RSMeans

Plumbing Estimating Methods

Third Edition

Includes:

- *Standard Plumbing & Fire Protection Systems*
- *Special Systems Including Medical Gas & Glass Piping*

Joseph J. Galeno and Sheldon T. Greene

WILEY

RSMeans

Plumbing Estimating Methods

Third Edition

Includes:

- *Standard Plumbing & Fire Protection Systems*
- *Special Systems Including Medical Gas & Glass Piping*

Joseph J. Galeno and Sheldon T. Green

This book is printed on acid-free paper.

Published by John Wiley & Sons, Inc., Hoboken, New Jersey

Published simultaneously in Canada

The editors for this book were Melville Mossman and Andrea St. Ours. The managing editor was Mary Greene. The production manager was Michael Kokernak. The production coordinator was Marion Schofield. The electronic publishing specialist was Sheryl Rose. The proofreader was Robin Richardson. The book and cover were designed by Norman R. Forgit.

For general information on our other products and services, or technical support, please contact our Customer Care Department within the United States at 800-762-2974, outside the United States at 317-572-3993 or fax 317-572-4002.

Wiley publishes in a variety of print and electronic formats and by print-on-demand. Some material included with standard print versions of this book may not be included in e-books or in print-on-demand. If this book refers to media such as a CD or DVD that is not included in the version you purchased, you may download this material at http://booksupport.wiley.com. For more information about Wiley products, visit www.wiley.com.

Library of Congress Cataloging-in-Publication Data:

ISBN: 978-0-87629-704-9

10 9 8 7 6 5 4 3 2 1

Table of Contents

Preface

Plumbing Estimating Methods, Third Edition was written to help readers become competent plumbing estimators. It should also serve as a valuable reference in fulfilling the everyday assignments of the job.

Unfortunately, an individual wishing to make estimating his career may find it difficult to locate colleges or technical schools that offer practical, comprehensive estimating programs. The courses that are offered usually lack depth and are intended only to show the future architect or engineer the role that estimating plays in the construction industry. There are many excellent reference manuals published today, but these books are, for the most part, written for the seasoned plumbing estimator rather than the beginner.

The obvious question then is: "How does one become a plumbing estimator?" The usual procedure consists of on-the-job training with a plumbing contractor or construction consultant. The estimator trainee is placed under the guidance of a senior estimator and taught the fundamentals of taking off and pricing a job. The problem with this method is that the training process is constantly interrupted by the everyday business of the firm. The education of an estimator trainee can become a long, tedious, and often costly effort for all concerned.

This book explains and illustrates plumbing estimating procedures step-by-step. It assumes a basic knowledge of plumbing systems and the ability to read architectural/engineering drawings. By the time the reader completes the last chapter, he will have the basic knowledge required to function as a competent plumbing estimator.

About The Authors

Joseph J. Galeno, MBA, is Chief Executive Officer of Projex, Inc., a construction management firm in New York City. He has more than 30 years' experience in construction management, claims, cost engineering, and scheduling. His work has covered virtually all types of commercial, industrial, utility, institutional, and medical facilities in the public and private sectors. Mr. Galeno is the author of the *Plumbing Estimating Handbook* (Van Nostrand Reinhold, 1976) and of numerous articles on claims and other construction topics. He is a member of the American Association of Cost Engineers, the American Society of Sanitary Engineers, and the Building Contractors Association, and has served as a construction arbitrator for the American Arbitration Association.

Sheldon T. Greene, Senior Construction Consultant on plumbing/mechanical estimating, construction management, and value engineering for Projex, Inc., has over 40 years' experience as a construction professional. His work has focused on managing projects in the areas of cost control, contract negotiations, and field supervision. He is a coauthor of the *National Plumbing Estimator* (O.T.S., Inc. 1983), and of the *Building Cost File* (Construction Publishing Co., Inc., 1974). Mr. Greene serves as an arbitrator for the American Arbitration Association, is an Assistant Adjunct Professor at New York University, and is on the board of directors of the American Society of Sanitary Engineering. He is a member of the American Society of Plumbing Engineers, the American Society of Professional Estimators, the American Association of Cost Engineers, and the American Arbitration Association.

Melville J. Mossman, PE, is the Chief Mechanical Engineer at RSMeans. He has been senior editor of Means *Mechanical* and *Plumbing Cost Data* for more than 29 years. He is a contributing editor on all other Means cost products; is responsible for the development, maintenance, and update of the mechanical database; responds to all mechanical technical cost inquiries; and provides technical support on consulting projects. A registered Professional Engineer, Mr. Mossman has taught Means' Mechanical/Electrical Cost Estimating seminars for many years and has also lectured at ASHRAE events.

Introduction

The third edition of *Means Plumbing Estimating Methods* contains the same essential information on plumbing systems and estimating as earlier editions, but now includes Means latest union and non-union wage rates, and city and historical cost indexes. Also included is a greatly expanded Chapter 8, which provides a thorough explanation of how to utilize *Means Plumbing Cost Data* together with numerous sample pages from a recent edition. A new Chapter 9 provides an in-depth look at a complete unit price estimating example. Sources of code information and price guides has also been completely updated.

Part I of the book,"The Plumbing System,"provides an in-depth view of the materials, components, and subsystems that together form a plumbing system. The text, charts, and illustrations cover plumbing fixtures, piping systems, valves, pipe fittings, and equipment. This section also includes discussions on"thin wall"steel sprinkler piping, combination hubless DWV systems, chemical fire suppression systems, grooved pipe mechanical joint systems, and an expanded section on medical gases. Charts of standard symbols used by engineers on design drawings have been included in the Appendix.

Part II,"Plumbing Estimating,"consists of Chapters 3 through 9, which deal exclusively with estimating, from the initial preparation stage to the completion of a final contract bid estimate. Included is a discussion of the estimating tools and reference manuals that are necessary and helpful. Sample takeoff and estimate forms are provided, along with step-by-step instructions on how to take off a plumbing job. There is also a sample takeoff, job specification, drawings, and an estimate. The estimate is based on comprehensive labor-hour production tables for fixtures, devices, equipment, and piping found in the Appendix, along with a chart showing the average prevailing hourly wage rates for plumbers in 30 major U.S. and Canadian cities.

It should be noted that the costs included in the sample takeoff and estimate in Chapters 5 and 6 serve only to illustrate plumbing estimating methods. They are not intended as a pricing reference for creating estimates. Regardless of the year or project location, the procedures will be identical, the material and labor costs will vary geographically, and productivity will vary according to the crew, supervision, and the plumbing contractor's efficient use of tools, labor, and scheduling.

Chapter 6 discusses the various markups necessary to complete the estimate, along with the summary sheet for the sample job takeoff. Chapter 7 covers different methods and types of estimating, such as change order analysis, estimating for additions and alterations, budget estimating, and a systems approach to plumbing costs, as well as a section on computerized estimating systems. In Chapter 8, the proper use of *Means Plumbing Cost Data* is explained along with sample pages from the book.

Part I

The Plumbing System

Part I

Introduction

The word *plumbing* is derived from the Latin word *plumbum*, meaning lead, which was the basic metal plumbers worked with until about sixty years ago. History tells us that plumbing was viewed as a necessity by the earliest known ancient civilizations. However, plumbing as a complete system providing water for various needs, such as bathing, drinking, and waste removal, is a relatively new idea developed in the United States in the early part of the nineteenth century. One can define plumbing today as the system of piping and other apparatus for conveying water, liquid wastes, sewage, storm water, and certain special gases or liquids within or adjacent to any building. A complete plumbing system, much like a complete electrical system, consists of smaller subsystems and components.

Before one can successfully estimate plumbing costs, one must have at least a working knowledge of what a plumbing system is and how it functions. Part I of this book is designed to convey that knowledge to less experienced plumbing estimators, and to serve as a review for those who are well versed in plumbing installations.

Part I consists of Chapter 1, which describes the basic materials and methods used in various subsystems, and Chapter 2, which covers individual subsystems and components that together form a complete plumbing system. Part II follows with seven chapters detailing plumbing estimating procedures.

Chapter 1

Basic Materials and Methods

A plumbing system is composed of a number of smaller subsystems. Many of the same materials and methods are used to construct various subsystems. For example, copper tubing may be used in both domestic water systems and medical gas systems. Excavation and backfill is used on both sanitary and storm systems, as well as water or gas systems. Familiarity with the basic systems and materials is essential to the plumbing estimator. Therefore, we begin with an in-depth discussion of this subject.

Pipe and Pipe Fittings for Interior Plumbing

Pipe is available in a wide variety of sizes and materials. The size depends on the characteristics and quantity of the substance to be conveyed, and the service requirements. We will learn in Chapter 2 that service factors include the materials to be conveyed, hot or cold water, sanitary waste, storm water, natural gas, or acid waste. Pipe materials are also selected based on whether they will be used above or below grade, system pressure, external loads and problems of corrosion. Costs, too, are an important consideration.

Pipe is classified by its nominal size. Pipe materials are ferrous (containing iron), non-ferrous, and non-metallic. Examples of ferrous pipe are cast iron and black or galvanized steel. Brass and copper are commonly used non-ferrous materials. Non-metallic pipe materials include plastics and glass for use in building plumbing systems. For site work, pipe is also made of such other non-metallic materials as concrete and vitrified clay. The different types of pipe are categorized further into certain classes or schedules.

Pipe materials and sizes used for each type of plumbing subsystem will be described in Chapter 2. The next section will begin with piping terms, how pipes are joined, and pipe fittings.

Steel Pipe (Ferrous)

Steel pipe is manufactured as either black or galvanized (zinc-coated) with plain, threaded, or grooved ends, and can be purchased in random lengths up to 21′. The following grades of steel pipe are the most commonly used for plumbing and process systems.

1. Schedule 5 (Thin Wall)
2. Schedule 10 (Thin Wall)

3. Schedule 40 (Standard Weight)
4. Schedule 80 (Extra Strong)
5. Schedule 160 (Double Extra Strong)

These schedules are based on wall-thickness. Of the schedules mentioned, Schedule 40 threaded pipe is the one most commonly found in plumbing work. Although rather susceptible to corrosion, steel pipe is strong, comes in a wide range of sizes, and is less expensive in comparison to other metals used for similar applications.

Joints and Fittings

Although steel pipe may be welded using carbon steel weld fittings, this method is rarely used for plumbing installations other than large diameter gas lines. Tapered threads and grooved mechanical joints are the most common forms of joining steel pipe for plumbing and gas work.

A recent steel piping innovation that has gained industry-wide acceptance is **thin wall steel pipe** for use on sprinkler systems. Thin wall pipe provides the industry with a properly engineered lower cost alternative for use in fire protection applications. Further discussion on thin wall pipe can be found in Chapter 2 under Fire Protection Systems.

Pipe threading is done by manual or power-operated threading devices equipped with hard tempered steel dies. Dies are available in all size ranges necessary to produce tapered pipe threads. Threads are gauged according to standards set forth by the American Standards Association. Figure 1.1 illustrates a threaded joint.

In joining threaded pipe, the plumber first cleans the male threads of the pipe and the female threads of the fitting with a hard wire brush, then applies pipe joint compound to the male end. The pipe and fitting are then aligned and tightened by hand. Finally, the plumber completes the joint by tightening with a pipe wrench. In some systems, Teflon tape is used in lieu of pipe joint compound.

Grooved mechanical joint piping systems have steadily gained in popularity in recent years. The system eliminates many of the time-consuming installation requirements used in welding, threading, and flanging pipe, thereby resulting in labor savings. Grooved mechanical

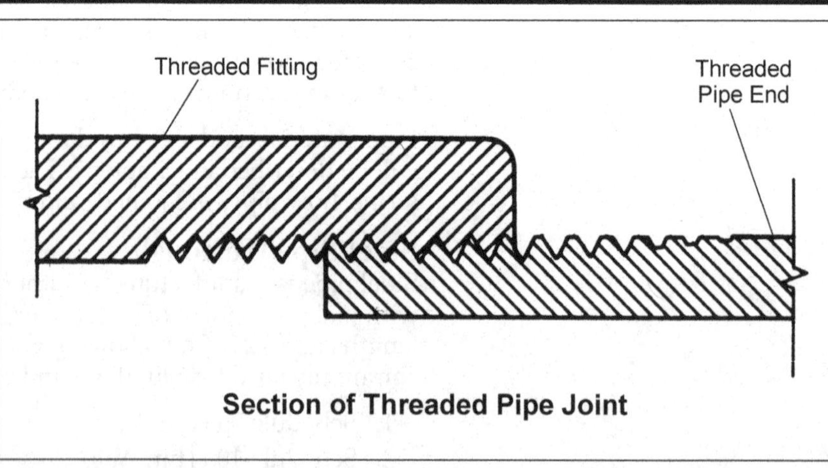

Section of Threaded Pipe Joint

Figure 1.1

joints combine the advantages of low cost installation, complete design integrity and simple, economic maintenance. This system and its associated fittings are used in the fire protection field, building service systems (i.e., storm and sanitary drains), potable, heating, cooling, and condenser water systems (not available for steam or gas piping). The fittings are designed for use with plain end, cut grooved, roll grooved, or bevel end steel pipe (see Figure 1.2), and are available plain or galvanized finish. This kind of fitting is produced for use with lighter schedules of piping, as well as with standard weight (Schedule 40). The wide variety of sizes and configurations of grooved mechanical joint fittings allows a complete system to be assembled using only these fittings. Special tools are required for field or shop cutting, rolling of grooves, and drilling outlets in the pipe wall where hook-type or other mechanically joined outlets are substituted for tees or welded nozzles.

Figure 1.2

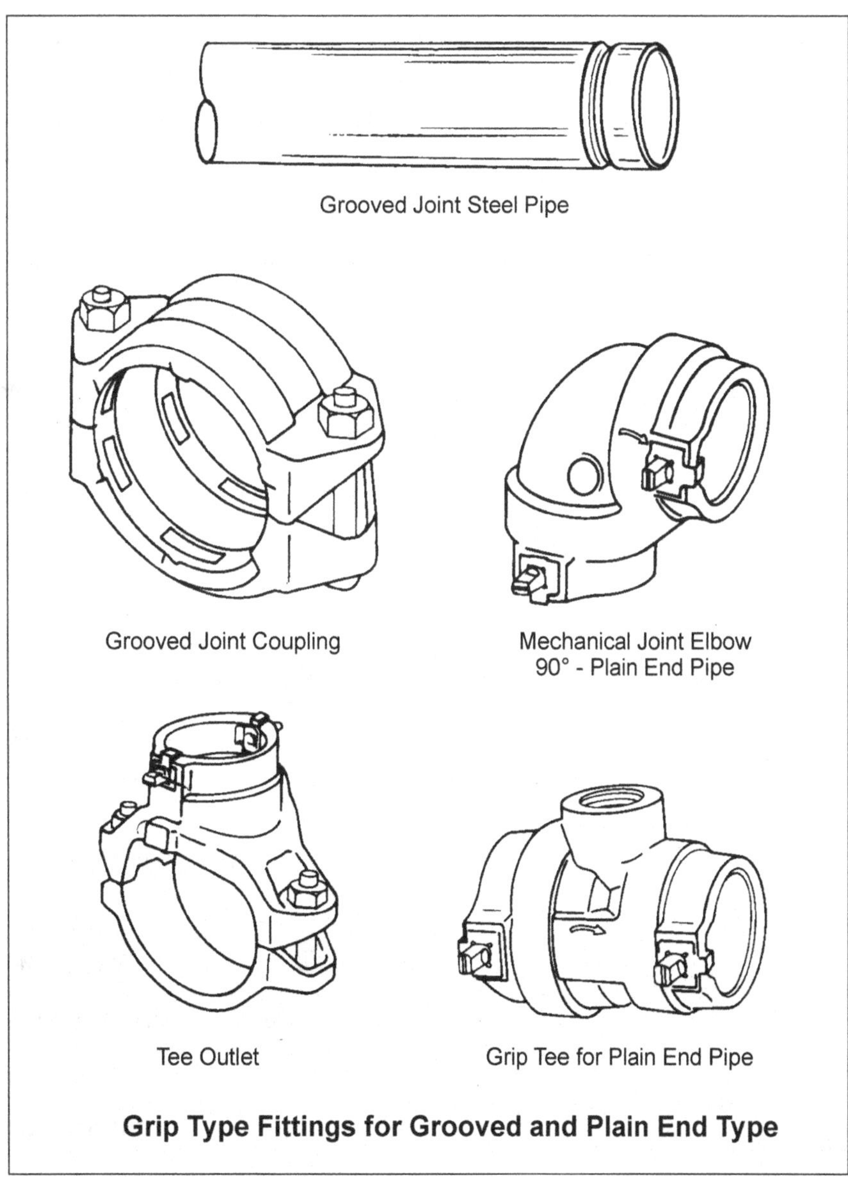

Grip Type Fittings for Grooved and Plain End Type

Depending on the particular service, fittings used to join steel pipe may be cast iron, malleable iron, and cast iron drainage pattern, all of which are available in black or galvanized. Pressure ratings indicate the recommended maximum basic pressure under which the fitting will properly function.

Where short pieces (1″ to 6″ long) of threaded pipe are required to connect fixtures and equipment, pipe nipples are used. Pipe nipples are manufactured in sizes ranging from 1/2″ to 12″. Pipe nipples longer than 12″ can be obtained on special order. Figure 1.3 illustrates typical threaded iron pipe fittings and nipples. See Figure 1.4 for a complete list of fittings in this category.

Ductile Iron Water Pipe (Ferrous)

Ductile iron water pipe is a steel-iron alloy, manufactured with flanged, mechanical, or push-on joints, and can be purchased in 18′ lengths. The class of pipe can be specified by pressure or by wall thickness. Pressure

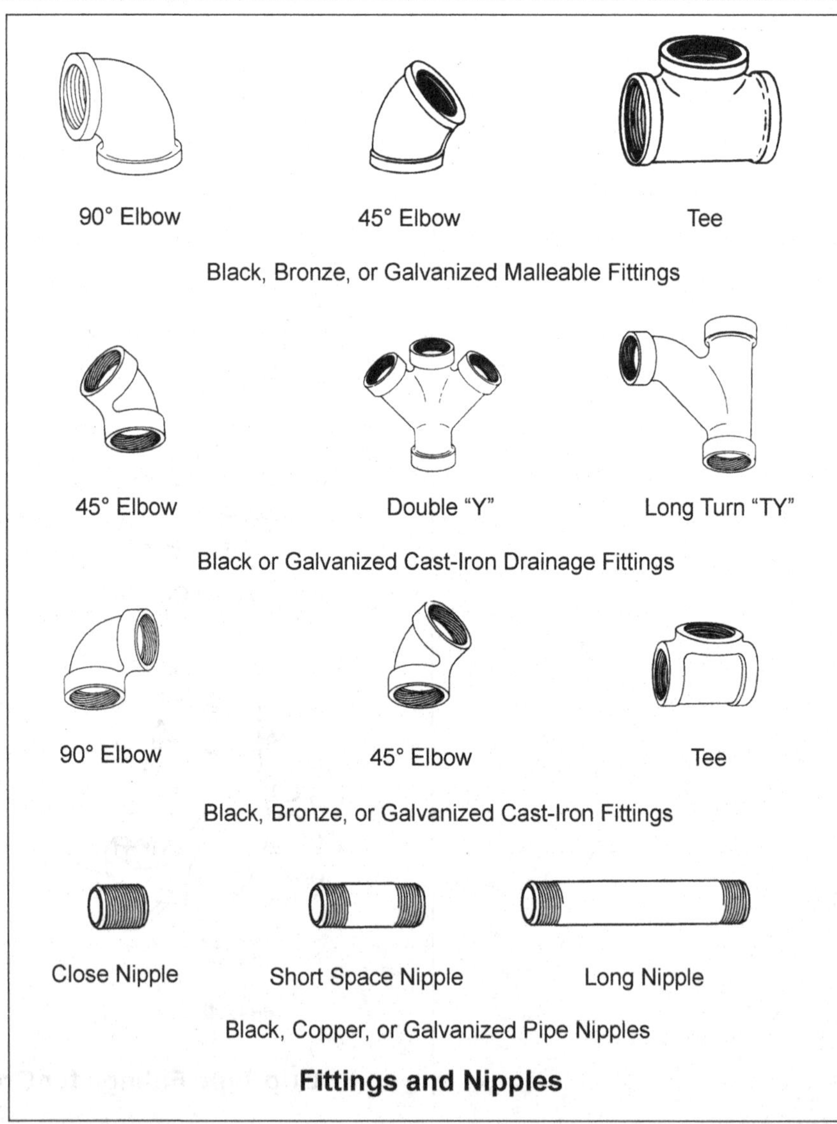

Fittings and Nipples

Figure 1.3

MANUFACTURED BLACK OR GALVANIZED MALLEABLE IRON FITTINGS (THREADED) (1/8" TO 12" PIPE DIAMETER)	
90° Elbows	Return Bends
90° Reducing Elbows	Side Outlet Elbows
90° Street Elbows	Female Drop Elbows
90° Reducing Street Elbows	Side Outlet T
45° Elbows	Couplings
45° Street Elbows	Caps, Plugs
T's	Street T's
Reducing T's	Reducing Street T's
Crosses	'Y' Branches
Unions	

MANUFACTURED BLACK OR GALVANIZED CAST-IRON DRAINAGE FITTINGS (THREADED) (1½" TO 12" PIPE DIAMETER)	
90° Short Turn Elbows	Short Turn 'TY's
90° Short Turn Reducing Elbows	Long Turn 'TY's
90° Long Turn Elbows	Reducing Short Turn 'TY's
45° Long Turn Elbows	Reducing Long Turn 'TY's
45° Short Turn Elbows	Short Turn Double 'TY's
90° Extra Long Turn Elbows	Long Turn Double 'TY's
60° Elbows	Reducing Short Turn Double 'TY's
Three-Way Elbows	Reducing Long Turn Double 'TY's
22½° Elbows	45° 'Y' Branches
Three-Way Reducing Elbows	45° Reducing 'Y' Branches
11¼° Elbows	45° Reducing Double 'Y' Branches
90° Elbows w/2" Side Outlet	60° 'Y' Branches
90° Elbows w/2" Heel Outlet	'P' Traps
90° Street Elbows	60° Reducing 'Y' Branches
45° Degree Street Elbows	Running Traps
Basin T's	Bath 'P' Traps
Hub Roof Connections	'S' Traps
T's	Half 'S' Traps
Increasers	Tucker Connections
Couplings	
Reducing T's	

MANUFACTURED BRONZE THREADED, CAST-IRON THREADED, PLASTIC, SOLDER JOINT AND BRAZED FITTINGS (1/8" TO 12" PIPE DIAMETER)	
90° Elbows	Couplings
90° Elbows (Reducing)	Couplings (Reducing)
90° Elbows (Long Turn)	45° 'Y' Branches
45° Elbows	45° 'Y' Branches (Reducing)
90° Elbows (Street)	Return Bends
T's	Adapters
T's (Reducing)	Caps
Crosses	Plugs
Crosses (Reducing)	Bushings
Unions	

Figure 1.4

classes are 50, 100, 150, 200, 250, 300, and 350 psi, the pressure under which the particular class of pipe is designed to operate.

Pipe wall thickness classes for ductile iron pipe are 50, 51, 52, 53, 54, and 56. A cement lining for the inside walls of the pipe is available for use in locations where soft water could cause an encrusting condition called *tuberculation*. Ductile iron offers excellent resistance to corrosion, and resists physical damage during backfill operations or other stressful occurrences.

Joints and Fittings

Flanged joints on ductile iron water pipe are made by threading flanges onto threaded pipe ends. A rubber or special application face gasket is placed between the two flanges to be joined, which are then bolted together. (See Figures 1.5 and 1.6.)

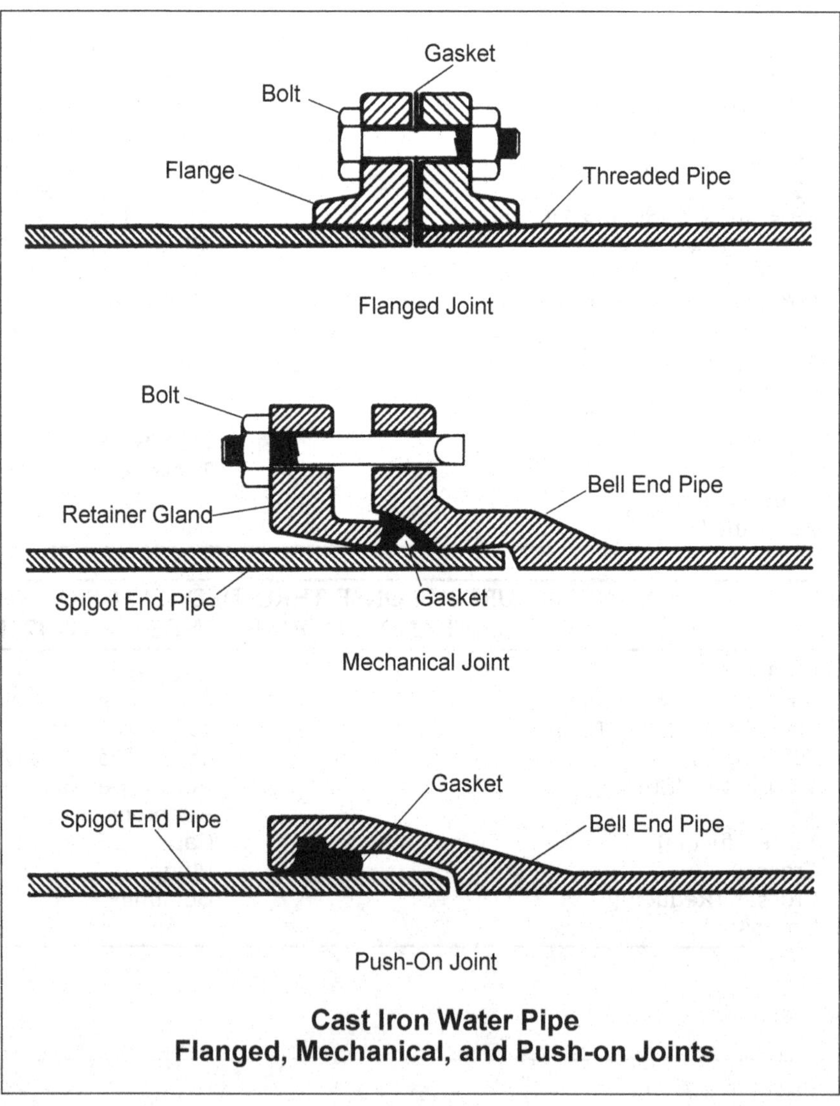

Cast Iron Water Pipe
Flanged, Mechanical, and Push-on Joints

Figure 1.5

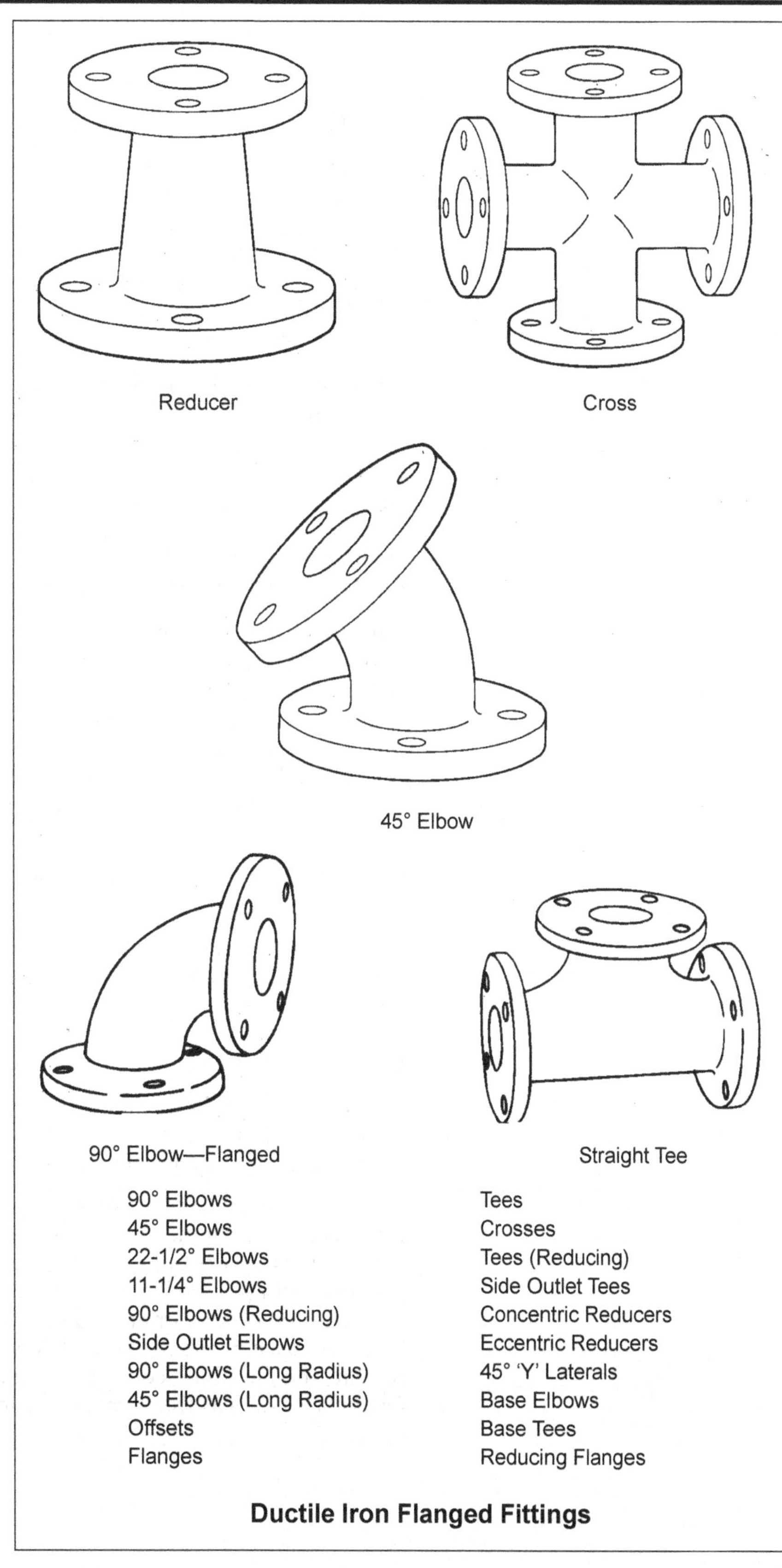

Ductile Iron Flanged Fittings

Figure 1.6

Push-on joints are compression-type joints with rubber or neoprene gaskets, which are inserted into the hub or bell end of the pipe. (See Figure 1.5.) A special joining tool is used to slip the spigot or plain end of the pipe into the bell end. Push-on joints allow speedy, low-cost installation. Push-on fittings are manufactured with bell ends.

A mechanical joint is actually a cross between a flanged and a push-on joint. The bell end of the pipe is cast with a flange facing; the spigot end of the pipe section to be joined is inserted into the bell end with a rubber ring gasket and an outer ductile iron retainer gland. When this operation is complete, the two sections are bolted together. (See Figure 1.5.) Bell-end mechanical joint fittings are available in a variety of patterns. See Figure 1.6 for a complete list of fittings in this category.

Cast Iron Soil Pipe (Ferrous)

Cast iron soil pipe is manufactured in 5′ and 10′ lengths and is available as either *service*, or *extra-heavy weight*, depending on the pipe-wall thickness. Joints are either lead and oakum (caulked), push-on, or clamp joint (hubless). Cast iron soil pipe offers excellent resistance to corrosion, to hydrogen sulphide gas action (an oxidation process creating sulphuric acid which can damage piping), and to abrasion from sand and gravel particles. (See Figure 1.7 for a list of soil pipe fittings.)

Joints and Fittings

The lead and oakum or caulked joint is made with molten lead and oakum packing. (See Figure 1.8.) The spigot end of the pipe or fitting is inserted into the bell end of the pipe or fitting to be joined. The oakum packing is then inserted in the bell with a caulking tool, followed by molten lead poured over the oakum. After the lead has cooled, it is driven into the joint with a caulking tool, and a tight, leakproof joint is formed.

The push-on joint is a compression joint similar in principle to that previously described for ductile iron water pipe (See Figure 1.5).

The clamp or hubless joint is a relatively new innovation for cast iron soil pipe and fittings. (See Figure 1.8.) The clamp joint is made by inserting a neoprene gasket around the hubless pipe and fitting joint, after which a stainless steel shield is placed around the gasket and tightened with two stainless steel bands.

Cast iron soil pipe fittings (Figure 1.9) are manufactured in a wide variety of design patterns for all three types of joints mentioned above. (See Figure 1.8.) See Figure 1.7 for a list of fittings in this category.

A recent innovation in hubless soil pipe systems is the combination hubless DWV system. The system, as designed, utilizes ordinary hubless pipe, fittings, and couplings along with multi-outlet and other special purpose fittings. (See Figure 1.10.) Various roughing combinations can be fabricated off-site or at the point of installation, with no wasted joints or space. Since the system is mainly designed to be fabricated off-site, the plumbing contractor can expect a better installation with few delays (such as those caused by inclement weather).

MANUFACTURED C.I. SOIL PIPE FITTINGS (BELL & SPIGOT LEAD OR PUSH-ON JOINTS) (2″ TO 15″ PIPE DIAMETER)	
Increasers	T w/Brass Cleanout
Quarter Bends	Fresh Air Inlet
Fifth Bends	Reducing 'Y'
Sixth Bends	Combination 'Y' & Eighth Bend
Eighth Bends	Sanitary Cross
Sixteenth Bends	Long Sanitary Cross
Quarter Bend w/Heel Inlet	Double 'Y'
Quarter Bend w/Inlet	'Y'
Long Quarter Bend	Cross w/Brass Cleanout
Long Eighth Bend	Upright 'Y'
Short Sweep	Tapped T
Long Sweep	Tapped Sanitary T
Long Sweep (Reducing)	Tapped 'Y'
Long Sweep (Increasing)	Long Tapped T
Double Quarter Bend	Reducing Sanitary Cross
Double Hub Quarter Bend	Double Comb. 'Y' & Eighth Bend
Double Hub Eighth Bend	Closet Fittings
Double Spigot Quarter Bend	Tapped Cross
Long Double Spigot Quarter Bend	Sanitary T w/Brass Cleanout
T	'Y' w/Brass Cleanout
Sanitary T	'Y' & ⅛ Bend w/Brass Cleanout
Inverted 'Y'	Offsets
T w/Inlet	Plugs
Sanitary Tee w/Inlet	Reducers
'Y' w/Inlet	Iron Body Cleanouts
Sleeves	
Traps	

MANUFACTURED HUBLESS C.I. SOIL PIPE FITTINGS (CLAMP JOINT) (1½″ TO 6″ PIPE DIAMETER)	
Quarter Bends	Sanitary Tap T's
Fifth Bends	Double Vertical San. Tap T's
Sixth Bends	Sanitary Tap T's w/90° Side Tap
Eighth Bends	'Y's
Sixteenth Bends	Upright 'Y's
Quarter Bends w/2″ Heel Outlet	Double 'Y's
Quarter Bends w/2″ Side Outlet	Tapped 'Y's
Double Quarter Bend	Comb. 'Y' & Eighth Bend
Tapped Quarter Bends	Double Comb. 'Y' & Eighth Bends
Short Sweeps	Sanitary Crosses
Long Sweeps	Sanitary Tapped Crosses
Long Quarter Bends	'P' Traps
Long Eighth Bends	Reducers
Sanitary T's	Increasers
Sanitary T's w/2″ Side Outlet	Plugs
San. Tees w/2″ Inlet Both Sides	Hub Adapters
Sanitary T's w/2″ 90° Inlet R&L Above Center	Test T's
Sanitary T's w/2″ 45° Inlet Above Center	Ferrules

Figure 1.7

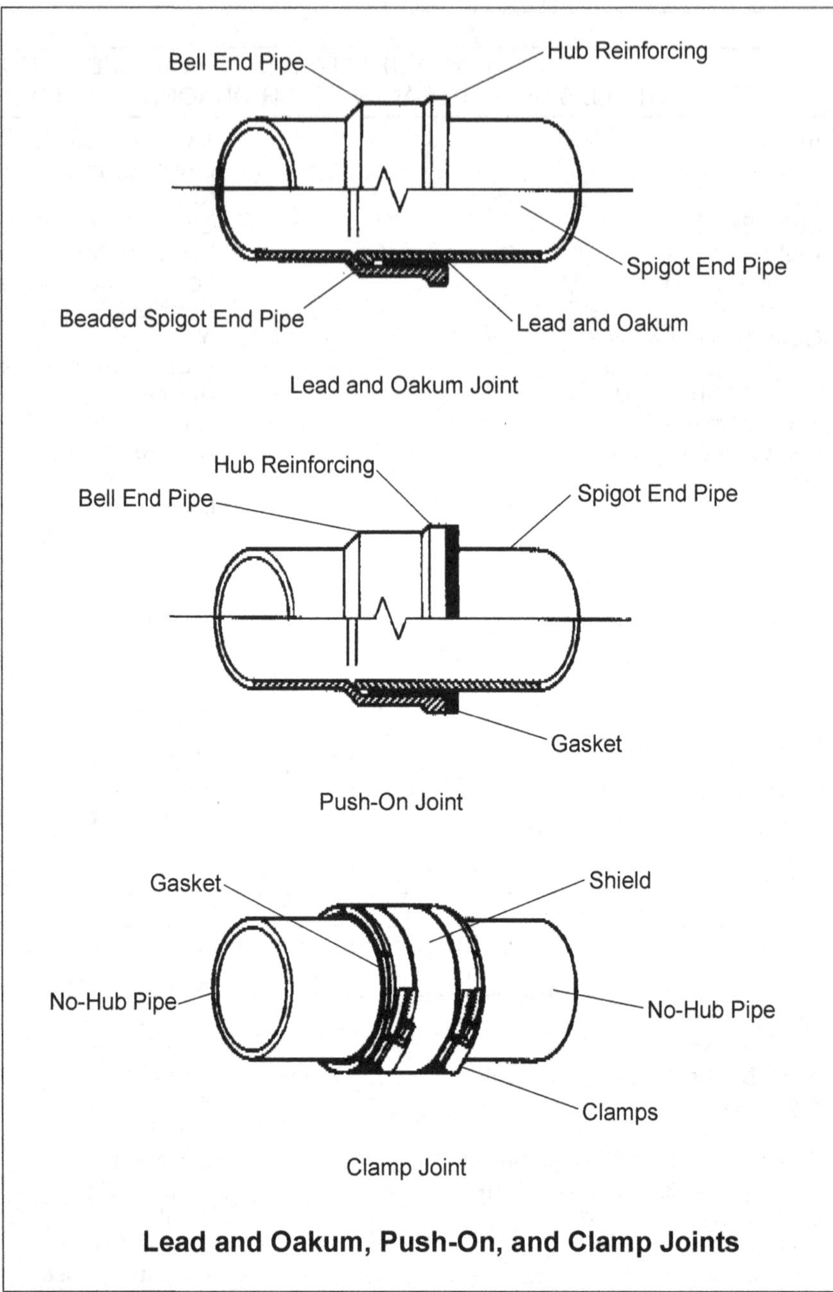

Lead and Oakum, Push-On, and Clamp Joints

Figure 1.8

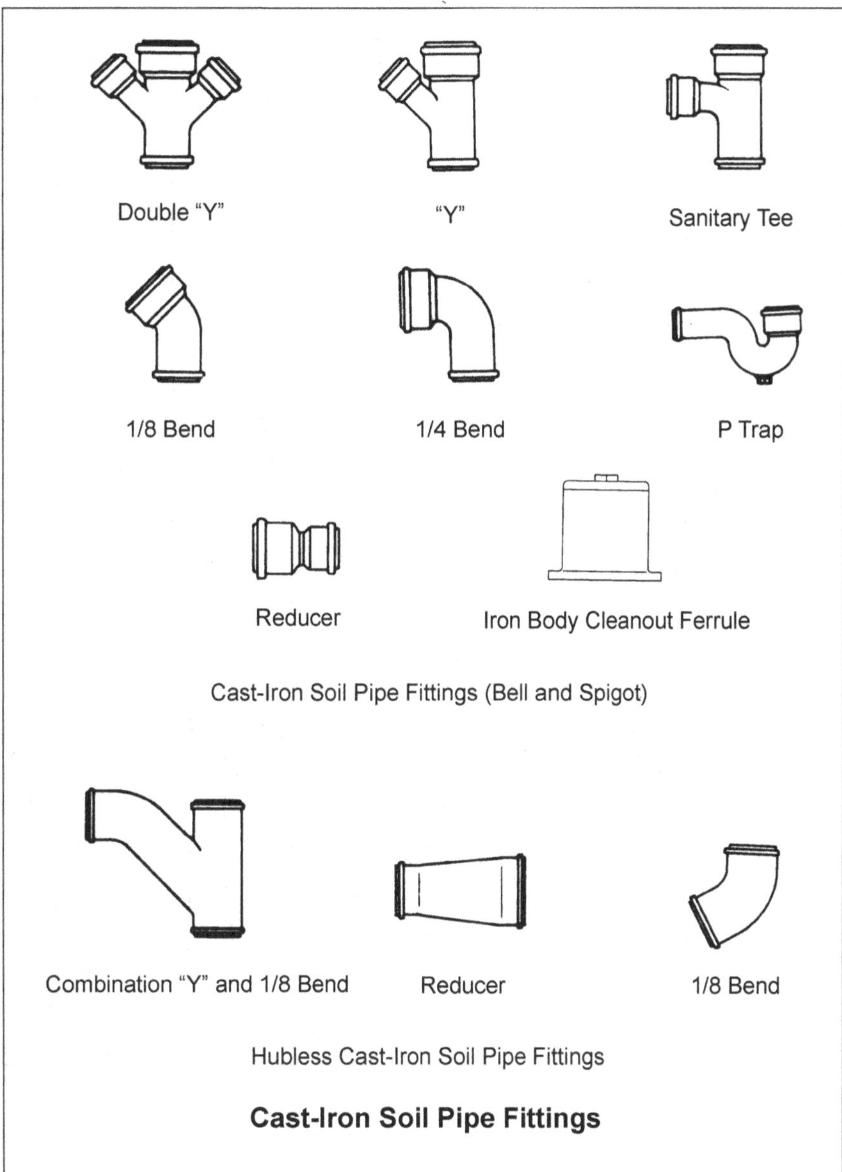
Double "Y"
"Y"
Sanitary Tee
1/8 Bend
1/4 Bend
P Trap
Reducer
Iron Body Cleanout Ferrule
Cast-Iron Soil Pipe Fittings (Bell and Spigot)
Combination "Y" and 1/8 Bend
Reducer
1/8 Bend
Hubless Cast-Iron Soil Pipe Fittings
Cast-Iron Soil Pipe Fittings

Figure 1.9

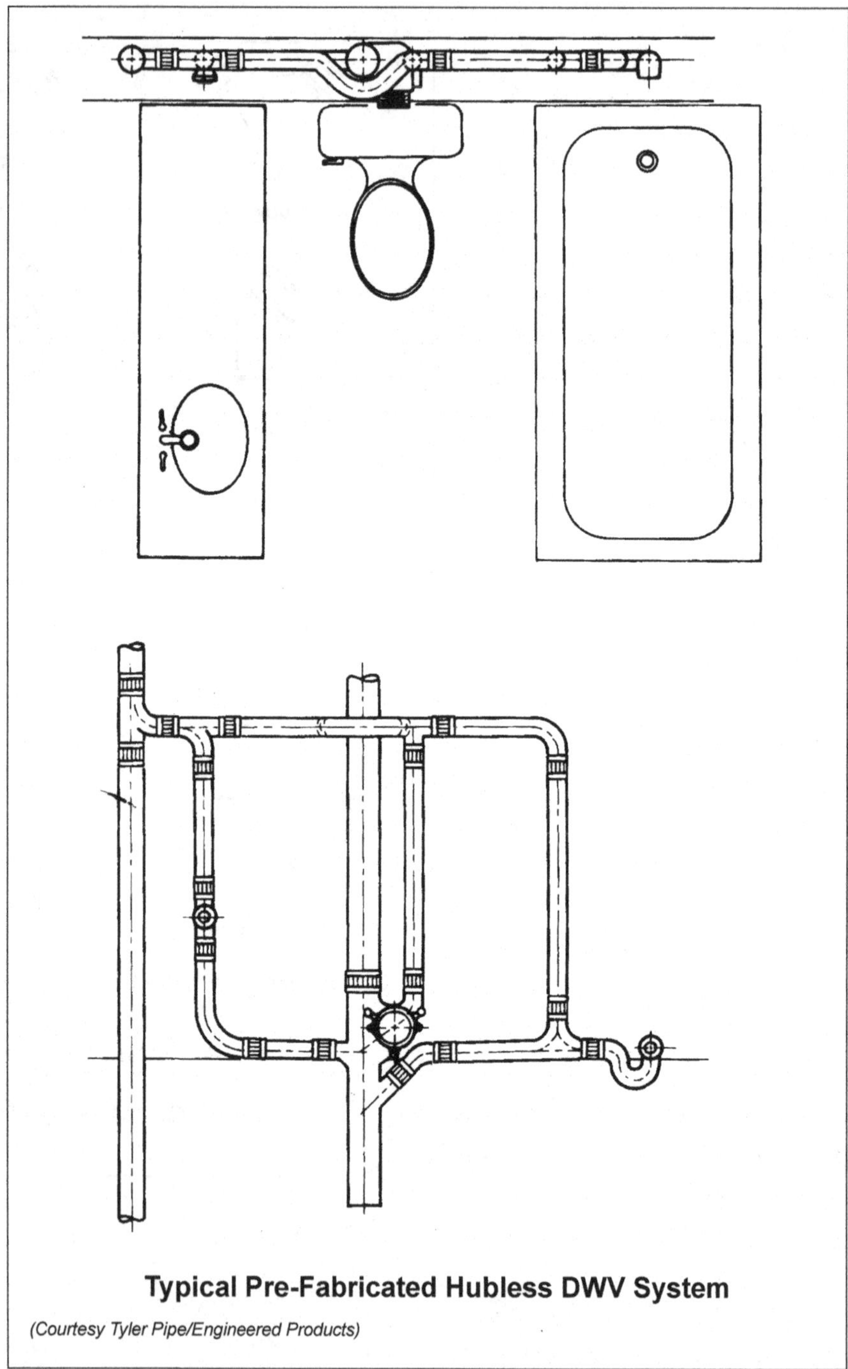

Typical Pre-Fabricated Hubless DWV System

(Courtesy Tyler Pipe/Engineered Products)

Figure 1.10

It should be noted that the plumbing contractor must investigate local union regulations to determine whether it is permissible to prefabricate units off-site. In addition, it is important to state that the optimal benefit realized by using this system is on multiple/typical roughing installations.

Red Brass Pipe (Non-Ferrous)

Red brass pipe is made from an alloy of 85% copper and 15% zinc. Brass pipe can be purchased in lengths of 12′ to 16′ and is produced with threaded ends. Brass offers excellent resistance to corrosion.

Joints and Fittings

Threaded cast-bronze fittings or flanged cast-bronze fittings may be used to join brass pipe. The threading operation is similar to that for steel pipe.

For special requirements, threadless fittings are produced for brazing to pipe. Brazing is a joining method using brazing filler metals, which melt at temperatures ranging from 1100°F to 1500°F. The filler metals are alloys containing silver or copper and phosphorous. In making a brazed joint, burrs on pipe must be removed and the pipe thoroughly cleaned. Brazing flux is applied to remove any traces of oxides. Heat is then applied to the pipes being joined, using an oxyacetylene flame. Once the proper temperature is achieved, the brazing filler rod or wire is placed at the joint and will enter the socket of the fitting, completing the joint. (Figure 1.4 lists a complete line of fittings in this category.)

Plastic Pipe (Non-Ferrous)

Plastic is a material that has been growing in popularity in recent years. Plastic pipe is actually a broad title for many different types of pipe of plastic composition. Some of the most popular plastic piping materials manufactured today are:

PVC-Polyvinyl chloride
CPVC-Chlorinated polyvinyl chloride
PE-Polyethylene
PB-Polybutylene
PP-Polypropylene
FRP-Fiberglass reinforced plastic
ABS-Acrylonitrile-butadiene-styrene

Most of the above pipe is manufactured in 20′ lengths, and each has its own distinctive applications and characteristics. PVC is stronger than most thermoplastics and has excellent chemical resistance to corrosive fluids, but it may be damaged by certain hydrocarbons. CPVC has the capacity to handle corrosive fluids at a temperature 40°F to 60°F above other plastics. CPVC also will not sustain combustion. PE offers excellent chemical resistance and can be used at temperatures below 130°F, but its bending strength is less than most other thermoplastics. PP is the lightest plastic piping material, yet is stronger and has greater corrosion and chemical resistance than PE. Unlike other plastic materials, FRP is reinforced with fiberglass, which gives it very high strength. ABS is light and easy to handle, yet strong and corrosion-resistant.

With respect to systems applications, PVC pipe is the most commonly used material on drain, waste, and vent (DWV) and cold water systems. PVC competes with acrylonitrile butadiene styrene (ABS) piping, especially in DWV applications. Polypropylene is another competitor in DWV applications. Chlorinated PVC (CPVC) pipe has been developed for use in hot water systems, polyethylene in cold water systems, and polybutylene in both hot and cold water applications. CPVC and polybutylene are approved for use in wet type fire protection systems, while polyethylene is commonly used in gas and water services.

Plastics, on the whole, are non-toxic, lightweight, corrosion-resistant, low in cost, and easily installed.

Joints and Fittings

There are at least three different methods of joining plastic pipe, depending on the particular application. The most popular methods are threading, solvent cement, and heat fusion. Threading is usually restricted to Schedule 80 or heavier walled pipe. The threading operation is the same as for steel pipe.

A cement joint is made by applying solvent cement to both the pipe end and fitting socket. Prior to the setting of the cement, the two pieces are joined and rotated. Heat fusion is a method in which the joint surfaces of both the pipe and the fitting are heated at the same time. An electrical 120 volt, 4 ampere fusion power unit is used. A prefabricated fusion coil is placed on the pipe, then both the coil and pipe are inserted into the fitting. A compression clamp is placed over the fitting shoulder and the power unit attached to the clamp. The wire tail piece of the fusion coil is attached to a terminal post on the clamp, and the power unit is then turned on for approximately 90 seconds to create the fused joint.

Additional fitting types are compression, socket weld, butt weld, brass insert, plastic insert, clamp, grip, flared, and flanged.

Plastic fittings for both pressure and drainage applications are manufactured in a wide variety of sizes and patterns. (See Figure 1.11 for illustrations of pressure type fittings. Drainage fittings generally conform to the same shape as copper drainage fittings. See Figure 1.13 for a complete list of fittings in this category.)

Glass Pipe (Non-Ferrous)

Glass pipe is used in plumbing work to convey acid or acid-bearing wastes. Glass pipe is manufactured in regular or heavy schedule, and may also be wrapped with fiberglass and polyester resin for underground installations. Glass pipe is manufactured in either 5′ or 10′ lengths. This is one of the best materials for resisting corrosion. Glass pipe is also used for high purity water systems.

Joints and Fittings

Beaded-end glass pipe and fittings are the most common joint method used in plumbing work. A metal compression coupling with a corrosion-resistant Teflon gasket is designed to receive one beaded end and one plain cut end. The coupling is tightened with two bolts, which are part of the metal housing of the coupling. Couplings designed to receive two beaded ends are also available. A complete line of drainage fittings such as elbows, 45's, tees, wyes, and other shapes, are manufactured. See Figure 1.12 for a complete list of fittings in this category.

Conical joint glass piping is coupled via flanged, gasketed joints.

Pipe and Pipe Fittings — Site (Outside)

In addition to ductile iron pipe discussed earlier in this chapter, there are certain types of pipe used exclusively for outside or buried applications. In sewer and site drainage, the most common types of pipe found are concrete, vitrified clay, and galvanized corrugated metal.

Concrete Pipe

Concrete pipe is manufactured in two classes: *reinforced* and *non-reinforced*. The reinforced class is broken down further into five separate classes: I, II, III, IV, and V. These are based on the D-load (test load expressed in pounds per linear foot per foot of diameter) necessary to produce a crack in the pipe. The two joining methods used for concrete pipe are the single rubber ring gasket joint and the cement mortar joint. Concrete fittings are stock manufactured or can be custom made. However, where large diameters are involved, manholes are often used at the junction point instead of fittings.

Vitrified Clay Pipe

Vitrified clay pipe is manufactured in both standard and extra-heavy strengths. The pipe is molded of clay and glazed to make it more impervious to contamination. Clay pipe is manufactured in 2-1/2' and 3' lengths. The most popular joining method used today is "O" ring gasket joints. Fittings are manufactured for all size ranges.

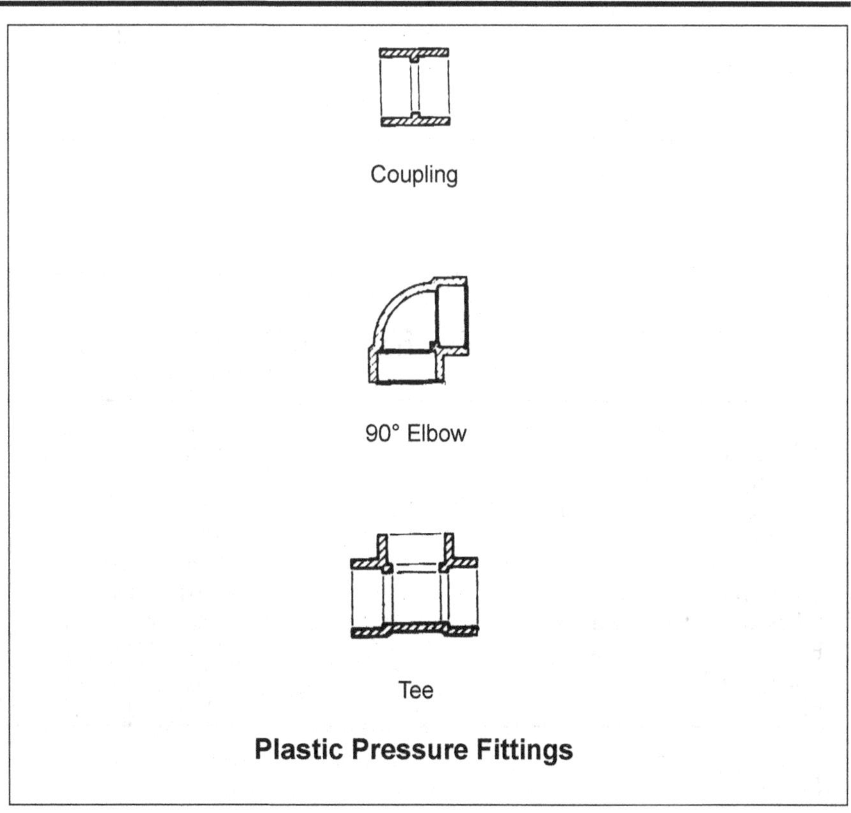

Plastic Pressure Fittings

Figure 1.11

Galvanized Corrugated Metal Pipe

Galvanized corrugated metal pipe is classified according to metal gauge, (16, 14, 12, 10, and 8 gauge). Corrugated metal pipe is heavily galvanized. A bituminous coating can be applied for added protection. The corrugated design increases the beam strength of the pipe. The pipe walls are joined by a number of methods: riveted seams, spot welded seams, helical lock seams, or bolted seams. Fittings are available and can also be custom fabricated.

Tube and Tube Fittings

Any student of plumbing estimating should understand the basic differences between *pipe* and *tubing*. Tubing generally has a thinner wall than pipe and is never threaded. Tubing is manufactured in both coils and lengths; coils are possible because of its greater bending flexibility.

MANUFACTURED GLASS DRAINAGE FITTINGS 1½" TO 6" PIPE DIAMETER	
Quarter Bends	Offsets
Sixth Bends	Return Bends
Eighth Bends	Cleanout and Test T's
Sixteenth Bends	Cleanout Plugs
Quarter Bend Reducers	P Traps
Sanitary T's	Swivel 'P' Traps
Compact Sanitary T	Swivel 'S' Traps
Drainline 'Y's	Running Traps
Compact Drainline 'Y's	Swivel 'P' Drum Traps
Combination 'Y's/Eighth Bends	Swivel 'S' Drum Traps
Compact Combination'Y's/Eighth Bends	Drum Traps
Upright 'Y's	Universal Traps
Vent Stack Increaser	Trap Cleanouts
Reducers	Cup Sinks
Double Quarter Bend	Couplings

Figure 1.12

MANUFACTURED D.W.V. COPPER & PLASTIC FITTINGS (1¼" TO 8" PIPE DIAMETER)	
Adapters Slip Joint	Elbows 11¼°
Adapters Soil Pipe	Elbows 60°
Adapters Threaded Female	Plugs
Adapters Threaded Male	Sanitary T's
Bushings	T's
Closet Flanges	'Y's
Couplings	Test Caps
Elbows 90° Short Radius	Vent Increasers
Elbows 90° Long Radius	Long Turn 'T's
Elbows 45°	Short Turn 'TY's
Elbows 22½°	Double Long Turn 'TY's

Figure 1.13

Copper water tubing is manufactured in three classes according to wall thickness; the heaviest, Type K; medium-grade, type L; and lightest-grade, Type M. Types K and L are available in soft or hard temper (hard temper tubing resists freezing very well). Soft temper copper is available in longer coils. The manufactured length of hard drawn copper tubing is 20'. Copper tubing is popular because it is lightweight, corrosion-resistant, requires no threading, and the soft temper bends easily.

Another type of copper tubing is DWV, designed exclusively for soil and waste drainage systems. The advantages of type DWV copper tubing over cast iron soil pipe are that it is manufactured in 20' lengths, thus requiring fewer joints; is easily cut; and when joined, is lightweight and space-saving due to its smaller outside diameter compared to bell end soil pipe.

Joints and Fittings

Copper tube is joined by either *brazing, soldering,* or *roll groove couplings*. Brazed joints are completed using the same method for threadless brass pipe described earlier in this chapter. Soldering is used for copper tubing where temperatures do not exceed 250°F. Solder joints create a capillary action, which draws the molten solder into the fitting. Flux is used as a wetting agent to assure a uniform spreading of the solder over the joint. Two types of solder are used to join copper tubing: 50-50 tin-lead solder for drainage and vent systems and 95-5 tin-antimony (or other lead-free materials) for potable water systems. The use of lead in solder for potable water systems has been banned in most areas of the country. Tin-antimony solder (95-5) can be used on moderate to high pressure temperature services and offers greater joint strength. In making up a solder joint, all burrs should be removed from the tubing and all surfaces to be joined should be cleaned. Flux is applied to the tube end and to the socket of the fitting. They are then joined. Using a propane torch, heat is then applied to the joint. When the metal is hot enough, the flame should be removed and the solder applied and melted at the joint. When cooled, the solder joint is completed.

Fittings used for copper water tube are wrought copper or cast bronze solder joint. Cast bronze solder joint drainage fittings and wrought copper joint drainage fittings are used for DWV copper. See Figure 1.14. (Figures 1.4 and 1.13 show complete lists of fittings.)

Valves and Control Devices

The contents of every piping subsystem must be controlled to ensure the proper system operation. The flow of liquids and gases, system pressures, and overall system operations are controlled with valves and devices. Valves are available in a wide variety of styles, materials, and sizes, with the style and material determined by the application and the size and on its capacity to control the liquid, gas, or pressure. Chapter 2 covers the valves and control devices found on various plumbing subsystems. This section introduces various valve and device categories, materials used, and terminology.

The major types of valves found in plumbing work are gate, globe, check, butterfly, and ball. These valves are manufactured for plumbing use in bronze, iron, and plastic.

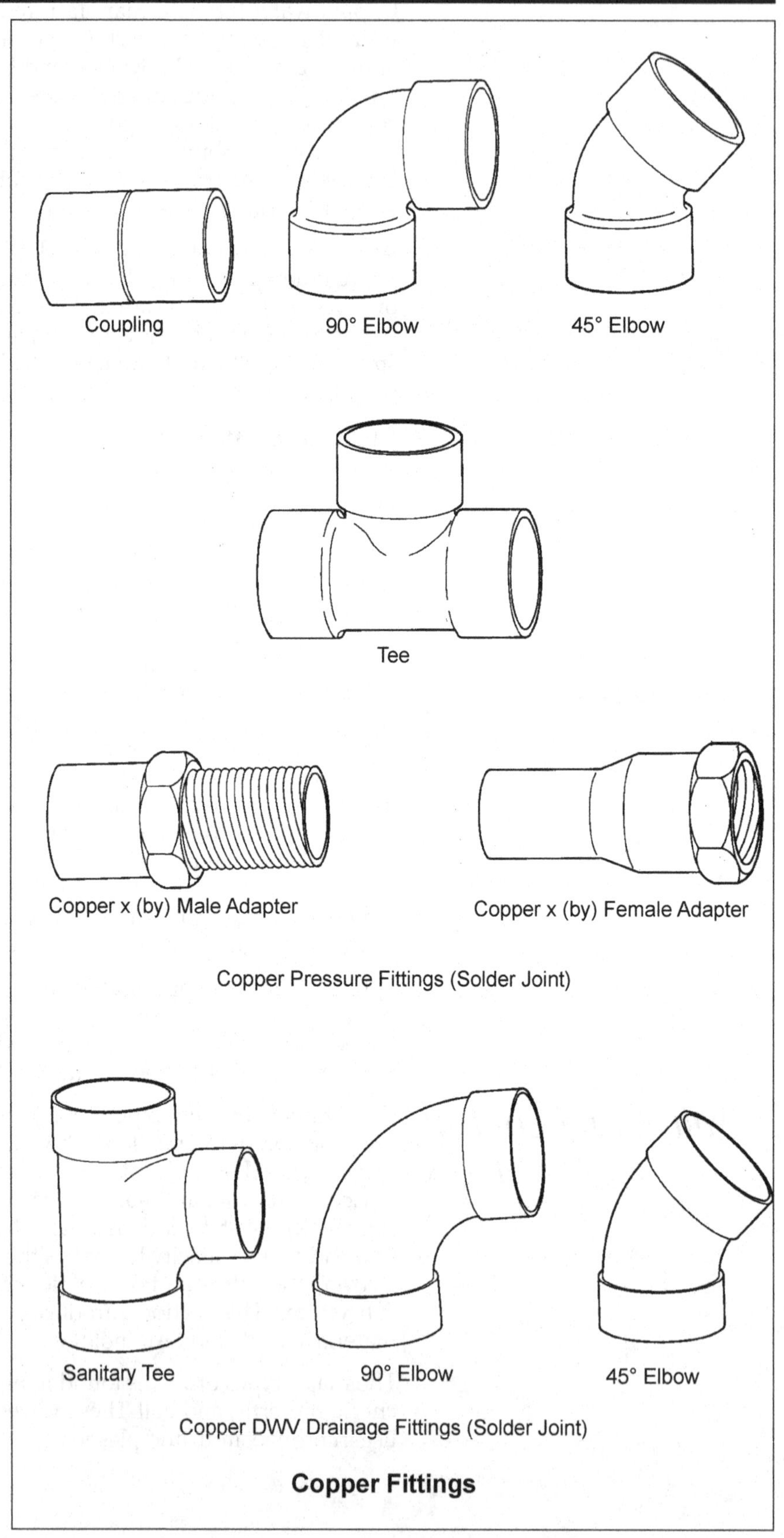
Coupling
90° Elbow
45° Elbow
Tee
Copper x (by) Male Adapter
Copper x (by) Female Adapter
Copper Pressure Fittings (Solder Joint)
Sanitary Tee
90° Elbow
45° Elbow
Copper DWV Drainage Fittings (Solder Joint)
Copper Fittings

Figure 1.14

Gate Valves (Bronze)

Bronze gate valves may be of either a split or solid wedge design, the wedge gate being the control piece regulating the flow. (See Figure 1.15.) Gate valves are required where full and unobstructed flow is necessary. Gate valves come equipped with rising stems or, where headroom is a factor, non-rising stems. Rising stems are particularly useful in determining the position of the wedge (open or closed). Another gate valve design is the outside screw and yoke type, desirable where frequent maintenance and lubrication is required. OS & Y valves also indicate by stem position whether the valve is opened or closed. Working pressures range from 125 psi to 300 psi for steam, and 200 psi to 600 psi for water, oil, and gas. Gate valves are manufactured with flanged, threaded, soldered, socket, or grooved ends.

Gate Valves (Iron-Body)

Like bronze valves, iron-body gate valves are of the wedge design. Their iron body construction keeps costs down to a minimum, making them desirable where large sizes are required. Outside screw and yoke designs are available in iron-body gate valves. Iron-body gate valves that are to be used for fire protection systems must be made to Underwriter and Factory Mutual specifications. Most iron-body valves start at 2″I.P.S. However, there are one or two designs available in smaller pipe sizes. Valve ends are either threaded or flanged.

Gate Valves (Plastic)

Polyvinyl chloride gate valves are available from 1/2″to 6″I.P.S. with socket, threaded, or flanged ends. They are manufactured with a working pressure rating of 125 psi. Plastic gate valves are normally limited in use to plastic piping systems only.

Globe Valves (Bronze)

Bronze globe valves are of the composition or plug disc design. (See Figure 1.16.) Globe valves are desirable where close flow control is important. Outside screw and yoke, angle, and hose end designs are also available. Working pressures range from 150 to 300 psi for steam

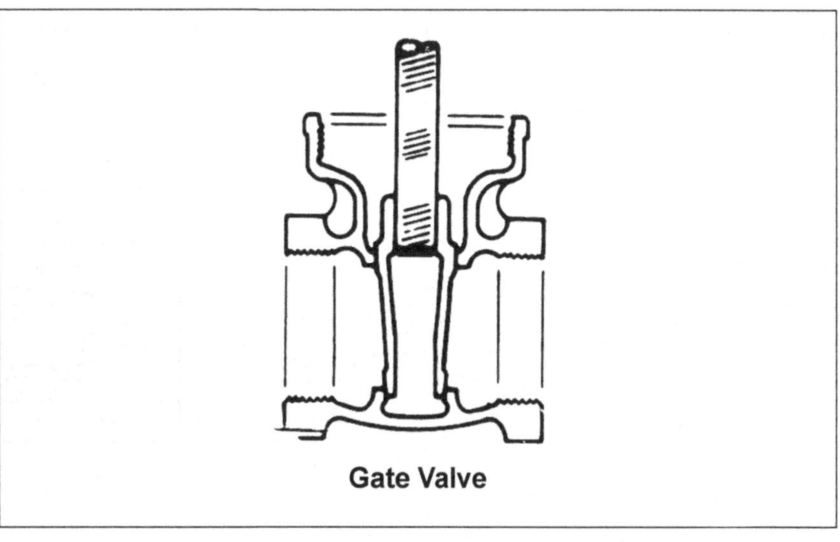
Gate Valve

Figure 1.15

service, and 150 to 600 psi for water, oil, and gas. Globe valves are manufactured with flanged, threaded, soldered, or socket ends.

Globe Valves (Iron-Body)

Iron-body globe valves, like bronze globe valves, are desirable where close flow control is important. However, like iron-body gate valves, their iron-body construction keeps costs down and is desirable where large sizes are required. Valve ends are either threaded or flanged. Most sizes begin at 2″ I.P.S., with a few designs available starting at 1/2″ I.P.S.

Check Valves (Bronze)

Bronze check valves are of the lift or swing check design. (See Figure 1. 17.) Check valves are required to prevent reverse flow in a pipeline. They are opened by pipeline pressure and closed by back pressure or the weight of the swing check mechanism. Working pressures range from 125 to 300 psi for steam, and 175 to 300 psi for water, oil, and gas. Check valves are manufactured with threaded, flanged, soldered, or socket ends.

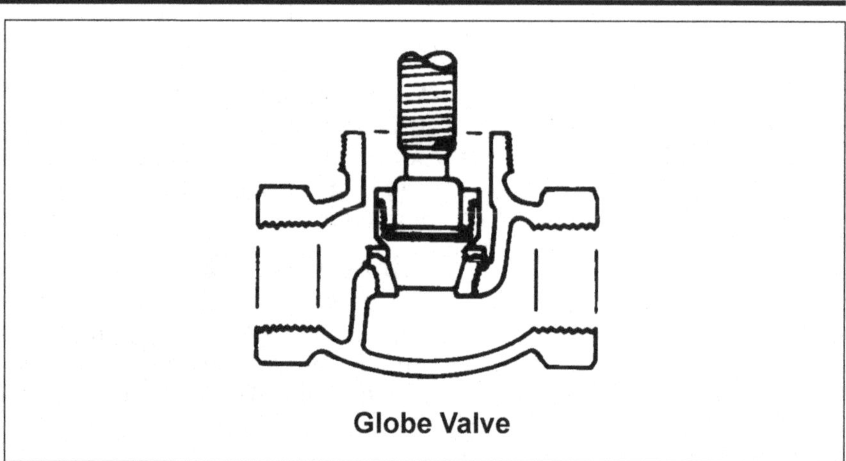

Globe Valve

Figure 1.16

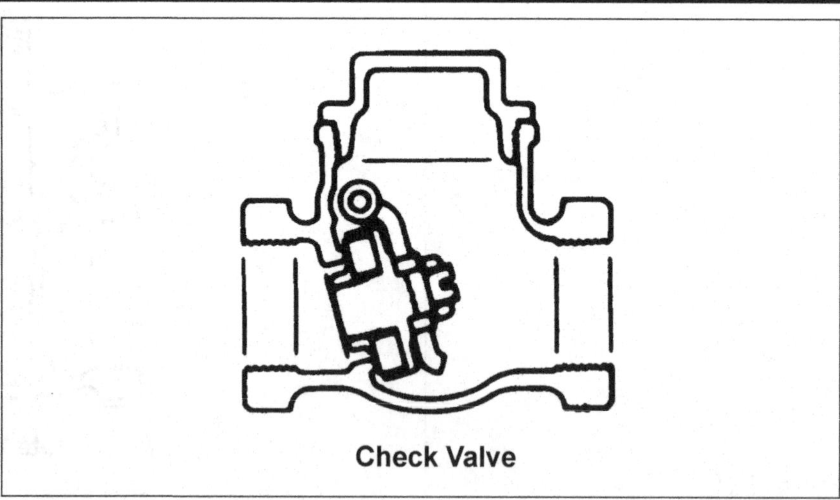

Check Valve

Figure 1.17

Check Valves (Iron-Body)

Iron-body check valves are similar to bronze check valves in that they are either of the lift, swing check, or non-slam spring-loaded design. As with iron-body gate and globe valves, the iron-body construction is less costly than bronze, making them desirable for use in large size installations. The operating principle is the same as that for bronze check valves. Iron-body check valves are manufactured with either threaded or flanged ends.

Ball Valves (Bronze)

Ball valves are popular, especially for control of medical gases. Ball valves are compact, have little flow restriction, and are relatively low in cost. Working pressures are 150 psi for steam and 400 psi for water, oil, and gas. Sizes range from 1/4″ to 2″ I.P.S. Figure 1.18 illustrates a typical ball valve.

Ball Valves (Plastic)

Plastic ball valves are manufactured in polyvinyl chloride (PVC), chlorinated polyvinyl chloride (CPVC), and polypropylene (PP). Working pressure is 150 psi, and sizes range from 1/4″ to 4″ I.P.S. Plastic ball valves are available with socket or threaded ends and should only be used on plastic piping systems.

Butterfly Valves (Iron-Body, Carbon Steel, Bronze)

Butterfly valves are a low cost, lightweight, and compact piping component. The valves are quickly and easily installed, utilizing flanged or grooved ends. Butterfly valves are designed to ensure bubble tight shut-off and to provide low pressure drop through the valve. Figure 1.19 illustrates a typical butterfly valve.

Control Devices

Control devices include any devices on a piping subsystem (other than standard valves), which aid the control of one or more subsystem functions. Examples of control devices are pressure reducing valves, temperature and pressure relief valves, backflow preventers, strainers, vacuum breakers, shock absorbers, expansion joints, and hose bibs. (See Figure 1.20.) The next chapter covers in greater detail their functions and applications.

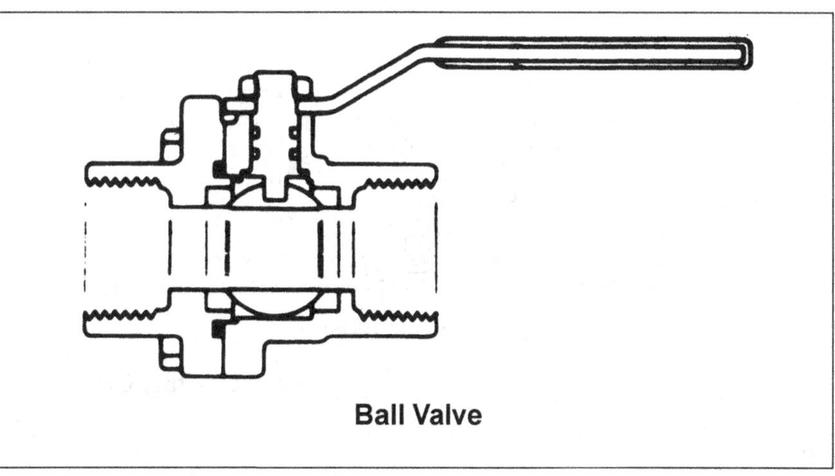

Ball Valve

Figure 1.18

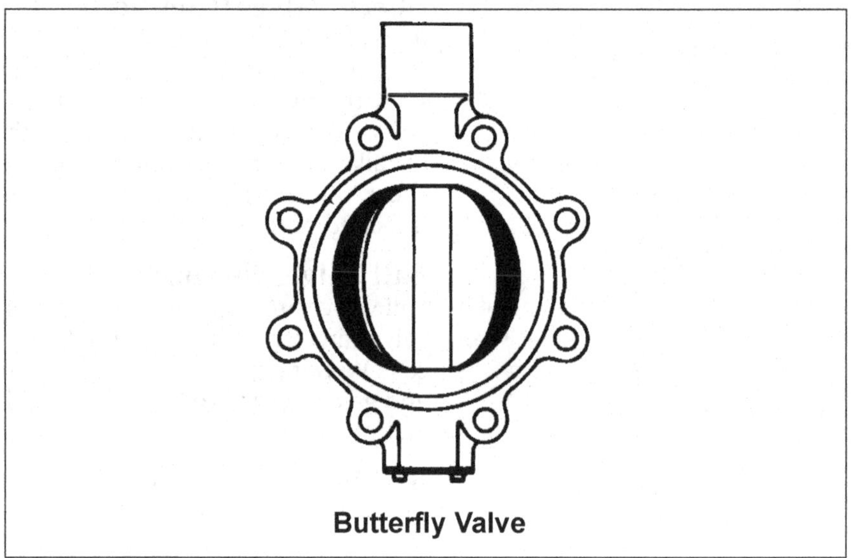

Butterfly Valve

Figure 1.19

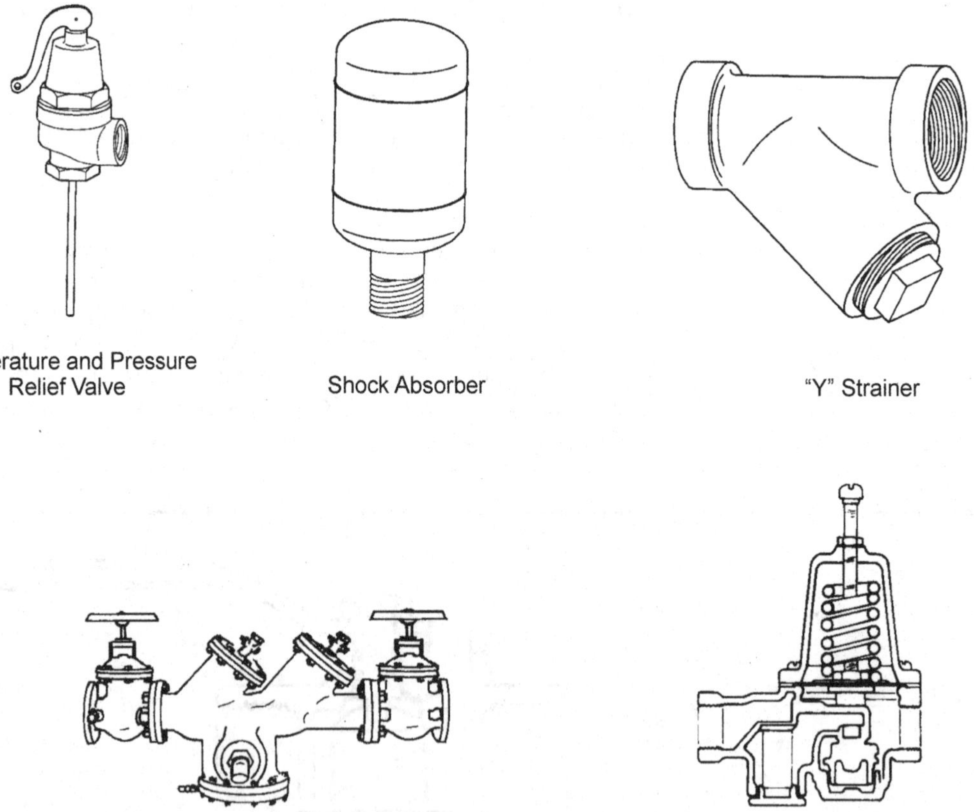

Temperature and Pressure Relief Valve

Shock Absorber

"Y" Strainer

Backflow Preventer

Pressure Reducing Valve

Typical Control Devices

Figure 1.20

Pipe Support Systems

Pipe supports fasten and support piping systems to walls, ceilings, floor slabs, or structural members within a building. (See Figures 1.21a, b, c, and d.) Some supports carry single pipelines, such as band hangers, clevis hangers, single or double rod roll hangers, and riser clamps. Other supports carry multiple pipe runs, such as trapeze hangers and pipe racks. Hanger or support assemblies used in plumbing are usually of the band, clevis, roll, or trapeze design. Clevis hanger assemblies consist of a clevis hanger, insert and threaded rods with nuts. Depending on the type of floor construction used in a building, various methods are used to anchor hangers. For a metal deck with concrete fill, all inserts are set before concrete is poured. After the pour is complete and the inserts imbedded in the cured slab, the plumber installs the hanger rods and clevis hangers to make a complete installation.

Pipe clamps are used as supports for vertical risers and are usually installed at every other floor. Pipe clamps are manufactured in two pieces. The clamp is secured around the pipe with two bolts and rests on the floor slab. Trapeze hangers are anchored much like clevis hangers, except that two threaded rods are required for each assembly. Sleeves are used within a building for individual piping lines passing through walls and floors. Fire-rated sleeves are designed to prevent the penetration of flames during fire. Watertight sleeves are used for piping lines passing through walls and floors exposed to outside elements, such as foundation walls and floor slabs on grade. A pipe sleeve is made of steel or plastic and is usually two sizes larger than the pipe passing through it (or the outside diameter of the pipe insulation).

If the estimator does not have recommended intervals for pipe supports indicated in his job specifications or on the drawings, Figure 1.22 can be used as a guide.

Pipe Insulation

Pipe insulation or covering is installed on piping systems to retard heat loss and prevent condensation. In plumbing, all concealed hot, cold, and recirculating water piping should be insulated. It is also good engineering practice to insulate all horizontal offsets or storm drainage piping in hung ceilings to prevent condensation.

Insulation is available in rigid or flexible form and is manufactured from fiberglass, cellular glass, rock wool, polyurethane foam, closed cell polyethylene, flexible elastomeric, rigid calcium silicate, phenolic foam or rigid urethane. However, fiberglass is by far the most common material used to insulate plumbing lines. Insulation thickness for hot, cold, recirculation, and storm lines is usually 1/2″ and 1-1/2″ for piping lines exposed to freezing conditions. Pipe insulation comes with a vapor barrier jacket made of aluminum foil between layers of kraft paper. Fiberglass insulation should not have a thermal conductivity exceeding .22 BTU per square inch per hour at a mean temperature of 75°F. Fittings and valves should be insulated with hydraulic setting cement insulation of the same thickness specified for piping. Jacketing is also required for fittings and valves and should be coated with white vapor barrier lap cement.

The plumbing contractor usually subcontracts the pipe insulation to an insulation contractor. The insulation contractor will, in turn, submit his marked-up price to the plumbing contractor to be included in the plumbing estimate.

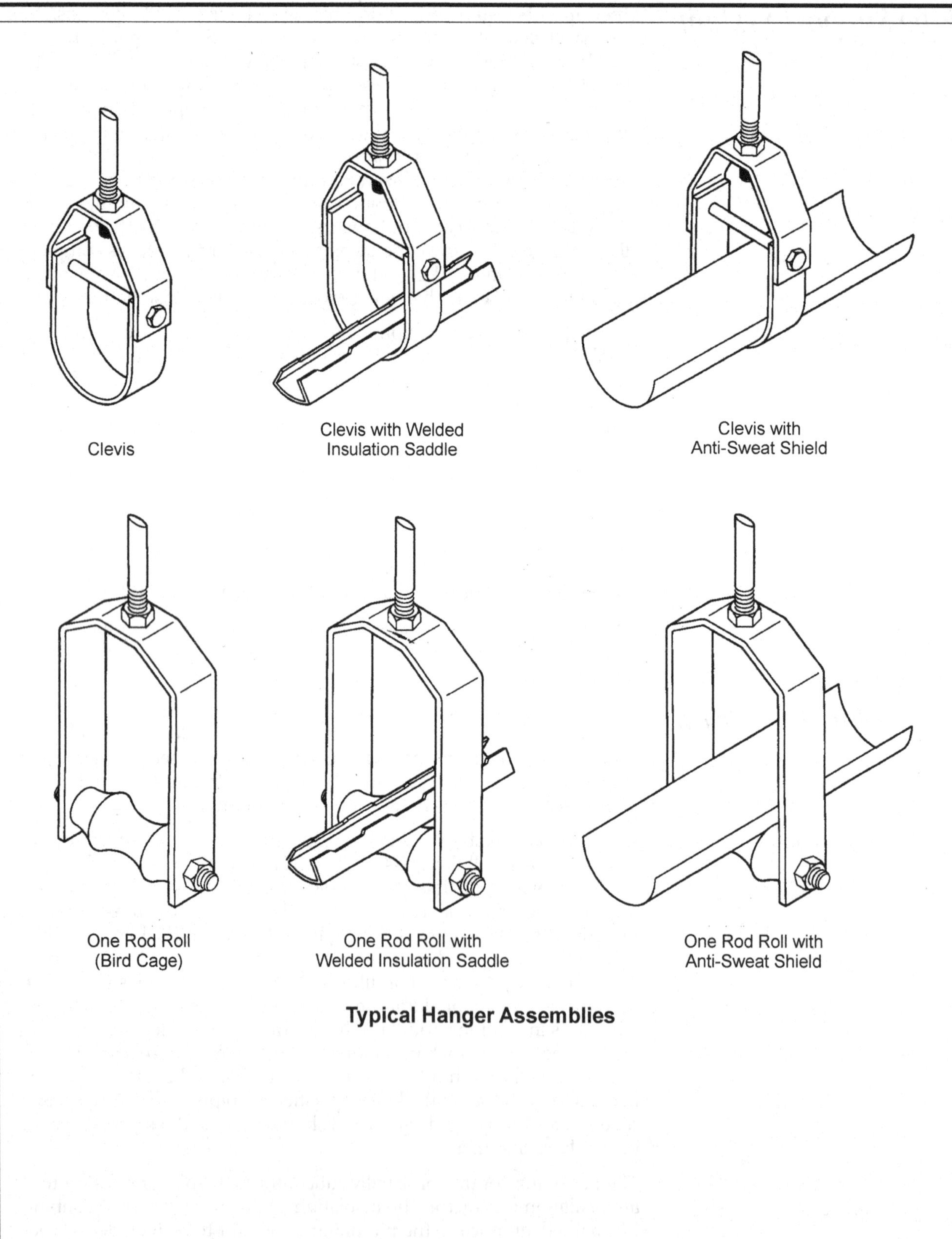
Clevis
Clevis with Welded
Insulation Saddle
Clevis with
Anti-Sweat Shield
One Rod Roll
(Bird Cage)
One Rod Roll with
Welded Insulation Saddle
One Rod Roll with
Anti-Sweat Shield
Typical Hanger Assemblies

Figure 1.21a

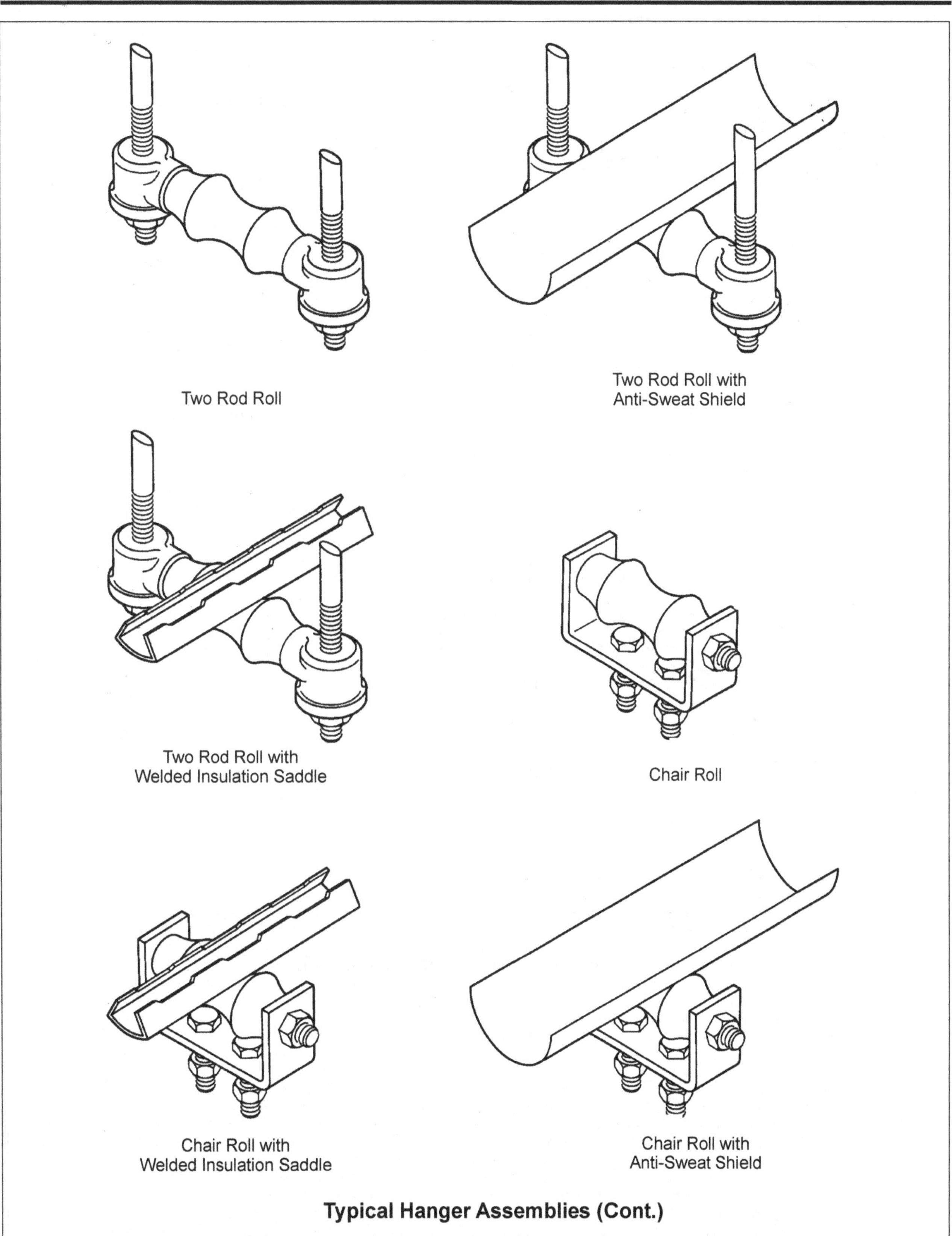

Typical Hanger Assemblies (Cont.)

Figure 1.21b

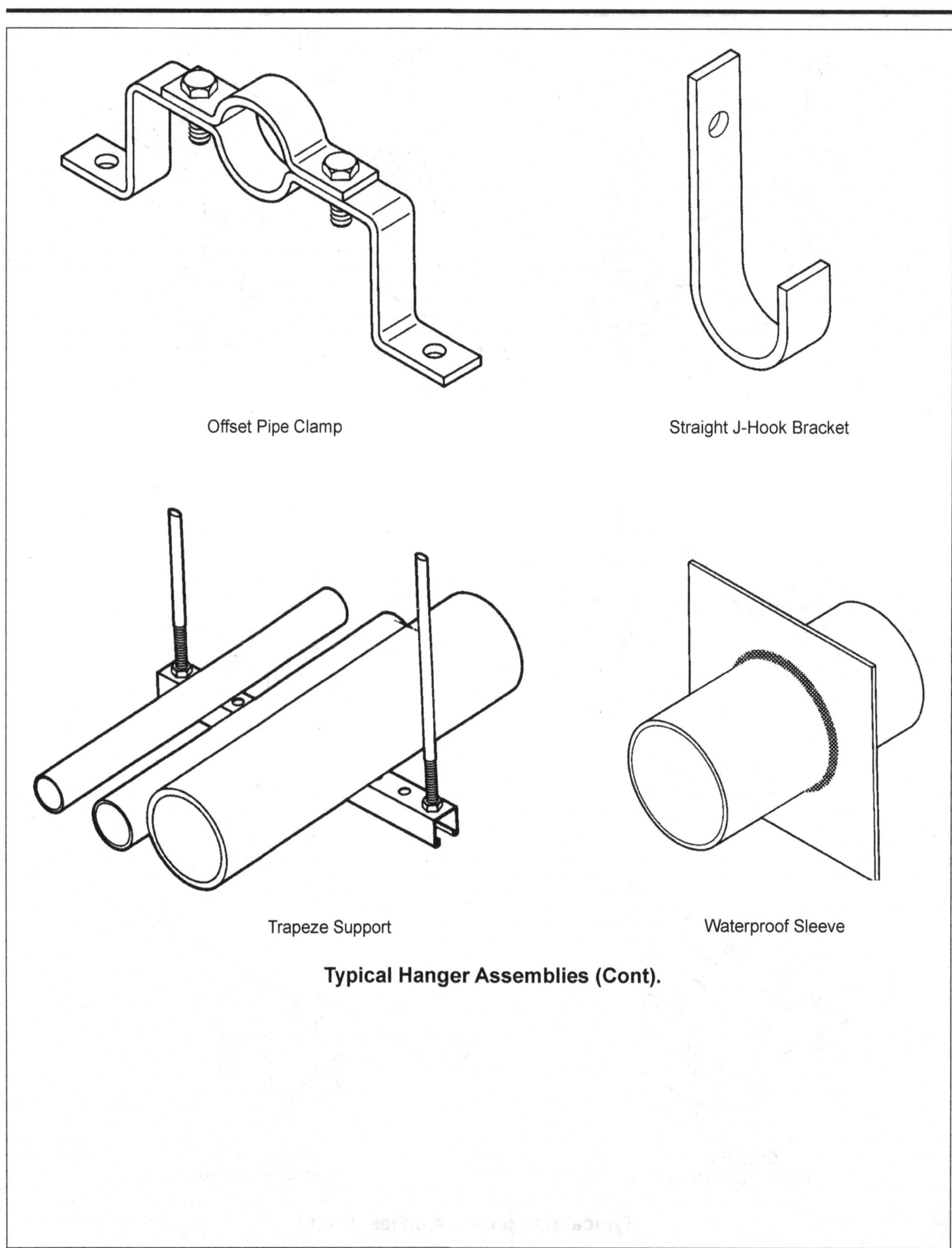

Typical Hanger Assemblies (Cont).

Figure 1.21c

Painting and Identification

Painting of piping may or may not be part of the plumbing contract, although it is very often a part of the plumbing estimate and is usually a subcontracted item. Exposed piping systems are painted according to color codes set forth by the architect or engineer in the job specifications. The types of primer, paint, and number of coats required will also be stated in the specifications. Painting of pipe is generally estimated per lineal foot.

Pipe system identification is usually accomplished by one of two methods:

1. Spray-on stenciling
2. Manufactured pipe markers

Location and type of pipe markers are outlined in the job specifications. Valve tags are a manufactured item and are usually attached to valves with metal "S" hooks. In addition to tags, the contractor must turn over to the owner a valve chart clearly showing the location and function

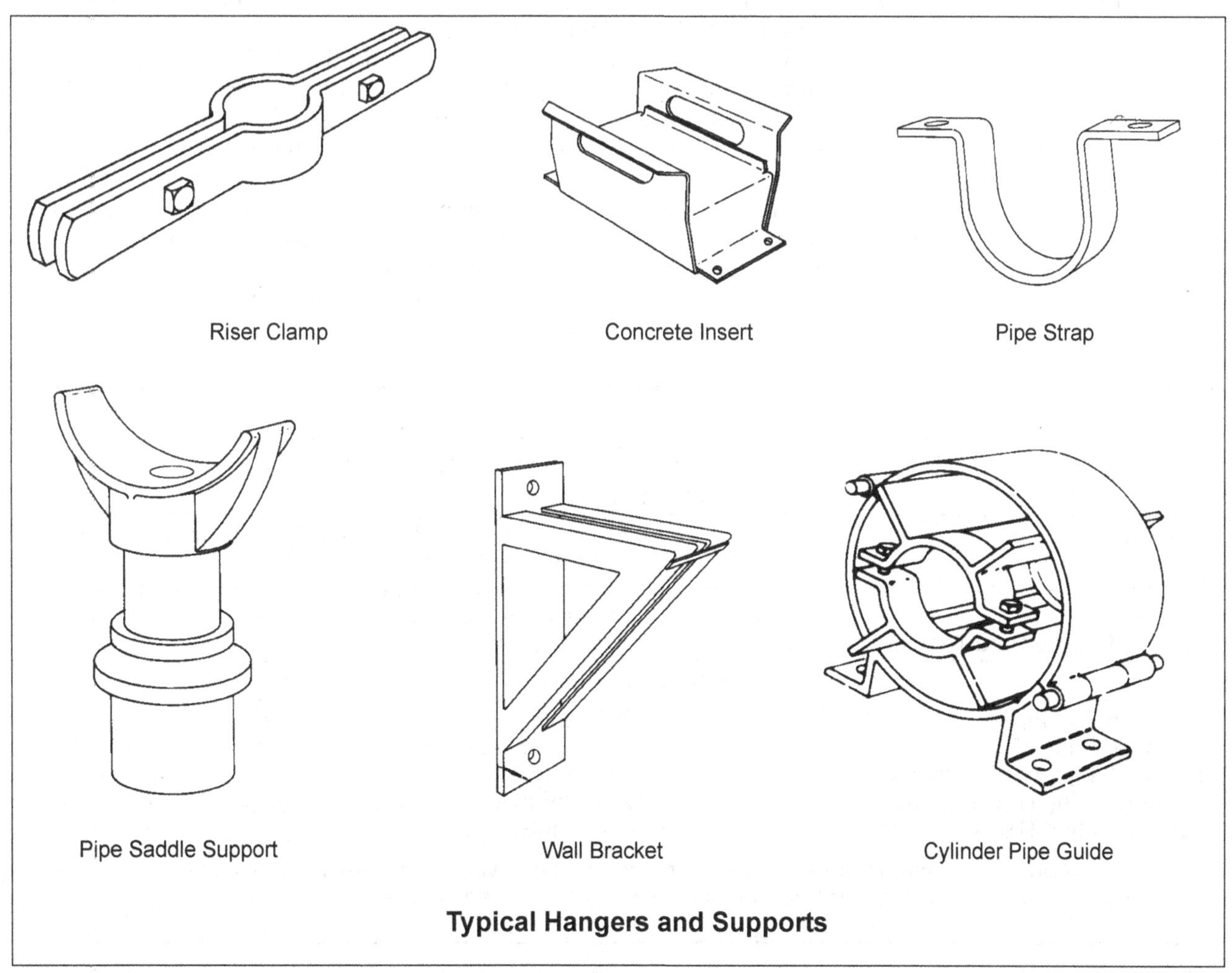

Typical Hangers and Supports

Figure 1.21d

of all strategic valves. Markers and tags are taken off according to the specified number of required locations.

Clearly marked piping systems and valves are essential to an owner so that maintenance crews can readily identify a system or valve in case of emergency.

Rigging

Rigging is the labor and equipment needed to hoist or place heavy items. The rigging work plumbers generally encounter on a project is the hoisting of heaters, pumps, and heavy fixtures such as bathtubs onto various floors of multi-story buildings. On small projects up to eight stories high, the plumbing contractor may decide to rent his own hoist with a qualified engineer present, if required by local union regulations.

On high-rise projects, it might be beneficial for the plumbing contractor to hire a union rigger if the general contractor's hoist is not available. Average costs for both methods are usually similar, differing only by a few dollars. If rigging is required on projects, the estimator should obtain a quote from a licensed rigger during the bidding stage.

Concrete Pads and Vibration Isolation

Concrete pads and vibration isolators are usually required for all large, motorized pieces of plumbing equipment such as circulating/pressure pumps, fire pumps, and air compressors. Concrete pads and vibration isolators ensure proper structural load distribution and noise abatement. Depending on the job specifications, the responsibility to furnish and install these pads may or may not be part of the plumbing contract. The size of the concrete pad is usually governed by the equipment manufacturer's recommendations. If the concrete pad is furnished by others, it is still the plumbing contractor's responsibility to provide manufactured vibration isolators to the installing contractor. If the concrete pad is the responsibility of the plumbing contractor, he will usually subcontract this work to a concrete contractor, who furnishes a price for inclusion in the plumbing estimate. Since pad

MINIMUM RECOMMENDED INTERVALS FOR PIPE SUPPORTS	
Vertical Piping	
Cast-Iron Soil Pipe	At base and at each story height, but in no case at intervals greater than 20′.
Threaded Pipe	At every other story height but in no case at intervals greater than 25′.
Copper Tubing (Hard Temper)	At each story height.
Horizontal Piping	
Cast-Iron Soil Pipe	At 5′ intervals and behind every hub.
Threaded Pipe (1″ or less)	At 8′ intervals
Threaded Pipe (1¼″ or more)	At 12′ intervals
Copper Tubing (1¼″ or less)	At 6′ intervals
Copper Tubing (1½″ or more)	At 10′ intervals

Additional support must be considered where large valves, strainers, meters, etc. contribute added weight to the piping system.

Figure 1.22

sizes vary according to the piece of equipment, the plumbing estimator should request a quote for the particular job from a concrete contractor. However, if it is not possible to get a quote, some rule-of-thumb costs for concrete pads—including all formwork, bolts and vibration isolators, based on selected weights of equipment—are furnished in Appendix D.

Excavation and Backfill

Excavation for pipelines should be treated on an individual job basis. Due to the variety of soil conditions on different projects, soil classifications are generally broken into four classes:

> **Class I (Good):** Soft clay, loose medium sand, stiff clay, loose coarse sand, sand-gravel mixtures, loose gravel, compact fine sand, compact coarse sand, and gravel.
>
> **Class 2 (Fair):** Soft broken bedrock, compact partially cemented gravels and sand, and hardpan.
>
> **Class 3 (Difficult):** Foliated rock such as schist and slate, and sedimentary rock such as shale and sandstone.
>
> **Class 4 (Poor):** Massive bedrocks such as granite, diorite, and gneiss.
>
> Class 3 may require blasting, and Class 4 almost surely will. Blasting should be treated as a separate contract by the plumbing contractor.

Most excavation for plumbing work is accomplished with a trencher or backhoe and backfilled with a bulldozer. In tight areas or existing structures, hand excavation may be necessary.

Regardless of the class of soil or method employed, all excavation is estimated by the cubic yard. Figure 1.23 is a guide for calculating the total cubic yards of excavation required to install piping. It should be noted, however, that the table does not take into consideration sheeting, shoring, extreme depths, or multiple pipelines.

Sheeting and shoring should be included in excavation costs where piping is deeper than 6′ or where unstable soil conditions exist. The cost per cubic yard of excavation and backfill will have to be determined by soil conditions, depths, and the particular method used. The recommended bottom trench width for various pipe diameters is shown in Figure 1.24. Costs can be found in the current version of *Means Plumbing Cost Data*.

Tests and Adjustments

All piping systems in a plumbing installation, whether in the building or on the site, must be tested to uncover any material defects, improper connections, and poor workmanship. There are many methods used to test piping systems: soil, waste, vent, drain, water, and gas piping systems are tested using either water, smoke, air, or oil of peppermint. There are generally two construction periods during which tests are performed. The first takes place after all roughing has been completed. The plumber caps all lines, with the exception of the opening from which the test material will enter the system. The second test is performed after the installation of all fixtures.

In the air test, compressed air is forced into the system until a pressure of 3 psi is attained. If any leaks occur, there will be a pressure drop.

Condition 'A'

Condition 'B'

Condition 'C'

WIDTH IN FEET	DEPTH IN FEET																
	2	3	4	5	6	7	8	9	10	11	12	13	14	15	16	17	18
1	.07	.11	.15	.19	.22	.26	.30	.33	.37	.41	.44	.48	.52	.56	.59	.63	.67
2	.15	.22	.30	.37	.44	.52	.59	.67	.74	.81	.89	.96	1.04	1.11	1.18	1.26	1.33
3	.22	.33	.44	.56	.67	.78	.89	1.00	1.11	1.22	1.33	1.44	1.56	1.67	1.78	1.89	2.00
4	.30	.44	.59	.74	.89	1.04	1.18	1.33	1.48	1.63	1.78	1.92	2.07	2.22	2.37	2.52	2.67
5	.37	.56	.74	.93	1.11	1.30	1.48	1.67	1.85	2.04	2.22	2.41	2.59	2.78	2.96	3.15	3.33
6	.44	.67	.89	1.11	1.33	1.55	1.78	2.00	2.22	2.44	2.66	2.89	3.11	3.33	3.55	3.77	4.00
7	.52	.78	1.04	1.30	1.55	1.81	2.07	2.33	2.59	2.85	3.11	3.37	3.63	3.89	4.14	4.40	4.67
1	.14	.28	.45	.65	.89	1.17	1.48	1.83	2.22	2.65	3.10	3.61	4.15	4.72	5.33	5.98	6.67
2	.22	.39	.60	.83	1.11	1.43	1.77	2.17	2.59	3.05	3.55	4.09	4.67	5.27	5.92	6.61	7.33
3	.29	.50	.74	1.02	1.34	1.69	2.07	2.50	2.96	3.46	3.99	4.57	5.19	5.83	6.52	7.24	8.00
4	.37	.61	.89	1.20	1.56	1.95	2.36	2.83	3.33	3.87	4.44	5.05	5.70	6.38	7.11	7.87	8.67
5	.44	.73	1.04	1.39	1.78	2.21	2.66	3.17	3.70	4.28	4.88	5.54	6.22	6.94	7.70	8.50	9.33
6	.51	.84	1.19	1.57	2.00	2.46	2.96	3.50	4.07	4.68	5.32	6.02	6.74	7.49	8.29	9.12	10.00
7	.59	.95	1.34	1.76	2.22	2.72	3.25	3.83	4.44	5.09	5.77	6.50	7.26	8.05	8.88	9.75	10.67
1	.22	.44	.74	1.11	1.55	2.07	2.67	3.33	4.07	4.89	5.77	6.73	7.77	8.89	10.06	11.32	12.67
2	.30	.55	.89	1.29	1.77	2.33	2.96	3.67	4.44	5.29	6.22	7.21	8.29	9.44	10.65	11.95	13.33
3	.37	.66	1.03	1.48	2.00	2.59	3.26	4.00	4.81	5.70	6.66	7.69	8.81	10.00	11.25	12.58	14.00
4	.45	.77	1.18	1.66	2.22	2.85	3.55	4.33	5.18	6.11	7.11	8.17	9.32	10.55	11.84	13.21	14.67
5	.52	.89	1.33	1.85	2.44	3.11	3.85	4.67	5.55	6.52	7.55	8.66	9.84	11.11	12.43	13.84	15.33
6	.59	1.00	1.48	2.03	2.66	3.36	4.15	5.00	5.92	6.92	7.99	9.14	10.36	11.66	13.02	14.46	16.00
7	.67	1.11	1.63	2.22	2.88	3.62	4.44	5.33	6.29	7.33	8.44	9.62	10.88	12.22	13.61	15.09	16.67

Guide for Calculating the Total Cubic Yards of Excavation Required to Install Piping

Figure 1.23

The oil of peppermint test is done by mixing an oil of peppermint solution with hot water and pouring it into the vent terminals on the roof. If any leaks are present, an odor of peppermint will be detected. The water test is done in a similar fashion, and if leaks are present, water will appear at the defective joint.

The estimator should include testing costs (labor) and a percentage for possible material replacement. Testing costs are generally estimated as a percentage of the total direct job cost. The plumbing estimator can generally assume a range of 2% to 4% of the direct cost. This percentage includes all material, labor, and test equipment.

RECOMMENDED BOTTOM TRENCH WIDTH FOR VARIOUS PIPE DIAMETERS

DIAMETER OF PIPE	BOTTOM TRENCH WIDTH
2″ to 12″	2 feet
14″ to 18″	3 feet
20″ to 24″	4 feet
27″ to 36″	5 feet
42″ to 48″	6 feet
54″ to 60″	7 feet

Figure 1.24

Chapter 2

Plumbing Subsystems and Components

Conversant with the various materials and methods used on a plumbing system, next we will study the various subsystems and components that together form the complete plumbing system. Depending on the type of building or facility, a *complete* plumbing system must be designed to meet the needs of the project. For example, a medical gas system would most certainly be required in a hospital, while it may or may not be a consideration in a nursing home.

Plumbing Fixtures and Trim

A plumbing fixture (sink, water closet, or shower) is probably the most basic plumbing component within a plumbing system. A plumbing fixture is any receptacle, device, or appliance that is supplied with water, or receives or discharges liquids or liquid-borne wastes. Each fixture performs an individual function, such as a shower, lavatory, washing machine, etc. Depending on the type of building, different types of fixtures will be encountered by the plumbing estimator, such as clinical service sinks, flushing rim sinks in hospitals and nursing homes, or wash fountains in schools and factories. There are also different styles of fixtures performing the same function, like wall-hung water closets with flush valves, as compared to water closets with floor-mounted tank and bowl combinations.

Plumbing fixtures are manufactured of various types of materials, most commonly *vitreous china, cast iron, enameled steel, stainless steel,* and *fiberglass.* Vitreous china possesses certain desirable qualities, such as resistance to chipping, scratching, and household acids. Cast iron offers the same benefits and is also exceptionally durable. Enameled steel, though not as durable as cast iron, is an excellent choice where budget-priced, attractive fixtures are desired. Stainless steel is very durable and, with no plating or coating, it is virtually chip-proof. However, the use of stainless steel is generally limited to sinks, drinking fountains, and certain special prison fixtures. Fiberglass is a material that has gained in popularity in recent years. Fixtures such as toilets, lavatories, bathtubs, shower stalls, and combination bath/shower/wall "surround" systems are now commonplace. The light weight of fiberglass reduces installation time considerably. Fiberglass tub and shower walls and bathtub eliminate the need for tiled walls.

Floor, roof, and area drains are generally constructed of cast iron, which is resistant to corrosion and exceptionally strong.

Figures 2.1 through 2.4 illustrate examples of typical fixtures installed by plumbing contractors.

Every fixture needs trim, which consists of devices that control the flow of water between the piping system and the fixture. Trim is normally exposed and chrome-plated. (See Figure 2.5.)

Examples of trim are flush valves, faucets, vacuum breakers, cup strainers, traps, supply fittings, and stop valves. (Some are supplied with the plumbing fixture; others are purchased separately.) Certain trim items may be rough brass, or, where exposed to the eye, chrome-finish.

Figure 2.6 shows a representative plumbing fixture and trim schedule. The fixtures correspond to those shown in Figures 2.1 through 2.3. Some of the listed trim is shown in the fixture illustrations.

This schedule is by no means complete, since there are hundreds of different types of plumbing fixtures on the market today. However, the schedule gives the estimator a cross-section of common fixtures. In an actual project, the estimator would consult the engineer's specifications (discussed in Chapter 4) for the specific manufacturers and types of fixtures and trim required.

Bathroom Accessories

Accessories are actually neither plumbing fixtures nor trim. However, since some accessories are functionally related to certain types of fixtures or areas requiring fixtures, we have included them in this section of the book. Unless specified or because of a local union regulation, lavatory accessories are not usually the responsibility of the plumbing contractor, but this is not always the case. For some projects, these accessories are specified to be supplied and installed by the

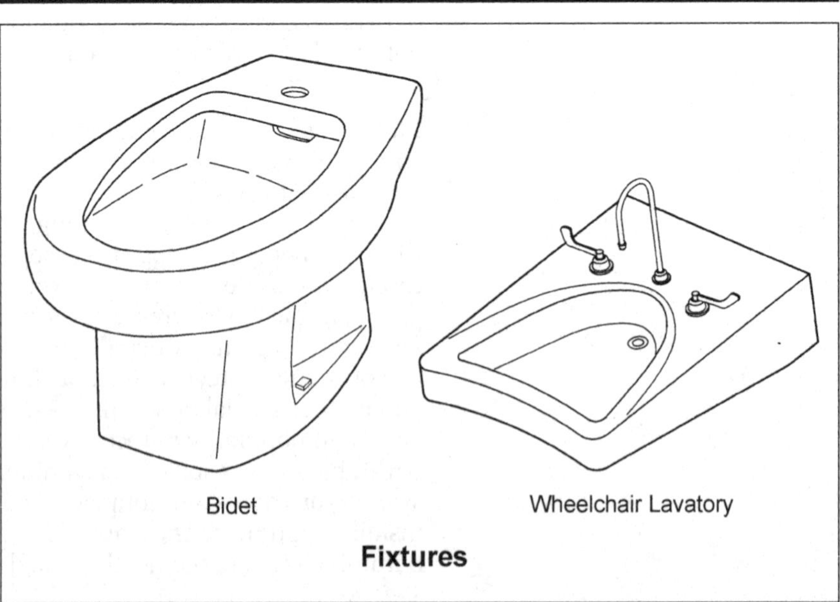

Fixtures

Figure 2.1

plumbing contractor. In others, accessories are supplied by the general contractor and installed by the plumbing contractor.

Therefore, the plumbing estimator should be aware if these items are to be included in the price.

A list of toilet and bath accessories would include:

Toilet paper holders
Soap dispensers
Towel dispensers
Waste receptacles
Sanitary napkin dispensers
Sanitary napkin disposers
Facial tissue dispensers
Mirrors
Medicine cabinets
Grab bars (handicapped)
Soap dishes
Tumbler and toothbrush holders
Towel bars
Pail and ladder hooks
Shower curtain rods
Broom and mop holders
Shelving robe hooks

Figure 2.2

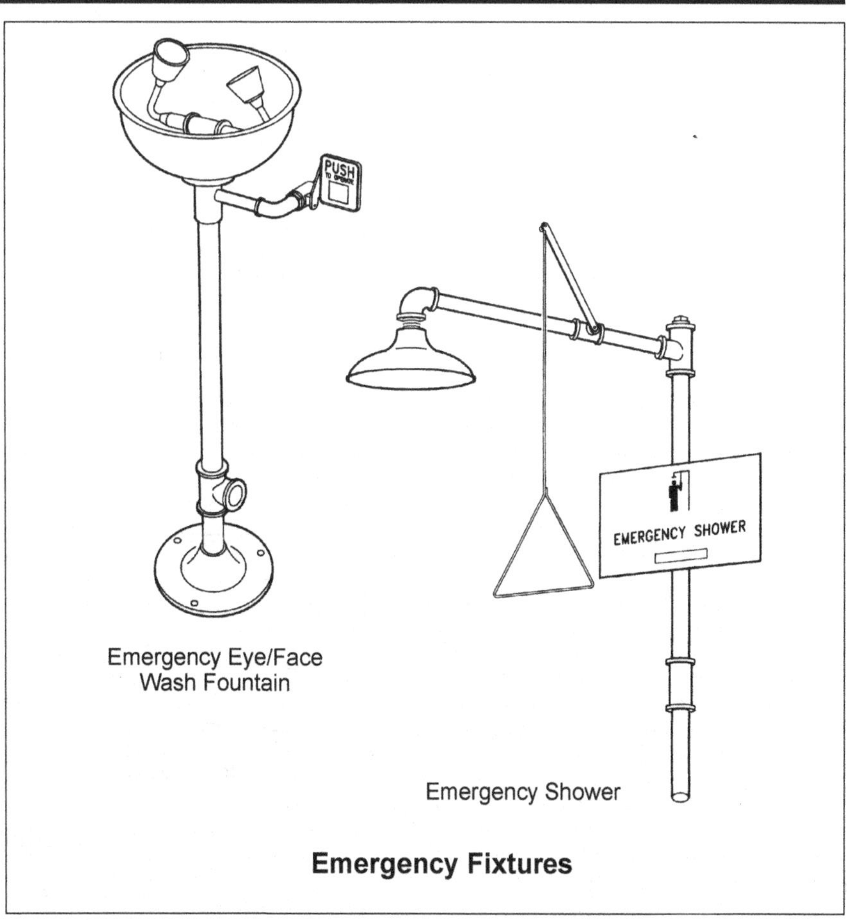

Emergency Fixtures

Toilet and bath accessories are generally constructed of stainless steel, chrome, or anodized aluminum. If the accessories are part of the plumbing contract, the estimator, before attempting a quantity takeoff, should obtain a set of drawings showing exact locations of the specified accessories, and should verify these locations with the architect.

Hook-Up of Fixtures and Trim Supplied by Others

Depending on the engineer's specification, certain plumbing fixtures or equipment may be owner- or general contractor-supplied. However, the rough-in and installation of these items are usually the responsibility of the plumbing contractor. The estimator should consult the architect and/or specifications to learn exactly which items, supplied by others, the plumbing contractor must install.

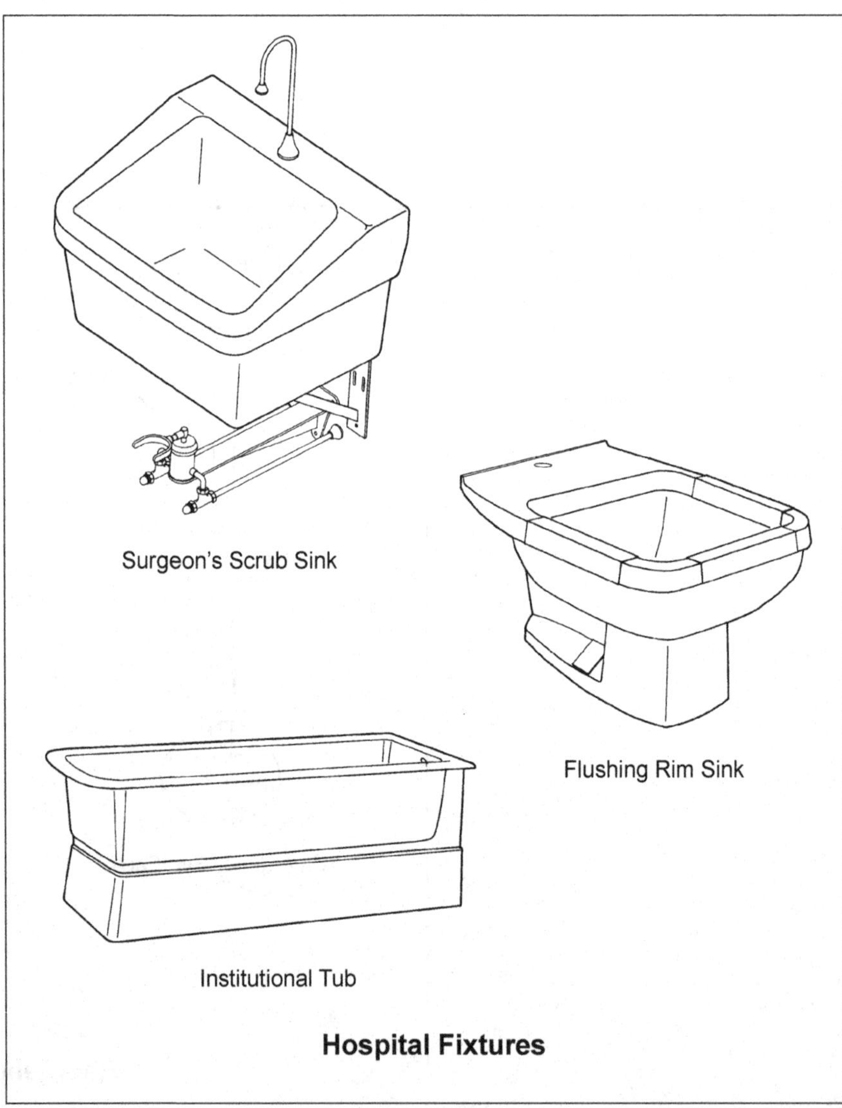

Hospital Fixtures

Figure 2.3

Owner- or general contractor-furnished shed fixtures are usually found in, but not limited to, the following areas:

Kitchens and Cafeterias

- Sinks
- Steam kettles
- Ranges
- Griddles
- Coffee urns
- Dishwashers
- Ice makers
- Refrigerators

Laboratories and Hospitals

- Lab sinks
- Fume hoods
- Cup sinks

Figure 2.4

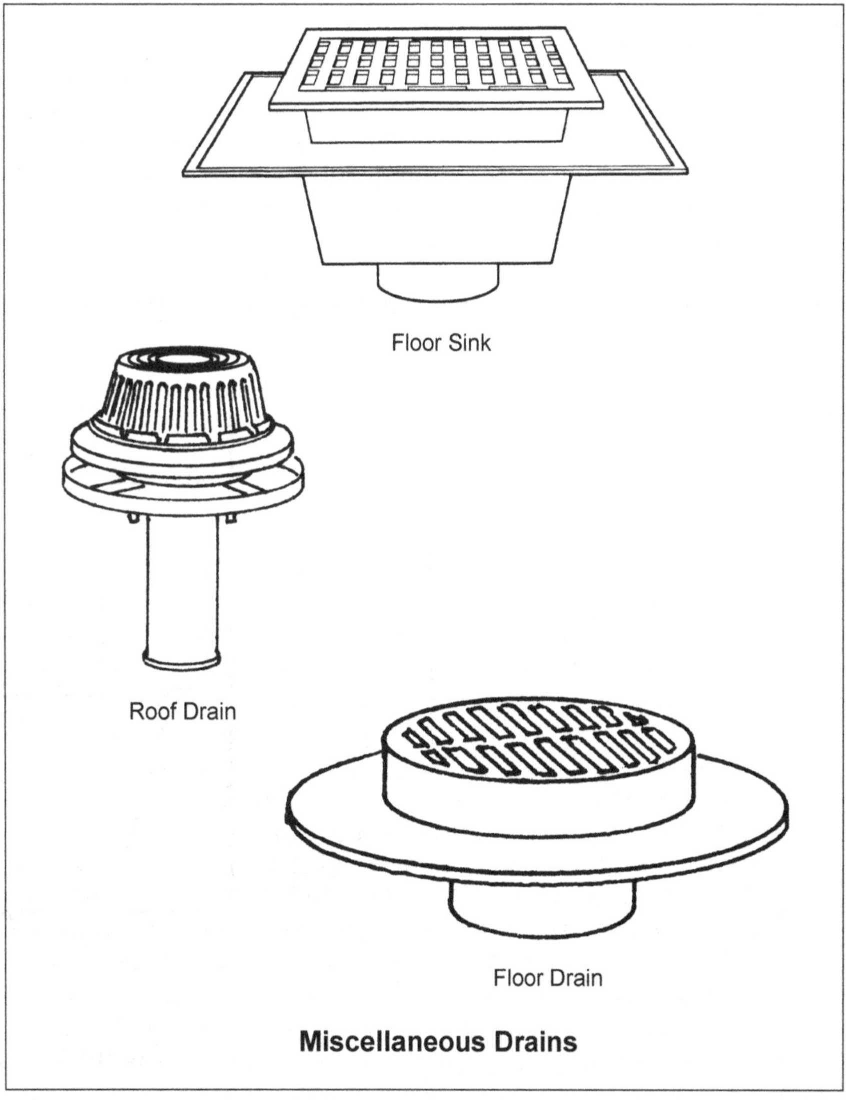

Miscellaneous Drains

Hydro-therapy baths
Autopsy tables
Laboratory tables
Sterilizers
Bedpan sterilizers

There are generally accepted standards of pipe sizes for connecting fixtures to hot and cold water, vent, and waste lines, as shown in Figure 2.7. While this table can be used as a guide, the estimator should refer to local plumbing codes for exact requirements.

Domestic Hot and Cold Water Systems

The domestic hot and cold water system is a common piping system found on virtually every plumbing job. The domestic hot and cold water system in a building consists of:

—Water distribution piping (mains, risers, branches, runouts)
—Fittings
—Control valves and devices
—Hot water generators
—Pumps
—All appurtenances used for conveying water to the plumbing fixtures

The cold water supply in a building begins once it enters the building wall. It is usually connected to a water meter (see Figure 2.8a) at this point of entry, and from the meter begins its distribution to the necessary fixtures and equipment. (Note: Water meters can also be

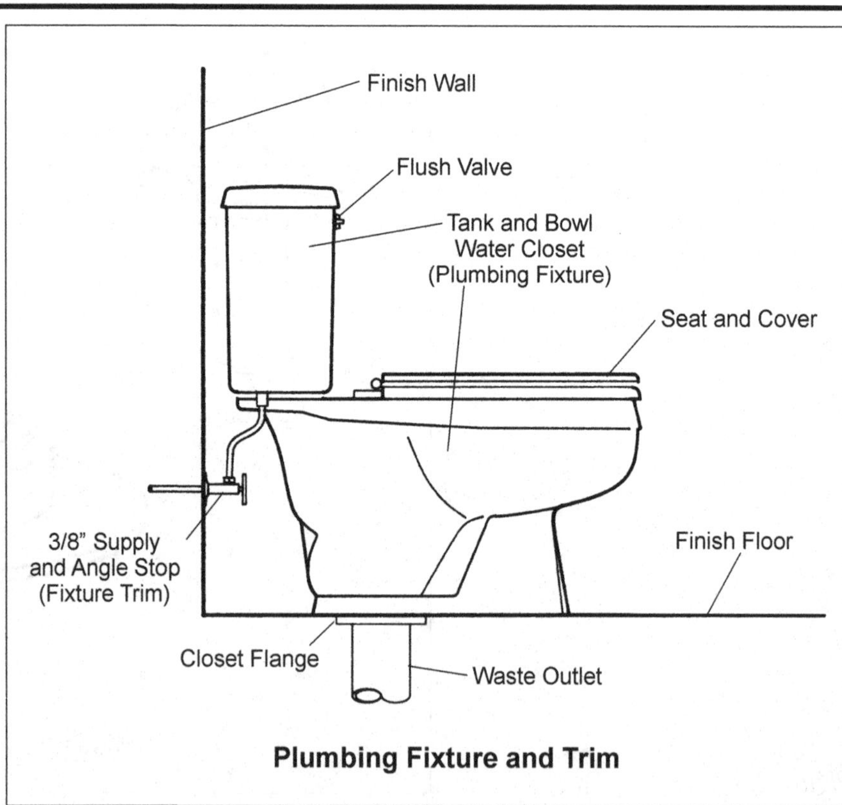

Figure 2.5

TYPICAL PLUMBING FIXTURE AND TRIM SCHEDULE			
FIXTURE	TRIM	TRIM SUPPLIED WITH FIXTURE	
		Yes	No
Water Closet (Flush Valve Type, Wall-Hung) (Vitreous China) *Note*: Carrier fitting included with rough piping.	Toilet seat	*X	
	Flush valve		X
Water Closet (Tank Type, Floor-Mounted) (Vitreous China)	⅜″ supply/angle stop and escutcheon		X
	Wax gasket/closet or lead bend		X
	Toilet seat	*X	
	Bolt caps/closet flange	X	
	Water control/flush/trip lever	X	
Lavatory (Vitreous China)	⅜″ faucet/1¼″ pop-up drain		X
	1 pair ⅜″ supplies/angle stops and escutcheons		X
	1¼″ "P" trap/tailpiece		X
Service Sink (Vitreous China)	½″ service sink faucet		X
	Rim guard, vacuum breaker, bucket hook, integral stops (all optional)		X
	3″ trap with strainer		X
Sink (Kitchen) (In-counter Type) (Enameled Cast Iron) *Note*: Stainless Steel Counter Frame not Provided by Plumber.	⅜″ faucet with spray		X
	1½″ crumb cup strainer/tailpiece		X
	1½″ cast brass "P" trap		X
	1 pair ⅜″ straight stops		X
Bathtub (with Shower) (Enameled Cast Iron)	Shower head, arm and escutcheon		X
	Bath spout		X
	Bath/shower diverter valve with handles		X
	1½″ remote pop-up bath drain with tailpiece and outlet tee		X
Electric Water Cooler	Drain strainer and bubbler	X	
	⅜″ self-closing valve	X	
	1½″ x 1¼″ cast brass "P" trap/ tailpiece and ⅜″ supply		X
Dishwasher	⅜″ angle stop and supply		X
	Special 1¼″ x ½″ "Y" tailpiece cast brass		X
Clothes Washer	(1) Drain hose (2) Fill hoses with female threaded connections and vacuum breakers	X	
	1 pair ½″ hose end control valves		X

*Seat not always supplied with fixture, check model number

Figure 2.6

TYPICAL PLUMBING FIXTURE AND TRIM SCHEDULE			
FIXTURE	TRIM	TRIM SUPPLIED WITH FIXTURE	
		Yes	No
Wash Fountain (54" Diameter) (Granite, Stainless Steel)	Spray head, supporting tube, strainer, foot valve, operating mechanism, foot levers, rail, mixing and volume control valves, combination stop, check valves, and soap dispenser	X	
	2" cast brass "P" trap		X
Drinking Fountain (Wall-Hung, Semi-Recessed) (Vitreous China)	Cross handle supply valve, brass strainer plate, ⅜" service shut-off valve, 1¼" tailpiece waste-outlet	X	
	Mounting frame		X
	1½" x 1¼" cast brass "P" trap		X
Shower (Stall) 36" x 34" (Terrazzo) Receptor In lieu of tile walls, metal wall panels may be specified (Check specification)	4" grid drain with 2" caulked outlet	X	
	Shower head, arm and escutcheon		X
	1 pair angle valves or (1) mixing valve		X
	2" cast brass "P" trap		X
Shower (Column Type) (Wall Pack)	Units are supplied with necessary trim. Piping for water supplies, vents, and drain are connected directly to units.	–	–
Urinal (Flush Valve Type, Wall-Hung) (Vitreous China) *Note*: Carrier, if required, is included in rough piping.	Flush valve		X
Flushing Rim Sink (Wall-Hung, Integral Trap) (Vitreous China)	Flush valve with vacuum breaker		X
	Rim guard		X
	½" supply fitting with screwdriver stops, 6" elbow control handles, spout with bucket. Hook and fork brace		X
Institutional Tub (Cast Iron) *Note*: Thermostatic Mixing Valve necessary unless tempered water system used.	Cast iron enamel base. Built-in over-rim bath filler with compression		X
	Bypass valves, four-arm handles, and escutcheons. Built-in spray fitting with compression valves		X
	Elevated vacuum breaker, nozzle with hook, 3' rubber hose with spray four-arm handles and escutcheons		X
	Lever-operated pop-up drain with 2" tailpiece		X

Figure 2.6 (cont'd.)

TYPICAL PLUMBING FIXTURE AND TRIM SCHEDULE			
FIXTURE	TRIM	TRIM SUPPLIED WITH FIXTURE	
		Yes	No
Bidet (Floor-Mounted) (Vitreous China)	Supply fittings, vacuum breakers, bolt caps 1¼″ cast brass 'P' trap	X	X
Scrub Sink (Wall-Hung) (Vitreous China)	½″ supply fitting with 2″ spray screwdriver stops, 6″ elbow control handles		X
	Drain plug with metal grid and 1½″ tailpiece		X
	1½″ cast brass 'P' trap		X
	1½″ trap nipple with cast escutcheon		X
Floor Drains (Integral Trap) (Cast Iron)	Clamping device		X
	Strainer	X	
Roof Drain (Cast Iron)	Dome strainer	X	
	Deck clamp, drain receiver		X
Wall Hydrant (Box Type)	Wall clamp, nikaloy finish, cylinder lock, extra deep box, nozzle type vacuum breaker, elbow union assembly		X
Emergency Shower	1″ self-closing valve, deluge head, 32″ chain and pull-ring	X	
Eye Wash Fountain	½″ supply w/ball valve, 1½″ drain pipe, 1½″ drain tee, stainless steel bowl and aerators, floor stand, 1½″ cast brass 'P' trap	X	
Area Drain (Cast Iron)	Strainer, hinged top	X	
Trench Drain (Cast Iron)	Strainer	X	
	Additional sections		X
Floor Sink	Strainer (vitreous enameled)	X	
	Sediment bucket, flashing ring, trap primer		X

Figure 2.6 (cont'd.)

located outside the building in concrete vaults.) The hot water supply starts from the hot water generation equipment and from there begins its distribution to the fixtures. In some buildings, a hot water circulation main is desired to provide continuous hot water circulation throughout the entire system.

The domestic water system in a building is the potable water system, which means it is suitable and used for human consumption, being sufficiently free of impurities that could cause disease. However, there

PLUMBING FIXTURE AND MINIMUM PIPING CONNECTION SCHEDULE				
FIXTURE	WASTE	VENT	HOT WATER	COLD WATER
Water Closet (Flush Valve)	4″	2″	–	1″
Water Closet (Tank Top, Low Flush)	4″	2″	–	½″
Lavatory	1¼″	1½″	½″	½″
Service Sink	3″	2″	¾″	¾″
Sink (All-purpose)	1½″	1½″	½″	½″
Bathtub	1½″	1½″	½″	½″
Electric Water Cooler	1½″	1½″	–	½″
Kitchen Sink	1½″-2″	1½″	½″	½″
Dishwasher	1½″	1½″	½″	–
Clothes Washer	1½″	1½″	½″	½″
Wash Fountain 54″ Dia.	2″	1½″	¾″	¾″
Drinking Fountain	1½″	1½″	–	½″
Shower (Stall)	2″	1½″	½″	½″
Shower (Column Type)	3″	1½″	1″	1″
Shower (Wall-Pack)	3″	1½″	¾″	¾″
Urinal (Flush Valve)	2″	1½″	–	¾″
Flushing Rim Sink	3″	2″	½″	(1)1″ (1)½″
Clinical Sink	3″	2″	½″	1″
Institutional Tub	2″	1½″	(2)½″	(2)½″
Scrub Sink	1½″	1½″	½″	½″
Bidet	1½″	1½″	½″	½″
Floor Drain	2″-3″-4″	1½″	–	–
Roof Drain	2″-3″-4″-5″-6″	–	–	–
Wall Hydrant	–	–	–	¾″
Emergency Shower	–	–	–	1″
Eye Wash Fountain	2″	1½″	–	¾″
Area Drain	3″-4″-5″-6″	–	–	–
Trench Drain	3″-4″-5″-6″	–	–	–
Floor Sink	2″-3″-4″	1½″	¾″	¾″

Figure 2.7

is increasing concern that the water we drink is not pure enough. As a result, more sophisticated filtration systems are becoming popular, as is the use of backflow preventers. Backflow preventers provide greater protection from cross connections in industrial, commercial, and institutional buildings. A typical example of a backflow preventer is shown in Figure 2.8b.

The standard piping materials used in domestic water systems are type K, L, and M copper tubing, and in limited applications, brass and galvanized steel pipe. On cold water, polyvinyl chloride (PVC) plastic pipe is gaining a wider acceptance in certain areas of the country.

All of these are used for aboveground installations. In cases where large diameter pipe is called for, especially around pumps and equipment, flanged ductile iron or cast iron water pipe may be used. Ductile and cast iron are especially desirable for these applications because of their relatively low cost in comparison to copper or steel (which become very expensive for sizes 4" in diameter and larger). Type K soft or hard copper tubing is generally used inside the building or for below-grade or buried installations. Of the five types of materials mentioned for aboveground use, type L copper tubing is the most commonly used material today.

The types of fittings and joining methods used for hot and cold water systems are solder-joint, or roll grooved pressure fittings for copper tubing; threaded, or flanged cast brass fittings for brass pipe; and

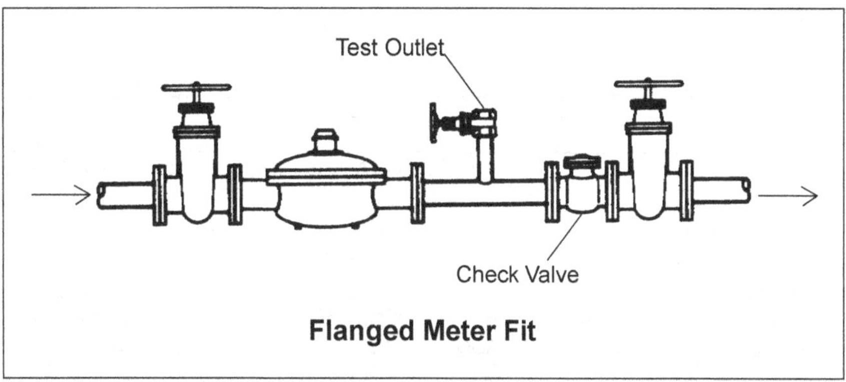

Flanged Meter Fit

Figure 2.8a

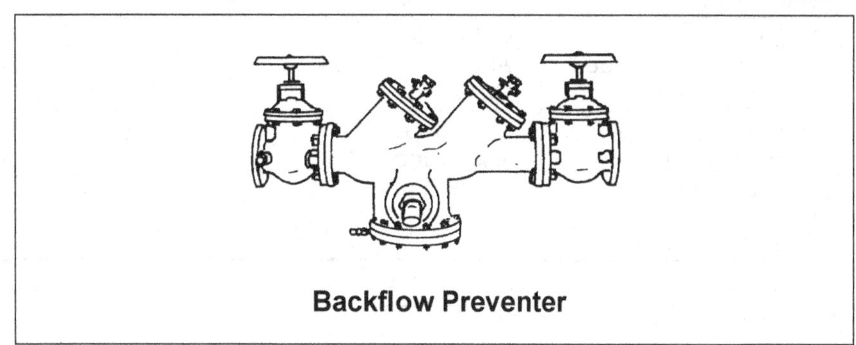

Backflow Preventer

Figure 2.8b

threaded galvanized cast iron fittings for steel pipe; and threaded- or cement-joint fittings for PVC pipe. They are available in many sizes and a wide variety of types, as shown in Figures 1.4 and 1.7.

Figure 2.9 is a concise guide to the various types of pipe, fittings, recommended service, and sizes generally used on domestic hot and cold water systems. The flow and operation of a hot and cold water system is controlled through various valves and other devices. (See Figures 1.5, 1.6, 1.7, 1.8 and 1.9.) Gate valves are the standard method of controlling water in a system, and are used where full and unobstructed flow is required. Globe valves become more desirable when operation is frequent and close, and more critical flow control is essential. Check valves are designed to prevent reverse flow in the piping system.

Valves are located in critical areas: around equipment, at fixture branches, and at runouts to risers to ensure isolation, if necessary. The various types of valves and control devices encountered by the estimator on a hot and cold water system are described later in this chapter, in Figure 2.10.

Hot water in a building is obtained through the use of hot water/steam heaters or generators. (See Figures 2.11 and 2.12.) There are five basic water heater classifications:

- Storage
- Instantaneous
- Semi-instantaneous
- Steam and water-mixing valve
- Water blending valve (hot/cold)

DOMESTIC WATER PIPING AND FITTING SCHEDULE			
PIPE	SERVICE	SIZE	FITTINGS
Type L Copper Hard-Drawn	Hot and Cold Water Above Grade	4″ and Under	Cast Bronze—Solder Joint Wrought Copper—Solder Joint
Brass Pipe Threaded	Hot and Cold Water Above Grade	4″ and Under	Cast Bronze—Threaded Joint
Galvanized Steel Pipe Threaded Schedule 40	Hot and Cold Water Above Grade	2½″ and Above	Galvanized C.I. Threaded Fittings 125 lb.
Type K Copper Hard-Drawn	Hot and Cold Water Below Grade	3″ and Under	Wrought Copper—Solder Joint
PVC Plastic Pipe Schedule 40 Threaded or Cement Joint	Cold Water Above and Below Grade	4″ and Under	PVC Cement Joint Fittings Schedules 40 and 80
Ductile-Iron Water Pipe (Flanged, Mechanical or Neoprene Joint) Class 150 or 250	Cold Water Above and Below Grade	4″ and Above	D.I. Fittings Flanged, Mechanical or Neoprene Joint

Figure 2.9

DOMESTIC WATER SYSTEM VALVE AND CONTROL DEVICE CHART		
VALVE OR DEVICE	DESCRIPTION	APPLICATION
Globe Valve (Straight Pattern)	Bronze Construction, Threaded ⅛"-3", Flanged 1"-3", 150 p.s.i. Pressure Rating, Composition Disc, Screw-Over Bonnet. 125 lb. Working Pressure.	General purpose globe for water system, beneficial where close flow control is essential.
Globe Valve (Angle Type)	Bronze Construction, Threaded ⅛"-3", 150 p.s.i. Pressure Rating, Composition Disc, Screw-Over Bonnet. 125 lb. Working Pressure.	Same as above except used where angle connection is desired.
Globe Valve (Solder End)	Bronze Construction, Solder Joint ¼"-3", 300 p.s.i. Pressure Rating (Water), Composition Disc. 125 lb. Working Pressure.	Standard globe valve used on domestic water lines. Solder ends for joining to type K, L, and M copper.
Gate Valve (Straight Pattern)	Bronze Construction, Threaded ¼"-3", Flanged ½"-3", 150-300 p.s.i. Pressure Rating, Inside Screw Non-Rising Stem, Solid Wedge. Screw-In Bonnet. 125 lb. Working Pressure	General purpose gate for shut-off control, is particularly useful where headroom is limited.
Gate Valve (Solder End)	Bronze Construction, Solder Joint ⅜"-3", 200 p.s.i. Pressure Rating (Water), Inside Screw Non-Rising Stem, Solid Wedge. 125 lb. Working Pressure.	Standard gate valve used on domestic water lines. Solder ends for joining to type K, L, and M copper.
Gate Valve (Solder End)	Bronze Construction, Solder Joint ⅜"-3" 200 p.s.i. Pressure Rating (Water), Traveling Stem, Solid or Split Wedge. 125 lb. Working Pressure.	Similar to valve above except equipped with traveling stem which indicates wedge position.
Check Valve (Swing Type)	Bronze Construction, Threaded ¼"-3". Bronze Disc. Screw-In Cap. 300 p.s.i. Pressure Rating. 125 lb. Working Pressure.	General purpose check valve, for prevention of backflow on water lines.
Check Valve (Swing Type Solder End)	Bronze Construction, Solder Joint ⅜"-3". Bronze Disc. Screw-In Cap. 300 p.s.i. Pressure Rating. 125 lb. Working Pressure.	Ideal check valve for water lines, providing non-return control service. Solder ends for joining to type K, L, and M copper.
Globe Valve (Iron Body Straight Pattern)	Iron Body Construction, Threaded 2"-4", Flanged 2"-8", 125 p.s.i. Pressure Rating, Composition Disc, Outside Stem and Yoke, Renewable Bronze Seat Ring. 125 lb. Working Pressure.	Utilized where large pipe sizes occur, iron body construction keeps cost at a minimum. Globe valve ideal where close flow control is desired.
Gate Valve (Iron Body Straight Pattern)	Iron Body Construction, Threaded 2"-4", Flanged 2"-20", 150 p.s.i. Pressure Rating, Inside Screw, Non-Rising Stem, Bronze Trim, Solid Wedge. 125 lb. Working Pressure.	General purpose gate for large pipe sizes, low-cost iron body construction. Ideal where full and unobstructed flow is essential.
Ball Valves (Bronze, Plastic)	Bronze or Plastic Construction Threaded, Solder Cement Joint. 150 lb. Working Pressure.	General purpose valve for water system, few flow restrictions.
Butterfly Valves (Iron Body, Carbon Steel, Bronze)	Iron Body, Bronze and Carbon Steel Construction. Flanged or Grooved Joints. 125-200 lb. Working Pressure.	General purpose, low cost. Ideal for equipment hook-up, due to size and serviceability.

Figure 2.10

The heat sources for these generators can be steam, hot water supply (boiler), electric, and gas- or oil-fired. The tanks are constructed of steel and are ASME-rated pressure vessels. (The American Society of Mechanical Engineers founded this committee—Boiler and Pressure Vessels—to develop standards for construction of steam boilers and other pressure vessels.) Linings for the tanks can be cement, copper silicone, or glass. Capacity, size, and type of heater vary depending on the hot water demand for the particular building, and are usually specified by the mechanical engineer.

In buildings where a hot water circulation line has been designed, it becomes necessary to furnish a hot water circulating pump. These pumps are usually small, in-line fractional horsepower pumps of bronze or iron-body construction.

DOMESTIC WATER SYSTEM VALVE AND CONTROL DEVICE CHART		
VALVE OR DEVICE	DESCRIPTION	APPLICATION
Check Valve (Iron Body Swing Type)	Iron Body Construction, Threaded 2"-4", Flanged 2"-12". 125 p.s.i. Pressure Rating, Bronze Seat Ring and Disc, Bolted Cover. 125 lb. Working Pressure.	Low-cost iron body construction. Used to prevent reverse flow in piping system.
Pressure and Temperature Relief Valves	Bronze Construction, Temperature Relief 210°F., Pressure Ranges 75 to 175 lbs. ½" to 2" Pipe Size.	Used to prevent high water temperature and pressure on a hot water tank or heater, which can cause flashing of steam at faucets and tank damage.
Pressure Reducing Valves	Bronze Body and Trim or Iron Body with Bronze Trim. Threaded ⅜"-2", Flanged 1"-12". Reduced Pressure Ratings 2-150 p.s.i.	Used to regulate downstream pressures on water system.
Backflow Preventers	Bronze Body and Trim or Iron Body with Bronze Trim ¾"-10" Screwed or Flanged.	Used on direct water connections where the possibility of non-toxic contamination from backflow exists.
Expansion Joints	Piston or Bellows Type, Threaded, Flanged or Solder Joint. ½"-20". 150-250 p.s.i.	Used to control expansion and contraction of pipe lines due to temperature changes.
Strainers (Y-Type)	Bronze Construction, Threaded ¼"-4", Iron Body, Flanged ½"-14". 150-600 p.s.i. Pressure Ratings.	Used to strain foreign matter from pipe lines.
Shock Absorbers	Precharged Nitrogen Water Hammer Arresters, Stainless Steel Construction. Plumbing and Drainage Institute Sizes A through F. Fixture Ratings from 1 through 330.	Used to prevent water hammer caused by quick closing valves or lever faucets at fixtures.
Vacuum Breakers	Bronze Construction, Threaded ¼"-3".	Used to prevent back-syphonage of polluted water into potable water lines.
Hose Bibbs	Bronze Construction, ½"-3" Threaded or Solder Joint. 125 lb. Working Pressure. Threaded Hose Outlet Connection.	Used as hose connection for washdown or garden hose attachment.

Figure 2.10 (cont'd.)

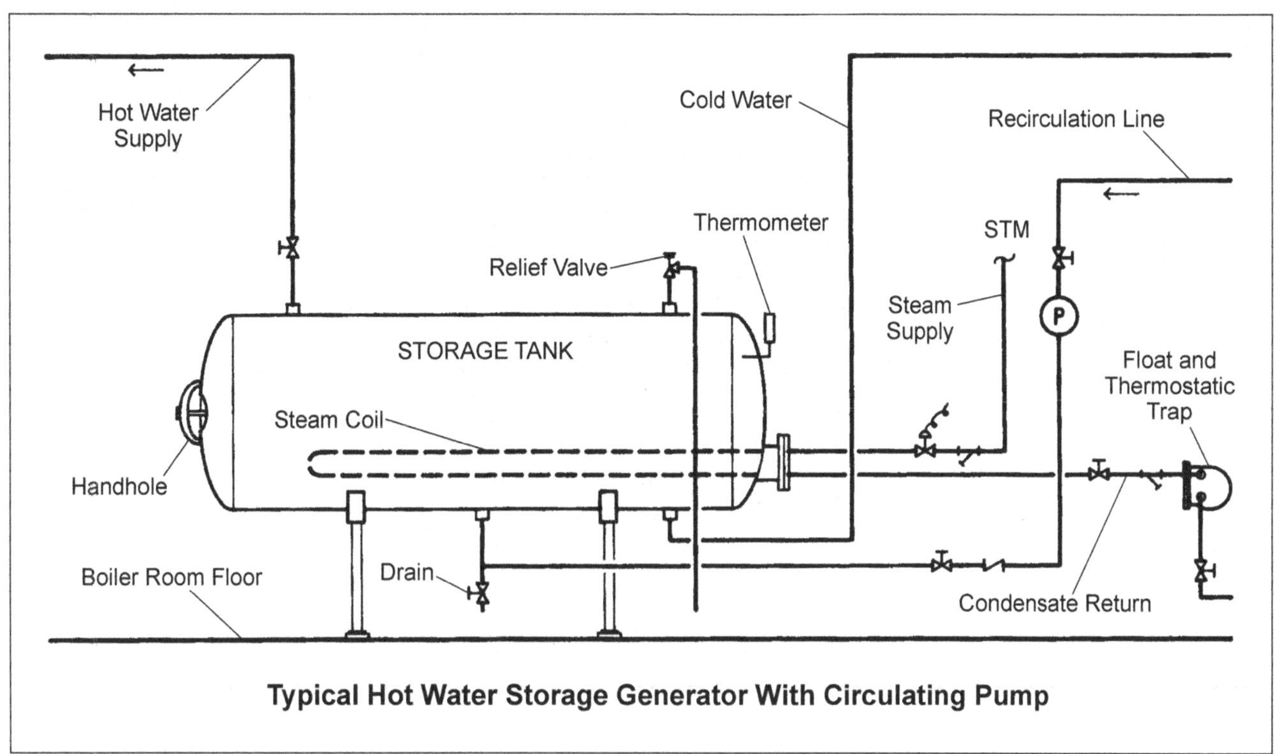

Typical Hot Water Storage Generator With Circulating Pump

Figure 2.11

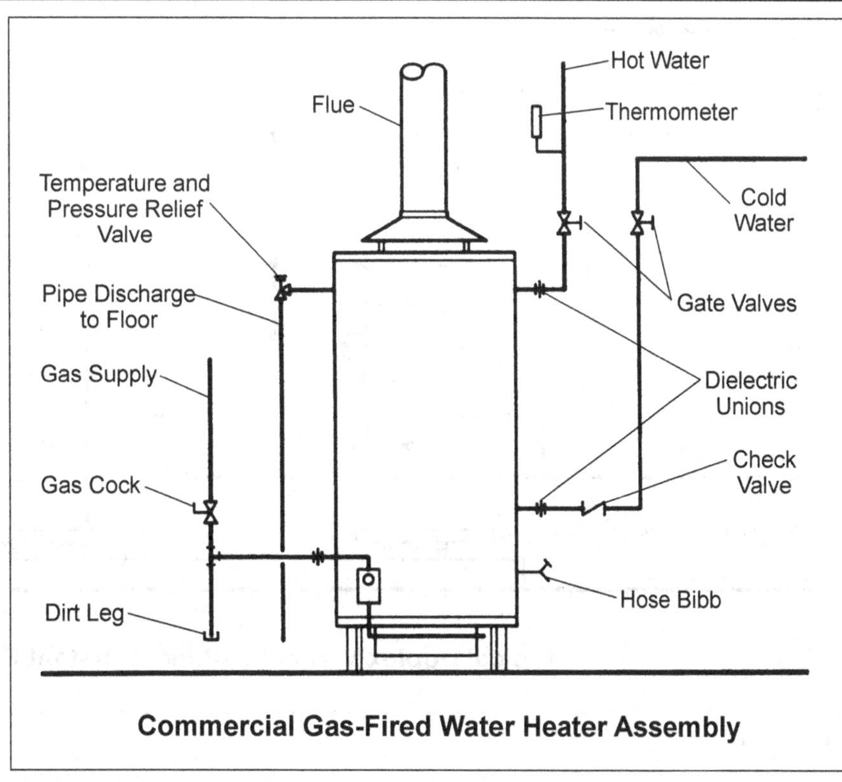

Commercial Gas-Fired Water Heater Assembly

Figure 2.12

In high-rise buildings or buildings where outside street water pressure is inadequate, it becomes necessary to furnish a cold water booster or constant pressure pumps. (See Figure 2.13.) Booster pumps are usually duplex or triplex in design and are sized according to the required water demand. These are rather large pumps ranging from 5 to 25 horsepower and are of the vertical turbine or centrifugal design. Some engineers may specify a roof tank system for buildings requiring a large water demand. The water is pumped up to the roof tank, which serves a dual purpose as a domestic water reservoir and fire system reservoir. In case of a pump breakdown, the building's water supply needs can still be met for a short period through means of gravity. Roof tanks are usually constructed of wood (fir, cypress, or redwood), and are subcontracted to contractors specializing in their construction.

Local municipalities usually require water meters in most buildings to accurately determine water consumption. Meters generally range in size from 3/4" to 8" but, unlike gas meters, are furnished and installed by the plumber. (See Figure 2.8a.)

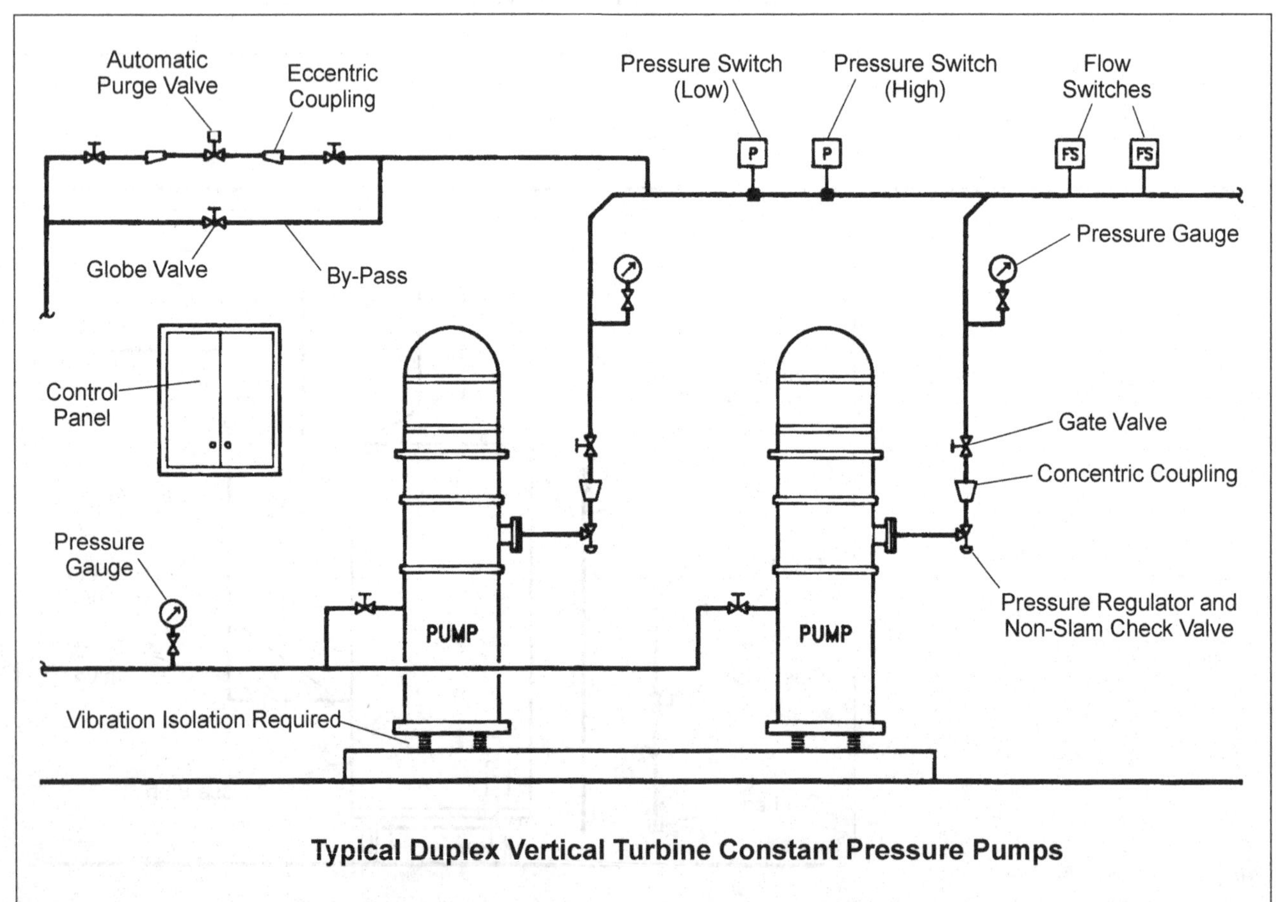

Figure 2.13

Sanitary Waste and Vent Systems

The sanitary waste and vent system in a building carries sewage only and excludes storm, surface, and ground water. The system consists of three areas:

1. Building house drain (sanitary)
2. Soil/waste piping and stacks
3. Vent piping and stacks

Building House Drain

The building house drain is that part of the lowest piping of a drainage system that receives the discharge from the soil, waste, and other drainage pipes of the building and conveys it to the building house sewer by gravity.

The building house drain must have a house trap and fresh air inlet. The house trap should be located near the foundation wall of the structure and should be the same size as the building house drain. The purpose of the house trap is to prevent circulation of air between the house drainage system and the building house sewer. The fresh air inlet is connected to the building house drain immediately upstream from, and within four feet of, the house trap.

Soil/Waste Piping and Stacks

Soil piping is that part of the sanitary waste and vent system that conveys sewage that contains fecal (animal or vegetable wastes) matter to the building house drain. Waste piping, like soil piping, discharges into the building house drain. However, it conveys only liquid waste, free of fecal matter. Both soil and waste piping are shown in Figure 2.14.

Also shown is a soil stack, which is any vertical soil pipe riser extending floor to floor. Soil and waste lines from the plumbing fixtures are connected to the main soil stack.

Cleanouts (see Figure 2.15) should be installed on all soil and waste lines at every change in direction greater than 45° on all horizontal pipes of the drainage system. In addition, a cleanout should be provided at or near each vertical waste or soil stack.

In some cases, it may be necessary to install grease or oil interceptors. These devices are designed to separate and retain grease, oil, or any undesirable matter from normal wastes and permit normal sewage or liquid wastes to discharge into the disposal terminal by gravity. A grease trap can be seen in Figure 2.15.

If fixtures or floor drains are installed below the sewer level outside and cannot drain into the gravity system, a sewage ejector is necessary. (See Figure 2.16.) Ejectors are usually duplex in design and are installed in concrete or cast iron basins. The ejector has the capability to force the sewage up to the gravity system through a pump discharge line.

Vent Piping and Stacks

The vent piping shown in Figure 2.14 is installed as part of the sanitary system, primarily to provide:

- A flow of air to or from a drainage system.
- A circulation of air within such a system to protect trap seals from siphonage and back pressure.

The vent stack, also shown on Figure 2.14, is any vertical vent pipe extending floor to floor, connected to the vent system to provide circulation of air to and from any part of the drainage system. The *vent* stack ends at a minimum of two feet above the roof. This is called the vent *terminal* or *extension*. The vent terminal is flashed with lead or copper sheet flashing to make it weather-tight. In areas of the country where freezing poses a problem, vent terminals should be at least four inches in diameter to prevent frost or snow from clogging the vent opening.

In some cases, vent lines may be connected to the soil stack at a point above the highest fixture. This is called *stack venting*.

There are a number of different methods used to vent fixtures and drains on a sanitary system, some of which are listed below. The choice will depend on prevailing codes.

- Common vent and vent header
- Circuit vent
- Back vent
- Sovent

Figure 2.14

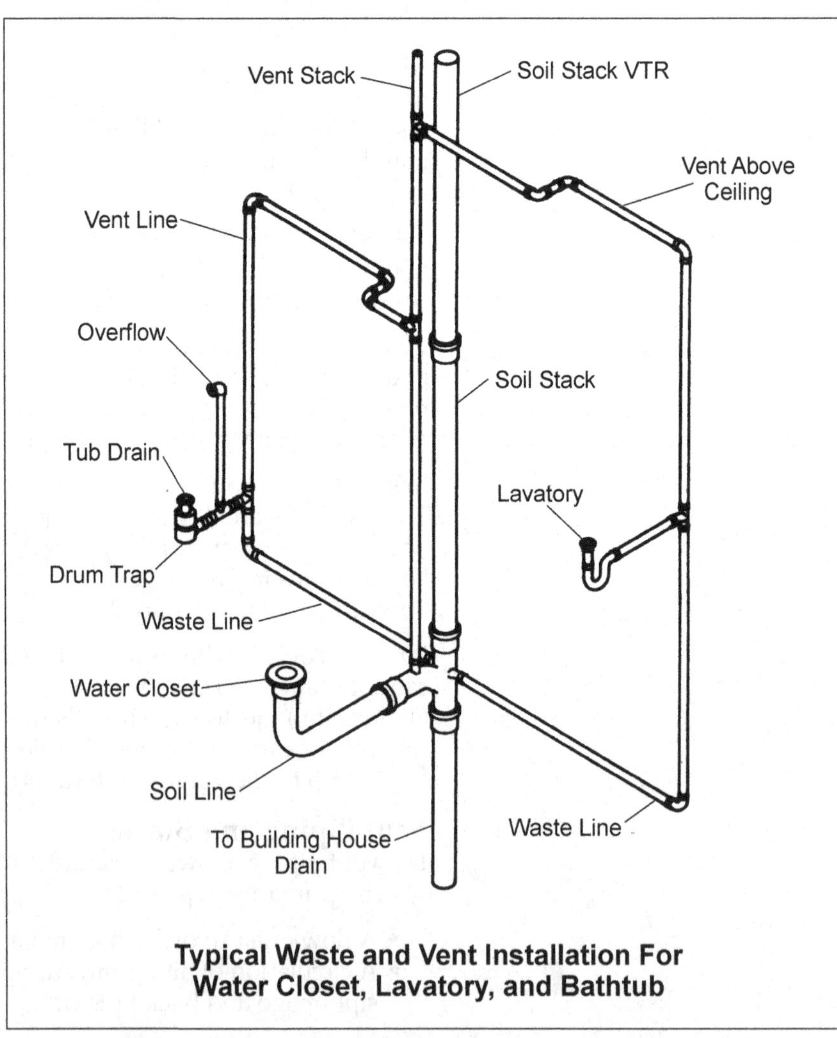

Typical Waste and Vent Installation For Water Closet, Lavatory, and Bathtub

Definitions of these venting methods appear in the Glossary.

There are a number of different piping materials suitable for installation on a sanitary waste and vent system:

- Cast iron soil pipe
- Galvanized steel pipe
- DWV copper
- DWV plastic (PVC, ABS)

Compatible fittings are available for all of the above pipe, as shown in Figures 1.4, 1.7, and 1.8. Gaining wider acceptance and popularity over bell and spigot lead joint pipe and fittings are the neoprene and hubless joint design, as described in Chapter 1. Both of these designs can be used throughout the whole system, with the exception that hubless joints cannot be used for underground installations. Combination hubless DWV fittings, as described in Chapter 1, may be used if appropriate to the design. Galvanized steel pipe with galvanized drainage fittings are also specified quite often on

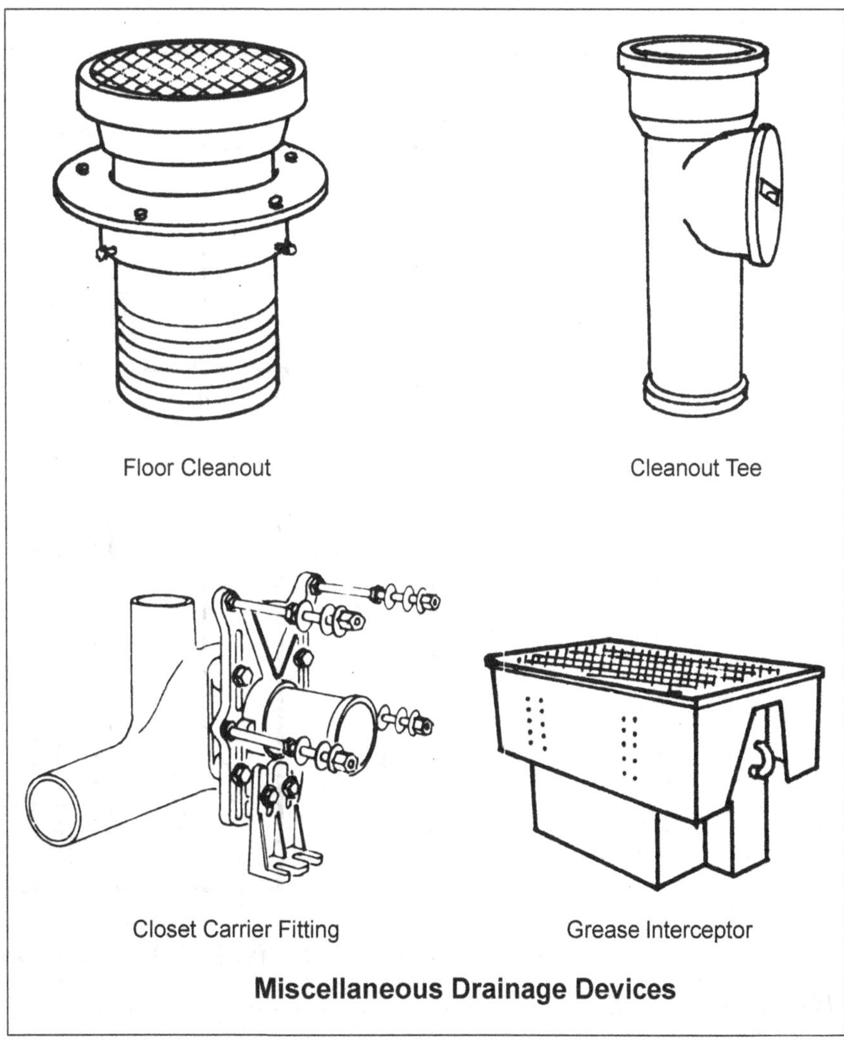

Miscellaneous Drainage Devices

Figure 2.15

aboveground installations and almost always on vent lines 2-1/2″ in diameter and below.

DWV copper and plastic are very popular because of the relative speed with which they can be installed due to light weights and ease of handling. However, these materials can only be installed in aboveground installations.

The above mentioned piping materials and associated fittings, which are found on sanitary waste and vent systems, are outlined further in Figure 2.17. Wall-mounted plumbing fixtures such as urinals and water closets are supported by carrier fittings (shown in Figure 2.15), which are an integral part of the sanitary system roughing. Carrier fittings are made of cast iron or steel. The carrier fittings are completely concealed in the wall and support the fixture by a face and base plate, which is anchored to the floor. Fixture supports are manufactured in various combinations and are readily adaptable to virtually any roughing layout.

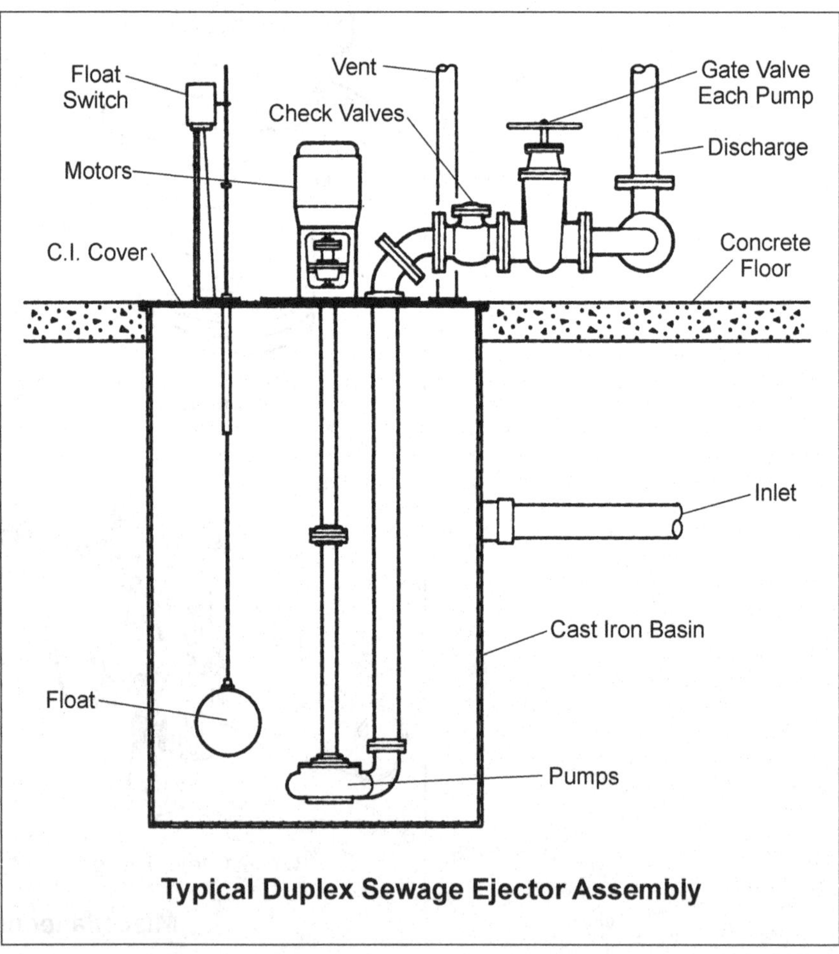

Typical Duplex Sewage Ejector Assembly

Figure 2.16

Storm Drainage System

A storm drainage system is required on virtually all buildings whose design includes a flat roof. The storm system components are:

1. Building house storm drain
2. Storm branch lines
3. Leaders

The building house storm drain is that part of the lowest piping of a storm drainage system that receives clear water drainage from leaders, surface run-off, ground water, subsurface water, condensate, cooling water, or other similar storm or clear water drainage pipes inside the walls of the building. This system conveys the drainage to the building house storm sewer by gravity.

The storm branch lines are all the horizontal drainage lines located above and connected to the building storm drain, the only exception being any horizontal pipe connected to a *single* roof or gutter drain which is considered part of the leader.

A leader is the vertical drainage pipe for conveying storm water from roof or gutter drains to the building storm drain, building house drain (combined), or other means of disposal.

The storm drainage system is similar to the sanitary system in a number of ways.

- House trap required
- Cleanouts required
- Types of pipe and fittings

These devices were fully explained in the previous section. All roof drains should be flashed with lead or copper sheet flashing for weatherproofing. Storm drainage that cannot drain into the gravity system because it is below the outside sewer invert will require a sump

SANITARY WASTE AND VENT PIPING AND FITTING SCHEDULE			
PIPE	SERVICE	SIZE	FITTINGS
Cast-Iron Soil Pipe Lead, Neoprene Joint	Soil, Waste, Vent lines above and below grade	2″-15″	Cast-Iron Soil Fittings Lead, Neoprene Joint
Cast-Iron Soil Pipe Hubless-Clamp Joint	Soil and Waste lines above grade	1½″-10″	Cast-Iron Soil Fittings Hubless-Clamp Joint
Galvanized Steel Pipe Threaded Schedule 40	Soil, Waste, Vent lines above grade	1¼″-12″	Galvanized C.I. Drainage Fittings for Soil and Waste Galvanized Malleable Iron Fittings for Vent
D.W.V. Copper Solder Joint	Soil, Waste, Vent lines above grade	1¼″-8″	D.W.V. Copper Solder Joint Fittings (Wrought or Cast Drainage)
D.W.V. PVC, ABS Cement Joint	Soil, Waste, Vent lines above grade	1½″-8″	D.W.V. Plastic PVC Cement Joint

Figure 2.17

pump. (See Figure 2.18.) Sump pumps are simple or duplex in design and are installed in either cast iron or concrete basins.

Fire Protection Systems

The three fire protection systems that will be discussed in this section are fire standpipe systems, automatic sprinkler systems, and FM200 systems. While fire as a source of damage and destruction will never be totally eliminated, the efforts of professionals should mitigate the tragedy it causes. Recently, major fires have resulted in multi-billion dollar lawsuits involving not only owner and owner agencies, but materials specifiers, architects, engineers, and even code writing certification groups. With all of these parties becoming involved, fire safety has become a major issue and a war is being waged against destructive fires.

Fire safety begins with the proper design and construction of fire-safe structures. Incorporated into that design should be a total systems approach to fire protection. There are three basic components of an engineered fire protection system: detection, alarm, and suppression. When integrated with architectural considerations such as separation walls, fireproof or fire-resistant materials, and a comprehensive fire education/evacuation program, the highest degree of reliability and life safety can be achieved.

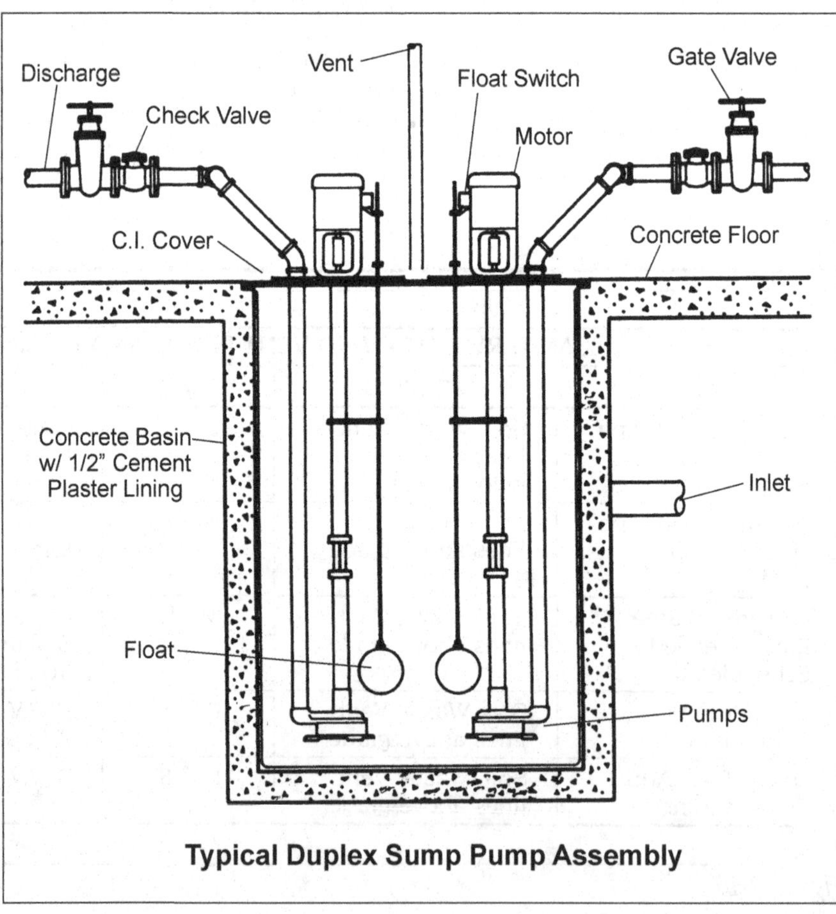

Figure 2.18

Fire Standpipe Systems

In most areas of the country, the fire standpipe system is part of the plumbing contractor's work, when not connected to or part of the sprinkler system. On the other hand, sprinkler systems are not always in the scope of plumbing work. For example, all sprinkler work in New York City is done by sprinkler contractors belonging to the steamfitters union. In many other municipalities across the country, however, sprinkler systems are part of the plumbing contract, even if the plumbing contractor subcontracts his work to a separate sprinkler contractor. In many areas fire protection contractors employ sprinkler fitters.

A fire standpipe is a system of piping for fire-fighting purposes (see Figure 2.19), consisting of a water supply that serves one or more hose outlets. Standpipe systems and their locations are governed by the various local codes. However, a standpipe system is usually required in buildings exceeding three stories in height, or two or more stories in height that have a net floor area of 10,000 square feet or more on any floor.

There are two types of standpipe systems: *wet* and *dry*. A wet standpipe system is one in which all piping is filled with water under pressure that is immediately discharged upon the opening of any hose valve. A dry standpipe system is one in which all piping is filled with air (whether compressed or at atmospheric pressure) to prevent winter freeze up. Water enters the system through a control valve actuated either automatically by the reduction of air pressure within the system or by the manual activation of a remote control located at each hose station. The hose stations are usually located within each stair enclosure or adjacent to the entrance of such enclosure. Fire hose stations consist of a 2-1/2" fire department valve with 75′, 100′, or 125′ of hose on racks or in cabinets. Certain municipalities require a fire reserve roof tank capable of delivering 3,500 gallons of water to each standpipe zone. In buildings where outside water main pressure is inadequate to handle the prescribed water demand, a fire pump is required. Siamese connections are fittings connected to a fire standpipe system, and are installed on the outside of a building with two hose inlets for use by the fire department, to furnish or supplement the water supply to the system. Siamese connections are required on all standpipe systems. One Siamese connection should be installed for every 300′ of exterior building wall.

The standard piping materials for a standpipe system are black steel Schedule 40 or 80 pipe with 350 lb. malleable iron fittings and 150 to 500 lb.-rated fire underwriter gate and check valves. The joining methods used are usually threaded or grooved. In addition to steel, and where code permits, chlorinated PVC piping is used on fire standpipe installations.

Sprinkler Systems

The sprinkler system is the system of piping, valves, and sprinkler heads for fire-fighting purposes, connected to one or more sources of water. According to the National Fire Protection Association (NFPA), sprinklers are the most effective means of automatically controlling fire in buildings. Sprinkler systems are versatile since they trigger the alarm, react immediately, and concentrate directly on the fire with continuous operation until the fire is completely extinguished.

The NFPA also states that sprinklers are 96.2% effective. The 3.8% that might be considered "failures" were due to improper water supply or failure by the facility to modify the system to account for an increased fire hazard.

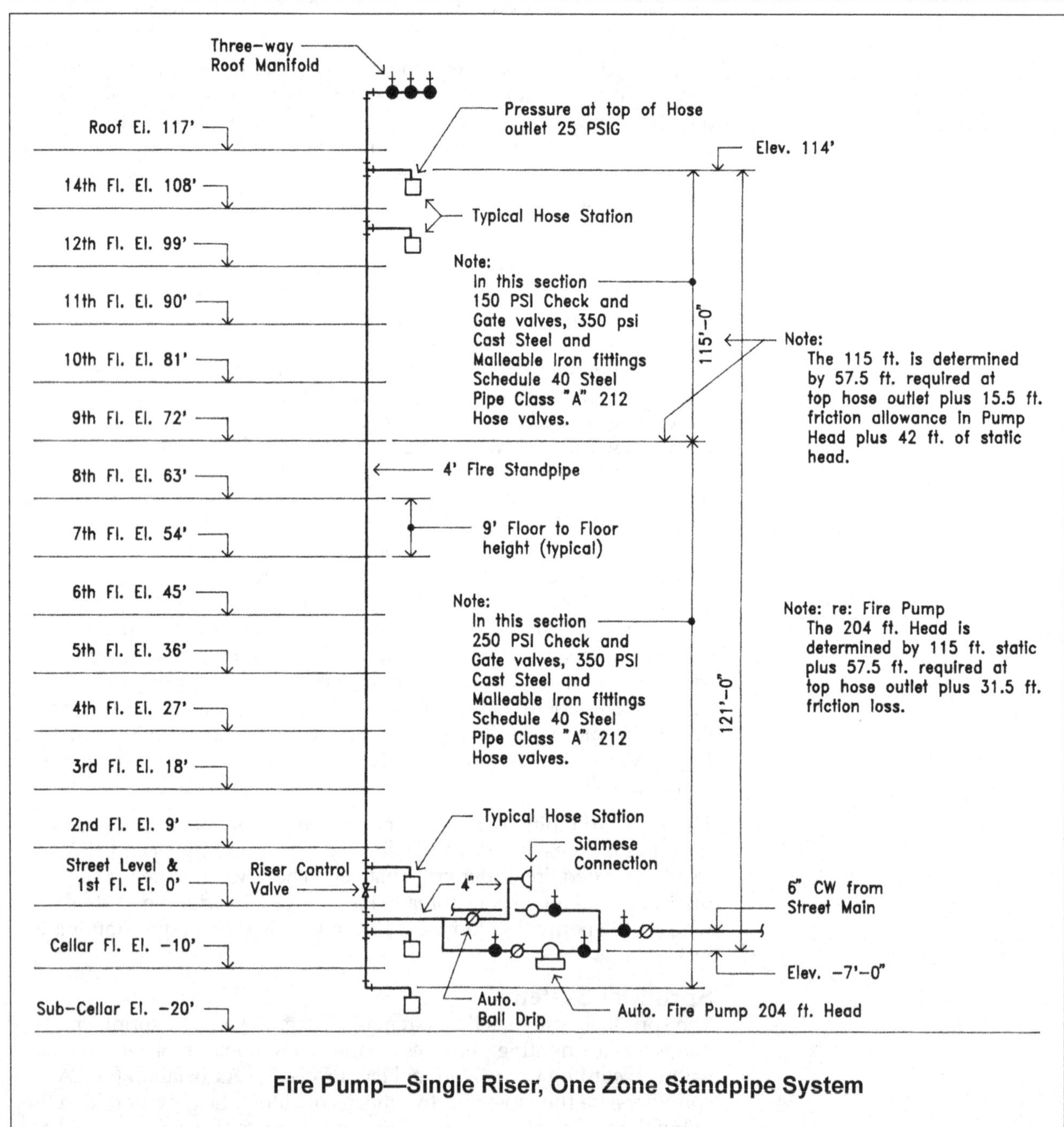

Fire Pump—Single Riser, One Zone Standpipe System

Figure 2.19

No other fire extinguishing agent functions as thoroughly as water. It cools and smothers the fire, while diluting and emulsifying the combustibles. Water damage is kept to a minimum by sprinkler systems for two reasons. First, only sprinkler heads directly over the blaze open up and discharge water. Second, when compared to a standard 2-1/2" diameter firefighting hose, the sprinkler at 75 psi delivers approximately 50 gallons of water per minute, as compared to the hose's 400 gallons per minute. Water consumption by sprinklers is minimal. For example, 37.4% of all reported sprinkler fires were controlled by one head; 73.4% were controlled by five or fewer heads; and 85% were controlled by 10 or fewer heads.

Lastly, and perhaps more important, records show that there have been only a dozen or so instances in which occupants have been killed by fire in a fully sprinklered building. The victims were killed by the fire itself, and were not "drowned" by the sprinkler system.

There are basically five types of automatic sprinkler systems: wet pipe, dry pipe, pre-action, deluge, and firecycle. A sprinkler design engineer normally selects the proper system for the facility he or she is designing, based on its hazard classifications. Buildings are evaluated and given a hazard classification based on two criteria: building construction, and building occupancy.

While there are numerous municipal building and fire codes, the most widely used and recognized minimum standards are the 16 volumes of the National Fire Code published annually by the National Fire Protection Association. In the NFPA Code 13, hazard classifications (light, ordinary, extra) are given to various types of buildings based on their occupancy. See Figure 2.20 for NFPA listings.

The five types of sprinkler systems mentioned above are designed for specific applications.

1. **The Wet Pipe System:** This system is the most common sprinkler used. The wet pipe system employs automatic sprinklers attached to piping containing water under pressure at all times. When the heat from a fire exceeds the temperature rating of the sprinkler head, it melts a fusible link, and water contained within the piping is discharged immediately onto the fire. This system is generally used wherever there is no danger of the pipes freezing. See Figure 2.21a for a typical wet pipe system.
2. **The Dry Pipe System:** Dry pipe sprinkler systems are used in unheated areas, such as parking garages. The system has automatic sprinklers attached to piping containing air under pressure. When a sprinkler head opens, pressurized air is exhausted or reduced through the heat until water begins to flow. Dry systems cost more than wet systems because of the necessary larger pipe sizes, additional equipment (accelerator, air compressor, etc.), and they require additional maintenance. See Figure 2.21b for a typical dry pipe system.
3. **The Pre-Action System:** Pre-action systems prevent accidental water discharge from defective sprinkler heads and/or fittings. The dry system is normally used in areas that contain costly equipment, such as computer rooms. Pre-action sprinklers operate

faster and reduce both fire and water damage in comparison with conventional dry pipe systems. Because piping is dry, it is nonfreezing and therefore applicable to conventional dry pipe service. The system piping supervision is provided by low pressure air, leakage of which sounds an alarm without tripping the main valve. The water supply valve is actuated independently of the opening of sprinklers, i.e., the valve is opened by the operation of an automatic fire detection system and not by the fusing of the sprinklers. Pre-action systems are limited to 1,000 sprinklers.

4. **Deluge System:** Although different in application, deluge and preaction sprinkler systems are very similar in design. Both use automatic fire detection systems, and both use the same main valve. The primary difference is that deluge systems use open sprinklers, while pre-action systems use closed sprinklers. A deluge system is designed to discharge water simultaneously from every sprinkler head in the system. Unlike the other three

FIRE HAZARD CLASSIFICATION BY OCCUPANCY

LIGHT HAZARD

- AUDITORIUMS
- CHURCHES
- CLUBS
- EDUCATIONAL
- HOSPITALS
- INSTITUTIONAL
- LIBRARIES (EXCEPT LARGE STACK ROOMS)
- MUSEUMS
- NURSING HOMES
- OFFICES
- RESIDENTIAL
- RESTAURANTS
- SCHOOLS
- THEATERS

ORDINARY HAZARD

- AUTOMOTIVE GARAGES
- BAKERIES
- BEVERAGE MANUFACTURING
- BLEACHERIES
- BOILER HOUSES
- CANNERIES
- CEMENT PLANTS
- CLOTHING FACTORIES
- COLD STORAGE WAREHOUSES
- DAIRY PRODUCTS MANUFACTURING
- DISTILLERIES
- DRY CLEANING
- ELECTRIC GENERATING STATIONS
- FEED MILLS
- GRAIN ELEVATORS
- ICE MANUFACTURING
- LAUNDRIES
- MACHINE STOPS
- MERCANTILES
- PAPER MILLS
- PRINTING & PUBLISHING
- SHOE FACTORIES
- WAREHOUSES
- WOOD PRODUCT ASSEMBLY

EXTRA HAZARD

- AIRCRAFT HANGARS
- CHEMICAL WORKS
- EXPLOSIVES MANUFACTURING
- LINOLEUM MANUFACTURING
- LINSEED OIL MILLS
- PAINT SHOPS
- SHADE CLOTH MANUFACTURING
- SOLVENT EXTRACTING
- VARNISH WORKS
- OIL REFINERIES
- VOLATILE OR FLAMMABLE LIQUID MANUFACTURING AND USE

Figure 2.20

systems, the sprinkler head has no fusible link and, therefore, remains open. When a detection system is activated, the deluge valve opens and water floods the area. Deluge systems are used where flammable liquids or other materials can spread quickly and endanger life safety. Deluge systems are limited in size to 225 sprinklers. See Figure 2.22 for typical pre-action and deluge systems.

5. **Firecycle System:** Technically speaking, firecycle is an on-and-off cycling pre-action sprinkler system. Through the use of heat detectors and an electrical control panel, firecycle systems have the capability of continued on-and-off cycling while controlling fire,

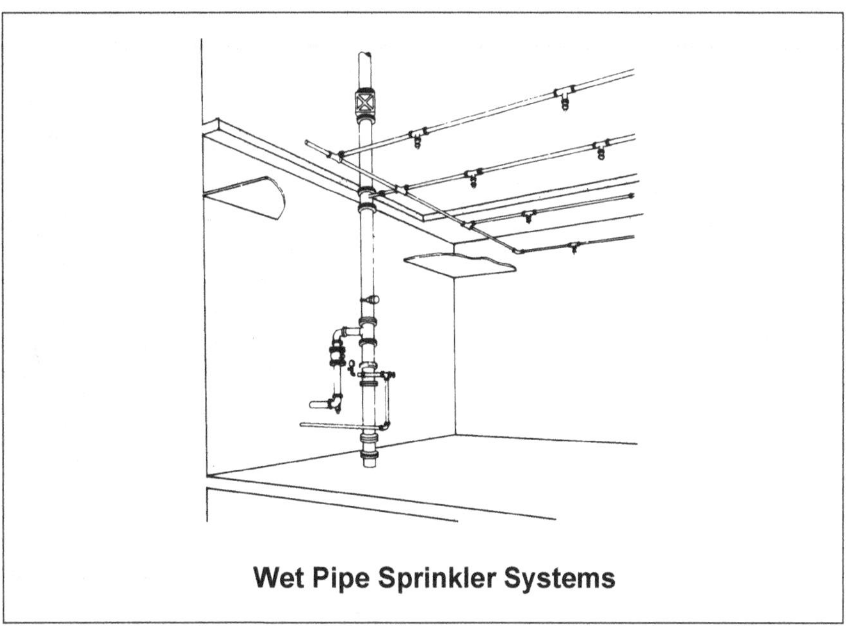

Wet Pipe Sprinkler Systems

Figure 2.21a

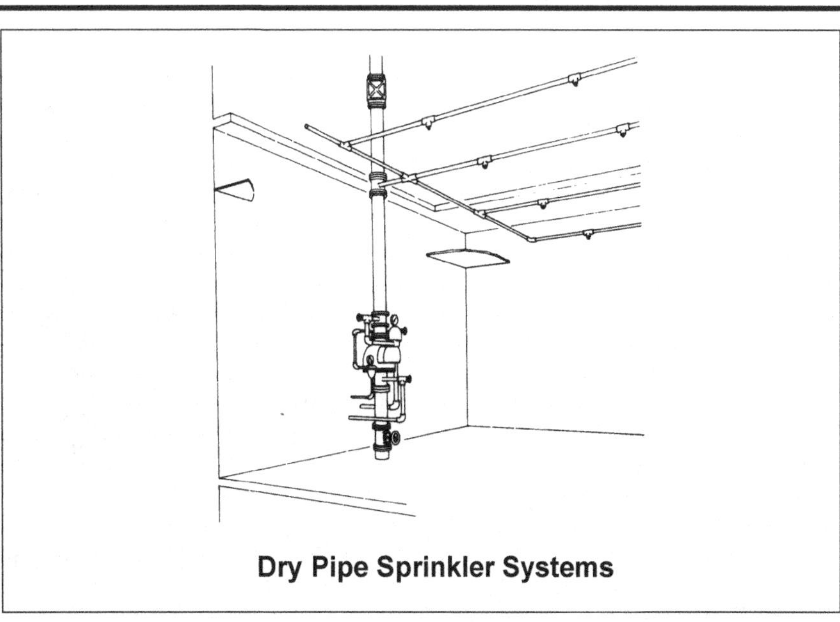

Dry Pipe Sprinkler Systems

Figure 2.21b

and shutting off the flow of water when the fire is extinguished. The system has been widely used in areas where water damage must be minimized. Because water cannot flow unless initiated by signal from the heat detection system, firecycle also eliminates the possibility of water damage stemming from accidents to sprinklers or piping. See Figure 2.23 for a typical firecycle system.

Sprinkler booster pumps may also be required if outside street pressure is inadequate. Depending on local jurisdiction, fire lines may be required to have fire detector meters. Piping materials, valves, and fittings for sprinkler systems are similar to those specified for fire standpipe systems.

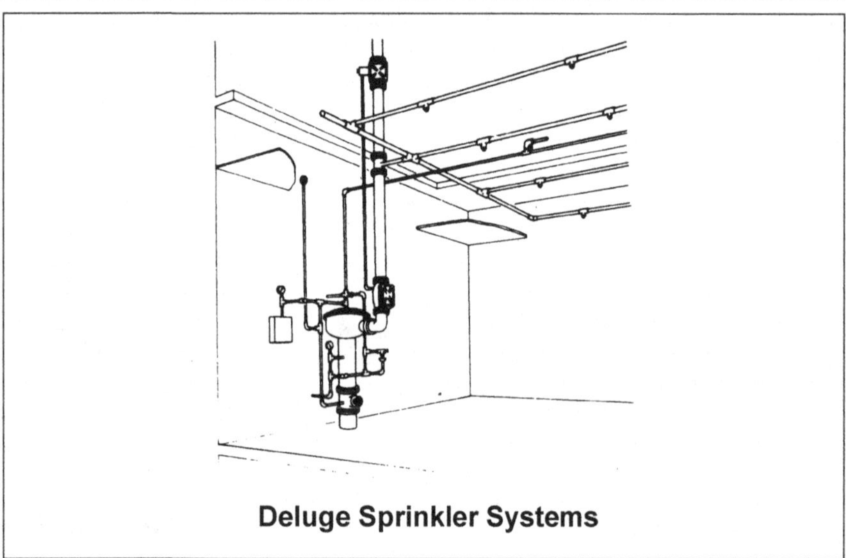

Figure 2.22

Deluge Sprinkler Systems

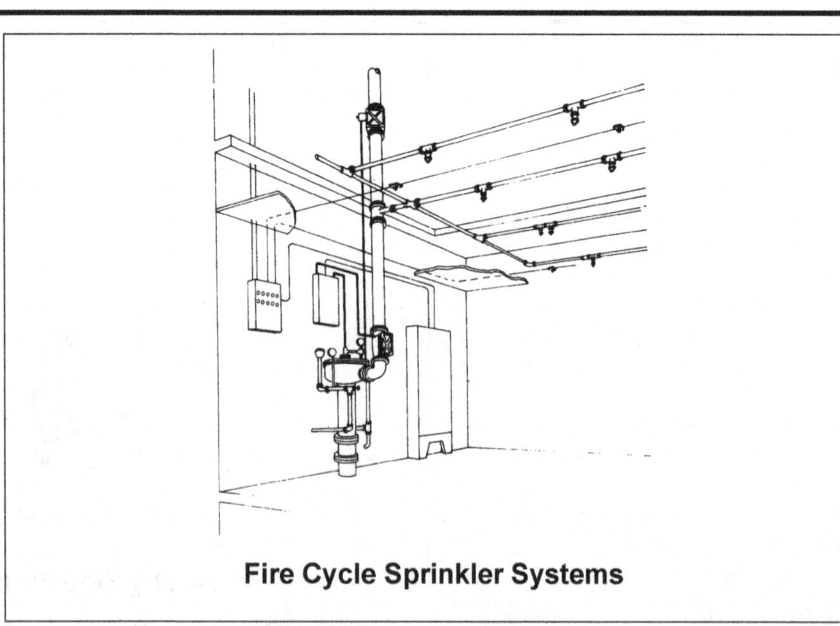

Figure 2.23

Fire Cycle Sprinkler Systems

No discussion of sprinkler systems is complete without a detailed look at the various types of sprinkler heads offered by manufacturers for specific applications. All automatic sprinkler heads are engineered to spread a blanket of finely divided water over an immediate fire area. They act quickly, before the fire gets out of control. They perform positively without fail.

Sprinkler heads are produced to operate at a variety of temperature ranges. The heads have their water opening securely capped. The cap is held in place with lever arms or rods, which, in turn, are held by a fusible link. In case of fire, the solder in the link fuses. Instantly the lever arms or rods are thrown clear by the spring pressure of the frame. The water cap is thrown out by water pressure. Through the orifice, water strikes the deflector, then distributes the water in a uniform pattern. See Figure 2.24 for various types of sprinkler heads.

Fire extinguishers are another of the fire-fighting devices that may fall under the plumbing, sprinkler, or general contract. They are mentioned here simply to note that they may have to be included in the estimate costs for fire protection.

Chemical, Foam, and Gas Fire Suppression

Chemicals, foams, and gasses of various types delivered through portable or fixed spray systems are used to extinguish fires in special applications. Fixed spray systems include foams and gasses, primarily carbon dioxide, FM200, or equivalent that extinguish fires by suffocation. Portable spray systems include foams and gasses, as well as a variety of dry chemicals and pressurized water.

Total Flooding Fire Suppression Systems

The primary chemical used for automatic fire suppression for many years was HALON 1301. Unfortunately, this chemical belongs to the group that is responsible for depleting the ozone layer. While the United States Environmental Protection Agency (EPA) mandated that production of HALON 1301 be stopped by December 31, 1993, it may still be found in older installations. All new systems, and older ones that have to be recharged, utilize a substitute chemical, one of the most common being FM200 (heptafluoropropane).

Figure 2.24

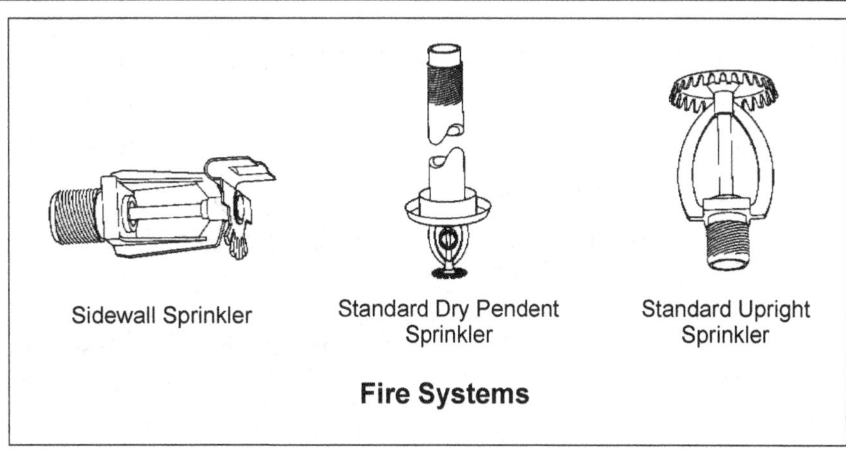

Fire Systems

FM200 and similar systems are called "clean," because they leave no residue that must be cleaned up or that could contaminate valuable items (such as records or electronic equipment). It is a nonconductor of electricity and, when used, floods the working area and penetrates cabinets or other electric or electronic enclosures where chemical powders cannot. It does not leave a residue upon evaporation. A "complete" system includes both detection components and controls with provision for both pre-alarm and automatic agent release. FM200 has been evaluated by toxicologists and found to be safe for use when people are present. Exposure to normal extinguishing concentrations does not cause health problems.

Fire Suppression Systems should be in compliance with NFPA Standard 2001 (Standard on Clean Agent Fire Extinguishing Systems), while the detectors themselves should be in accordance with NFPA 72. The detection system usually utilizes photoelectric and ionization smoke detectors in addition to heat detectors (U.L. Listed and F.M. approved). The first detector to be activated initiates an alarm signal generator. When a second detector is actuated, the system will generate a pre-discharge signal and start the pre-discharge condition.

A "control panel" that includes a battery standby supervises the total system, alarm signals, agent-releasing output, and manual and automatic control functions. Release of the chemical agent is accomplished by activating solenoids on the storage containers. To meet specifications, these systems are designed to completely discharge within ten seconds and provide a uniform minimum extinguishing

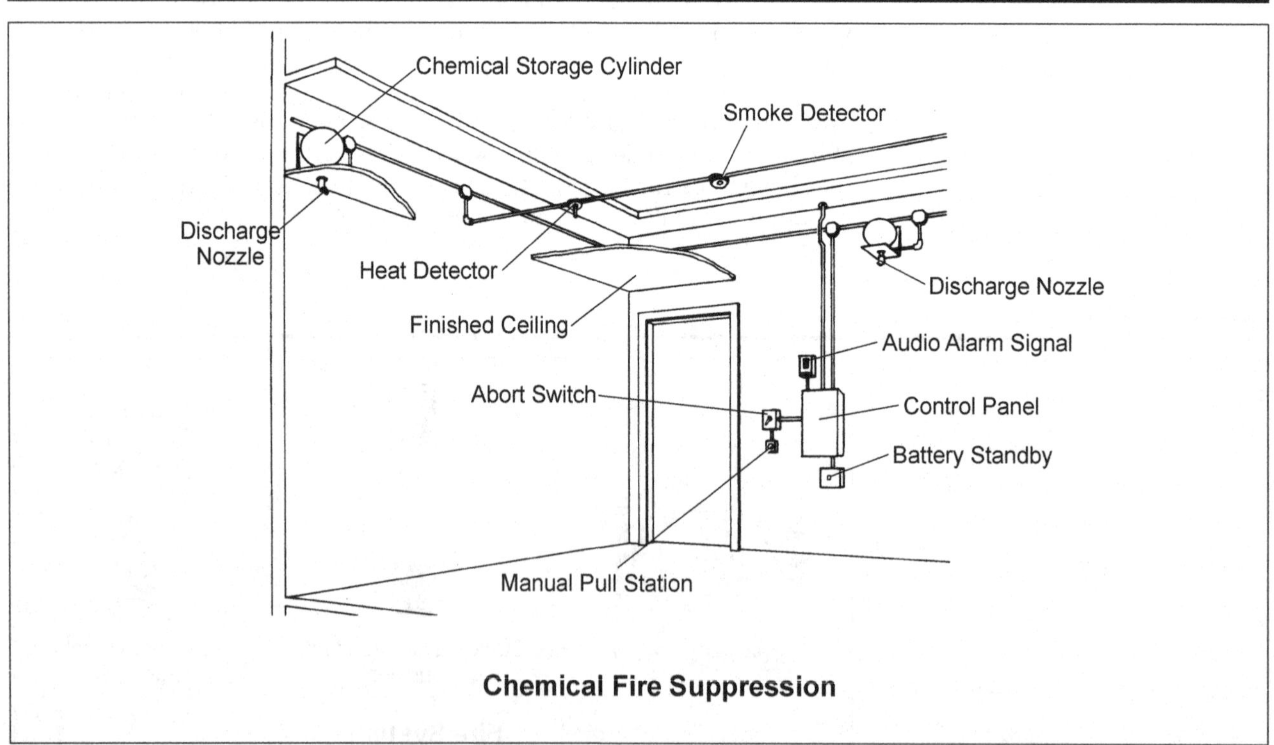

Chemical Fire Suppression

Figure 2.25

concentration of 7% or higher as recommended by the system/agent manufacturers.

Detection and actuation are critical requirements for fast extinguising, to eliminate not only fire damage, but also the accompanying risks of smoke, heat, carbon monoxide, and oxygen depletion. Fire and smoke detectors are wired to a control panel that activates the alarm systems, verifies or proves the existence of combustion, and releases the extinguishing agent, all in a matter of seconds.

The FM200 fire supression system is most effective in an enclosed area. The control system may also have the capability of closing doors and shutting off exhaust fans. These systems have been commonly used in the following places.

- Aircraft (both cargo and passenger)
- Tape and data storage vaults (rooms)
- Telephone exchanges
- Laboratories
- Radio and television studios
- Transformer and switchgear rooms
- Libraries and museums
- Bank and security vaults
- Electronic data processing
- Flammable liquid storage areas

Units of Measure: Self-contained or modular fire suppression systems are taken off and priced as each. For larger "pre-engineered" or "engineered" systems (central storage configuration) components should be taken off individually as each. Pipe should be taken off by the linear foot.

Material Units: A piping system (or other fixed spray system) is similar to a sprinkler piping system in material costs; hangers and supports are determined from the piping totals and cylinder locations. Dispersion nozzles (rather than heads) are utilized, although not as frequently as heads are used in sprinkler systems. Other materials to be taken off and priced are release stations detectors, deflectors, alarms, annunciators, abort switches, cylinders (containers), and controllers.

Labor Units: Before labor units can be estimated, discharge nozzles and detectors should be located and a piping route laid out from the nozzles back to the storage cylinders. The quantities should be taken off to determine fabrication labor. Steel T&C (threaded and coupled) piping and grip type connectors are most often used in these systems.

Takeoff Procedure: The takeoff should begin with the nozzles and other piping components, and proceed to the pipe and fitting takeoff. Any valves used in these systems will be found at the storage cylinders or manifolds.

Other Fire Extinguishing Systems

In addition to the methods of fire protection already covered in this chapter, the plumbing contractor may be required to furnish portable extinguishers, and the sheet metal contractor may be required to install or furnish kitchen hoods with built in fire dampers, spray assemblies, and fan disconnect apparatus.

Other spray systems similar to halon are used in the form of portable extinguishers or fixed pipe installations. These spray systems include foam used primarily for fuel fires and other flammable liquids, and carbon dioxide. This type of system extinguishes fires by smothering. Because carbon dioxide is slightly toxic, it is limited in use to fires in classifications B or C (listed below). It should also be noted that carbon dioxide dissipates and may allow re-ignition.

A third type of extinguisher covers a variety of dry chemicals and powders which are available for the range of fire classifications. Care should be taken to match the proper system with the expected hazard. To facilitate proper use of extinguishers on different types of fires, the NFPA Extinguisher Standard has classified fires into the following four types.

- **Class A** fires involve ordinary combustible materials (such as wood, cloth, paper, rubber, and many plastics) requiring the heat-absorbing (cooling) effects of water, water solutions, or the coating effects of certain dry chemicals which retard combustion.
- **Class B** fires involve flammable or combustible liquids, flammable gasses, greases, and similar materials where extinguishment is most readily secured by excluding air (oxygen), inhibiting the release of combustible vapors, or interrupting the combustion chain reaction.
- **Class C** fires involve live electrical equipment where safety to the operator requires the use of electrically nonconductive extinguishing agents. (Note: When electrical equipment is de-energized, the use of Class A or B extinguishers may be indicated.)
- **Class D** fires involve certain combustible metals (such as magnesium, titanium, zirconium, sodium, potassium, etc.) requiring a heat-absorbing extinguishing medium not reactive with the burning metals.

Some portable fire extinguishers are of primary value on only one class of fire. Some are suitable for two or three classes of fire; none is suitable for all four classes of fire.

Color coding is part of the identification system, and the triangle (Class A) is colored green, the square (Class B) red, the circle (Class C) blue, and the five-pointed star (Class D) yellow.

Units of Measure: Portable fire extinguishers and related accessories are taken off and priced as each. Fixed pipe systems, such as FM200 systems, are taken off as each or by component, depending on system configuration.

Material Units: Material units for portable fire extinguishers are relatively simple. Individual wall-hung extinguishers, when specified, will be furnished with mounting brackets. These same brackets may be used for mounting within certain types of extinguisher cabinets as well.

Specifications for cabinets may require one or more extinguishers or hose and valve combinations.

The cabinets would have the same material units as previously outlined under standpipe systems. Cabinet doors and trim may be painted steel, aluminum, or stainless steel. The cabinet front may be solid panel, glass, wire glass, plexiglass, or a combination of solid panel and glass. Decals, blankets, spanner wrenches, axes, and alarms are among the safety and operating options available.

Fixed or permanent systems will have piping and heads or nozzles similar to the apparatus in the FM200 systems. Kitchen exhaust hoods would probably arrive on the job with the piping assembly in place. Unique fuel loading areas might require a piping system and components.

Labor Units: Labor considerations for portable extinguishers include the following:

- Anchoring the wall bracket to the building structure
- Securing the extinguisher in its bracket

Pressurized water type extinguishers are shipped empty and must be charged in the field. Any extinguishers that have been set in place for standby use during construction will have to be removed, cleaned, and inspected or recharged before final acceptance.

Recessed cabinets are built into the walls by the General Contractor. The finished frames and doors are installed by the plumber as the extinguishers are placed. This is similar to the procedures previously outlined for fire hose cabinets.

Takeoff Procedure: Portable fire extinguishers and special built-in permanent spray systems are often not shown on the mechanical drawings. However, the specifications will indicate the types and locations required. The architectural floor plans and elevations may indicate each extinguisher station. The plumbing drawings will indicate whether or not extinguishers will be housed with fire hoses (in the same enclosure). Spray systems for kitchen exhaust hoods should be found either on the kitchen or HVAC equipment drawings or both. Piping arrangements for fuel storage and piping systems will be indicated on the drawings.

The estimator, in beginning the takeoff, should separate the extinguishers by type and floor. Permanent spray installations should have a pipe and fittings takeoff. The spray heads and any valves, pumps, or storage cylinders should be recorded and priced separately from any other piping system.

The project documents should be carefully read to determine who is responsible for supply and/or installation of these special items.

Special Piping Systems

Plumbing estimators are apt to encounter a variety of specialized piping systems in the course of a career. A book could be written about these special systems alone, so we will cover only the more common specialized systems in this section. They include:

- Natural gas systems
- Medical gas systems
- Acid waste systems
- Swimming pool and filtration systems

Natural Gas System

Natural gas serves as an energy source for items such as boilers, burners, hot water generators, unit heaters, etc. Gas is generally supplied to customers via gas mains supplied and maintained by local gas companies. In most cities, the gas service and meter to individual buildings is also provided by the gas company. However, the estimator should check with the local utility for a listing of the regulations for each particular project.

The gas system within a building consists of all the piping, valves, and devices starting from the gas meter. The plumber begins the distribution of piping within the building after the gas meter, supplying a gas regulator, if required. Gas regulators are usually required if the gas supply is at a pressure in excess of 1/2 psi. Pressures of up to 3 psi are permitted upon approval in certain cases, such as commercial or industrial buildings. Depending on the type of building, gas is used as a fuel for gas-fired equipment, or as in the case of laboratories, as a fuel for laboratory tables and equipment.

The most common piping material for a gas system within the building is Schedule 40 black steel pipe with 150 lb. black malleable iron fittings.

The flow of gas is controlled through the use of approved brass or iron-body gas cocks for pipe sizes 1/2″ to 2″, and through the use of lubricated plug valves for sizes 2-1/2″ and up. All pieces of equipment, such as boilers, gas ranges, unit heaters, and laboratory equipment should be individually controlled using these valves.

Medical Gas Systems

The medical gas system in a hospital is made up of five medical gas subsystems: oxygen, nitrogen, compressed air, nitrous oxide, and a vacuum subsystem. (See Figure 2.26.)

Oxygen is used for respiratory therapy. Oxygen supply in a hospital should always be plentiful, with outlets located in strategic and convenient areas. Oxygen is supplied to the system by one of two methods—either through a bulk oxygen storage tank on the site (furnished by a medical gas supplier) or through a gas manifold with individual cylinders of oxygen connected to it. Manifolds are either automatic or semi-automatic in design, but automatic manifolds are suggested for large installations since they require no maintenance other than changing depleted cylinders and routine inspection.

Nitrogen is a gas used to drive surgical tools, and its location within the hospital is usually limited to surgical suites. Nitrogen is supplied to the system by a manifold similar to the one used for oxygen.

A vacuum subsystem serves a variety of functions throughout the hospital. In the area of patient treatment, it is used to remove fluids during surgery and post-operative drainage. In the laboratory, it is used for transferring liquids from one container to another and for filtering and cleaning apparatus. The vacuum is created by rotary-oil and watersealed vacuum pumps.

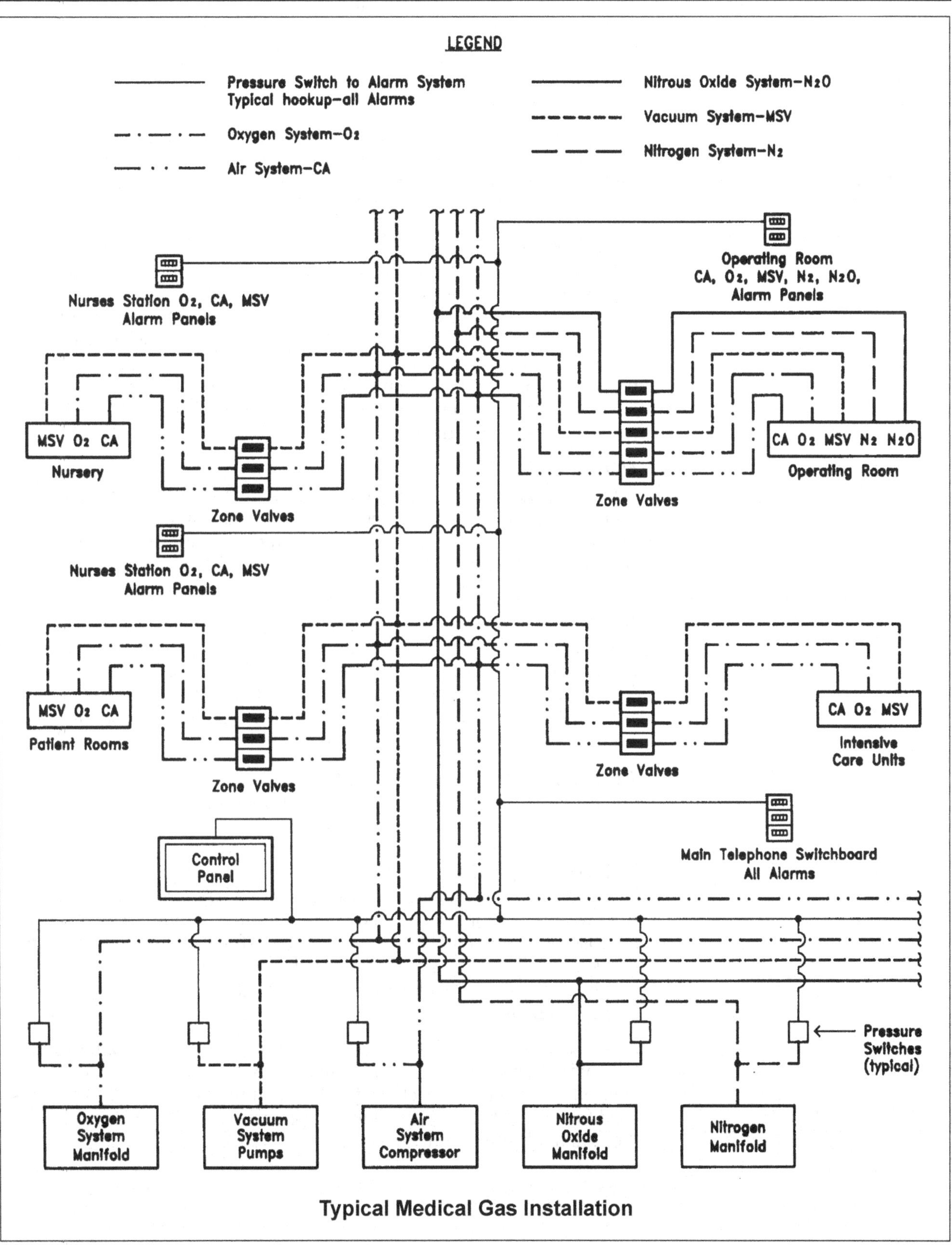

Typical Medical Gas Installation

Figure 2.26

Compressed air is used for respiratory therapy. Compressed air is usually specified for intensive care units, nurseries, emergency rooms, out-patient clinics, surgical suites, and laboratories. Compressed air is supplied to the system by reciprocating air-cooled or oil-free air compressors.

Nitrous oxide, the analgesic gas that produces a loss of sensitivity to pain, is supplied to the system by manifolds similar to the types used for oxygen and nitrogen. Oxygen and nitrous oxide are two gases that support combustion and should be adequately controlled in all areas.

System Control

All medical gas systems are controlled by zone control valves that are capable of isolating each of the medical gases in individual areas. For example, there would be separate zone valve assemblies for operating rooms, patient rooms, nurseries, etc. (See Figure 2.26.) Access to medical gases is obtained through medical gas outlets. Outlets are manufactured for various applications, such as individual outlets in patient rooms, hose-reel ceiling outlets for overhead services, and surgical ceiling columns in surgical suites.

Alarms

All medical gas systems have audio-visual monitoring alarm systems. Sudden pressure drops are picked up by pressure switches, which relay the signal to an alarm panel located at strategic points, such as nurses' stations. Local alarms are used in areas where the supply of gas may be critical to life support.

Type L or K copper tubing is the common piping material used for all medical gases. The fittings used on medical gas systems are wrought or cast bronze solder joint. Control valves are either globe or ball valves (solder joint). The copper tubing used for medical gases is washed with either a hot solution of sodium carbonate or trisodium phosphate. Tubing can be purchased prewashed or can be washed after installation.

Acid Waste and Vent System

The acid waste system is a special drainage system of pipe and fittings used to convey, from laboratory sinks or other receptacles, acid-bearing wastes prior to their dilution and discharge into the sanitary house drain. (See Figure 2.27.) Acid waste systems are generally located in laboratory areas of hospitals, schools, research laboratories, and colleges. Acid wastes cannot drain directly into the house sanitary system because of potentially dangerous or noxious fumes being released into that system. Damage to standard cast iron, copper, or steel drain lines is also likely to occur if acid wastes are allowed to drain into the sanitary system undiluted.

Acid wastes are diluted in individual or central acid neutralizing sumps filled with limestone chips. Individual sumps are usually located adjacent to or beneath each acid sink or fixture. If a central acid sump is used for several fixtures, it should be located at the lowest story above the sanitary house drain.

Consulting engineers usually specify one of three types of piping materials for use on acid waste systems:

- Glass pipe and fittings
- Polypropylene pipe and fittings
- Cast iron high-silicone pipe and fittings

Glass pipe is the most popular of the three materials. Although mainly used for above grade applications, glass can be buried below grade if heavy schedule pipe is encased with a special covering to give it greater loadbearing strength.

Polypropylene pipe is gaining wider acceptance because of its light weight and ease of handling during installations, but its use is limited to above and below ground installations.

Cast iron high-silicone pipe has been around for some time and is excellent for use on both above and below ground installations.

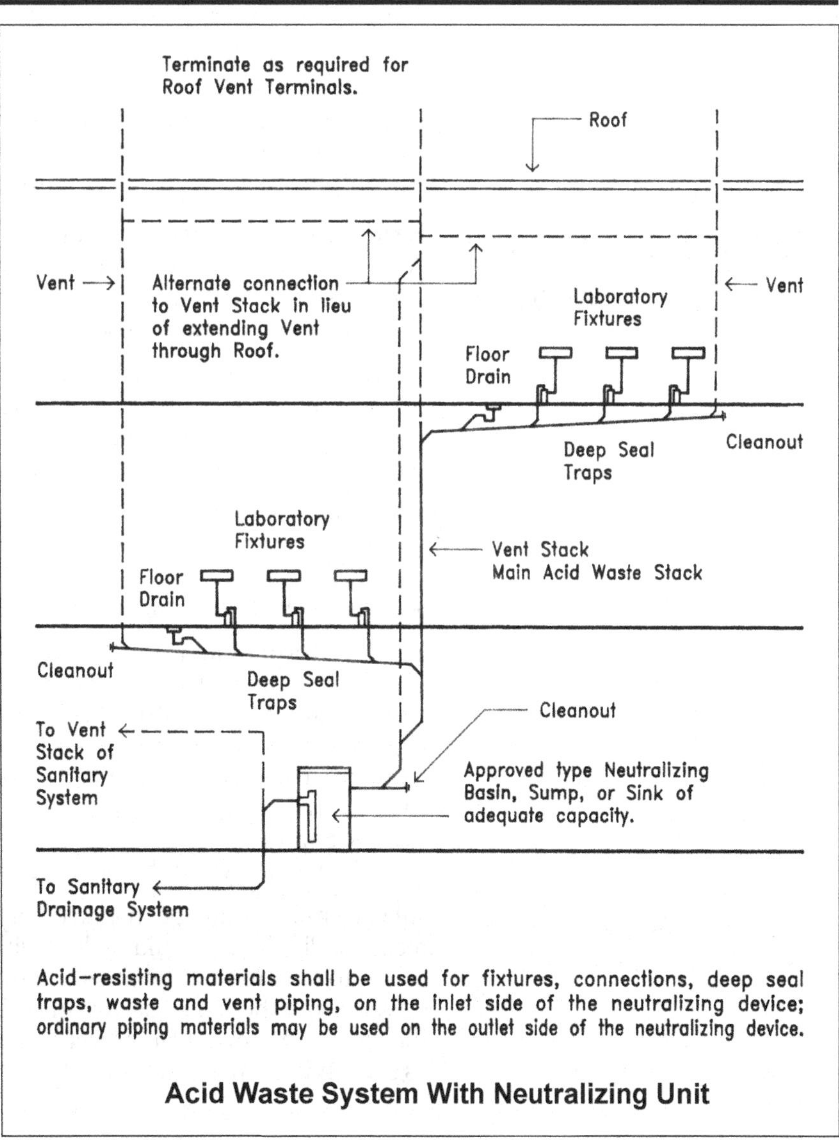

Acid Waste System With Neutralizing Unit

Figure 2.27

Swimming Pool and Filtration Systems

On a project where a swimming pool is to be installed, it is usually the responsibility of the plumbing contractor to furnish and install all the piping and filtration equipment. The plan of a typical swimming pool is shown in Figure 2.28, and the pool filter system is illustrated in Figure 2.29. The following is a list of components that are generally found in every pool system and should be included in the plumbing estimate, unless otherwise noted in the engineer's specifications.

- Filter
- Circulating pump
- Pool heater (if required)
- Soda ash and hypochlorinator
- Alum feed
- Filtration pipe, fittings, and valves
- Vacuum piping and fitting
- Gutter drain piping and fittings
- Main drain piping and fittings
- Pool supply piping and fittings

The estimator should consult the engineer's specifications for the type of filtration system and obtain a quote for such equipment from a pool equipment manufacturer. Depending on the system and specifications, piping found in pool installations is usually polyvinyl chloride, galvanized steel, copper, or cast iron.

Plumbing Site Work Systems

Sanitary and Storm Systems

At some sites, large drainage installations are the responsibility of the general contractor, but in many cases site drainage work will fall under the plumbing contractor's scope of work.

We will discuss the site sanitary and storm systems together, because the estimator often encounters a combined sanitary and storm system. Secondly, the method of installation and types of materials specified are often either the same or similar in many respects.

The sanitary sewer system on the project site is considered to be all the piping, manholes, and appurtenances conveying sewage from the building sanitary drain to a public sewer, private sewer, individual sewage disposal system, or other point of disposal.

The site storm drainage is the system of piping, manholes, catch basins, yard drains, dry wells, and all other appurtenances that convey clear storm water drainage from the building storm drain and from all other site areas requiring drainage, such as roads, parking lots, and plazas, to a public storm sewer, combined sewer, or drainage basin. A combined sewer is any sewer conveying sewage, storm water, and other clear water wastes. Sanitary, storm, and combined site sewer systems begin at a point five feet beyond the building wall.

All sewer lines must be installed very carefully on the site for various reasons. Pipelines should be laid with the proper pitch (1/8″per foot or more) to ensure that the lines will meet the invert of the existing public sewer or manhole. The sewer lines should enter the public sewer above the center to prevent the sewer backing up into the house sanitary drain as shown in Figure 2.30.

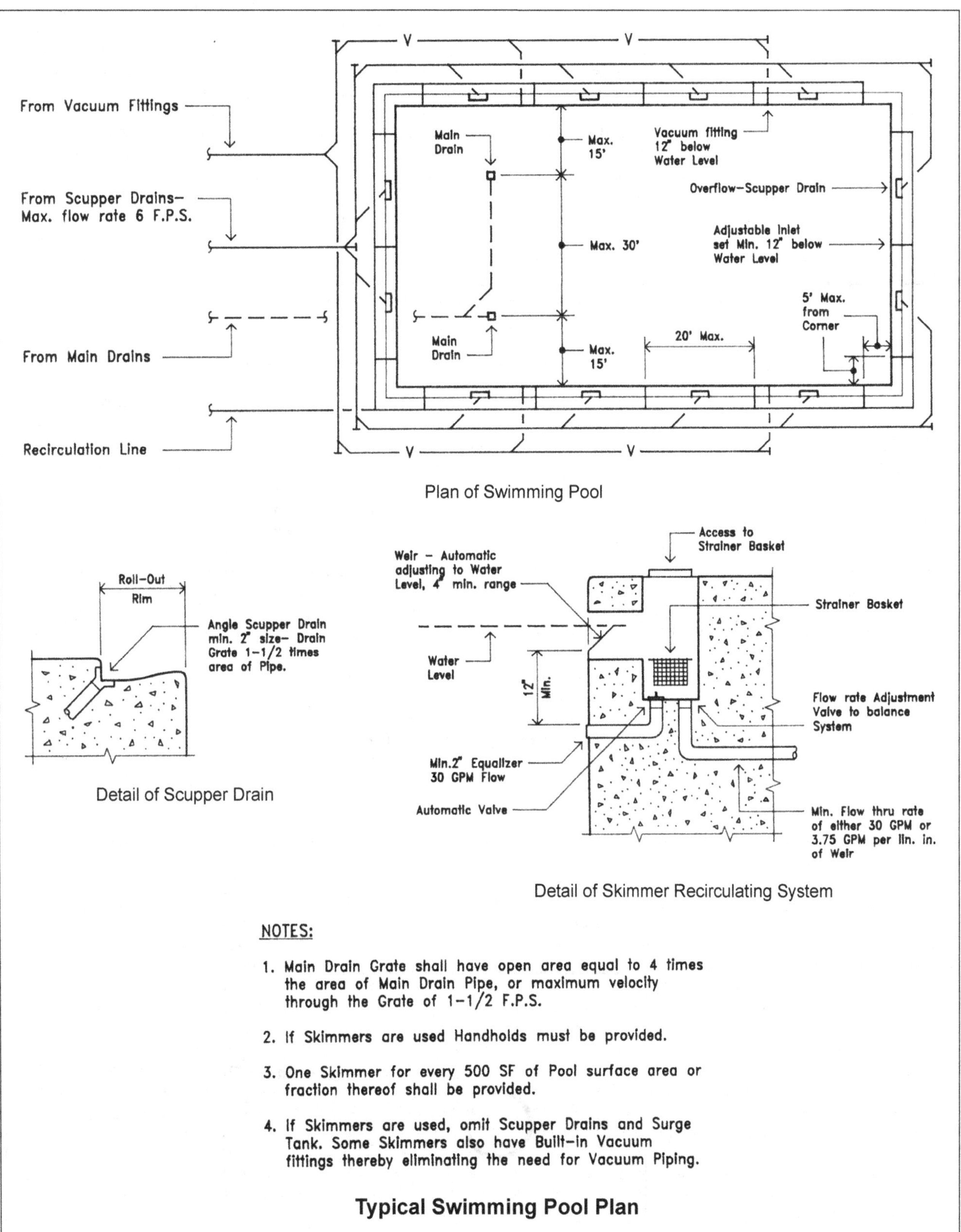
From Vacuum Fittings
From Scupper Drains– Max. flow rate 6 F.P.S.
From Main Drains
Recirculation Line
Main Drain
Max. 15'
Vacuum fitting 12" below Water Level
Overflow–Scupper Drain
Max. 30'
Adjustable Inlet set Min. 12" below Water Level
5' Max. from Corner
20' Max.
Main Drain
Max. 15'
Plan of Swimming Pool
Roll-Out Rim
Angle Scupper Drain min. 2" size– Drain Grate 1-1/2 times area of Pipe.
Detail of Scupper Drain
Weir – Automatic adjusting to Water Level, 4" min. range
Access to Strainer Basket
Strainer Basket
Water Level
12" Min.
Flow rate Adjustment Valve to balance System
Min.2" Equalizer 30 GPM Flow
Automatic Valve
Min. Flow thru rate of either 30 GPM or 3.75 GPM per lin. in. of Weir
Detail of Skimmer Recirculating System
NOTES:
1. Main Drain Grate shall have open area equal to 4 times the area of Main Drain Pipe, or maximum velocity through the Grate of 1-1/2 F.P.S.
2. If Skimmers are used Handholds must be provided.
3. One Skimmer for every 500 SF of Pool surface area or fraction thereof shall be provided.
4. If Skimmers are used, omit Scupper Drains and Surge Tank. Some Skimmers also have Built-in Vacuum fittings thereby eliminating the need for Vacuum Piping.

Typical Swimming Pool Plan

Figure 2.28

If soil conditions on the site are unstable, the plumbing contractor should take the steps necessary to prevent sagging of the lines, which could break and restrict the flow of wastes. Two methods of correcting sagging are wood piling supports placed at every hub or connection, or installing a crushed stone bed beneath the pipe.

Manholes are a very important part of any sewer system. In addition to their use for inspection and cleaning services, they are used as connection points where two or more sewer lines join together. Manholes are usually constructed of brick or are purchased in precast concrete sections. A pre-cast manhole is illustrated in Figure 2.31.

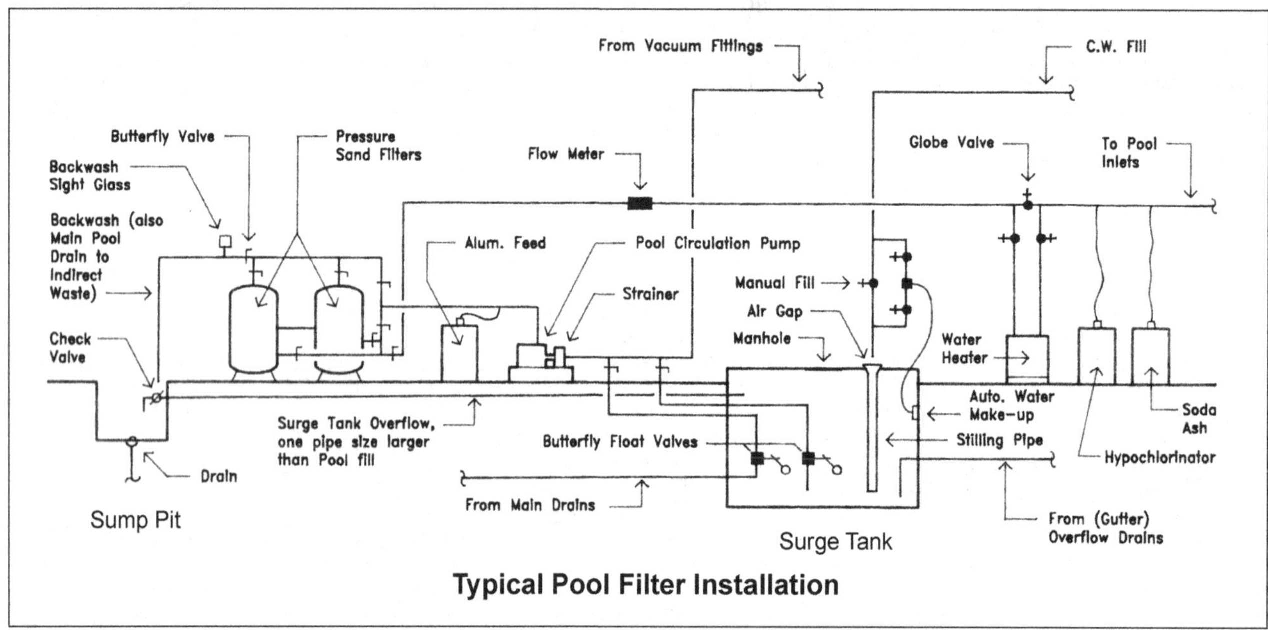

Figure 2.29

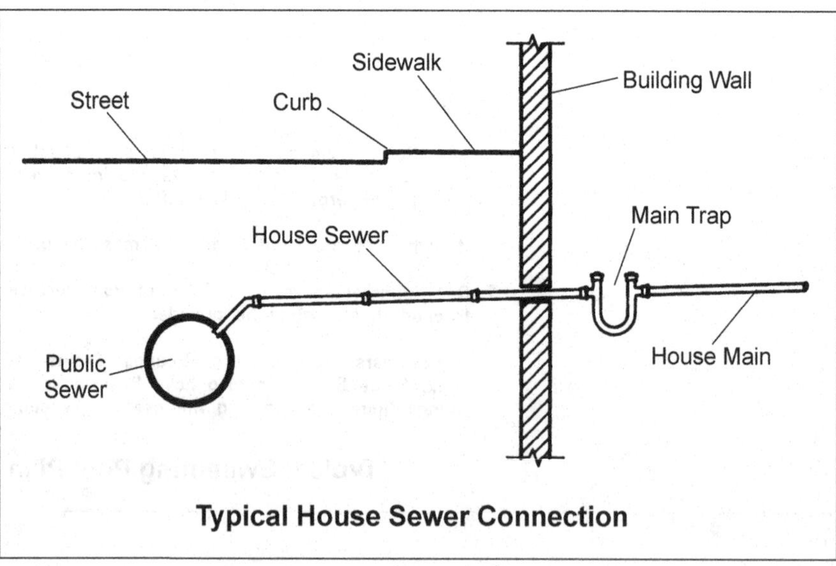

Figure 2.30

Proper drainage of areas such as parking lots and roads is essential to prevent flooding conditions. In many cases, surface drainage is sufficient to handle any site drainage problems. Surface drainage is simply the system of allowing rainwater to run off a properly graded site and seep back into the earth. However, surface drainage is usually inadequate where paved areas (such as parking lots or roads) are located on low or depressed areas of the site. Parking lots and roads are usually drained through a system of catch basins or yard drains which are connected to the storm drainage sewer or independent dry wells.

It should be noted here that the connection of the new sewer system to the existing public sewer is performed by the plumbing contractor. Permits for the sewer connection and for any street breaking have to be obtained and paid for by the plumbing contractor, who should include these costs in his estimate.

There are many different types of piping materials used for site drainage systems. The kind of piping material used by the contractor on a sewer project is usually governed by local codes or highway departments. Figure 2.32 gives the reader an excellent cross-section of commonly used drainage system piping materials used on the site.

Water Service Piping

The site water service piping is that portion of the water supply system extending from the street water main to the house control valve or meter. The procedure for installing any water main begins at the existing live main in the street. The street main is tapped by a special procedure performed by the local utility company. This procedure is called a wet

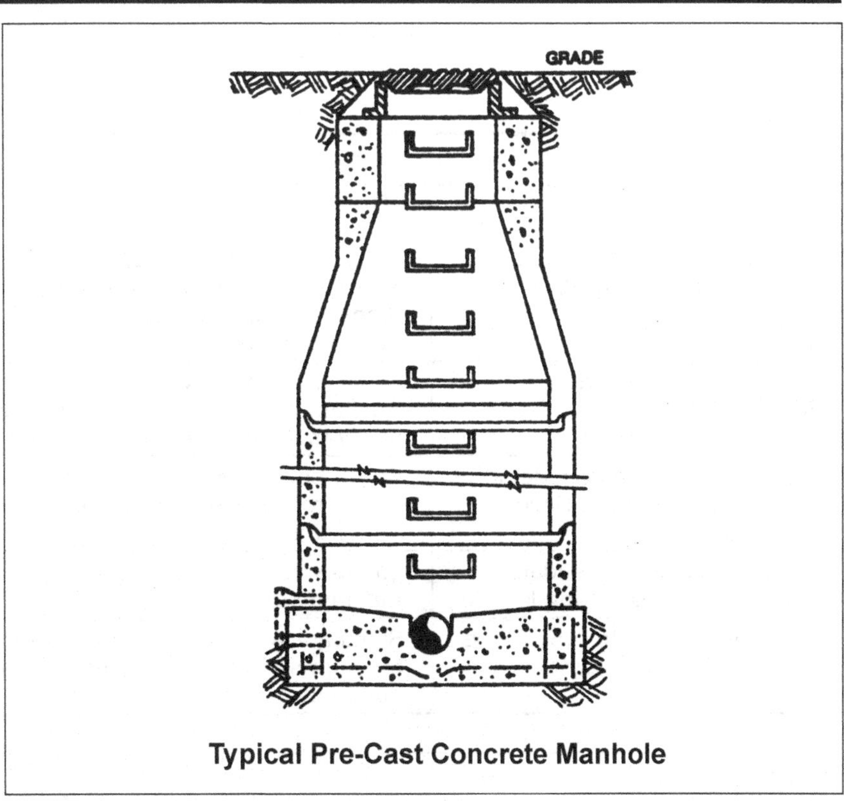

Typical Pre-Cast Concrete Manhole

Figure 2.31

connection for buildings requiring services of 3″ or larger, or wet taps for services under 3″. They are so called because the connection to the street main is made while water under pressure is present in the main. Wet taps and connections avoid shutdown of the main and probable interruption of service.

The utility companies leave a street valve in the closed position for the plumber to begin his portion of the installation, for which the plumbing contractor must secure a permit. A water main is usually buried at least 4′ from the top elevation of the pipe. This is usually the frost line in most areas of the country. The plumbing contractor then runs the main to the building, providing a sidewalk valve in a service box for easy access. Similar to sewer regulations, a permit for street cutting must be secured by the plumbing contractor. Depending on local regulations, the plumbing contractor may be charged to restore the street to its previous condition, and almost all municipalities require the plumbing contractor to install at least temporary or bagged pavement. The materials used for installing water mains on the site are outlined further in Figure 2.33.

Site Fire Protection System

Very often the plumbing estimator will encounter designs with extensive fire mains throughout the site. Buildings are often erected on large sites, and the nearest fire hydrant may be thousands of feet away from the structure. This makes it necessary for the engineer to design a complete fire loop system with fire hydrants strategically located throughout the site.

Fire hydrants are usually of the compression type design with various nozzle arrangements, as shown in Figure 2.34. Each fire hydrant is valved separately, with the valve located in a valve box. For site fire

STANDARD SITE DRAINAGE PIPING MATERIALS		
PIPE	RECOMMENDED SERVICE	SIZE RANGES
Reinforced Concrete Pipe Classes 1, 2, 3, 4 and 5 Ring Joint	Storm and Sanitary Sewers requiring large diameter pipe and exposed to heavy traffic or buried to extreme depths.	12″-96″
Unreinforced Concrete Pipe Ring Joint	Storm and Sanitary Sewers requiring small to medium diameter pipe and not exposed to heavy traffic or extreme depth pressure	6″-24″
Galvanized Corrugated Metal Pipe 16, 14, 12 and 8 Gage Bolted Joints	Storm Sewers requiring large diameter pipe where resistance to external loads is important.	8″-72″
Vitrified Clay Pipe Standard and Extra-Strength Ring Joint	Sanitary and Storm Sewers requiring small to medium size pipe.	4″-36″
Extra Heavy Cast Iron Soil Pipe Lead or Neoprene Joint	Sanitary and Storm Sewers requiring small to medium size pipe but requiring maximum durability.	2″-15″

Figure 2.32

protection systems, it is always of the utmost importance for fire crews to see if certain valves are opened or closed. The use of indicator posts, as shown in Figure 2.34 remedies this situation. Indicator posts are placed over the buried valve, with windows indicating if the valve is open or shut. A wrench supplied with the indicator post is used to open or close the valve when and if necessary.

STANDARD SITE WATER PIPING MATERIALS		
PIPE	RECOMMENDED SERVICE	SIZE RANGES
Cement-lined Ductile Iron Pipe Classes 150 and 250 Lead, Neoprene or Mechanical Joint	Services requiring water mains larger than 3" in diameter.	4"-48"
Type 'K' Copper Tubing Solder Joint	Services requiring water mains 3" in diameter or less.	½"-3"

Figure 2.33

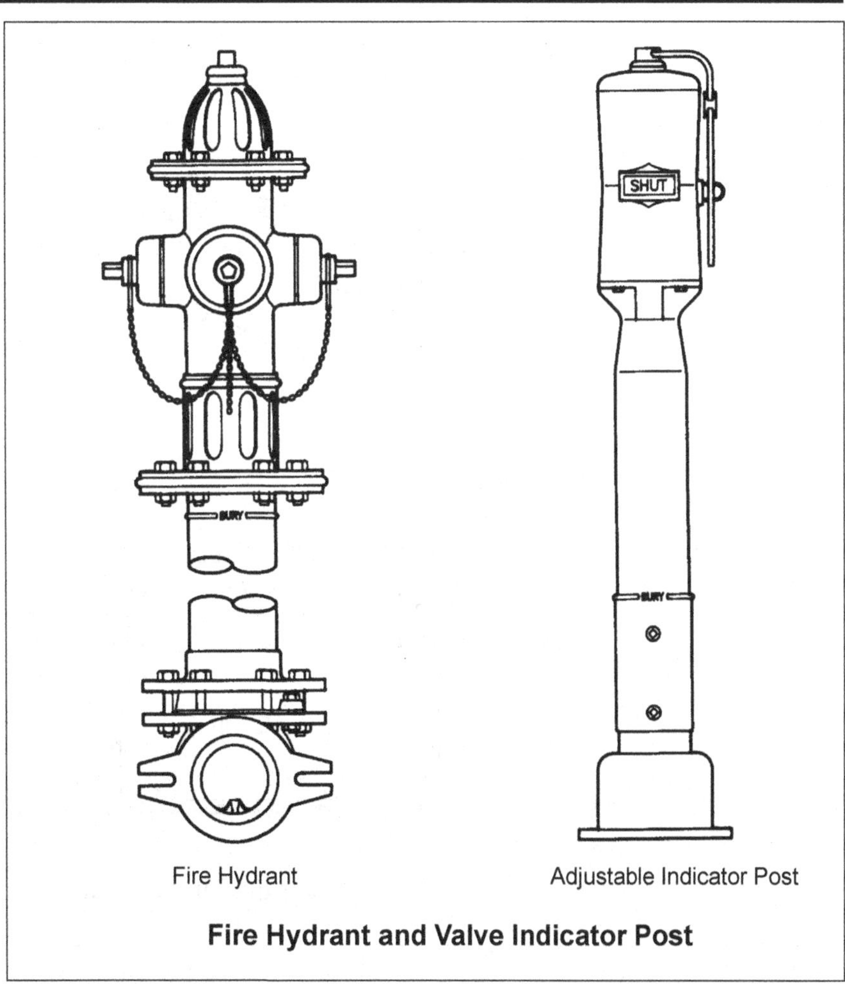

Fire Hydrant and Valve Indicator Post

Figure 2.34

Septic Tank System

In areas where no public sewer facilities are available, human wastes must be disposed of in some other fashion. The septic tank system of disposal has long been the accepted practice for dealing with this problem. The septic tank system is basically a three-step operation. Wastes from the building enter the septic tank, which is designed to separate solids from liquids. The solid wastes settle to the bottom of the septic tank, while the liquid effluent is discharged to a distribution box and then into a leaching area of perforated, split-tile pipe, or seepage pits. (See Figure 2.35.) Both the solid and liquid wastes decompose through bacterial action and other natural processes. Septic tanks and distribution boxes are usually constructed of precast concrete and can be purchased in varying capacities. It is essential that septic tanks be of sufficient capacity to handle the incoming sewage. Septic tanks must be cleaned periodically, but there are chemicals, which help decomposition, thus making the cleaning intervals less frequent.

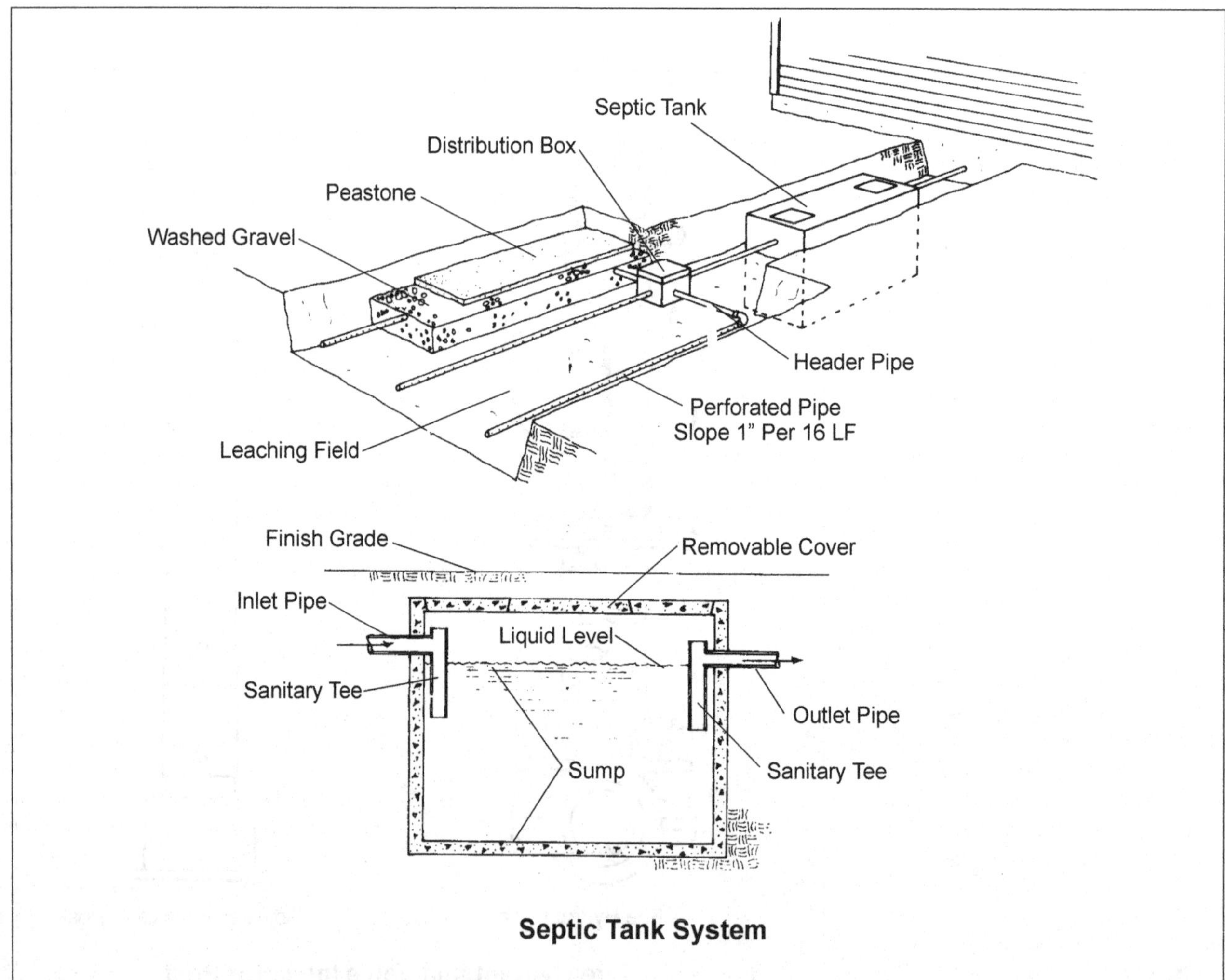

Septic Tank System

Figure 2.35

Lawn Sprinkler System

Lawn sprinkler systems have gained steadily in popularity and are being included more frequently in the design criteria of a project. This popularity stems from the efficiency with which lawn sprinklers can perform their function. Buildings with large lawn areas such as schools, colleges, and industrial parks cannot depend on maintenance men with garden hoses to water lawns, which tends to be a more costly and less efficient method of operation.

Lawn sprinkler systems can function both manually and automatically and, if properly maintained, require little maintenance. Lawn sprinklers are designed to give an even distribution of water with complete lawn coverage. Sprinkler heads are placed in patterns so that the spray from each head will slightly overlap. This also compensates for some loss in water pressure, which is bound to happen from time to time. The sprinkler system is zone-designed, and each zone is controlled by a manifold of valves (either manually or automatically) with timer devices. Sprinkler heads are usually of the pop-up design, which lift up into the spray position by water pressure. The piping, usually Schedule 80 PVC plastic pipe, is buried in the ground 6" to 12" inches deep. The piping should have a slight pitch so that the system can be drained in the winter months. This is achieved by installing a drain valve at the end of each zone.

Part II

Plumbing Estimating

Part II

Introduction

An estimate is defined as an approximate judgment of the dollar value of a project. However, looking beyond the strict definition of the word, one will see a procedure that is neither all art nor all science, but perhaps a little of each. An estimate is an art in that it takes skill and talent to perform, and a science, as it deals with facts, laws, and systems.

Chapter 3 discusses the preparation required to perform a plumbing estimate. Taking off, writing up, pricing, and completing the estimate are covered in Chapters 4 through 6.

Chapters 7, 8, and 9 present other forms of plumbing estimating and provide information useful throughout an estimating career.

Chapter 3

Preparing for the Estimate

This chapter describes some basic preparation necessary before one can put pen to paper to begin taking off and pricing an estimate. The estimator must gather the proper tools, forms, and reference materials, and might want to use a checklist of these items to get organized. We will begin with a list of basic tools.

Estimating Tools

Like other craftsmen, an estimator needs the right tools to achieve professional results. The three basic tools an estimator should have on the worktable before beginning any estimate are measuring devices, a calculator, and pencils.

Plumbing drawings, like architectural drawings, give all dimensions in feet and inches. These dimensions are reduced in size by use of an appropriate scale. The most commonly used scale on drawings is 1/8″ equals 1′. Other scales used are 1/4″= 1′, 1/2″= 1′, 1″= 1′, 3/8″= 1′, and 3/4″= 1′. The larger scales are usually used for details and equipment rooms, or other areas requiring greater clarity on the drawings. There are a number of measuring devices manufactured for reading the above scales.

An architectural scale rule resembles an ordinary ruler, but it has fractional scale divisions embossed on it. There are four edges to a flat architectural scale rule, with two scales embossed on each edge, making a total of eight scales. The scales run in opposite directions on each edge. For example, the 1/8″ scale is read from left to right, while the 1/4″ scale is read from right to left. Triangular rules with six scales are also in current usage. Various types of scales appear in Figure 3.1.

A tape measure is a retractable steel tape, equipped with only two fractional scale divisions: 1/8″ on one side and 1/4″ on the other side. (See Figure 3.1.) This device is especially desirable for piping takeoff because large measures can be accumulated at one time, as the 1/8″ scale has the capacity to measure up to 480′, and the 1/4″ scale can measure up to 240′.

The rotometer, or wheel, is the measuring device best suited to a large piping takeoff. (See Figure 3.1.) The rotometer is a precision instrument for measuring curved or straight distances in feet. It has three scale markings: 1/8″, 1/4″, and 1/2″. Most important, it has a cumulative

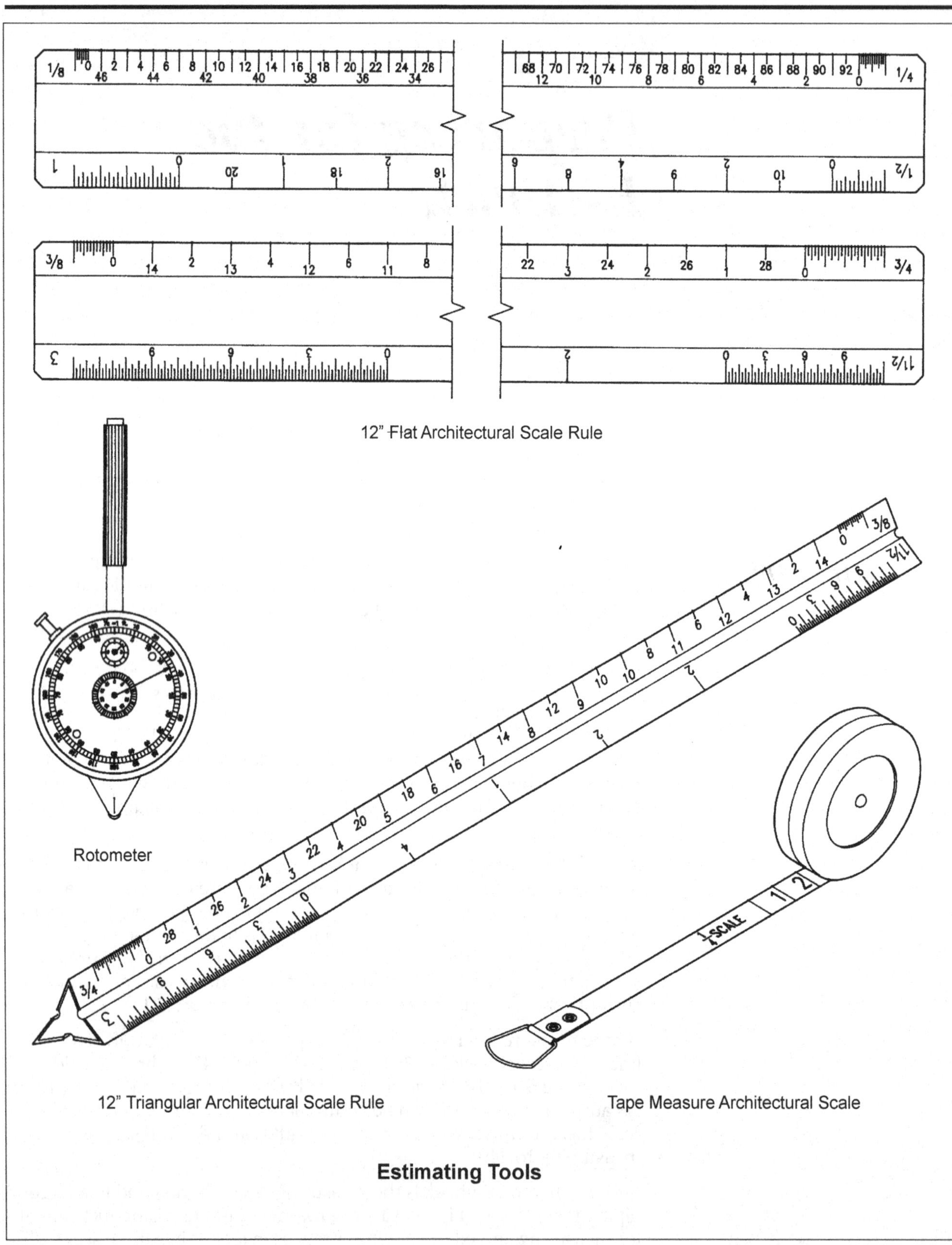

12" Flat Architectural Scale Rule

Rotometer

12" Triangular Architectural Scale Rule

Tape Measure Architectural Scale

Estimating Tools

Figure 3.1

register dial, which enables the plumbing estimator to take off a maximum of 2,400′ on the 1/8″ scale, 1,200′ on the 1/4″ scale, and 600′ on the 1/2″ scale. As with any tools, these function only as well as the person using them. The estimator must take great care in reading the scales properly. Many mistakes are made because of carelessness, and contractors can lose a great deal of time and money as a result.

A calculator is an essential item to the estimator. Of the many types on the market today, the electronic desk calculator and the pocket calculator are among the most popular. A model with a printed tape is recommended, rather than relying solely on calculators with lighted numerical displays. Estimators can ill afford to reach a final cost estimate without being able to recheck figures. A calculator (or adding machine) with a printed tape enables the user to review calculations. The calculator used by the estimator should have at least the following features.

- Printed tape function
- Full four function capability (+, –, /, x)
- Numerical capacity of 8 digits

Additional calculator features that are not essential but are valuable aids when estimating a project are:

- Full memory
- Percent key
- Square and square root key

In addition to standard lead pencils, the estimator should have a variety of colored pencils. The purpose of colored pencils is for color coding of all fixtures, equipment, and piping on the drawings. Color coding is a very important part of the takeoff procedure. Color-coded items are readily identified, should the plumbing estimator or an assistant have to check back on the drawings for progress or to take over the takeoff. Color coding also provides the estimator with a simple method of checking for completeness. It is recommended that the color codes be consistent for all the jobs estimated in the same office. The reader, of course, can create his or her own color code, but once a certain color code is established, it should remain the standard for all estimates within a firm—both in the shop and in the field. A sample color code follows:

Sample In-House Plumbing System Color Code

• Plumbing fixtures and trim		Red
• Equipment		Green
• Cold water		Blue
• Hot water		Red
• Sanitary waste and vent		Yellow
• Storm system		Brown
• Fire standpipe system		Orange
• Sprinkler system		Light green
• Medical gas systems:	Oxygen	Purple
	Nitrogen	Dark blue
	Nitrous oxide	Light blue
	Vacuum	Violet
	Compressed air	Dark green
• Natural gas system		Lavender
• Acid waste and vent system		Gold

Takeoff Forms (Quantity Sheets)

Takeoff forms or quantity sheets can save the estimator valuable time. All takeoff or quantity sheets should have certain basic information at the top of the sheet:

- Title of quantity sheets (plumbing fixtures, piping, equipment, valves, fittings, etc.)
- Project title and job number
- Estimator's name
- Checked by
- Date
- Page number

Many firms prefer to design their own takeoff and estimating forms or use or adapt published forms. The sample takeoff and estimate in Chapters 1-6 of this book are recorded on forms that the authors have designed. The forms used for the Means illustrated estimate are from *Means Forms for Building Construction Professionals*.

They are:

- Plumbing fixture quantity sheet
- Equipment quantity sheet
- Piping quantity sheet
- Fitting quantity sheet
- Valve and device quantity sheet

Plumbing Fixture Quantity Sheet

This form (shown in Figure 3.2) is usually the first in the series and allows for recording the number of fixtures by floor. The "Total" column to the right indicates the total number of a particular fixture in the entire building. The total column at the bottom shows the total fixtures of all types per floor.

Equipment Quantity Sheet

On this sheet (shown in Figure 3.3), the plumbing equipment from each system is listed with enough information included for proper identification. This sheet becomes very useful when ordering equipment.

Piping Quantity Sheet

The piping in a particular system is broken down by diameter and linear feet of pipe. The figures for linear feet of pipe are scaled off from the drawings. The total linear feet of each pipe size appears at the bottom of each piping quantity sheet (Figure 3.4). The column to the right is divided into six boxes. In each box, the total is entered under the appropriate heading for various operations or materials that might be required for that particular system. The divisions are:

- CY excavation: Total cubic yards of required excavation
- Lbs. of flashing: Total lbs. of roof or drain flashing material
- Solder/flux: Total lbs. of solder and flux required to join pipe
- Lbs. of lead: Total lbs. of caulking lead required to join pipe
- Lbs. of oakum: Total lbs. of packing oakum required to make up the lead joints
- Gas cylinders: Total number of gas cylinders required to melt lead or solder

PLUMBING FIXTURE QUANTITY SHEET	PROJECT	JOB NO.	ESTIMATOR	CHECKER	DATE	SHEET OF

FIXTURE TYPE	FLOOR										
	BASEMENT	FIRST	SECOND	THIRD	FOURTH	FIFTH	SIXTH	SEVENTH	EIGHTH	ROOF	TOTAL
TOTAL											

Figure 3.2

EQUIPMENT QUANTITY SHEET	PROJECT	JOB NO.	ESTIMATOR	CHECKER	DATE	SHEET OF

ITEM	QTY.	SYSTEM	CAPACITIES, DESCRIPTION AND OTHER INFORMATION

Figure 3.3

PIPING QUANTITY SHEET			SYSTEM				PROJECT			JOB NO.			ESTIMATOR		CHECKER	DATE	SHEET OF
PIPE SIZE																	
1/8"	1/4"	3/8"	1/2"	3/4"	1"	1-1/4"	1-1/2"	2"	2-1/2"	3"	4"	5"	6"	8"	10"	12"	C.Y. EXC.
																	Lbs. of flashing
																	Solder/flux
																	Lbs. of lead
																	Lbs. of oakum
																	Gas cylinders
																	TOTAL

Figure 3.4

Of course, each box may not apply to every system. Where this is so, the estimator should simply enter "not applicable" in that particular box as a reminder that it has not been omitted.

Fitting Quantity Sheet

Like pipe, fittings are taken off by system and by pipe size. These are counted individually (see Figure 3.5), with their totals shown in the narrow columns to the right of each wide column.

Valve and Device Quantity Sheet

Valves and devices such as shock absorbers, vacuum breakers, strainers, etc., are listed by pipe size for a particular system and, like the fittings, are totaled in the narrow columns to the right. (See Figure 3.6.)

Estimate Forms

Estimate forms aid an estimator in compiling and pricing the estimate neatly and concisely by properly organizing the information. These sheets are the formal presentation of an estimate. They contain work classifications, itemized quantities, material costs, and labor costs. Like quantity takeoff sheets, estimate forms should contain certain basic information on the top of the sheet:

- Firm name and address
- Classification
- Project title and location
- Date
- Architect/Engineer
- Type of estimate
- Estimate number
- Sheet number
- Estimator
- Checker

Each major system is subdivided into individual components. In the actual estimate to follow, each system is given a number (2, 3, 4, etc.), which is entered in the item column, and each segment of that system is given a consecutive decimal number to identify it (e.g., .01, .02, .03, etc.). The total quantity of each component is transferred from the quantity sheets to be priced by material and labor costs. The *Labor* column is used to enter labor-hours, production units and totals, while the *Labor Cost* column is used to enter the labor dollar rate, along with the total labor cost.

A company's own historical cost data is, of course, the most accurate source for cost data, but published cost indexes, such as *Means Plumbing Cost Data*, are good sources of information that may be lacking from a firm's own data.

Estimate forms can also be obtained from published collections, and used as is, or adapted to the specific needs and approaches of an individual firm. The sample estimates in this book are shown on forms designed by the authors for use in their firm. Figure 3.7 is a blank Estimate Sheet, which will be shown completed in Chapter 6.

Summary Sheet

Every estimate contains a form called a Summary Sheet. On this form, the estimator summarizes all totals arrived at on the estimate forms, so the final estimated cost can be calculated. The Summary Sheet contains the totals of the applicable systems broken out by material and labor, a listing of job overhead costs, the contractor's profit, and finally, the total

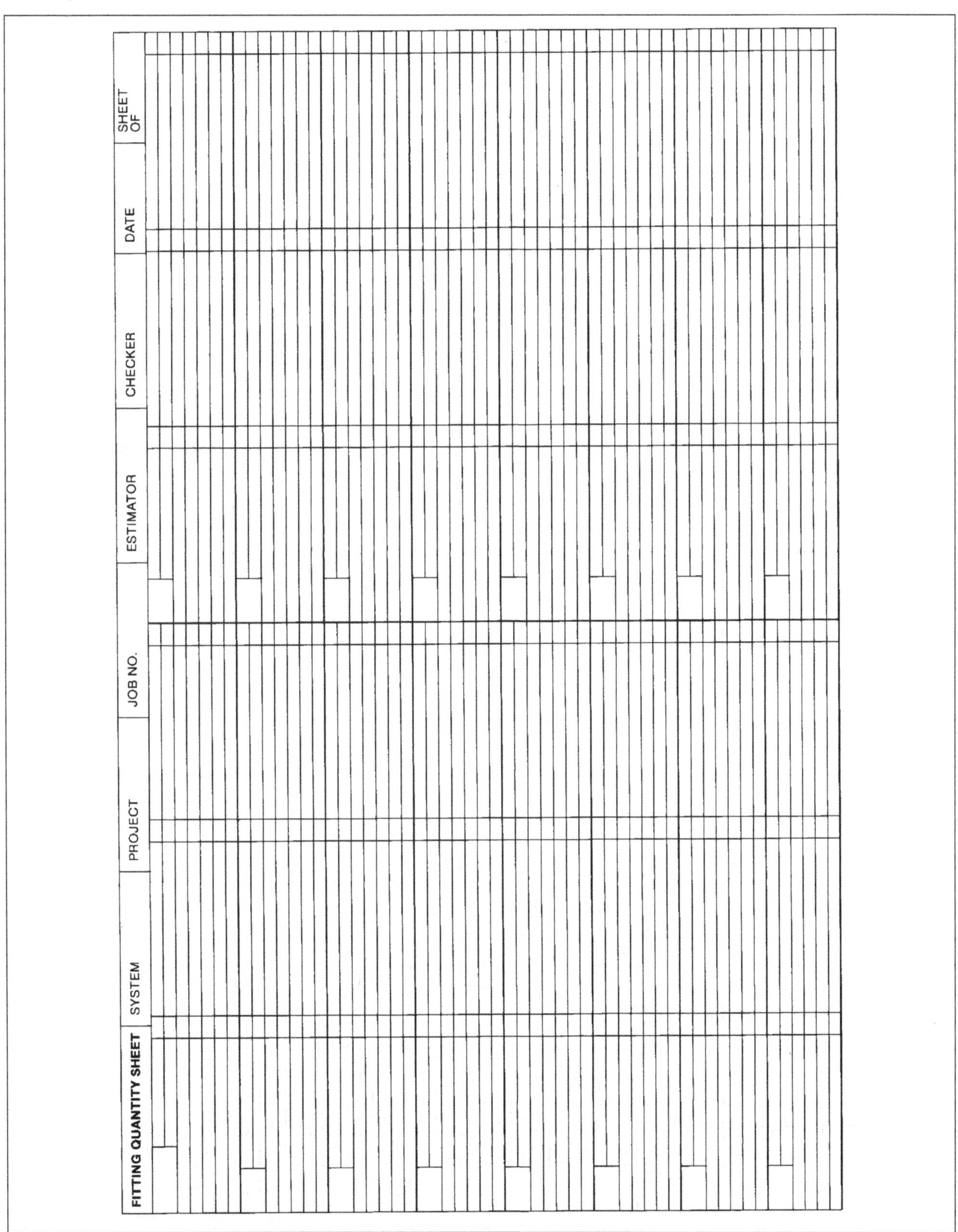

FITTING QUANTITY SHEET
SYSTEM
PROJECT
JOB NO.
ESTIMATOR
CHECKER
DATE
SHEET
OF

Figure 3.5

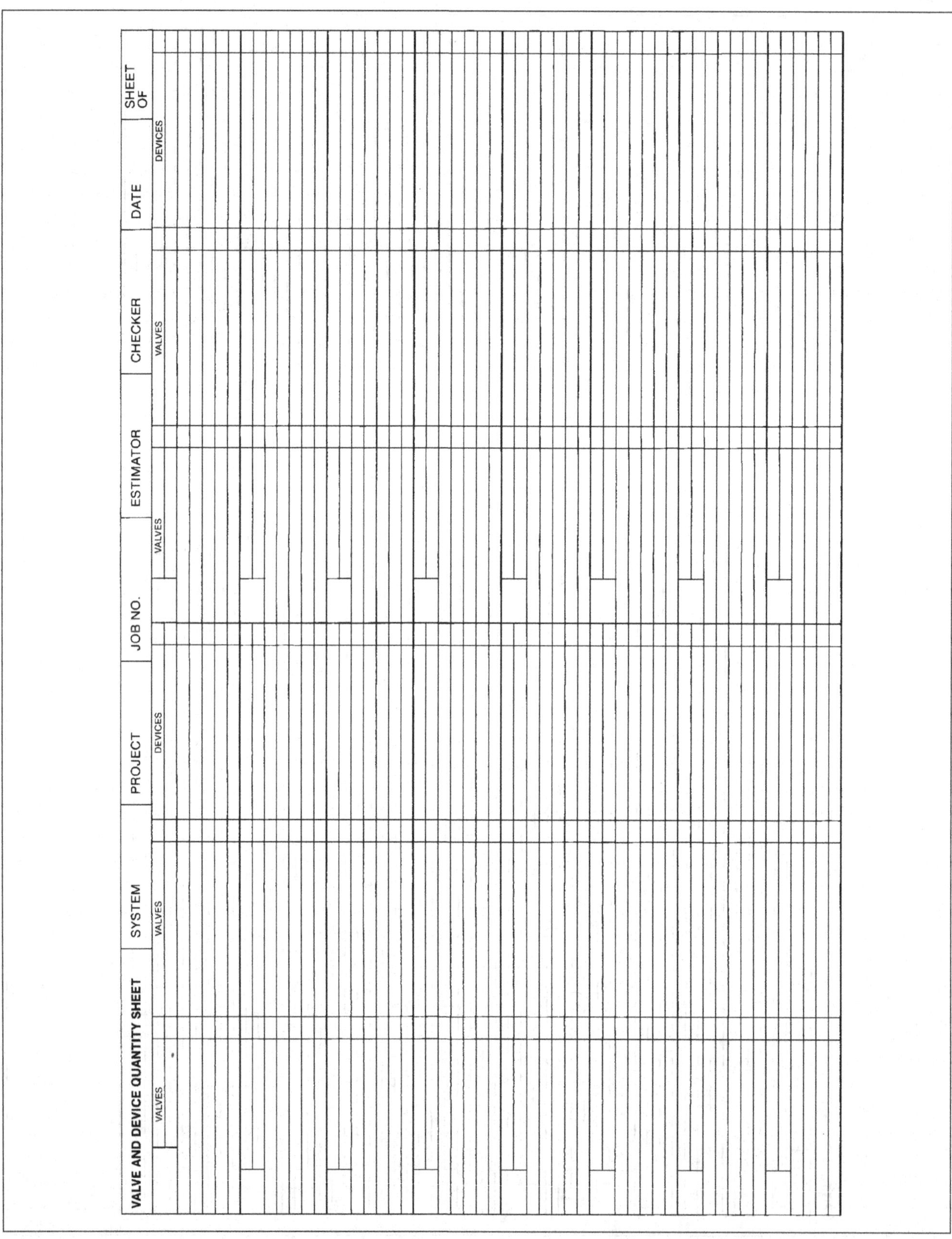

VALVE AND DEVICE QUANTITY SHEET	SYSTEM	PROJECT	JOB NO.	ESTIMATOR	CHECKER	DATE	SHEET OF
VALVES	VALVES	DEVICES		VALVES	VALVES	DEVICES	

Figure 3.6

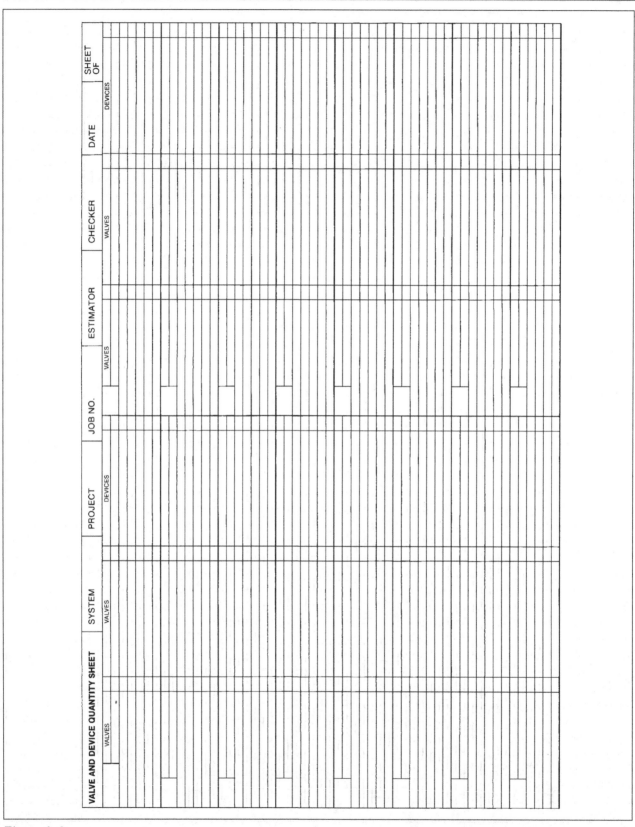

SHEET NO.

PROJECT

ESTIMATE NO.

LOCATION

CLASSIFICATION

DATE

ARCHITECT

ESTIMATOR

PRICED BY

CHECKED BY

DESCRIPTION	QUAN-TITY	UNIT	MATERIAL		LABOR				TOTAL COST MAT./LABOR
			UNIT COST	TOTAL COST	UNIT LH	TOTAL LH	UNIT RATE	TOTAL COST	

Figure 3.7

job cost. A sample of a Summary Sheet can be seen in Figure 3.8. The process of filling out estimate forms and summary sheets is discussed further in Chapters 5 and 6.

Plumbing Codes

Every project is designed according to a particular plumbing code. The plumbing codes are written to protect the public health by setting minimum standards and procedures to be followed in plumbing system design. Large cities, such as New York and Baltimore, write their own official plumbing codes, but in rural areas and small towns where there might not be a local plumbing code, an existing one created elsewhere is generally adopted. There are a number of plumbing codes published that a town or municipality may want to use, such as:

- **International Plumbing Code**: International Code Council (ICC).
- **Uniform Plumbing Code**: International Association of Plumbing and Mechanical Officials (IAPMO).
- **Fire Codes**: National Fire Protection Association, Inc. (NFPA).

Addresses for obtaining the above codes appear in Appendix D.

The plumbing code is of vital concern to a plumbing estimator for a number of reasons. Every estimator should have a copy of the plumbing code that has jurisdiction over the project. In this way, the plumbing estimator can verify that the plans and specifications are designed and written according to that code. It they are not, and the estimator figures the project as designed and is awarded the contract for work, he or she may still be required to install the material according to the respective code, at a possible additional cost. In addition, the plumbing estimator can use the code as a guide to properly estimate projects with incomplete drawings or specifications.

Vendor Catalogs

Manufacturers of plumbing products publish catalogs describing the items they manufacture. These vendor catalogs usually contain sketches, item descriptions, and model numbers of the manufacturers' full line of products. In addition, catalogs usually contain valuable engineering information and, in some cases, prices. The plumbing estimator should have a complete up-to-date library of vendor catalogs for reference purposes. Vendor catalogs are always free of charge and can be obtained by simply mailing or phoning in a request to the respective manufacturer. These catalogs are regularly updated, and estimators can request the latest product information on an ongoing basis.

Trade Price Sheets

To price a project correctly, an estimator needs up-to-date material costs. The prices of some materials change every day, and a price sheet even one month old may be completely outdated and useless. Items such as plumbing fixtures, valves, piping, and fittings usually appear on printed trade price sheets for use by the plumbing estimator. These price sheets are typically distributed by plumbing wholesalers and manufacturers, and may contain the discount allowed the plumbing trade. When pricing items subject to frequent fluctuation, one should include an appropriate margin. Large items, such as hot water generators and pumps, rarely appear on published price sheets. For these items, the estimator would call the manufacturer's sales representative for the price of the particular piece of equipment or materials specified. The plumbing estimator without current material prices leaves the firm open to the unnecessary risk of a mispriced

Summary Sheet

ABC PLUMBING COMPANY, INC.
CONTRACTORS
ANYWHERE, USA

SHEET NO. ______

PROJECT ______ ESTIMATE NO. ______

LOCATION ______ CLASSIFICATION ______ DATE ______

ARCHITECT ______ **ESTIMATOR** ______ PRICED BY ______ CHECKED BY ______

DESCRIPTION	QUAN-TITY	UNIT	MATERIAL		LABOR				TOTAL COST MAT./LABOR
			UNIT COST	TOTAL COST	UNIT LH	TOTAL LH	UNIT RATE	TOTAL COST	

Figure 3.8

estimate, resulting in lost contracts and money. A good backup reference for estimators who are unable to obtain a current price is the current edition of *Means Plumbing Cost Data,* which is updated regularly. This publication is also useful as a labor guide for unfamiliar methods or material installations.

Working Areas

Activity in contracting offices usually resembles the pressroom of a newspaper at a deadline. Unfortunately, the estimator cannot always have peace and quiet while trying to estimate a project. However, there are certain things an estimator can do to make the situation more bearable. The estimator should have a well-lit office with a large drafting table and a reference table. Distractions, such as radios, should be avoided if at all possible. An estimator constantly exposed to distractions will make mistakes, again, at the possible cost of lost contracts.

Chapter 4

The Takeoff

This chapter first discusses the prerequisites for performing the takeoff, such as reading the specifications, making notes, and scanning the drawings. The reader will learn the proper procedures for taking off a job and compiling a complete list of quantities. Finally, a sample job is used for a step-by-step analysis of an actual sample takeoff, complete with specifications, drawings, and quantity sheets.

Reading and Digesting Specifications

Specifications are the written instructions prepared by the design engineer that the contractor must follow in estimating and installing his or her portion of the work. It should be noted here that specifications generally take precedence over drawings. If any conflict of information occurs between the drawings and specifications, the specifications are the final word, unless otherwise stated by the engineer or architect through a written addendum to the contract.

Making Notes About Specifications

Before attempting any actual work, the estimator should spend time reading and digesting the specifications to fully understand the scope of work, the types of materials required, and any other requirements of a bidding contractor.

When reading the job specifications, it is advisable to make notes regarding anything unusual. Examples of noteworthy items include:

- Vaguely stated scope of work
- Noncompliance with plumbing code having jurisdiction
- Missing model numbers or equipment capacities

Becoming Familiar with the Drawings

Every estimator should have an overall "feel" for the job, in addition to knowledge of items listed in the specifications. Before beginning the takeoff, the estimator should scan all of the drawings in order to determine the complexity of the job, the approximate hours required to perform the work, and the method of design and construction used.

Making Notes About the Drawings

As with specifications, the estimator should also make notes about the drawings while scanning them. All required plans must be accounted for in the drawing set, and the drawings must be complete with respect to areas requiring plumbing, pipe sizing, details, and schedules. Designs should be drawn according to code. If any of the three conditions is not met, the estimator should make the applicable notes and notify the design engineer for clarification.

Coordination of Drawings and Specifications

After reviewing the job specifications and drawings individually, the estimator should then verify that the two are compatible. Often, design firms will have one person writing specifications for a job, and another person designing it. For various reasons, such as rushing to meet deadlines or drafting oversights, specifications and drawings can conflict at times. For example, the drawings may show a sump pump, while there is no mention of it in the specifications. These situations are infrequent with bid documents, but they can happen, and the estimator should be aware of them.

Taking Off Quantities from Drawings

The estimator should establish his or her own effective sequence for taking quantities off from drawings, and stick with it throughout his or her estimating career. The takeoff sequence the author has found best is shown below. This method is used on the sample takeoff that follows.

1. Plumbing fixtures
2. Equipment
3. Sanitary waste and vent system (pipe, fittings, devices) below grade
4. Sanitary waste and vent system (pipe, fittings, devices) above grade
5. Storm system (pipe, fittings, devices) below grade
6. Storm system (pipe, fittings, devices) above grade
7. Hot and cold water system (pipe, fittings, valves, devices)
8. Natural gas system (pipe, fittings, valves, devices)
9. Fire standpipe system (pipe, fittings, valves, devices)
10. Site work

Note: The sample project in this book does not have a sprinkler system or any other special system other than natural gas. The estimator will, of course, encounter projects that do include other systems. Special systems are taken off individually in the same manner shown for other systems.

Plumbing Fixtures

Plumbing fixtures are taken off by marking the fixture with a colored pencil on the drawings and then entering the fixture type onto the fixture quantity sheet. (See Chapter 3, Figure 3.2.) Fixture types are usually coded by the engineer to correspond with the codes in his specification, for example, P-1 water closet, P-2 lavatory, etc. Plumbing fixtures should be taken off separately for each floor. The quantity sheet includes columns to identify the floors each fixture type may appear on. This method aids the estimator in checking the fixture count, by totaling the columns down and the row across, and entering the single grand total in the lower right-hand corner, as shown on the sample takeoff. If a mistake was made, the estimator will arrive at two different totals, and should recheck the takeoff.

When taking off fixtures from the drawings, the estimator is automatically taking off fixture trim, since all trim required is called out in the specification under each fixture type. The counting of fixtures, as well as valves, fittings, and other devices throughout the job, is done by the following "slash" method. For example, if four P-1 water closets are shown on the basement plan, they are entered on the quantity sheet as ////. If five 1/2" gate valves are shown on the basement plan for the water system, they are entered on the valve and device quantity sheet for that system as ~~////~~. This method eliminates the totaling of many one and two-digit numbers at the end of the takeoff. The estimator simply counts up each full block of slashes as 5 (~~////~~). This recording method is also known as the "picket fence" method.

Equipment

Due to the small amounts of equipment used for the plumbing portion of a project, equipment is taken off simply by marking the piece of equipment in colored pencil on the drawings and entering it on the equipment quantity sheet (Chapter 3, Figure 3.3). The equipment quantity sheet has *item of equipment, quantity, system,* and *capacity* columns, each to be filled out as the individual piece of equipment is taken off. It is important that the estimator fill in the system column, because each piece of equipment will be transferred onto the estimate form to be priced with its appropriate system. Capacities for equipment can be found either on the drawings or in the specifications.

Sanitary Waste and Vent System (Below Grade)

Before taking off any piping, the estimator should take off all fittings and sleeves, such as "Y"s, 1/8 bends, sweeps, cleanout deck plates, etc. Fittings are taken off by placing a line through them on the drawings with a colored pencil and entering them by size onto the fitting quantity sheet (Figure 3.5 in Chapter 3). Most below-grade sanitary piping in a building is comprised of cast iron soil pipe and fittings. The smallest size pipe manufactured in cast iron service is two inches. Therefore, there will not be any fittings less than two inches in diameter on the underground takeoff.

After the takeoff of fittings is complete, totals are entered in the columns provided on the right side of each size category, as shown on the sample takeoff forms.

The estimator is now ready to take off the underground sanitary piping by size and linear foot, passing through all of the fittings. By measuring through the overall length of the piping run, rather than taking off only the piping between fittings, the estimator allows for some waste, which is necessary for any contingency that may occur during installation of the piping system. As the pipe is being taken off, the estimator marks off the section of pipe on the drawing with a colored pencil and then enters the footage on the piping quantity sheet for below-grade sanitary under the appropriate size of pipe, as shown on the sample takeoff.

If excavation and backfill is specified to be performed by the plumber, the estimator is now ready to take off excavation. Underground pipe below a building is rarely deeper than 4'. Drawings that are considered final documents generally furnish the estimator with two measurements: a *finished floor elevation* and a *pipe invert,* which is the lowest elevation or bottom-most portion of the pipe. If these two measurements are given, the estimator can arrive at the depth of the pipe by subtracting the pipe invert from the finished floor elevation.

With the pipe depth known, the estimator can calculate the cubic yards to be excavated by using the chart and method shown in Figures 1.23 and 1.24. If finished floor elevations and pipe inverts are not given, the estimator should assume a 4′ burial. The number of calculated cubic yards of excavation is then placed in the extreme right-hand column of the piping quantity sheet, as shown on the sample takeoff at the end of this chapter. These quantities are later transferred to the estimate forms.

If the cast iron soil pipe is of lead-joint design, the estimator must determine the total pounds of lead, oakum (jute packing), and gas necessary to complete the installation. To do this, he or she must first have completed the piping and fitting takeoff in order to know the total number of joints required. Cast iron soil pipe is manufactured in 5′ and 10′ lengths. For estimating purposes and for figuring lead quantities, the 5′ length is used. The amount of lead consumed per joint is determined by the diameter of pipe. (See Figure 4. 1.)

Excavation and backfill are often performed by the prime or general contractor. The plumbing estimator should check the specifications and General Conditions to define responsibility for this part of the work and avoid costly duplication.

The following example will demonstrate how to estimate quantities of lead, oakum, and gas for cast iron soil pipe.

Let us assume that we have taken off the following quantities of soil pipe and fittings for a project:

6″ XH cast iron soil pipe	–15 L.F.
4″ XH cast iron soil pipe	–10 L.F.
6″ × 4″ ″Y″	–1 ea.
4″ 1/8 bend	–1 ea.
4″ sweep	–1 ea.

LEAD REQUIRED TO CAULK CAST IRON SOIL PIPE JOINTS					
Pipe & Fitting Diams. Inches	Lead Ring Depth Inches	Service Weight		Extra Heavy Weight	
		Cu. Ins.	Wt. Lbs.	Cu. Ins.	Wt. Lbs.
2	1	2.81	1.15	2.91	1.19
3	1	3.90	1.60	4.17	1.71
4	1	4.98	2.04	5.25	2.15
5	1	6.06	2.49	6.24	2.56
6	1	7.15	2.93	7.42	3.04
8	1.25	15.06	6.17	15.49	6.35
10	1.25	18.90	7.75	19.34	7.93
12	1.25	25.53	10.47	26.02	10.67
15	1.5	43.09	17.67	43.38	17.8

Figure 4.1

The estimator proceeds to calculate the lead, oakum, and gas as follows. (Note that each hub, rather than each opening, is the basis for the number of joints.)

15 L.F. of 6″pipe (5′lengths) = three 6″joints × 3.04 lbs. per joint		
		= 9.12 lbs. of lead
10 LF. of 4″pipe (5′lengths) = two 4″joints × 2.15 lbs. per joint		
		= 4.30 lbs. of lead
6″× 4″″Y″ (one 6″joint & one 4″joint)		
	3.04 + 2.15	= 5.19 lbs. of lead
4″1/8 Bend (one 4″Joint)		= 2.15 lbs. of lead
4″Sweep = (one 4″joint)		= 2.15 lbs. of lead
Total		= 22.91 lbs. of lead

Oakum is usually estimated at one-tenth the weight of lead. Therefore, 22.91 lbs. of lead/10 = 2.29 lbs. of oakum.

Gas consumption is approximately one (instopropane) cylinder per 200 lbs. of lead.

After calculating the lead, oakum, and gas, the estimator should enter all quantities in the columns provided on the piping quantity sheet to be transferred later to the estimate forms.

Note: The quantities of lead, oakum, and gas entered on the piping quantity sheet include all fittings taken off. The estimator must work with both the fittings and pipe quantity sheets to perform the exercise.

If the cast iron soil pipe is of the neoprene joint or clamp (hubless) design, it is suggested that the estimator:

1. Count the number of joints to arrive at a quantity of neoprene gaskets or clamps.
2. Price them based on an average size.

Sanitary Waste and Vent System (Above Grade)

Using the same takeoff procedure as the below-grade sanitary system, the estimator should begin taking off all fittings on a separate fitting quantity sheet for above-grade sanitary. (See the sample fitting quantity sheet, Figure 3.5.) If hubless cast iron fittings are used, they are taken off in the same manner as bell and spigot pipe fittings.

After all fittings are taken off, the estimator can proceed with the above-grade sanitary piping. The above-grade piping takeoff procedure is similar to that previously explained for the underground. However, the estimator will have to take off the piping in two separate steps: (1) horizontal runs and branches, and (2) risers and drops to fixtures and equipment.

Riser diagrams or floor plans often provide the estimator with dimensions showing heights floor-to-floor. With this information, the estimator can estimate the height of risers and approximate the lengths of drop pieces down to fixtures.

Horizontal piping and risers are taken off in separate operations, because the estimator will most likely refer to two or more drawings, and the takeoff will be easier if done separately. However, the estimator can combine both the horizontal and riser quantities on one piping quantity sheet. If the drawings are not equipped with riser diagrams, the task becomes more difficult. All drawings are two-dimensional, and to take off risers and drop pieces, the estimator must try to visualize the drawings in a three-dimensional state.

Another item taken off for above-grade installations is *pipe hangers and supports*. After all the piping is taken off, the estimator can determine the quantities of hangers and supports by dividing the total linear footage of pipe in each size category by the recommended intervals stated in the pipe support section (shown in Figure 1.22), or as stated in the project specifications. This will give the estimator the approximate number of hanger assemblies required for the installation. The hanger quantities can be entered on the fitting quantity sheet until it is time to list them on the estimate forms.

Floor drains through a suspended slab and vents through the roof generally have to be flashed with either sheet metal or a membrane. If lead is used, four square feet of sheet lead per unit has to be carried in the estimate. Sheet lead can weigh either four or six pounds per square foot, depending on the engineer's specifications. To estimate the total pounds of sheet flashing required for a project, the estimator performs one of the following calculations:

Number of units × 4 lbs. × 4 SF = Total Pounds
or Number of units × 6 lbs. × 4 SF = Total Pounds

Once the sheet flashing quantity is calculated, the estimator enters the figure on the piping quantity sheet (Figure 3.4) under that heading. Other metallic flashing material is estimated similarly, but membrane flashing is calculated and priced by the square foot.

Hot and Cold Water System

On the hot and cold water system, fittings and sleeves are again the first items to be taken off. The fitting takeoff procedure is the same as outlined for the below-grade sanitary system. Copper water fittings in the reducing sizes are taken off as "× reducing." For example, if the estimator has a 3″ × 2″ × 2″ tee, it is taken off as a 3″ × reducing tee. (See the fitting quantity sheet, Figure 3.5.)

Valves and devices are the next item to be taken off. Valves are taken off according to type and size and are entered on the valve and device quantity sheet for hot and cold water, as shown on the sample takeoff. If items such as access panels are part of the contract, they should be taken off at this time and placed on the valve and device quantity sheet.

Water piping is taken off in a similar fashion as the sanitary piping above grade. The estimator first takes off mains and branches, and then proceeds with risers and drop pieces. The quantities are then entered onto the pipe quantity sheet for hot and cold water. (See the sample water piping quantity sheet at the end of the chapter.) Pipe hangers and supports are taken off by dividing the total linear footage of water piping in each size category (by the intervals recommended in Chapter 1, Figure 1.22) and are then entered on the water fitting quantity sheet, as shown on the sample takeoff.

Solder, flux, and gas—the materials used for joining water pipe and fittings—should now be estimated. Solder joint pipe and fittings are usually joined by soft (non-lead) solder. Solder, flux, and gas are difficult items to estimate, but using the chart in Figure 4.2, the estimator can arrive at a relatively accurate amount of solder and flux required for an installation. The estimator should count the number of joints required for each fitting and valve in its own respective size category. For example, a 2″ tee requires three 2″ joints and a 2″ ell requires two 2″ joints. The estimator should again refer to the chart in Figure 4.2 for the pounds of required solder per 100 joints. The total amounts of solder, flux, and gas can now be entered on the piping quantity sheet.

Once the estimator has determined the total linear footage of water piping, he knows the total linear footage of pipe that requires insulation. All hot and cold water piping is usually insulated, and all the estimator needs to do to estimate insulation is multiply the cost of insulation by the actual footage of water piping taken off, plus fittings and valves. Insulation is usually estimated and installed by an appropriate insulation subcontractor.

Storm System Below and Above Grade

The storm system below and above grade is taken off in the same manner as the appropriate sanitary waste and vent system. The only difference is that the estimator should keep horizontal pipe offsets in hung ceilings separate, since these sections of pipe are normally insulated to prevent condensation.

Fire Standpipe System and Natural Gas System

These two systems are taken off using the same procedure as the hot and cold water system. See the sample takeoff sheets for these systems.

Site Work

Site piping and fittings are taken off by size and system, and the quantities are entered on the respective quantity sheets. (See the sample site piping quantity takeoff sheets.) For drainage systems, the estimator takes off manholes and catch basins, and notes their depths. A top or rim elevation and a bottom or invert elevation are normally given at each manhole and catch basin. By subtracting the invert from the top elevation, the estimator will arrive at the depth of the manhole or catch basin. The estimator should also note connections of new sewers to existing sewers. Manholes, catch basins, and connections to existing sewers should be entered on the fitting quantity sheet. (See the sample site drainage fitting quantity sheets.)

ESTIMATED POUNDS OF SOFT SOLDER REQUIRED TO MAKE 100 JOINTS*

Size	3/8″	1/2″	3/4″	1″	1 1/4″	1 1/2″	2″
Pounds	.5	.75	1.0	1.4	1.7	1.9	2.4
Size	2 1/2″	3″	3 1/2″	4″	5″	6″	8″
Pounds	3.2	3.9	4.5	5.5	8.0	15.0	32.0

*Two oz. of flux will be required for each pound of solder. One tank of PRESTO gas will be required for every 500 joints.

Figure 4.2

Water service piping and fittings are taken off by size and entered on their respective quantity sheets. (See sample water piping quantity sheets.) Valves, or any wet connections or taps, are then taken off and entered onto the valve quantity sheet as shown on the sample site estimate. Valves are taken off by size. For wet connections, the size of both the existing main and the new connection should be noted. For example, a new 4″ main connected to an existing 8″ main is taken off as an 8″ x 4″ wet connection. Taps are taken off based on the size of the new service.

Excavation is calculated as shown in the excavation section of Chapter 1. The estimator can determine the appropriate depth of drainage piping by the manhole and catch basin rim and invert elevations given on the drawings. Water piping excavation is based on an assumed depth of 4′, unless otherwise noted on the drawing. The cubic yards of excavation are then entered on the respective piping quantity sheet for each system.

Note: Excavation, backfill, and even site work piping may be subcontracted to other appropriate specialists. However, the estimator must still provide a responsible estimate for this work. Careful reading of the General Conditions or specifications will determine whether the site work is to be carried by the plumber or general contractor.

Items Related to the Plumbing Estimate

The following items, not included in every project, must be evaluated by the plumbing estimator. Valve tags, rigging, and pipe markers, when required, are always the work of the plumber. The number of valve tags should be the same as the number of valves.

The number of pipe markers is specified, based on designated intervals per lineal foot of pipe.

Rigging, the handling of large equipment, will be performed either by the plumber's own workers, or may be subcontracted. (See Chapter 1 for more detail.)

Pipe painting and concrete pads are normally not part of the plumbing contract. However, the specifications may, in some instances, require the plumber to include concrete pads in his bid. In this case, prices would have to be solicited from the appropriate trade subcontractors. Pipe painting, other than stenciling, is not the work of the plumber. If, however, a cost is required for the bid, the appropriate subcontractor must be consulted. Concrete pads should be estimated as outlined in Chapter 1.

Once these figures are estimated, they can be entered directly on the estimate forms. After all quantities are taken off and totaled on the quantity takeoff sheets, the figures should be rechecked by another individual. Upon completion, the person performing the check should initial the takeoff sheet in the space provided.

Sample Job

Imagine that you have been asked to bid on the "Three-Story Service Center," our sample project located in a medium to large U.S. city. Assume that you have received the following drawings and specifications for the purpose of performing a takeoff and estimate.

The building is a three-story, slab-on-grade structure of approximately 45,000 square feet. The superstructure is of steel frame construction with upper floor framing of bar joist, metal deck/concrete fill design. The interior partitions are to be constructed of 5/8" drywall with metal studs. The roofing system is of standard built-up design. Site soil conditions are good, consisting of soft clay and loose, medium sand. Access to the site is excellent with required utilities along nearby roads.

You should now become familiar with the sample drawings in Figures 4.3 through 4.11. As mentioned in the Preface, it is assumed that you are familiar with architectural prints. Therefore, elementary explanations are not necessary. However, you may want to refer to Appendix A for clarification of the plumbing fixtures and piping, fitting, valve, and device symbols found on the drawings.

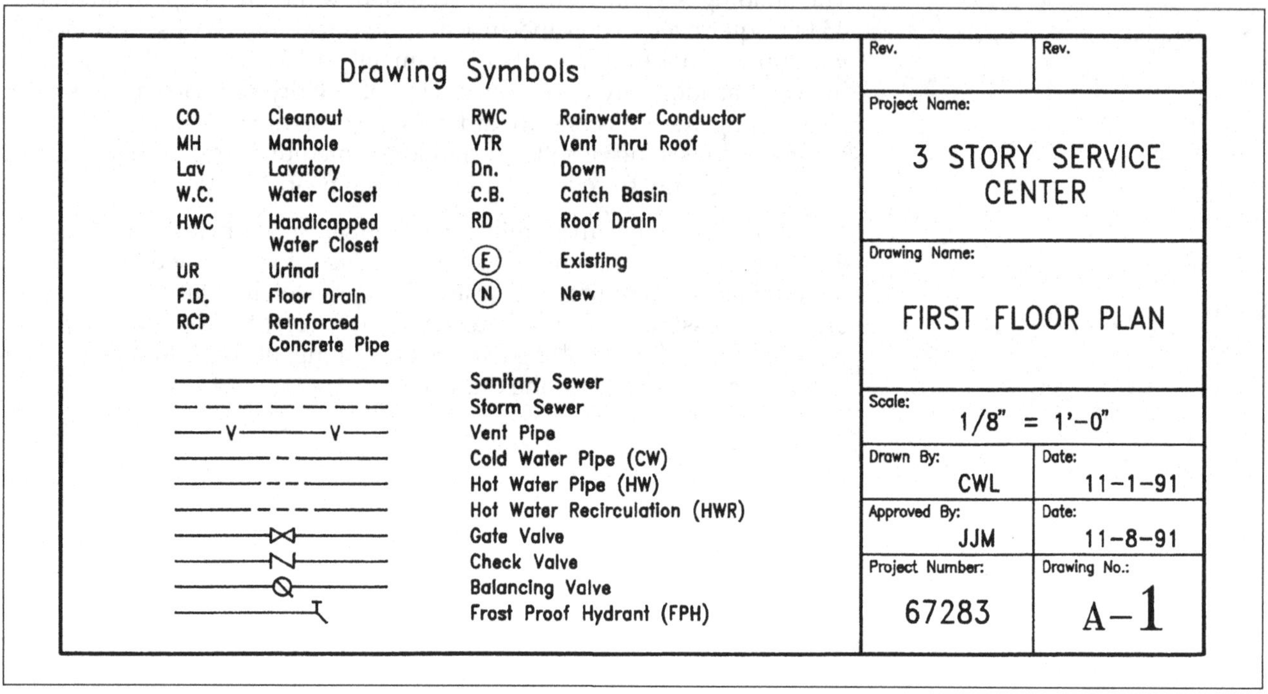

Drawing Symbols

CO	Cleanout	RWC	Rainwater Conductor
MH	Manhole	VTR	Vent Thru Roof
Lav	Lavatory	Dn.	Down
W.C.	Water Closet	C.B.	Catch Basin
HWC	Handicapped Water Closet	RD	Roof Drain
UR	Urinal	(E)	Existing
F.D.	Floor Drain	(N)	New
RCP	Reinforced Concrete Pipe		

Sanitary Sewer
Storm Sewer
Vent Pipe
Cold Water Pipe (CW)
Hot Water Pipe (HW)
Hot Water Recirculation (HWR)
Gate Valve
Check Valve
Balancing Valve
Frost Proof Hydrant (FPH)

Rev.	Rev.
Project Name: 3 STORY SERVICE CENTER	
Drawing Name: FIRST FLOOR PLAN	
Scale: 1/8" = 1'-0"	
Drawn By: CWL	Date: 11-1-91
Approved By: JJM	Date: 11-8-91
Project Number: 67283	Drawing No.: A-1

Figure 4.3

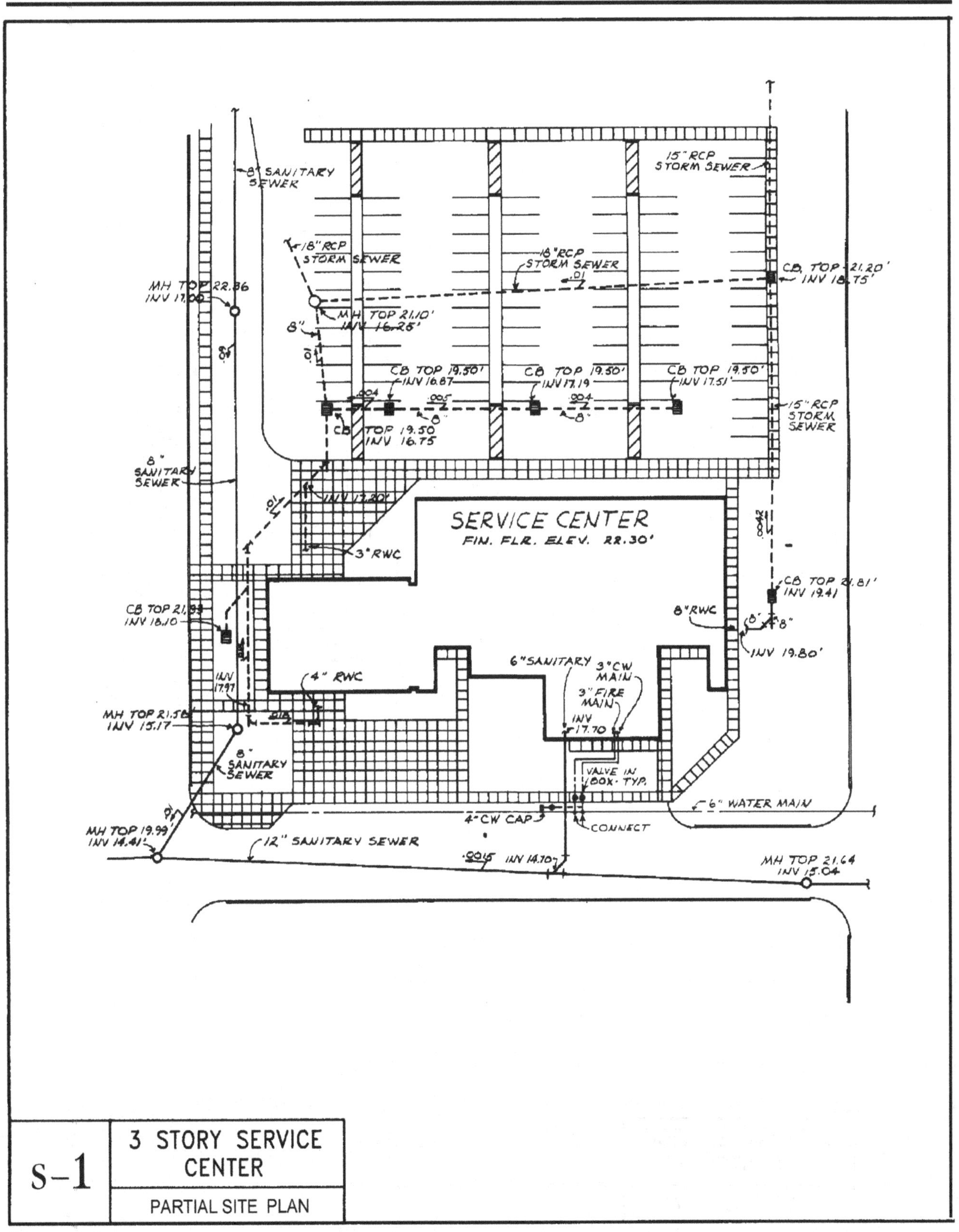

8" SANITARY SEWER
15" RCP STORM SEWER
18" RCP STORM SEWER
18" RCP STORM SEWER
.01
CB. TOP 21.20'
INV 18.75'
MH TOP 22.86
INV 17.00
MH TOP 21.10'
INV 16.25'
CB TOP 19.50'
INV 16.87
CB TOP 19.50'
INV 17.19
CB TOP 19.50'
INV 17.51'
CB TOP 19.50
INV 16.75
15" RCP STORM SEWER
8" SANITARY SEWER
INV 17.20'
SERVICE CENTER
FIN. FLR. ELEV. 22.30'
3" RWC
CB TOP 21.81'
INV 19.41
8" RWC
INV 19.80'
CB TOP 21.83
INV 18.10
4" RWC
6" SANITARY
3" CW MAIN
3" FIRE MAIN
INV 17.70
INV 17.97
MH TOP 21.58
INV 15.17
8" SANITARY SEWER
VALVE IN BOX. TYP.
6" WATER MAIN
4" CW CAP
CONNECT
MH TOP 19.99
INV 14.41'
12" SANITARY SEWER
.0015
INV 14.70
MH TOP 21.64
INV 15.04
S-1
3 STORY SERVICE CENTER
PARTIAL SITE PLAN

Figure 4.4

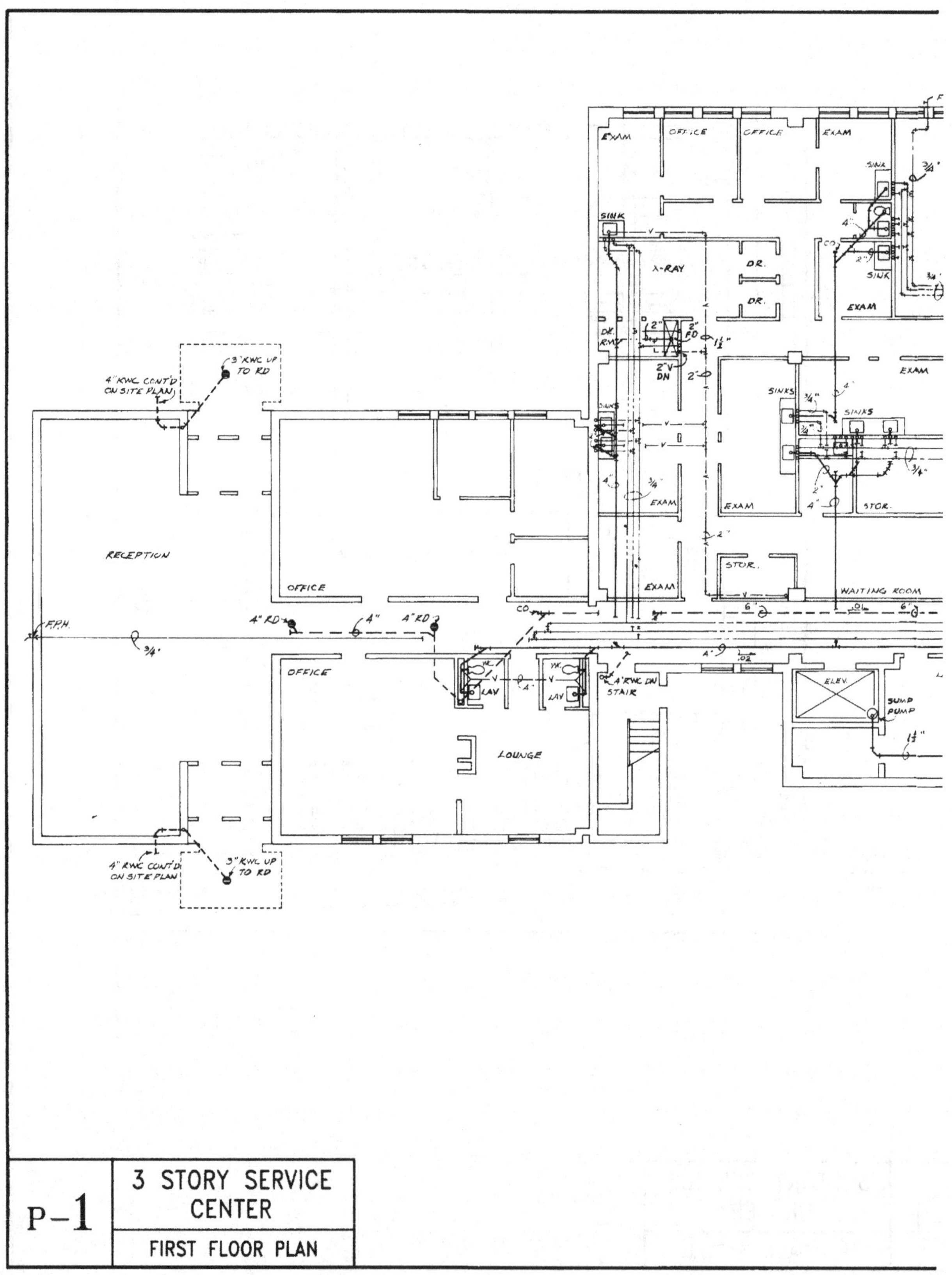
EXAM
OFFICE
OFFICE
EXAM
SINK
X-RAY
DR.
DR.
EXAM
DK. RM.
EXAM
SINKS
SINKS
EXAM
EXAM
STOR.
STOR.
EXAM
WAITING ROOM
RECEPTION
OFFICE
OFFICE
LOUNGE
LAV
LAV
STAIR
ELEV.
SUMP PUMP
4" KWC CONT'D ON SITE PLAN
3" KWC UP TO RD
4" RWC CONT'D ON SITE PLAN
3" KWC UP TO RD
F.P.H.
4" RD
4" RD
4"KWC DN
P-1
3 STORY SERVICE CENTER
FIRST FLOOR PLAN

Figure 4.5

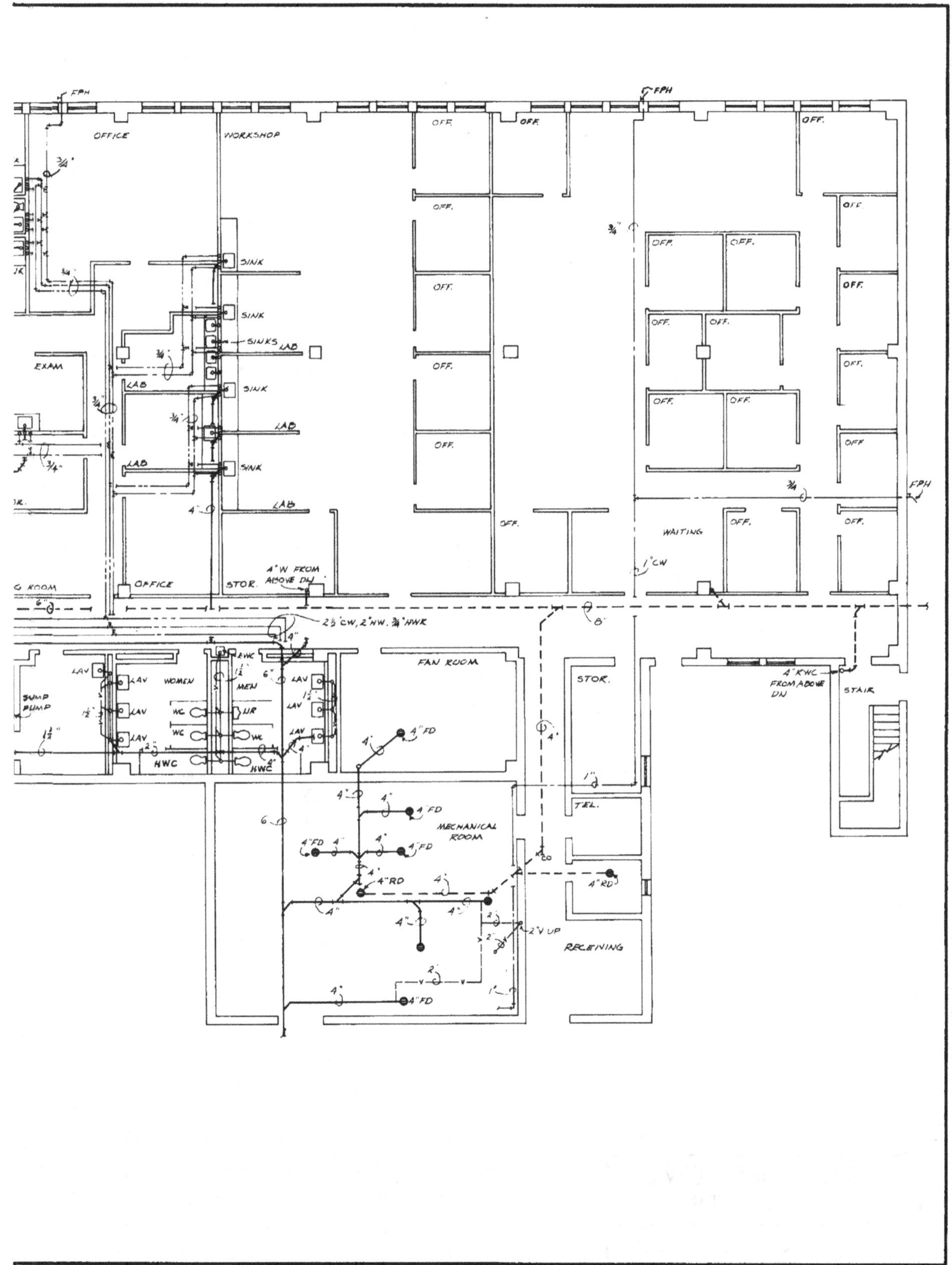

Figure 4.5 (cont'd.)

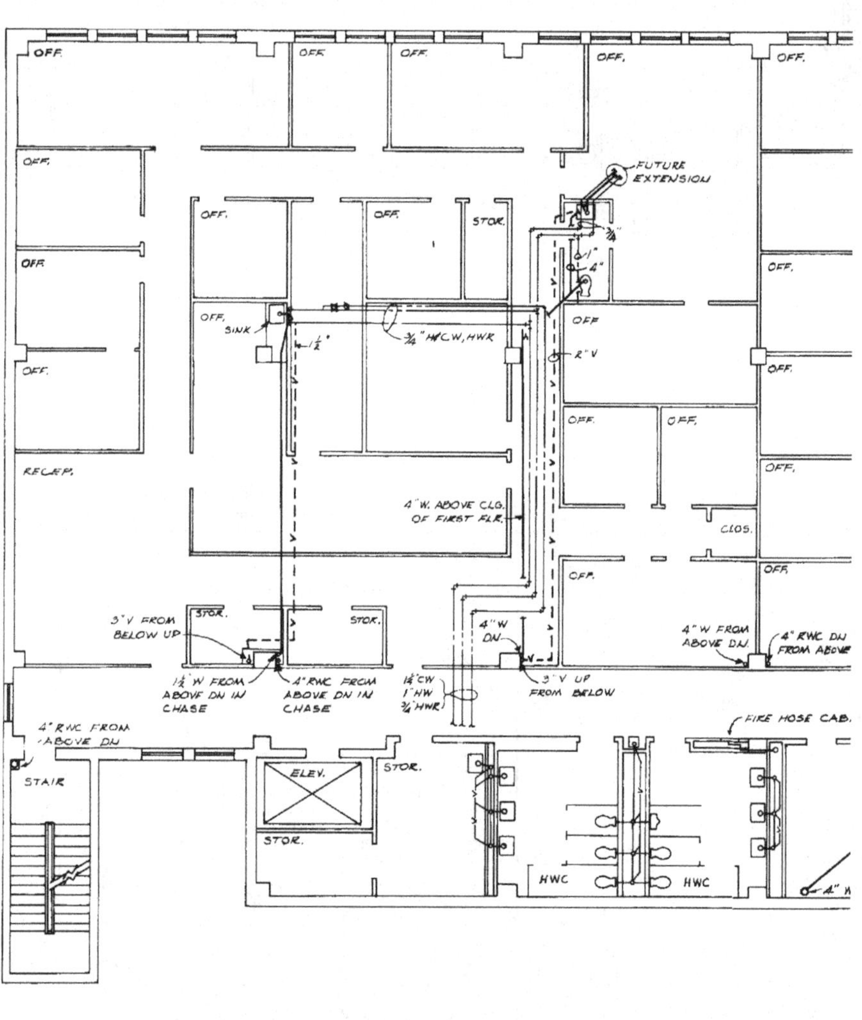

P–2	3 STORY SERVICE CENTER
	SECOND FLOOR PLAN

Figure 4.6

Figure 4.6 (cont'd.)

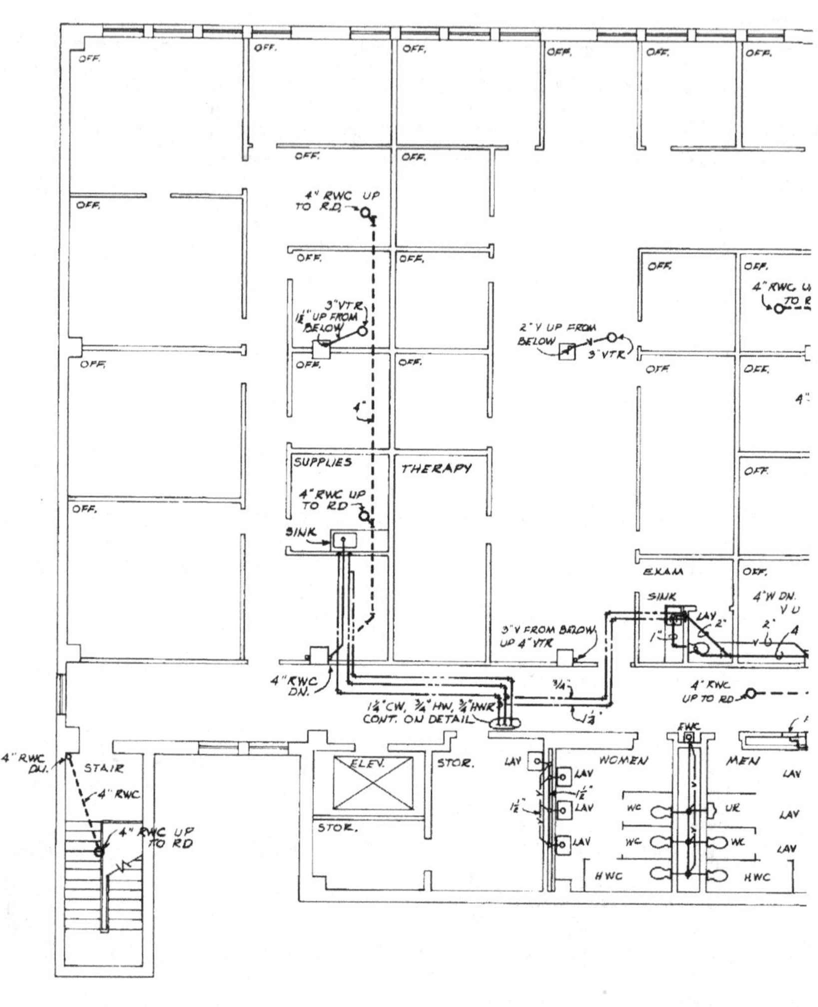

P–3	3 STORY SERVICE CENTER
	THIRD FLOOR PLAN

Figure 4.7

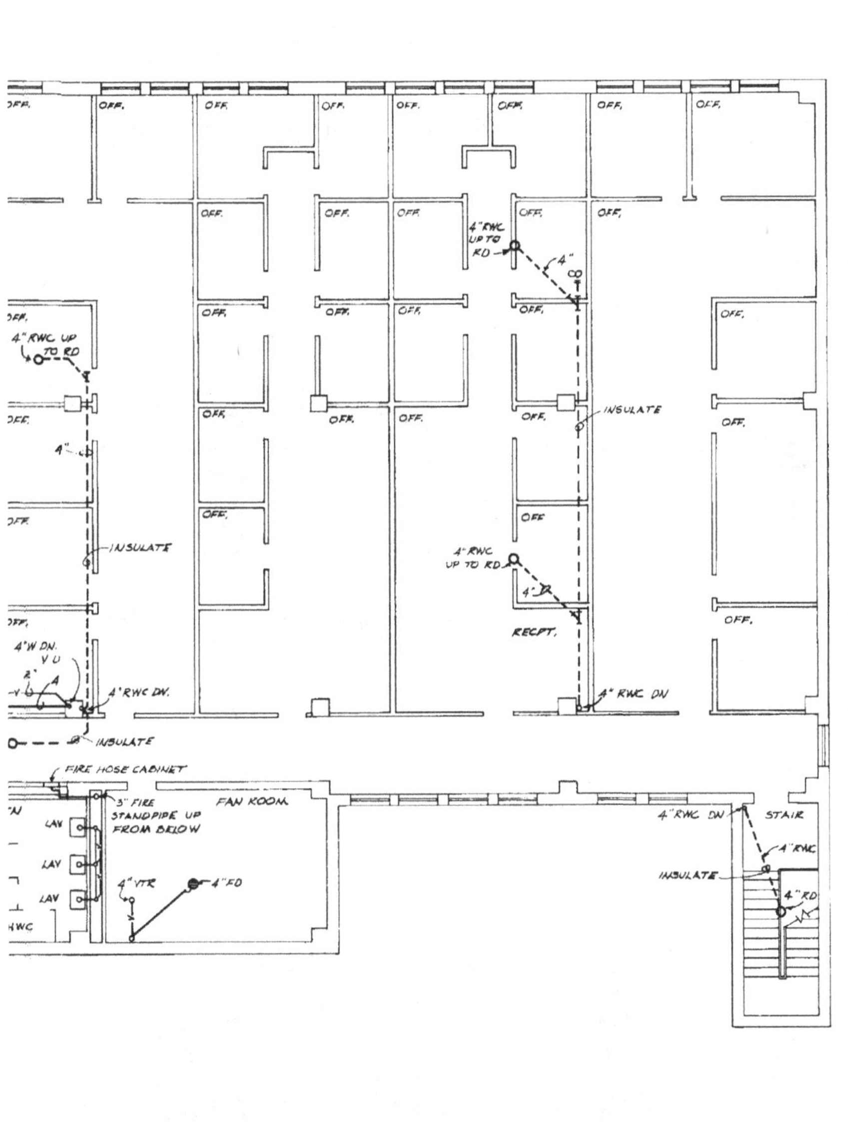

Figure 4.7 (cont'd.)

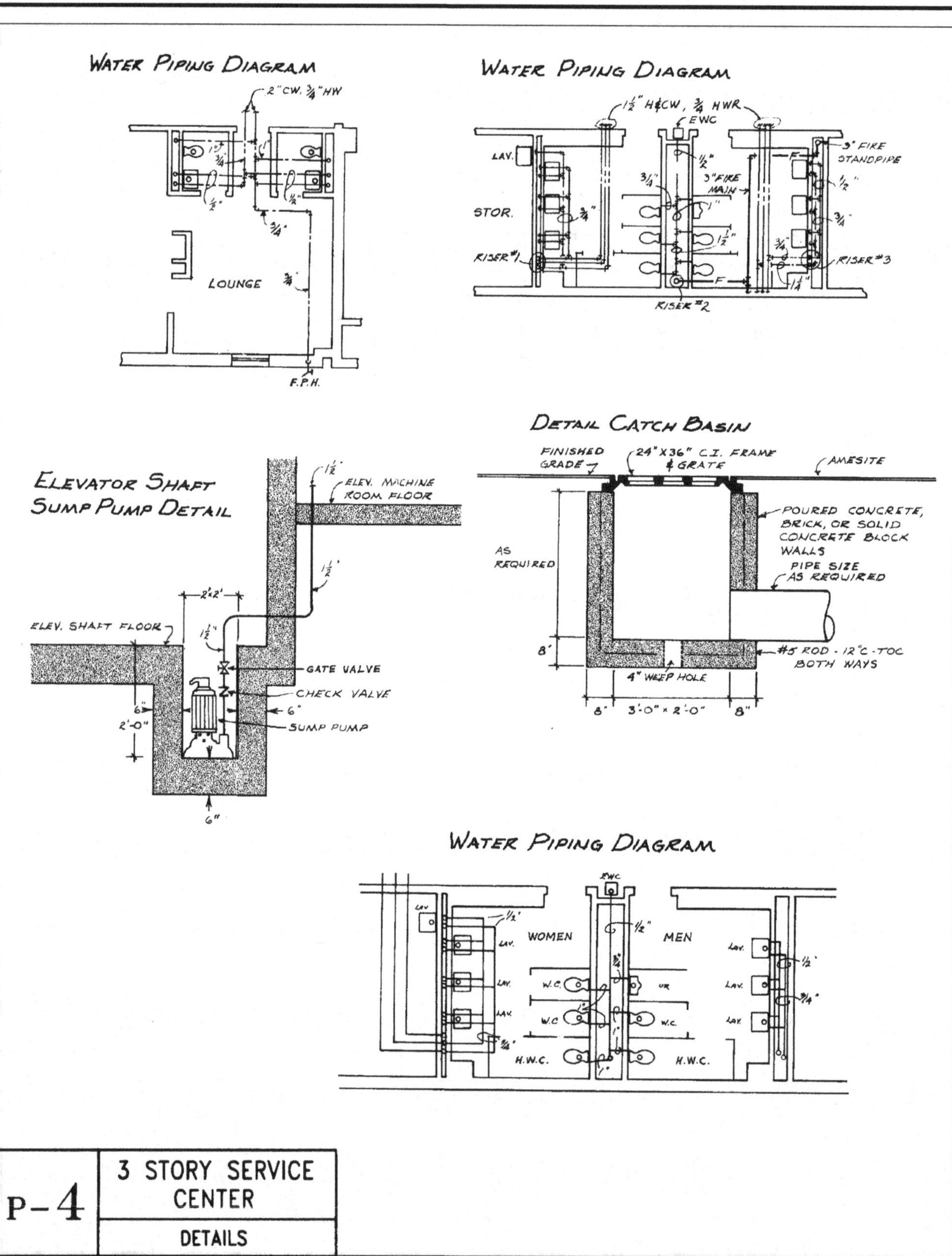
WATER PIPING DIAGRAM
2" CW. 3/4" HW
LOUNGE
F.P.H.
WATER PIPING DIAGRAM
1 1/2" H&CW, 3/4 HWR
EWC
LAV.
STOR.
RISER #1
3" FIRE MAIN
3" FIRE STANDPIPE
RISER #3
RISER #2
ELEVATOR SHAFT SUMP PUMP DETAIL
ELEV. MACHINE ROOM FLOOR
ELEV. SHAFT FLOOR
GATE VALVE
CHECK VALVE
SUMP PUMP
DETAIL CATCH BASIN
FINISHED GRADE
24"X36" C.I. FRAME & GRATE
AMESITE
POURED CONCRETE, BRICK, OR SOLID CONCRETE BLOCK WALLS
PIPE SIZE AS REQUIRED
AS REQUIRED
#5 ROD - 12"C-TOC BOTH WAYS
4" WEEP HOLE
3'-0" x 2'-0"
WATER PIPING DIAGRAM
EWC
WOMEN
MEN
LAV.
W.C.
UR
H.W.C.
P-4
3 STORY SERVICE CENTER
DETAILS

Figure 4.8

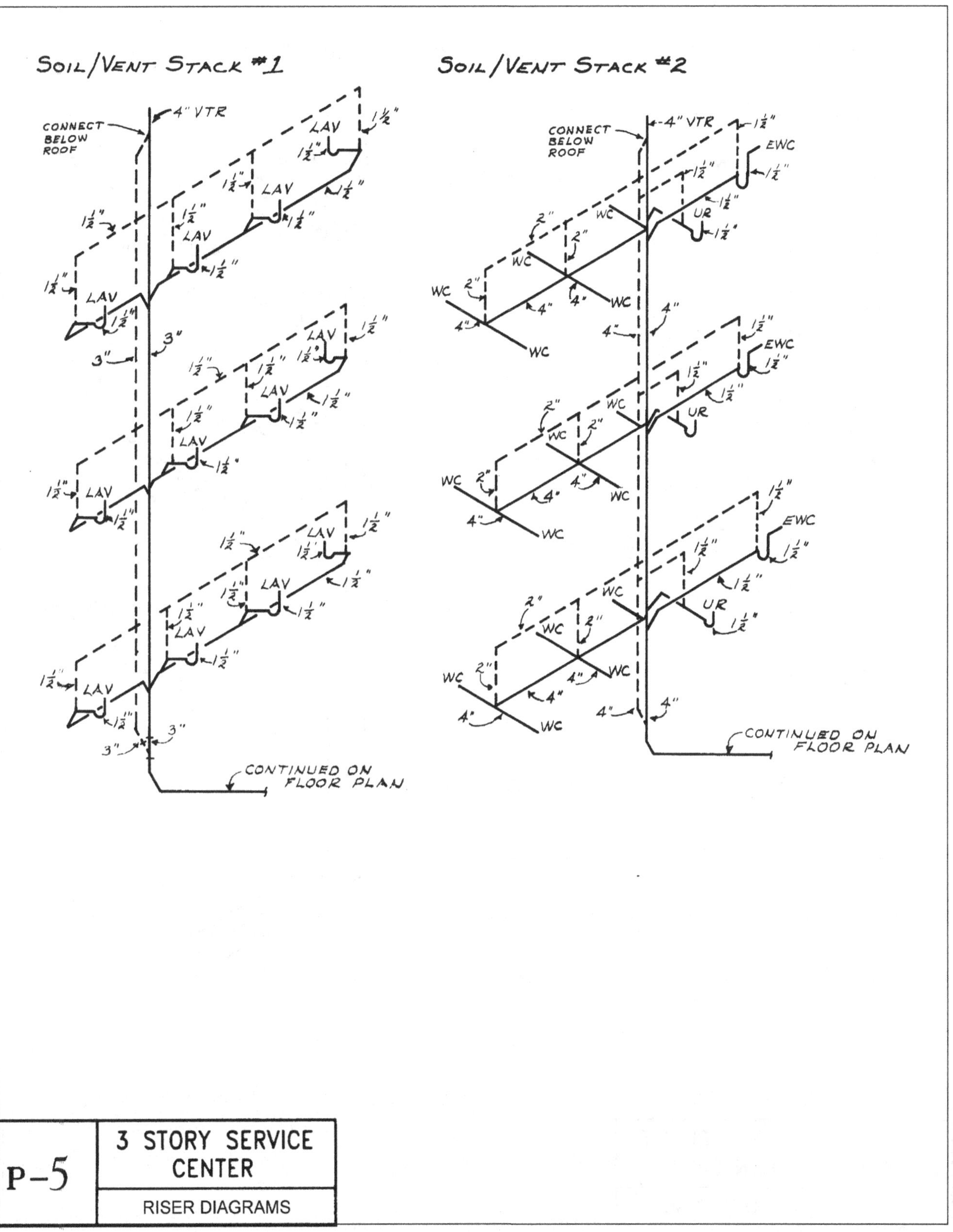

SOIL/VENT STACK #1
SOIL/VENT STACK #2
CONNECT BELOW ROOF
4" VTR
LAV
WC
UR
EWC
CONTINUED ON FLOOR PLAN
P-5
3 STORY SERVICE CENTER
RISER DIAGRAMS

Figure 4.9a

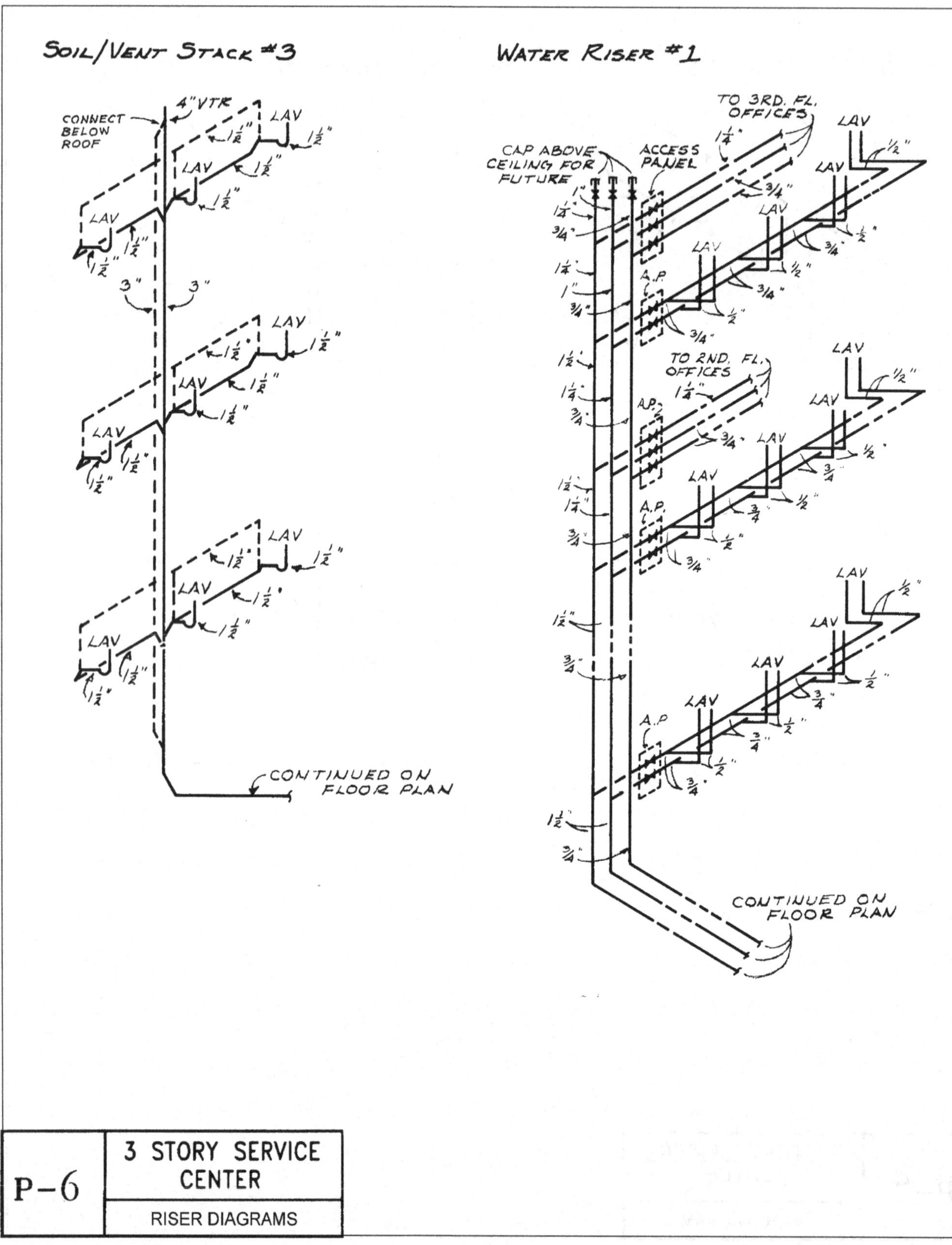
Soil/Vent Stack #3
4" VTR
CONNECT BELOW ROOF
LAV
3"
CONTINUED ON FLOOR PLAN
Water Riser #1
TO 3RD. FL. OFFICES
CAP ABOVE CEILING FOR FUTURE
ACCESS PANEL
A.P.
TO 2ND. FL. OFFICES
CONTINUED ON FLOOR PLAN
P-6
3 STORY SERVICE CENTER
RISER DIAGRAMS

Figure 4.9b

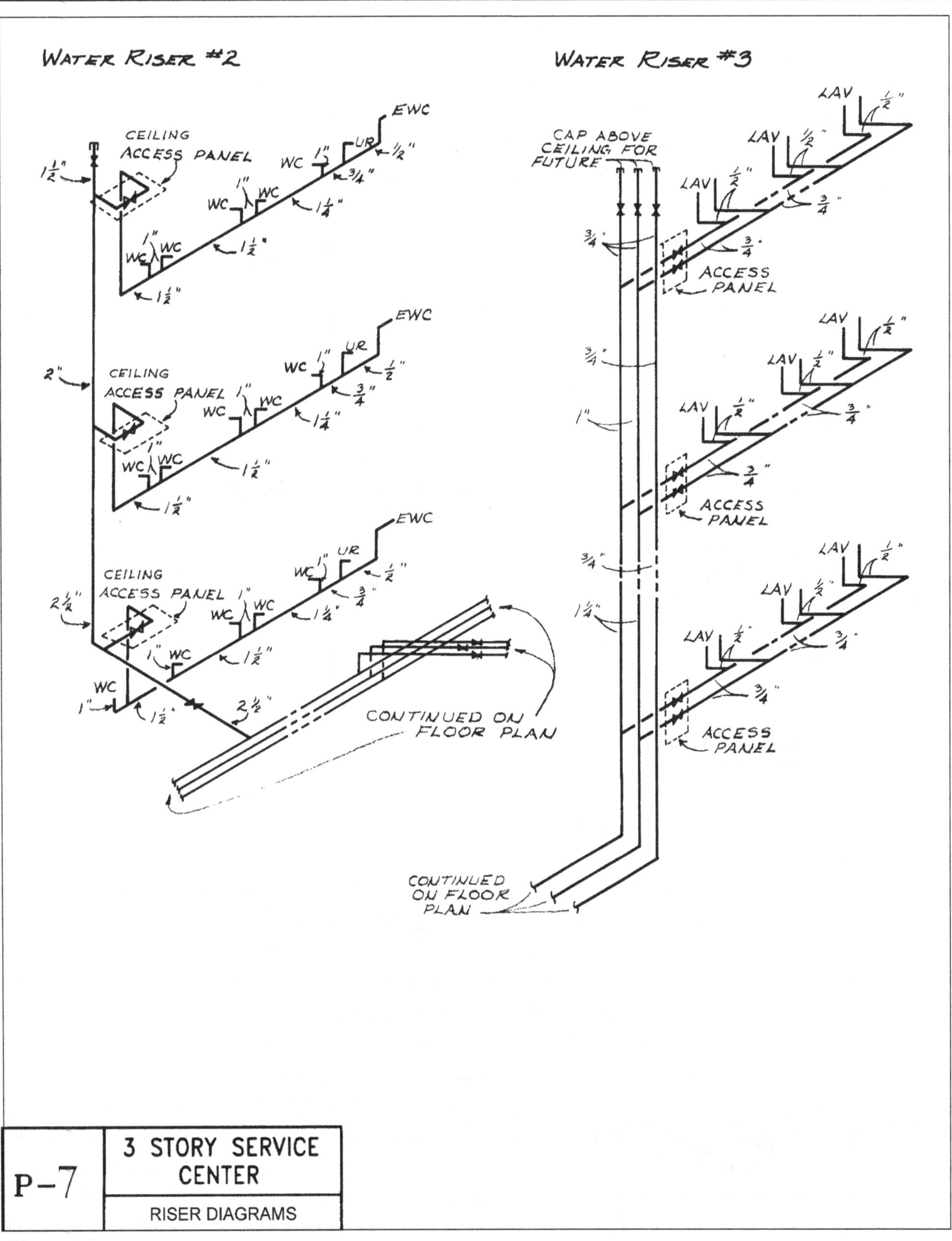
WATER RISER #2
WATER RISER #3
CEILING ACCESS PANEL
EWC
WC
UR
LAV
CAP ABOVE CEILING FOR FUTURE
ACCESS PANEL
CONTINUED ON FLOOR PLAN
P–7
3 STORY SERVICE CENTER
RISER DIAGRAMS

Figure 4.9c

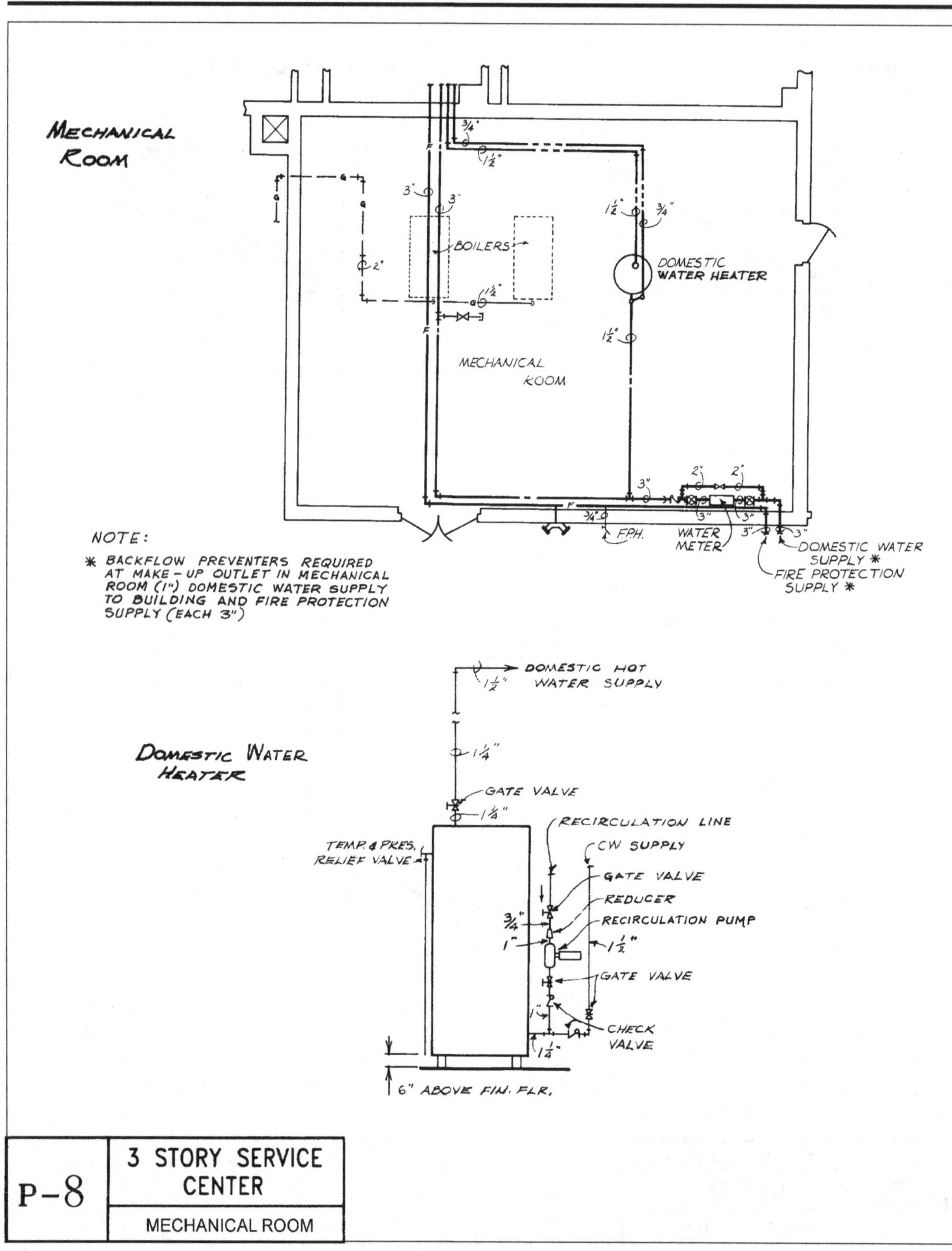
MECHANICAL ROOM
BOILERS
DOMESTIC WATER HEATER
MECHANICAL ROOM
NOTE:
* BACKFLOW PREVENTERS REQUIRED AT MAKE-UP OUTLET IN MECHANICAL ROOM (1") DOMESTIC WATER SUPPLY TO BUILDING AND FIRE PROTECTION SUPPLY (EACH 3")
F.P.H.
WATER METER
DOMESTIC WATER SUPPLY *
FIRE PROTECTION SUPPLY *
DOMESTIC HOT WATER SUPPLY
DOMESTIC WATER HEATER
GATE VALVE
TEMP. & PRES. RELIEF VALVE
RECIRCULATION LINE
CW SUPPLY
GATE VALVE
REDUCER
RECIRCULATION PUMP
GATE VALVE
CHECK VALVE
6" ABOVE FIN. FLR.
P-8
3 STORY SERVICE CENTER
MECHANICAL ROOM

Figure 4.10

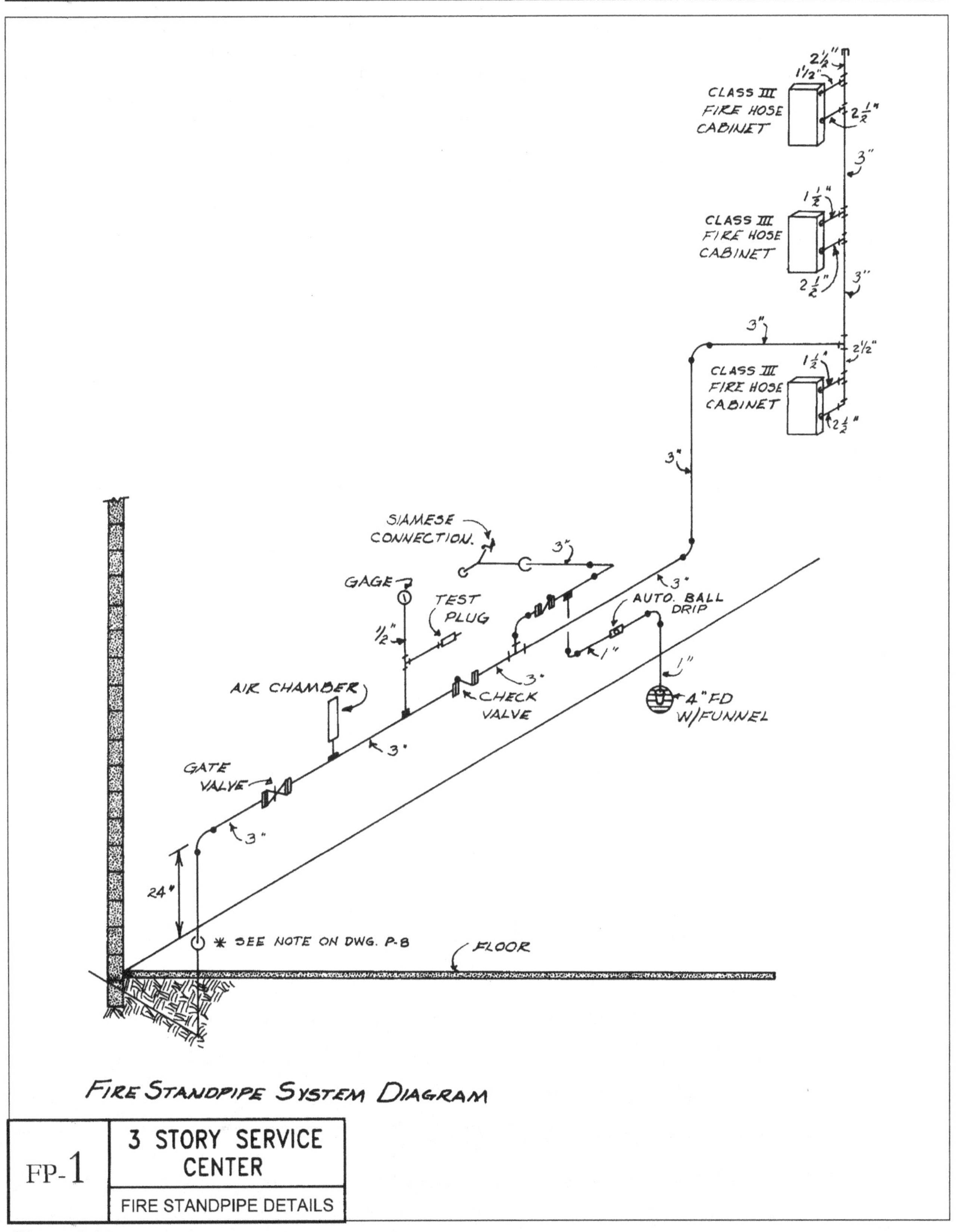
CLASS III
FIRE HOSE
CABINET
CLASS III
FIRE HOSE
CABINET
CLASS III
FIRE HOSE
CABINET
SIAMESE
CONNECTION
GAGE
TEST
PLUG
AUTO. BALL
DRIP
AIR CHAMBER
CHECK
VALVE
4" FD
W/FUNNEL
GATE
VALVE
24"
* SEE NOTE ON DWG. P-8
FLOOR
FIRE STANDPIPE SYSTEM DIAGRAM
FP-1
3 STORY SERVICE
CENTER
FIRE STANDPIPE DETAILS

Figure 4.11

Following the drawings, you will find the sample specifications for the Three-Story Service Center (Figures 4.12 through 4.57). In the sample specifications, there are numbered "specification notes" at the bottom of certain pages that correspond to the flagged and numbered sections within the body of the specification. The purpose of these notes is to extract and explain the most relevant portions of a specification.

SAMPLE JOB SPECIFICATIONS

Three-Story Service Center | Index

DIVISION 15 - PLUMBING
SECTION 1
BASIC MATERIALS AND METHODS

Figure 4.12

Three-Story Service Center PLUMBING 1-1

DIVISION 15 - PLUMBING

SECTION 1

BASIC MATERIALS AND METHODS

PART 1: GENERAL

1.01 NOTICE

A. General Conditions and Schedule of Drawings apply to and are hereby made part of this Section.

B. Contractor consult these Sections in detail as he will be responsible for and governed by conditions set forth therein and work indicated.

1.02 SCOPE

A. Work complete in all details including fixtures and equipment as hereinafter specified, with all appurtenances common to various systems generally consisting of piping, valves, hangers and supports, insulation, covering, structures, excavation and backfilling, cleaning, testing and such other material and work as is necessary, specified or required to form complete and properly operating systems as herein specified or indicated.

B. Following items are included in work required and are described hereinafter in detail:

1. Domestic water service, hot, cold and recirculating water piping.
2. Fire protection system water service and standpipe system.
3. Plumbing fixtures.
4. Building sanitary and storm water drainage.
5. Catch basins, storm water sewer and sanitary sewer.
6. Insulation.
7. Temporary Water.

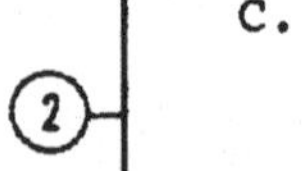

C. It is not intended that these Specifications or the accompanying Drawings show every detail; Contractor furnish all labor and install all material required for complete systems functioning as described, whether or not specifically called for or indicated.

SPECIFICATION NOTES

1. Estimator must take note of scope of work to see exactly what is required of him as a bidding contractor.
2. A note such as this should cause the estimator to work closely with the plumbing code having jurisdiction.

Figure 4.13

Three-Story Service Center PLUMBING 1-2

1.03 PLUMBING REFERENCES

A. Abbreviations

1. BTU - British Thermal Unit
2. cfm - cubic feet per minute
3. fpm - feet per minute
4. gal - gallon
5. gpm - gallons per minute
6. hp - horse power
7. lb - pound
8. psi - pounds per square inch
9. C - degree Centigrade
10. F - degree Fahrenheit
11. ft or ' - foot
12. gph - gallons per hour
13. in. or " - inch
14. wwp - water working pressure
15. sp - static pressure

B. Technical societies, trade organizations, governmental agencies.

1. AGA - American Gas Association
2. ASME - American Society of Mechanical Engineers
3. ASTM - American Society for Testing Materials
4. AWWA - American Water Works Association
5. NFPA - National Fire Protection Association
6. UL - Underwriters' Laboratories, Inc.
7. USA - USA Standards Institute

C. Definitions

1. "Provide" shall mean "furnish and install".
2. "Herein" shall mean "contents of a particular Division" where this term appears.
3. "Indicated" shall mean "indicated on Contract Drawings".
4. "Equal" shall mean "approved equal".
5. "Contractor" shall mean "Contractor or subcontractor for the work described".

Figure 4.14

Three-Story Service Center PLUMBING 1-3

1.04 SHOP DRAWINGS AND SAMPLES

A. Shop drawings

(3)

1. Furnish six (6) sets of shop drawings and pictorial or descriptive data.
2. Shop drawings marked with project designation.
3. Obtain approval within forty-five (45) days after signing respective Contract.
4. Data for respective trade submitted in separate folios.
5. Obtain approval of Architect before ordering.
6. Furnish performance curves showing efficiency, capacity, head and brake horsepower for all pumps, compressors and heat exchangers.
7. Furnish additional shop drawings and descriptive data, other than those listed herein, as requested by Architect.

B. Materials

1. Architect reserves right to require submission of samples of any or all articles or materials proposed to be used under these Specifications.

(4)

1.05 GOVERNING REQUIREMENTS

A. Mechanical installations comply with all applicable codes, ordinances, rules, regulations, and laws in effect.

B. Following considered minimum requirements:

1. State Plumbing Code.
2. National Plumbing Code.
3. Occupational Safety and Health Administration.

C. Construction comply with Department of Labor, Bureau of Labor Standards, Safety and Health Regulations for Construction.

(5)

D. Contractor obtain and pay for all necessary permits for work, including sanitary and storm sewers, as part of Contract.

E. Contractor arrange and pay for all required inspections and furnish required certificates of inspection to Owner.

(6)

1.06 TESTS

A. Arrange and pay for all tests required by Authorities specified.

B. Contractor notify Architect three (3) working days before tests are made.

3. Indicates extent of shop drawing work, and helps estimate engineering cost when marking up estimate.
4. Indicates codes estimator must follow in his development of the estimate.
5. Indicates the bidding contractor is responsible for the paying of permits; aids in applying a dollar value for permits and fees in cost estimate.
6. Indicates to estimator that dollars must be included in estimate for testing systems.

Figure 4.15

C. Repeat tests after defects are corrected.

D. Drainage system.

1. Test applied before pipe is covered.
2. System filled with water and subjected to not less than 10' of hydrostatic head.
3. Water remain in system not less than 15 minutes with no leaks or lowering of water level.
4. Air test of 5 psi for 15 minutes acceptable in lieu of water test.

E. Interior water supply system

1. Hydrostatic pressure test of not less than 125 psi.
2. Test pressure applied for not less than one hour with no leaks.

F. Gas piping

1. Gas piping tested as required by local Gas Light Co. and State Plumbing Code.
2. Tests meet requirements of NFPA.

G. For other tests, see particular equipment specified.

1.07 PAINTING

A. Equipment

1. Equipment furnished factory-finished with colors as selected by Architect.

2. Contractor refinish equipment with matching finish when damaged during shipment or construction.

B. Piping and insulation covering

1. Piping and covering in masonry walls, trenches and underground painted with two coats of cut-back-asphaltum paint, except cast-iron pipe may have factory-applied coating.

2. Exposed piping painted as specified in Painting Section of this Specification.

C. Hangers, supports and insulation covering painted as specified in Painting Section of this Specification.

D. All surfaces to be painted must be thoroughly cleaned of rust, scale, grease and foreign matter.

Figure 4.16

(7)

1.08 TEMPORARY FACILITIES

A. Water

1. Provide temporary water supply in locations directed by General Contractor for drinking water and construction purposes.
2. Water for above purposes will be separately metered and paid for by General Contractor.

1.09 AS-BUILT DRAWINGS

A. As work progresses, record on one set of plumbing Drawings all changes from the installation originally indicated.

B. Record final location of underground lines by depth from finished grade.

C. Record offset distances from buildings, curbs, or edges of walks.

D. Locate piping from interior walls and floors.

E. Submit to Architect for approval and record the above required information in colored pencil on blueprints of Contract Drawings.

1.10 INSTRUCTING ATTENDANT

A. Verbal Instruction

1. After all tests and adjustments, Contractor instruct attendant or Owner's representative in all details of operation of respective system.
2. Supply attendants to operate the systems until Architect is satisfied that the systems have been installed in accordance with these Drawings and Specifications and are functioning properly.
3. Provide services of equipment manufacturer's engineer to instruct representative of Owner in operation and maintenance of Mechanical equipment and controls.

B. Written Instructions

1. Provide two (2) copies of printed instructions and diagrams covering operation and maintenance of each item of equipment and controls.
2. Instructions furnished in bound covers and posted at locations designated by Architect.
3. Diagrams include performance curves for all pumps, minimum size 8-1/2" x 11".

7. Indicates to estimator that temporary facilities are to be installed under his contract.

Figure 4.17

Three-Story Service Center PLUMBING 1-6

1.11 GUARANTEE

A. Contractor leave entire system installed under respective contracts in proper working order.

B. Contractor responsible for specified performance of all equipment.

C. Contractor replace any work or material which develop defects, except ordinary wear and tear, or fail to perform satisfactorily, within one (1) year from the date of final acceptance.

D. Date of final acceptance or partial acceptance of system determined by Architect.

PART 2: MATERIALS

2.01 MATERIAL & EQUIPMENT REQUIREMENTS

A. Design

1. Materials and equipment conform to capacity, efficiency design and material specified.
2. Equipment must meet dimension and space requirements.
3. Sizes and capacities indicated or specified are minimum requirements; Contractor may use larger sizes provided space requirements are met and do not result in additional installation, maintenance or operating costs to Owner.
4. Materials and appliances of types for which there are UL Standard requirements, listings or labels have such listing of UL, be so labeled, and conform to their requirements.

B. Materials

1. Equipment or material of the same type or classification shall be product of same manufacturer.
2. Provide all new materials new, of best of their respective kind, and conforming with accepted standards of trade.
3. Equipment and accessories not specifically described or identified by manufacturer's catalog number designed in conformity with applicable technical standards and specifications of societies, organizations and/or agencies listed herein, suitable for maximum working pressure and have neat and finished appearance.
4. In all cases where device or part of equipment is herein referred to in singular number, it is intended that such reference apply to as many such items as are

Figure 4.18

required to complete installation.

5. Manufacturer's names and catalog numbers are given to describe and illustrate type, quality and design of material and equipment required.
6. Data on comparable material and equipment of other than listed manufacturers may be submitted with written request for approval of Architect; if approved, Architect will issue Addenda to Specifications.
7. No request for above approval will be considered later than ten (10) days before bids are due.

2.02 PIPE AND FITTINGS

A. Schedule 40, black steel pipe and fittings.

1. Provide pipe of uniform thickness with smooth cylindrical interior.
2. Pipe conform with USA Standard B36.10.
3. Provide standard weight black banded malleable iron fittings.
4. Fittings conform with ASTM Standard A-47.

(8)

B. Copper pipe and fittings.

1. Provide Type L hard copper pipe for pipe 1-1/2" and smaller above ground.
2. Provide Type K hard copper pipe for pipe larger than 1-1/2" above ground.
3. Provide Type L or K, as specified herein, soft copper pipe underground. (9)
4. Pipe conform to ASTM Standard B-88.
5. Copper pipe fittings wrought, sweat type; no joints underground. (10)
6. Fittings conform to ASTM Standard B-62.

C. Cast-iron soil pipe and fittings.

1. Provide cast-iron soil pipe of extra heavy weight, bell and spigot pattern, factory-coated inside and out with coal-tar pitch varnish, cylindrical and smooth, free from sand holes, cracks, and other defects.
2. Pipe and fitting conform with USA Standard 40.1 and ASTM Standard A-74.

(13)

3. Hubless pipe and fittings as manufactured by Tyler Pipe Industries, or equal, acceptable above ground. (11)

D. Terra cotta pipe and fittings.

1. Provide extra strength, hub and spigot pattern, vitrified, impervious clay sewer pipe.

8. Indicates to estimator the type of piping and fittings specified.
9. Notice that engineer has given bidding contractor a choice of materials. Estimator can use least expensive alternate and yet comply with specifications.
10. Estimator cannot figure joints for underground copper.
11. Engineer again has given bidding contractor a choice.

Figure 4.19

2. Pipe sound and well burned, with clear ring, smooth and free from blisters, cracks, or large chips.
3. Pipe and fittings conform to ASTM Standard C-200.

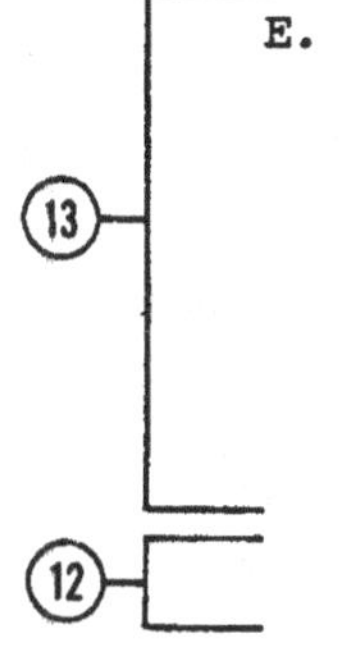

E. Class 250 ductile iron pressure pipe and fittings.

1. Provide pipe conforming to USA Standard A21 and A21.7.
2. Pipe of type with bolted mechanical joints.
3. Fittings conform to AWWA Specifications for Class D special castings or USA Standard A21.10 for Class 250 short-body fittings.
4. Pipe and fittings coated inside and out with coal-tar pitch varnish.

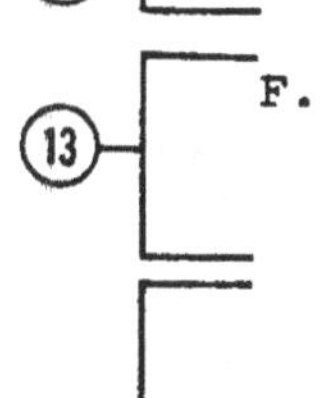

5. Provide cement lining as indicated conforming with AWWA Standard C104.

F. Galvanized Steel Schedule 40 pipe and fittings.

1. Provide pipe, of threaded, heavily and uniformly galvanized inside and outside.
2. Pipe conform to ASTM Standard A-53.
3. Provide galvanized cast iron recessed drainage fittings, screw pattern on waste and drain piping.
4. Provide galvanized, cast iron or malleable iron, flat banded pattern fittings on vent piping.

(14)

12. Cement lining is required for cast-iron water pipe. Estimator should figure accordingly.
13. Indicates to estimator type of piping and fittings specified.
14. Indicates to estimator type of piping and fittings specified.

Figure 4.20

Three-Story Service Center PLUMBING 1-9

G. Concrete pipe

(14)
1. Provide machine tongue and grooved reinforced concrete pipe with closed joints.
2. Pipe true circle, or uniform thickness and straight in direction of axis, sound, without cracks or large chips.
3. Pipe conform to ASTM Standard C76.

H. Valves

1. Provide valves of one manufacturer; Crane, Jenkins, Lunkenheimer or Walworth.
2. Valves designed for 125 psi wwp.
3. Provide gate valves for shut-off valves, 1" and larger.
4. Provide globe valves for all throttling valves.
5. Valves have manufacturer's name and working pressure cast integral.
6. Provide sleeve end type valves for solder joints on copper pipe.
7. Provide globe pattern valves for sizes 3/4" and smaller.
8. Gate valves conform to following:

   a. 2-1/2" and smaller: Bronze, inside screw non-rising spindle, solid or split wedge, screwed bonnet.
   b. 3" and larger: Bronze, outside screw and yoke, rising stem, bronze-mounted, solid or split bronze wedge, renewable seat rings and bolted bonnet, flanged.

(16)

9. Check valves conform to following:

   a. 2-1/2" and smaller: Bronze body, renewable composition disc, screwed top.
   b. 3" and larger: Bronze body, renewable seat ring, composition disc bolted at top, flanged.

10. Globe and angle valves conform to following:

   a. 2-1/2" and smaller: Bronze body, renewable composition disc, screwed bonnet.
   b. 3" and larger: Bronze body, bronze mounted with renewable seat ring and composition disc, outside screw and yoke, flanged.

11. Provide bronze body, bronze or brass mounted, double gate valve conforming to AWWA Standard Specifications, for valves in pipe underground; terminate valve stems in wrench nuts and furnish two (2) suitable keys.

12. Valves placed in accessible position and installed with stems vertical or as directed by Architect.

15. Indicates quality of valves specified.
16. Indicates types of valves, and types of valves for various size ranges specified.

Figure 4.21

13. Provide valve boxes for valves underground.

    a. Standard cast iron, adjustable shaft having minimum diameter of 5-1/4".
    b. Casting coated with two coats of coal-tar pitch varnish.
    c. Lids of all boxes bear the word "Water" or letter "W".

I. Unions.

1. Provide screwed pattern, galvanized malleable iron unions for sizes 2-1/2" and smaller.
2. Provide standard weight, flanged pattern, galvanized cast-iron unions sizes 3" and larger.
3. Unions suitable for 300 psi wwp and be of ground joint type with brass seat ring pressed into head piece.

J. Gaskets.

1. Provide full-faced rubber ring type, 1/16" thick gaskets for above ground; cast iron or malleable flanges.
2. Provide soft asbestos gaskets with graphited finish for fixture outlets to floor flanges.
3. Gaskets suitable for 125 psi wwp.

K. Nipples.

1. Provide nipples of same material as pipe or tubing on which they are installed.
2. Nipples extra strong when unthreaded portion is less than 1" long.
3. Running thread nipples are prohibited.

L. Cleanouts

1. Provide bodies conforming in thickness to that required for pipe on which installed.
2. Cleanouts extend not less than 1/4" above hub if installed on soil pipe.
3. Cleanouts have heavy brass plugs not less than 1/4" thick, provide with raised nut of not less than 3/16" height for removal of plug.
4. Cleanouts terminate 2" below finished floor in cast brass cleanout frame installed flush with floor.
5. Provide Josam, or equal Wade or Zurn, frame and cover in floor.
6. Cleanouts in wall terminate 2" inside finished surface in chromium-plated brass or stainless steel flush deck plate, Josam, or equal Wade or Zurn.

17. Indicates quality of cleanouts to be provided.

Figure 4.22

Three-Story Service Center PLUMBING 1-11

M. Traps.

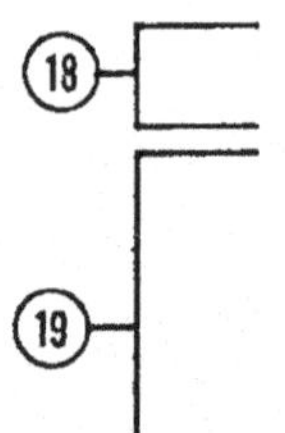

1. Provide traps, except on fixtures, cast iron and conforming to piping systems on which installed.
2. Provide 2-1/2" minimum and 4" maximum seal for traps.
3. Fixture traps, except water closets and urinals, have full size cleanout plugs below water line.
4. Traps self-cleaning type.
5. Running traps have vent hub on each side.
6. P-traps have vent hub on one side.
7. Traps below ground floor have cleanout extended to floor level.

N. Sleeves and Escutcheons.

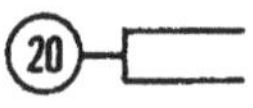

1. Provide steel or cast-iron pipe sleeves.
2. Provide one piece, chromium plated steel or brass, escutcheons.

O. Gauges.

1. Provide Ashcroft Series 10-10, or equal, crosby-Ashton Lonergron or Marsh gauges.
2. Gauges of bourdon tube type with bronze movement, white dial and black figures.
3. Provide aluminum case with close fitting friction ring, chromium plated, less back flange.
4. Scale approximately twice normal working pressure.

(21) 5. Gauges connected by means of brass pipe and fittings with brass shut-off cock.

P. Thermometers.

1. Provide Wexler Type AA5a, or equal, Moeller or Taylor thermometers.
2. Thermometer 9" scale red reading mercury type with brass, separable socket pattern, and 3/4" thread.
3. Provide 3-1/2" stem length inclined for visibility when located overhead.

Q. Joint Material.

1. Provide hot-poured joints between terra cotta and cast-iron pipe.

(22) a. Material not soften sufficiently to destroy effectiveness of joint at 160°F.
b. Material not become brittle at low temperatures.
c. Material not soluble in any wastes normally carried by drainage system.
d. Material adhere tightly to pipe with no injury to

18. Indicates all traps may not be shown on drawings, estimator must include them.
19. Indicates size and types of traps required.
20. Indicates estimator must figure sleeves and describes type required.
21. Indicates type and quality of gages and thermometers.
22. Indicates a certain joining method required.

Figure 4.23

joint.
e. No deterioration when immersed 5 days in 1% solution of hydrochloric acid or 5% solution of caustic potash.

2. Provide dry twisted jute packing or substitute approved by architect, for joints between terra cotta and cast iron.
3. Provide mortar of 1:2 Portland Cement-sand mixture for concrete sewer pipe joints.
4. Provide neoprene rubber type joints for cast-iron "No-Hub" pipe.
5. Provide pure, soft, best quality lead for cast-iron pipe joints.
6. Provide tarred or white oakum packing for cast-iron soil pipe below floor; white oakum only used above finished floor level.
7. Provide soft solder, composition 95/5 for joints on copper tubing; flux for solder joints non-corrosive type.

(23)

R. Flashings.
1. Provide sheetlead for flashing of all pipe extending through roof.
2. Cylindrical part of flashing fit snugly over pipe and extend over top and turn inside pipe.
3. 6" flange extended on sides of cylinder.
4. Roof drain provided with 36" square of 4-lb lead flashing.

(24)

2.03 MACHINERY VIBRATION ISOLATORS

A. Isolate all moving machinery from the building structure.

B. Isolating material guaranteed to effectively prevent noise or vibration being transmitted into building structure.

C. Design isolators to suit vibration frequency to be absorbed; isolator units of area and distribution to obtain proper resiliency under machinery load and impact.

D. Where equipment is bolted through isolators, isolate bolt head from equipment base with resilient washer; also use resilient bushing for bolt hole in base.

E. Cork type isolation.

1. Provide continuous cork layer under entire base or individual pads, as recommended by manufacturer.
2. Use pure natural corkboard with grains running horizontally, uniform thickness, reputable manufacturer; board made of pure cork granules, compressed and baked without foreign binder, free from defects, uniform thickness as manufactured by Armstrong Cork Co., or equal, also acceptable.

23. See Note 22.
24. Indicates flashing is part of bidding contractor's work.

Figure 4.24

Three-Story Service Center PLUMBING 1-13

F. Fabricated vibration isolator units.

1. Provide manufacturers standard catalog products with printed load ratings.
2. Each unit consist of steel top and bottom members with intermediate isolating material of cork, rubber, or spring steel as specified or approved for particular installation.
3. Provide Korfund Co. "LK" Vibro-Isolator, or equal, for steel spring type isolator.
4. Provide Vibration Eliminator Co., Korfund Co., or equal, for cork and rubber type vibration isolator.

G. Should objectionable noise or vibration in excess of design be produced and transmitted to occupied portions of the building by apparatus, piping, or other parts of mechanical installation due to improper installation, installing Contractor make necessary changes and additions as approved, without cost to Owner.

2.04 EQUIPMENT FOUNDATIONS

A. Equipment foundations inside and outside of building will be furnished by General Contractor.

(25)

1. Respective contractor shall furnish information and material required forming and imbedding, including anchor bolts, sleeves and washers to be built into foundation.
2. Respective contractor shall furnish for approval shop drawings showing the recommended foundation requirements of the equipment manufacturer.

2.05 ANCHOR BOLTS

A. Provide foundation bolts, sleeves, washers, nuts and templates to locate position of bolts for machinery on foundations.

B. Provide hook type anchor bolts of proper size and length to suit equipment.

C. Set bolts in steel pipe sleeves of approximately twice the bolt diameter and one-half imbedded length of bolts, set flush with top of rough concrete.

D. Contractor furnishing equipment is responsible for location of anchor bolts.

25. Indicates to estimator that equipment foundations are by general contractor. However, Items 1 and 2 must be furnished.

Figure 4.25

2.06 EQUIPMENT SUPPORTS AND STANDS

A. Provide supporting structures of strength to safely withstand stresses to which subjected and distribute properly load and impact over building areas.

B. Conform to applicable technical society's standards, and to codes, regulations of agencies having jurisdiction.

C. Provide structural steel or pipe frames and structural members rigidly braced, and secured with flanges bolted to the floor for floor mounted tanks set over four feet above the floor.

D. Provide suspended platform, bracket or shelf as indicated for ceiling or wall mounting.

1. Construction of structural steel members steel plates and rods as required.
2. Brace and fasten to building structure or to inserts as approved.

E. Locate supports for tanks so as to avoid undue strain on shell and interference with pipe connections to tank outlets.

1. Check support locations for clearance to pull tubes of tanks containing tubes.
2. Where saddles are indicated or specified for tank supports, use cast-iron or welded steel saddles of curvature to fit tank and not less than 150° support.

F. Respective Contractor provide required supports for equipment provided by them.

G. Submit detailed shop drawings of all supports; obtain approval before fabricating or constructing.

2.07 FLEXIBLE PIPE CONNECTORS

A. Provide flexible connections as indicated or required in water heating systems.

1. Connectors constructed of flexible, braided seamless copper tubing with brass ips male ends.
2. Connectors as manufactured by American Brass Corp., American Metal Hose Branch "Flexpipe" or Flexonics Corp.

B. Install connectors per manufacturer's written installation directions.

Figure 4.26

(26)

2.08 HANGERS AND INSERTS

A. Provide inserts constructed of cast iron or fabricated galvanized iron or steel.

1. Inserts of type to receive machine bolt head or nut after installation and permit adjustment of this bolt in one horizontal direction.
2. Inserts suspended from concrete floor, accurately locate before concrete is poured to be flush with concrete surface when forms are removed.
3. Where inserts are not provided fasten hangers by means of approved expansion bolts.
4. Wood plug not acceptable.

B. Provide hangers of malleable iron clevis or split ring type with machine thread and provisions for vertical adjustment.

(27)

1. Provide steel, copper or copper plated hangers for supporting copper pipe; where unplated steel hangers are used, copper piping cleaned and wrapped with insulating tape at point of contact of hanger.
2. Wire, band iron, chain and perforated strap iron not permitted for pipe or conduit supports.

(28)

3. Support horizontal steel piping and conduit on hangers in accordance with following schedule:

Pipe Size	Rod Diameter	Maximum Spacing
Up to 1-1/4"	3/8"	8'-0"
1-1/2" and 2"	3/8"	10'-0"
2-1/2" and 3"	1/2"	10'-0"
4" and 5"	5/8"	12'-0"
6"	3/4"	12'-0"
8"	7/8"	14'-0"

(29)

4. Support horizontal copper pipe and tubing on not more than 10' centers, using rod diameters in schedule above. above.
5. Support vertical risers with finished ring clamps approximately 8' from floor, runouts from ceiling 3' from risers and at intervals not greater than 8' apart, adjusted to maintain true and uniform grade of pipe.
6. No piping supported from other piping or conduit.
7. Piping in trenches supported on rods imbedded in concrete or having foot supports; spacing as specified for hangers.
8. Do not install hangers in cells of cellular floor intended for wiring.
9. Trapeze hanger may be used where several pipes can be installed in parallel at same level, provide piping is not heating piping exceeding 215°F.

26. Indicates types of pipe supports to be provided.
27. See Note 26.
28. Indicates spacing requirements of piping supports.
29. General pipe support instructions.

Figure 4.27

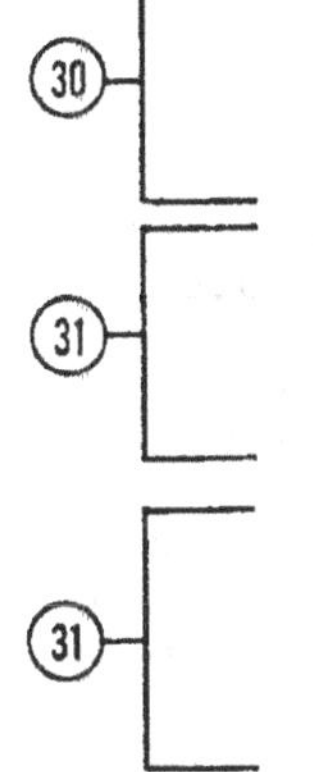

a. Hangers consist of two (2) horizontal steel channels bolted back to back with space between for hanger rods at each end, secured with washers and nuts.
b. Support from beam clamps on inserts.
c. Where provision for expansion and contraction in piping is required, use Fee & Mason Fig. 169 roller chairs, Fig. 160 pipe roller stands, or equal Grabler Mfg. Co. or Grinnell; fasten to trapeze channels.
d. Brace trapeze hangers as directed.

10. Support piping size 6" and larger by Grinnell Fig. 171 single roll type hangers with two rods and adjustable sockets, Grinnell Fig. 174 adjustable swivel pipe roll type hangers with one rod, or equal Fee & Mason Mfg. Co. or Grabler Mfg. Co., provide protection saddles.

11. All methods of supporting pipe work and equipment subject to approval of Architect.

2.09 MOTORS

A. Furnish, set and align all motors specified for driven equipment in Division 15 - Mechanical.

B. Motors conform with latest standards of IEEE and NEMA.

C. Motors designed for 60 Hz alternating current, voltage available, rated 40°C rise for continuous duty and equipped with alignment adjustment.

D. Connect motors 1/2 hp and larger for 3-phase operation unless otherwise indicated; connect smaller motors for 1-phase operation.

E. Motor load not less than 75% nor more than 100% of capacity when apparatus to which motors are connected is operating at specified capacity.

F. Nameplate voltage of motor same as voltage indicated and define upper limit; 5% less of nameplate voltage define lower limit.

2.10 MOTOR CONTROL EQUIPMENT

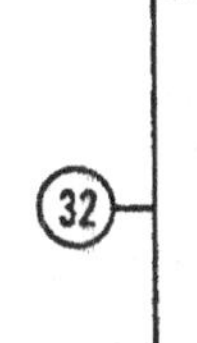

A. Motor control and starting equipment, including push-button and selector switches will be provided and connected by Electrical Contractor except as otherwise indicated or specified herein.

B. Mechanical Contractor check Division 16 - Electrical for motor control equipment requirements and meet all requirements specified therein for motor control equipment supplied by

30. Gives contractor option of using trapeze hangers for multiple lines; cost savings possible.
31. Indicates quality of hangers specified.
32. Indicates that estimator should coordinate with electrical drawings.

Figure 4.28

Mechanical Contractor.

2.11 ACCESS PANELS

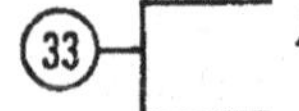

A. Provide access panels of required size for walls and ceilings where required for access to concealed valves, and controls.

1. Access panels suitable for flush mounting with smooth covers primed for painting, except baked enamel finish in tile walls, color as selected by Architect.
2. Panels constructed of 16-gauge steel with 14-gauge doors with concealed hinges and tamperproof cylinder lock of dual latch type; four (4) keys for locks furnished and delivered to Owner.
3. Access panels 12" x 12" minimum except where space requirements dictate smaller.

B. Provide 1/4" iron plate with tumbler lock, suitable lift device and flush hinge in angle iron frame for access panels in floor; panels mounted flush with floor.

C. Access panels in walls mounted with bottom 18" above floor or as directed by Architect.

D. Contractor responsible for installation of panels as work progresses, including furnishing of all anchors.

E. Access panels as manufactured by Inland Steel Products, Co., or equal.

2.12 TAGS, CHARTS AND IDENTIFICATION

A. Provide black, numbered and stamped 1-1/2" brass or aluminum tags fastened to valve by brass chain and S-hook.

B. Provide stencils on pipe at each valve in location easily read from floor; stencils indicate contents, size and flow direction in accordance with the following:

1. Domestic cold water - CW
2. Domestic hot water - HW
3. Domestic Recir. water - HWR
4. Size of letters in accordance with USA Standard A-13 or HPACCNA.

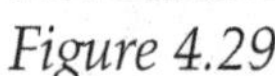

C. Provide brass or aluminum name plates, 2-1/2" x 3/4" as manufactured by Seton Name Plate Co., or equal, for all equipment.

1. Tags secured to equipment by rivets or screws.
2. Suitable adhesive may be used except in Boiler Room, subject to approval of Architect.
3. Provide duplicate name plates on motor controls.

33. Indicates access panels are provided by bidding contractor.
34. Indicates valve tag, chart, and pipe identification requirements for job.

Figure 4.29

D. Identify further materials and flow in Boiler and Equipment Rooms by W. H. Brady Co., Style 1235, or equal, direction arrows.

1. Secure arrows to piping or covering at changes in direction and at approximately 20' maximum intervals.
2. Identify mains at one readily visible central location for each service by W. H. Brady, Co., Style 2, or equal, pipe marker.
3. Color identification of arrows as follows:

34

1) CW - Green
2) HW - Blue
3) HWR - Purple

E. Provide 1/8" scale diagrams showing location, number and service or function of each tagged item.

1. Frame diagrams in approved metal frames with clear acrylic front, hinges and locks.
2. Secure to wall where directed.
3. Provide two (2) additional separate copies permanently covered and bound.

2.13 INSULATION

A. Pipe insulation

1. Provide first quality insulation and covering on all piping as specified herein.

2. Insulate domestic cold, hot and recirculating water lines with 1/2" thickness fiberglass preformed sectional insulation with vapor barrier jacket consisting of .001" aluminum foil between layers of kraft paper.
3. Preformed fiberglass insulation have thermal conductivity not to exceed .22 BTU per square in. per °F per hour at mean temperature of 75°F.
4. Flanges, bends, fittings and valve bodies on insulated piping to be insulated with hydraulic setting cement insulation to same thickness as pipe insulation.

   a. Coat cement with white vapor barrier lap cement wrapped with same jacket as piping insulation.
   b. Jacket coated with white vapor barrier lap cement.
   c. Jacket overlap adjacent jacket and be cemented thereto to maintain tight vapor barrier.

35. Indicates to estimator that all cold, hot and recirculating water piping must be insulated.

Figure 4.30

5. Leave uncovered all screwed or flanged unions, except on domestic cold water, or where exposed to freezing.

   a. Adjacent insulation beveled and covering applied in manner to provide access for wrenches without disturbing covering.
   b. Bring covering ends of insulation down over ends of insulation, terminate at least 1/2" from pipe union.

6. Insulate piping, fittings and valves in outside walls or wall chases, above ceiling or where otherwise exposed to freezing with 1-1/2" thickness insulation as specified herein.

7. Thoroughly clean all dirt, rust, dust, oil, paint or scale from all surfaces to be insulated.

8. Wrap piping in Boiler Room and where exposed with 40 lb. resin sized paper stapled in place.

   a. Paste 6 oz. canvas jacket over resin sized paper with Foster's No. 8142W adhesive, or equal.
   b. Where so finished, factory finish may be omitted and insulation secured with 16-gauge annealed wire on 12" centers or metal bands before applying paper.
   c. Cement canvas over fittings on smooth coat of cement.
   d. All exposed canvas covering finished glue sized for painting.

9. Apply all insulation in accordance with manufacturer's instructions and in workmanlike manner.

   a. Butt all sections of insulation tightly together to prevent open joints.
   b. Seal all jackets longitudinal and end overlap joints with white vapor barrier lap cement and, except where exposed, fasten with aluminum bands spaced not more than 18" apart, 3 bands per section.
   c. Seal joints between sections on vapor barrier jacket with factory supplied 4" wide sealing strips cemented as specified above.
   d. Trim ends of coverings true.
   e. Neatly finish ends with hydraulic setting cement insulation and canvas covering as specified for joints.
   f. No staples permitted in covering.
   g. Where insulation must be cut to clear joints, hangers or metal work, provide fiberglass between piping and hanger or metal work; fill opening with hydraulic setting cement to full diameter of covering.

Figure 4.31

h. All sections molded true to form and must be of proper size to fit snugly all pipes to which applied.

(36)

10. Insulate horizontal runs of storm water piping over finished ceilings with 1/2" thickness fiberglass, medium density flexible blanket insulation wrapped on piping.

11. Insulate all branch wastes from water coolers as specified for cold water piping.

2.14 PUMPS

(37)

A. Provide direct connected recirculating water pumps as indicated.

B. Provide pumps as manufactured by Bell & Gossett, Taco or Thrush.

C. Provide pumps 1 hp and larger to conform to the following:

1. Iron-body, bronze impeller, bronze-fitted with corrosion resistant steel shaft.

(38)

D. Provide pumps less than 1 hp to conform to the following:

1. All bronze construction.
2. Mount pump and motor in common housing designed for mounting in piping with flanged joints.
3. Single stage type.

E. All pumps conform to the following:

1. Pump and motor have sleeve bearings and mechanical seals.
2. Pump designed for 220°F water.
3. Finish pump, motor and base in manufacturer's standard factory finish.
4. Pump controlled as described under Automatic Temperature Control Section.

F. Obtain services of pump manufacturer for startup.

1. Obtain manufacturer's service representative to witness initial startup of pumps, check alignment, rotation and lubrication.
2. Representative check and report on pump suction and discharge pressures, current draw and rating of motor and ph of system water.
3. Submit report to Architect for approval.

36. Indicates estimator must figure on insulating horizontal storm piping offsets, and branch wastes from water coolers.
37. Indicates construction and manufacturer of recirculating pumps.
38. See Note 37.

Figure 4.32

PART 3: EXECUTION

3.01 INSTALLATION REQUIREMENTS

(39)

A. Contractors check all dimensions indicated immediately after award of contract, advise Architect promptly of discrepancies or interferences and obtain such measurements and information as may be required to satisfactorily install the work.

1. Before ordering any material or doing any work, Contractor verify all measurements and elevations at building site and be responsible for correctness of same.
2. Promptly submit any differences which may be found between field measurements and elevations and those indicated to the Architect for adjustment and approval before proceeding with work.

(40)

B. Contractors lay out their work and establish heights and grades in strict accordance with Drawings of building and finished site grades, and be responsible for accuracy of such layout.

C. Each Contractor consult with other concerned Contractors and Architect before any piping, ductwork, or conduit is installed.

1. Arrange work so that piping is kept as high as possible without interference.
2. Should Contractor fail to do so, or fail to agree, the decision of the Architect will be final.

D. Consider arrangement of piping, equipment and accessories approximate except where dimensioned.

1. Work installed generally as indicated and as directed by Architect.
2. Install piping as straight and direct as possible, parallel to or at right angles to building walls and other lines where exposed and properly spaced.
3. Install piping at uniform grade, supported on multiple hangers where practical, and adjusted to drop required.
4. Execute work in workmanlike manner so that it will present neat mechanical appearance when complete.
5. Right to make any reasonable changes in location, prior to rough-in or setting, to accommodate conditions arising during progress of work without additional cost to Owner, is reserved by Architect.

E. Align, level and adjust equipment for satisfactory operation; install so that connecting and disconnection of piping and accessories can be done readily, and so that all parts are easily accessible for inspection, operation and maintenance.

39. Indicates to estimator that close coordination with other trades necessary.
40. See Note 39.

Figure 4.33

Three-Story Service Center PLUMBING 1-22

F. Install material and equipment in accordance with manufacturer's written instructions and recommendations; submit such data to Architect prior to installation and consider this data part of these Specifications.

3.02 PIPING INSTALLATION

A. Cut piping accurately to measurements established at building and work into place without springing or forcing, properly clearing all windows, doors and supports.

1. Piping to have complete freedom of movement, except at anchor points, without causing stress or strain in pipe or any part of building during expansion and contraction.
2. Excessive cutting of building structure will not be permitted.
3. Use full-length pipe whenever possible.
4. Ream out pipe to full bore after cutting.
5. Where screwed joints are required, provide right hand, pipe standard, clean-cut, full-depth and tapering threads.
6. Make joints tight without caulking or use of lead or paint.

    a. Use of lubricant not permitted.
    b. Make up joints with "Teflon" tape.

7. Clean piping thoroughly before erection; clean after erection to remove foreign material from pipe.
8. Cap or plug open ends of piping during installation.
9. Copper piping and fittings mechanically clean, bright and fluxed.

B. Box unions, reducing bushings or caulking of joints will not be acceptable; use reducing fittings at all changes in pipe size.

C. Run piping as indicated.

1. Provide anchors and expansion bends as indicated or required.
2. Provide approved swing joints at mains and branch runouts to allow for expansion and contraction.
3. Do not project pipes beyond walls or limit lines more than necessary for installation.
4. Do not erect joints or fittings over any motor, switch-board, or other electrical equipment.
5. Adequately brace or clamp joints to prevent creeping or blowout.
6. Install piping in Boiler Room, Equipment Rooms and where exposed, close to ceiling; provide rise or drop in pipe for that purpose.
7. Run all piping, concealed, except in Boiler Room and Equipment Rooms.

Figure 4.34

8. Run mains carefully to insure unrestricted circulation and elimination of air pockets.

   a. Grade mains not less than 1" in 40'.
   b. Pitch pipe in direction indicated or required.

9. Architect reserves right to direct changes in run and details of piping as required by conditions encountered on site.

D. Use flange fittings in assembly of piping at equipment and in mains 3" and larger; provide union connections for pipe smaller than 3" and all connections to equipment.

E. Properly support pipe as specified herein; support piping underground by concrete or brick piers to prevent undue strain at joints, where disturbed earth is encountered.

(41) F. Provide sleeve on each pipe passing through walls, floors, partitions or ceilings.

1. Cut sleeve flush with surface, except as otherwise specified or required.
2. Provide sleeve one pipe size larger than pipe encased, except insulated pipes have sleeves of size to encase insulation.
3. Set sleeves in concrete or masonry during construction where possible.
4. Extend floor sleeves 1" above floor in toilets and other areas where water might be present.
5. Caulk sleeves embedded in concrete with graphite packing and an approved plastic and waterproofing compound.
6. Place escutcheon plate around pipe at exposed ends of sleeves in finished areas; secure with set screws.

G. After completion of installation, check piping for circulation and any excessive noise.

1. Promptly correct any defects.
2. Remove all concealed piping requiring repairs and have repairs to construction and finish made as required at respective Contractor's expense.

3.03 PROGRESS OF WORK

A. Mechanical and Electrical Contractors order progress of their work so as to conform to work of other trades.

B. Perform work underground as quickly as possible.

C. Complete all work within time specified for General Contract.

41. Indicates to estimator where sleeves are required.

Figure 4.35

Three-Story Service Center PLUMBING 1-24

3.04 CUTTING AND PATCHING

A. Install hangers, supports and pipe sleeves in floors, walls, partitions, ceilings, and roof slabs as construction progresses.

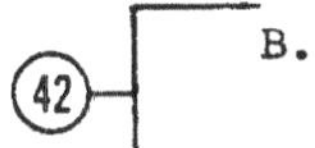

B. All cutting of concrete, brick or other material for passage of piping through floors, walls, partitions and ceilings will be done by General Contractor.

1. Respective Contractor deliver information to General Contractor for above purpose.
2. If respective Contractor delivers required information to General Contractor too late, or furnished information that is incorrect, or fails to order his work with General Construction, respective Contractor employ General Contractor to do cutting and patching required to install his work.

C. General Contractor will close openings around piping, ductwork and conduit with material equivalent to that removed.

D. Leave exposed surfaces in suitable condition for refinishing without further work.

E. Do not alter or cut any structural member without special permission of Architect.

3.05 CHASES, HOLES AND RECESSES

A. Mechanical and Electrical Contractors furnish to General Contractor, in advance of construction, exact details for provisions of chases, holes and recesses required for installation of all material and equipment furnished or installed by them.

B. Openings, recesses and chases required for such material and equipment will be provided by General Contractor.

3.06 DAMAGE TO OTHER WORK

A. Mechanical and Electrical Contractors responsible for damage to other work caused by their work or through neglect of their workmen.

B. Contractor who installed work do all patching and repairing, as directed by Architect; responsible Contractor pay cost of same.

42. Indicates to estimator that cutting is part of general contractor's work.

Figure 4.36

Three-Story Service Center PLUMBING 1-25

3.07 EXISTING SERVICES

A. Protect, brace and support existing active sewer, water, gas electric and other services where required for proper execution of work.

1. If existing active services are encountered that require relocation, make request in writing for determination; do not proceed with work until written directions are received.
2. Do not prevent or disturb operation of active services that are to remain.

B. Remove, cap or plug inactive services.

1. Notify utility companies or municipal agencies having jurisdiction.
2. Protect or remove the services as directed.

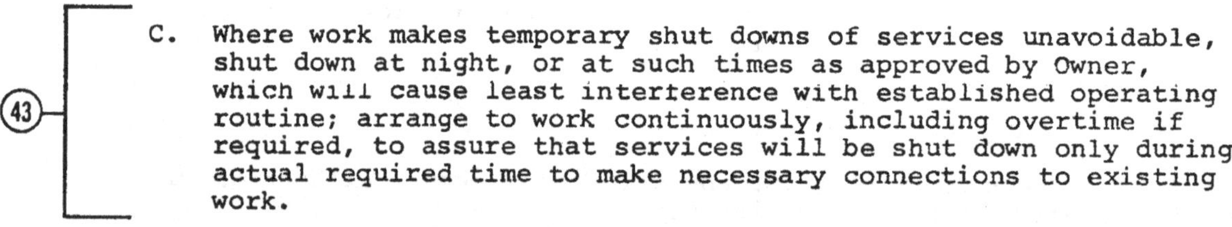

C. Where work makes temporary shut downs of services unavoidable, shut down at night, or at such times as approved by Owner, which will cause least interference with established operating routine; arrange to work continuously, including overtime if required, to assure that services will be shut down only during actual required time to make necessary connections to existing work.

3.08 EXCAVATION AND BACKFILL

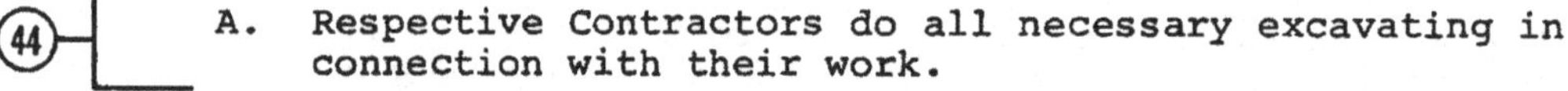

A. Respective Contractors do all necessary excavating in connection with their work.

1. Excavate bottom of trenches to exact depth and uniform grade in direction of flow so that pipes will be supported on solid bed of undisturbed earth, hubs, up grades, earth undercut at hubs so that each piece of pipe is supported throughout its entire length.
2. Refill excavation below required grade of piping with sand and firmly compact.
3. Excavate rock where encountered to grade of 4" below lowermost part of pipe grades specified.
4. Excavate principally with open-trench method.

   a. Excavate 12" wider than largest diameter of pipe, or tank, and to depths specified or required.
   b. Deposit materials excavated on sides of trenches and beyond reach of slides.
   c. Do not pile excavated material where it will interfere with traffic.
   d. Deposit all earth and other materials taken from trenches and not required for backfilling where directed.

43. Indicates to estimator that overtime costs may have to be included in his estimate.

44. Indicates to estimator that excavation and backfill is part of his contract.

Figure 4.37

e. Do not leave material where it will interfere with Owner's or other Contractor's operations.

(45) 5. Contractor provide shoring, bracing or sheet piling necessary to maintain banks of excavations.

a. Take out same as work progresses and filling in is accomplished, unless otherwise directed by Architect.
b. Arrangement of shoring must be such as to prevent any movement of trench banks and consequent strains on piping.
c. Provide shoring to prevent damage to work installed by other trades.

B. Exercise care to protect roots of trees to remain.

1. Do not cut any root greater than 1" in diameter.
2. Within branch spread of trees perform all trenching by hand.
3. Open trench only when utility can be installed immediately.
4. Prune injured roots cleanly.
5. Back fill as soon as possible after inspection and approval.
6. Perform all work under direction of Architect.

(46) C. Contractors do all pumping required to keep their excavation free of water.

D. Suitably protect trenches and openings for underground work by signs, barricades or enclosures, and flashing.

1. Display lights at night until completion of work.
2. Contractor responsible during construction for all damage to underground piping or structures.

E. Install underground tanks and piping below frost line, but not less than 3' below finished grade to top of tank or piping, except as indicated.

F. After work in trenches has been completed, fill with good fine earth, free from cinders, stones or brickbats.

1. Apply fill in 6" layers and carefully tamp compactly in place before next layer.
2. Fill containing stone not acceptable.
3. Backfill with concrete, as specified herein, trenches which pass under or within 18" of column footings or wall foundations.

a. Concrete backfill under footings or foundations run full width of such structure and at least 6" below bottom of piping.
b. Before backfilling pipes spirally wrap with two layers

45. If shoring and pumping are required estimator must include these items in his estimate.
46. See Note 45.

Figure 4.38

of 1" hair felt wound in opposite directions or sleeved to prevent direct bearing on pipe.

(47) G. General Contractor will restore streets, pavements, and other finished surfaces damaged by work to their original condition after suitable backfill.

(48) H. General Contractor will remove rock as directed by Architect where it is encountered.

3.09 OWNER'S EQUIPMENT

(49) A. Provide all required hot, cold and sanitary piping rough-in to equipment indicated such as laboratory equipment, dental equipment, etc.

B. Equipment will be furnished and placed under another part of this Contract; make all final connections to equipment as part of this Contract.

C. Detail drawings will be furnished to Mechanical Contractor showing rough-in dimensions for waste and water connections.

1. This contractor responsible for correct installation of these services, in accordance with Drawings furnished and his field check.
2. Check information with equipment manufacturer where necessary.
3. Should installation of rough-in be incorrect, due to failure to check in above manner or incorrect installation, make necessary changes in rough-in without cost to Owner.

3.10 GAUGE & THERMOMETER INSTALLATION

A. Provide in readily visible location gauges and thermometers where indicated and as follows:

1. Circulating water pumps suction and discharge, one (1) gauge each, 0-70 ft.
2. Chilled water supply and return, one (1) thermometer each, 30-120°F.
3. Condenser water supply and return, one (1) thermometer each, 30-120°F.

B. Provide gauges and thermometers of types specified herein.

47. Indicates that restoration of site is by general contractor.
48. Indicates rock excavation is by general contractor.
49. Indicates that estimator will most likely encounter equipment furnished by other, but must be installed by bidding contractor. Necessary materials and labor must be included in estimate.

Figure 4.39

Three-Story Service Center Index

DIVISION 15 - PLUMBING

SECTION 2

WATER SUPPLY SYSTEM

Figure 4.40

Three-Story Service Center PLUMBING 2-1

DIVISION 15 - MECHANICAL

SECTION 2

WATER SUPPLY SYSTEM

PART 1: GENERAL

(50)

1.01 SCOPE

A. Work under this Section includes following:

1. Water service and metering facilities.
2. Cold water distribution piping installed and connected from water service to plumbing fixtures, wall hydrants, hot water heater, heating and cooling systems.
3. Hot water distribution piping installed and connected from hot water heater to plumbing fixtures.
4. Hot water recirculation piping installed and connected from ends of branch piping to water heater.
5. Plumbing fixtures furnished, installed and connected to water supplies and sanitary drainage system.
6. Hot water heater.

(51)

B. It is not intended that Specifications or accompanying Drawings show every detail.

C. Contractor furnish all labor and install all material required for complete system functioning as described, whether or not specifically called for or indicated.

(52)

1.02 SHOP DRAWINGS

A. Furnish Shop Drawings as specified in Division 15 - Section 1.

B. Furnish Shop Drawings for following:

1. Plumbing fixtures, trim and roughing-in.
2. Wall hydrants.
3. Water heater.
4. Circulating pumps.
5. Insulation.

(53)

1.03 WATER SERVICE

A. Provide underground ductile iron pressure pipe water main of size indicated from main at curb to building.

50. Indicates scope of work for water supply system.
51. See Note 2.
52. Indicates which items require shop drawings.
53. Indicates type of piping specified for water main.

Figure 4.41

(54)

B. Provide all necessary accessories required by Water Company for their meter installation inside building.

1. Install gate valve at building entrance, valved bypass and spool piece for meter, as required by Water Company.
2. Arrange and pay for service, including service connection costs, meter and installation as part of Contract.

C. Provide tee in main as indicated for future connection of existing building.

PART 2: MATERIALS

(55)

2.01 WALL SUPPLY FITTINGS

A. Wall hydrants

1. Provide Josam Series 71000 , non-freeze type, wall hydrants for outside use.
2. Wall hydrant provided with 3/4", 45° angle nozzle, polished brass face, cast brass body, brass wall sleeve and pipe and loose key handle.

B. Inside hose bibbs

1. Provide Standard R7233 handle operated hose bibbs 3/4", all bronze, angle type.
2. Mount 18" above floor.

C. Approved manufacturers of wall supply fittings are American-Standard, Josam, Wade or Zurn.

(56)

2.02 WATER HEATER

A. Provide Jackson Model #GRE-120-T-H commercial, electric water heater.

1. Heater tank have minimum storage capacity of 120 gallons.
2. Heater designed for 150 psi wwp USA Standard.
3. Heater designed for 480-volt, 3-phase operation with 3 elements producing 18KW.
4. Provide glass- or stone-lined surfaces where exposed to water.
5. Provide all required fittings constructed of brass.
6. Casing constructed of "Bonderized" steel finished with baked-on enamel.
7. Provide fiberglass blanket type insulation between tank and casing.

54. Indicates water meter is part of bidding contractors work.
55. Indicates types and quality of wall hydrants and hose bibbs.
56. Indicates capacity, type and quality of hot water heater.

Figure 4.42

8. Heater have minimum recovery rate of 73.8 gph at 100° rise.
9. Water heater provided with the following:

   a. Adjustable thermostat with bulb immersed directly into water.
   b. High limit temperature control.
   (57) c. ASME spring loaded temperature-pressure relief valve, reset type set at 100 psi and 200°F.
   d. 3/4" drain valve.
   e. Magnesium anode rod.

10. Unit guaranteed for ten (10) years.
11. Entire unit approved by UL.

B. Approved manufacturers of water heater are Jackson, Ruud, A. O. Smith or Westinghouse.

2.03 PLUMBING FIXTURES

A. Provide fixtures of types hereinafter specified and of quantities indicated, making all required supply, waste, soil and vent connections with all fittings, supports, fastening devices, cocks, valves and traps leaving all in complete working order.

1. Provide carrier and fixture supports as required; install in walls as work progresses.
2. Contractor responsible for stability of all supports.
3. Wall-hung fixtures generally supported on fixture carriers.
(58) 4. Carriers as manufactured by Josam, Wade, or Zurn.

B. Trap all fixtures close to fixture so that discharge passes through not more than one trap before reaching house drain.

C. Provide fixtures clearly marked with manufacturer's name, trade-mark and quality or class of fixture; do not remove labels until work has been accepted.

D. Construction of fixtures conform to following except as specified otherwise:

1. White, twice-fired, vitreous china, non-absorbent, close grained, thoroughly vitrified, free from pores,unmarked, true and level.
2. Free from chips or flaws.
3. Fixture warranted not to craze, color or scale.
4. Exposed metal work chromium-plated, guaranteed against defects for one (1) year.
5. Strainers and screens of monel metal.
6. Fittings of best quality known to trade as No. 1 line.

57. See Note 56.
58. Indicates requirements and construction for plumbing fixtures.

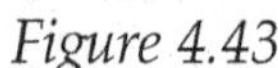
Figure 4.43

Three-Story Service Center   PLUMBING   2-4

7. Iron for enameled ware of best quality cast iron and of proper thickness to produce high grade quality.

E. Acceptable manufacturers for fixtures and trim are American Standard, Speakman, Crane Co., Eljer or Kohler.

(59)

1. American Standard and Speakman numbers are used herein to describe quality, type and finish desired.
2. Provide fixtures of one manufacturer for purposes of standardization.

F. Lavatories

1. Fixture - "Lucerne" No. 0350.132 for concealed arms, 20" x 18", 8" centers, front overflow, splash back; No. 0350.025 wall hanger with wall hanger where carrier installation is not possible.
2. Supply - Speakman S-4331 with "Autoflo" control, pop-up waste with 1-1/4" tailpiece and vandal-proof aerator; 2302.81 supplies 3/8" ips with key operated angle stops to wall, reducing couplings and escutcheons.
3. Trap - No. 4429.015 cast brass "P" trap with 1-1/4" inlet, 1-1/2" outlet and 4446.019 6" nipple.

(60)

G. Urinal

1. Fixture - "Washbrook" No. 6500.011 wall-hung, washout type with integral extend shields, flush spreader, trap with cleanout and top spud inlet.
2. Flush valve - Speakman No. K9082 BSP "Si-Flo" flush valve with key-operated angle stop and vacuum breaker, 1" x 4" nipple, escutcheon, chromium-plated nuts, bolts and washer.

H. Water Closet (Wall-Hung)

1. Fixture - "Afwall" No. 2477.016, wall-hung, syphon jet, elongated bowl, 1-1/2" top spud.
2. Flush valve - Speakman No. K-9000 BSP flush valve with key-operated angle stop, vacuum breaker, 1" x 4" nipple, escutcheon, chromium-plated nuts, bolts and washers.
3. Seat - Church American-Standard No. 5320.114 "Moltex" heavy duty solid plastic open front seat with stainless steel check hinge.

I. Water Closet (Floor-Mounted)

1. Fixture - "New Madera" No. 2222.016, floor-mounted, siphon jet, elongated bowl, 1-1/2" top spud, cast-iron floor flange, china caps and bolts.

59. Indicates quality of fixtures specified.
60. Indicates types of fixtures along with manufacturer's model numbers for both fixtures and trim.

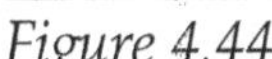

Figure 4.44

2. Flush valve - Speakman No. K-9000 BSP "Si-Flo" flush valve with key-operated angle stop and vacuum breaker.
3. Seat - Church American-Standard No. 5320.114 as specified for wall-hung water closet.

J. Service Sink

1. Fixture - "Lakewell" No. 7692.023, 22" x 18", wall hanger, plain black, enameled cast iron with 8379,018 rim guard.
2. Supply - Wall-mounted S-7115-ISVB double faucet with top brace, spout with bucket hook and hose end, integral stops, 5' of rubber hose.
3. Trap - Trap standard No. 7798.176 with cleanout.

K. Cabinet Sink

61
1. Fixture - Elkay No. LR-1918, single compartment, 18-gauge stainless steel, 19" x 18" self-rimmed with ledge.
2. Supply - Examination Rooms and Laboratories: Speakman S-7001, 3/8" gooseneck discharge nozzle, wrist action handles, 8" centers, cast drain plug, set screw plate and 1-1/4" tailpiece; other areas: Speakman S-4961 single control, brass swing nozzle, non-splash aerator indexed forged brass lever handle, 8" centers and Elkay LK99 drain and 1-1/4" tailpiece.
3. Trap - No. 4429.015 "P" trap with 1-1/4" inlet, 1-1/2" outlet, 4446.019 6" nipple and escutcheon.

L. Water Cooler

1. Fixture - Halsey-Taylor No. SW-8-A, or equal, surface wall-mounted, capacity of 8.0 gph at 90°F room temperature, 80°F inlet water and 50°F drinking water with 60% waste through precooler; include 1/5, 115-volt motor, hermetically sealed compressor, Freon-12 refrigerant, 10°F water temperature control adjustment, vinyl covering of color selected by Architect, steel housing, and electric cord with ground.
2. Mounting height as directed by Architect.
3. Unit include five (5) year warranty.

PART 3: EXECUTION

3.01 COLD WATER DISTRIBUTION

A. Install and connect cold water distribution from meter to water heater, plumbing fixtures, wall hydrants, hose bibbs, heating and cooling equipment, and as indicated.

61. See Note 60.

Figure 4.45

Three-Story Service Center PLUMBING 2-6

(62) B. Install valved branches to air conditioning equipment, hose bibbs, wall hydrants, fixtures and as indicated.

(63) C. All cold water lines installed with copper piping and fittings as specified herein, having nominal size indicated.

3.02 HOT WATER DISTRIBUTION

A. Hot water distribution piping installed and connected from water heater to plumbing fixtures and as indicated.

B. Install and connect hot water recirculation piping through circulation pump from water heater to valved connections at ends of distribution piping and as indicated.

(64) C. All hot water distribution lines installed with copper piping and fittings as specified herein, having nominal size indicated.

3.03 BRANCH SUPPLIES

A. Take branches and runouts to wall at all fixtures requiring water.

B. Provide branches from wall of size specified for fixture supplies.

(65) C. In no case provide branch runouts to fixtures less than following sizes:

Fixture	C.W.	H.W.
Water Closet	1"	-
Urinals	1"	-
Lavatories	1/2"	1/2"
Sinks	1/2"	1/2"
Water Coolers	1/2"	-

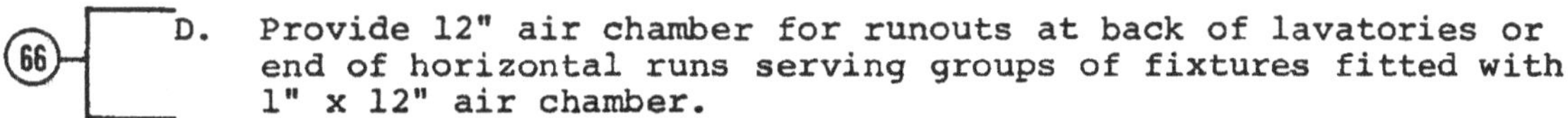

(66) D. Provide 12" air chamber for runouts at back of lavatories or end of horizontal runs serving groups of fixtures fitted with 1" x 12" air chamber.

(67) E. Place gate valves in all branches leading from mains to groups of fixtures and as indicated.

62. Indicates mentioned items require shut-off valves. These may not show up on drawings; however, estimator must include them.
63. Indicates pipe material to be used for cold water lines.
64. Indicates pipe material to be used for hot water lines.
65. Indicates minimum hot and cold water fixture connections.
66. Indicates air chambers to be used in lieu of mechanical shock absorbers; these are considerably less expensive.
67. Indicates all water branches are to be controlled with gate valves; these may not be shown on drawings; estimator must include them in estimate.

Figure 4.46

Three-Story Service Center Index

DIVISION 15 - PLUMBING

SECTION 3

SANITARY AND STORM WATER DRAINAGE SYSTEM

Figure 4.47

Three-Story Service Center PLUMBING 3-1

DIVISION 15 - PLUMBING

SECTION 3

SANITARY AND STORM WATER DRAINAGE SYSTEM

PART 1: GENERAL

(68)

1.01 SCOPE

A. Work under this Section includes the following:

1. Sanitary drainage and vent piping installed and connected from plumbing fixtures, floor drains and Owner's equipment.
2. Construction of sanitary sewer and connection to city sewer.
3. Storm water drainage piping installed and connected from roof drains to storm water sewer.
4. Construction of storm water sewer including yard basins, lampholes and connection to existing sewer.

(69)

B. It is not intended that Specifications or accompanying Drawings show every detail.

C. Contractor furnish all labor and install all material required for complete system functioning as described, whether or not specifically called for or indicated.

(70)

1.02 SHOP DRAWINGS

A. Furnish Shop Drawings as specified in Division 15 - Section 1.

B. Furnish Shop Drawings for following:

1. Cleanouts
2. Lampholes
3. Drains
4. Castings

1.03 CONNECTION OF SEWERS

A. Contractor arrange for connection of sanitary sewer with City agency.

(71)

B. Contractor pay all required fees and connection charges as part of his work.

68. Indicates scope of work for sanitary and storm system.
69. See Note 2.
70. Indicates which items require shop drawings.
71. See Note 5.

Figure 4.48

Three-Story Service Center PLUMBING 3-2

PART 2: MATERIALS

2.01 DRAINS

A. Provide drains connecting to the sanitary and storm water piping as indicated or specified herein.

B. Provide drains as manufactured by Josam, Wade or Zurn; Josam numbers are used herein to describe quality and type of drain required.

C. Floor Drains

1. Provide Series 32/00 cast-iron body floor drain for Boiler and Equipment Rooms; drain complete with loose set cast-iron anti-tilting grate with double drainage pattern and arranged for bottom outlet.
2. Provide Series 1465-U cast brass automatic trap seal valve with vacuum breaker for all floor drains.
3. Install trap seal valves with access panel and connected from nearest lavatory or sink to floor drains.
4. Provide all floor drains of inside caulk type.

D. Roof Drains

(72) 1. Provide Series 21540 cast-iron body roof drain with integral expansion joint with brass sleeve, removable low dome and sediment cup, non-puncturing flash clamp with integral gravel stop and arranged for bottom outlet.
2. Provide all roof drains with threaded bottom outlet.

2.02 LAMPHOLE COVER

A. Provide lamphole cover as manufactured by Josam, Wade or Zurn.

B. Provide Josam Series cast-iron lacquered lamphole cover complete with the following:

1. Serrated cut-off ferrule, rough brass raised head and screwed plug.
2. Rough cast brass round adjustable head have flanged rim and scoriated type tractor cover with vandal-proof screws.

2.03 SEWER MATERIALS

(73) A. Provide brick for sewer structures conforming to applicable requirements of ASTM STandard C62, Grade SW.

1. Clean, used paving or building brick which conforms to above, or radial brick or concrete meeting requirements given herein, may be used for sewer structures.
2. Precast storm sewer structures meeting standards and quality described herein will be acceptable.

72. Indicates type and quality of lampholes, floor and roof drains.
73. Indicates materials manholes and catch basins are to be constructed of.

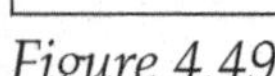

Figure 4.49

B. Provide concrete for sewer structures of 1:2:4 mix.

(74) C. Provide mortar for masonry in sewer structures of 1:3 cement-sand mix, provided that hydrated lime may be substituted for, not to exceed 10%, by weight, of cement.

D. Provide sewer structure castings as indicated.

1. Castings for sewer structures in roadways comply with State Highway Department Standards.
2. Provide tough, even grained, soft gray iron sewer structures for other areas.

(75) a. Structures free from burnt-on sand and other injurious defects.
b. Thoroughly clean and subject to hammer tests for soundness all castings before leaving foundry.
c. Castings given two (2) coats of coal-tar pitch varnish.
d. Provide locking devices for tops and gratings weighting less than 100 lbs.
e. Provide same design for like structures.

PART 3: EXECUTION

3.01 BUILDING DRAINAGE AND VENT LINES

A. Provide soil, waste, vent and drain piping and fittings of types indicated or listed below.

1. Cast iron pipe and fittings below ground.

a. Pipe and fittings below first floor, except as indicated.
(76) b. Pipe 2" and larger within the building and to 10' outside the building.

2. Hubless cast iron pipe and fittings.

a. All soil waste and vent piping 1-1/2" and larger above finished floor.
b. All vent piping 2-1/2" and smaller above finished first floor.

B. Building drains receive discharge of all stack, branch mains and runouts.

1. Arrange all horizontal mains, branches and runouts to present neat symmetrical appearance.
2. Provide support under vertical stacks.

C. Install mains at elevations and grades as indicated or directed.

74. See Note 73.
75. Indicates types of castings (manhole frames and covers) required.
76. Indicates type of piping required for sanitary and storm piping below grade in building, and for sanitary above grade in building.

Figure 4.50

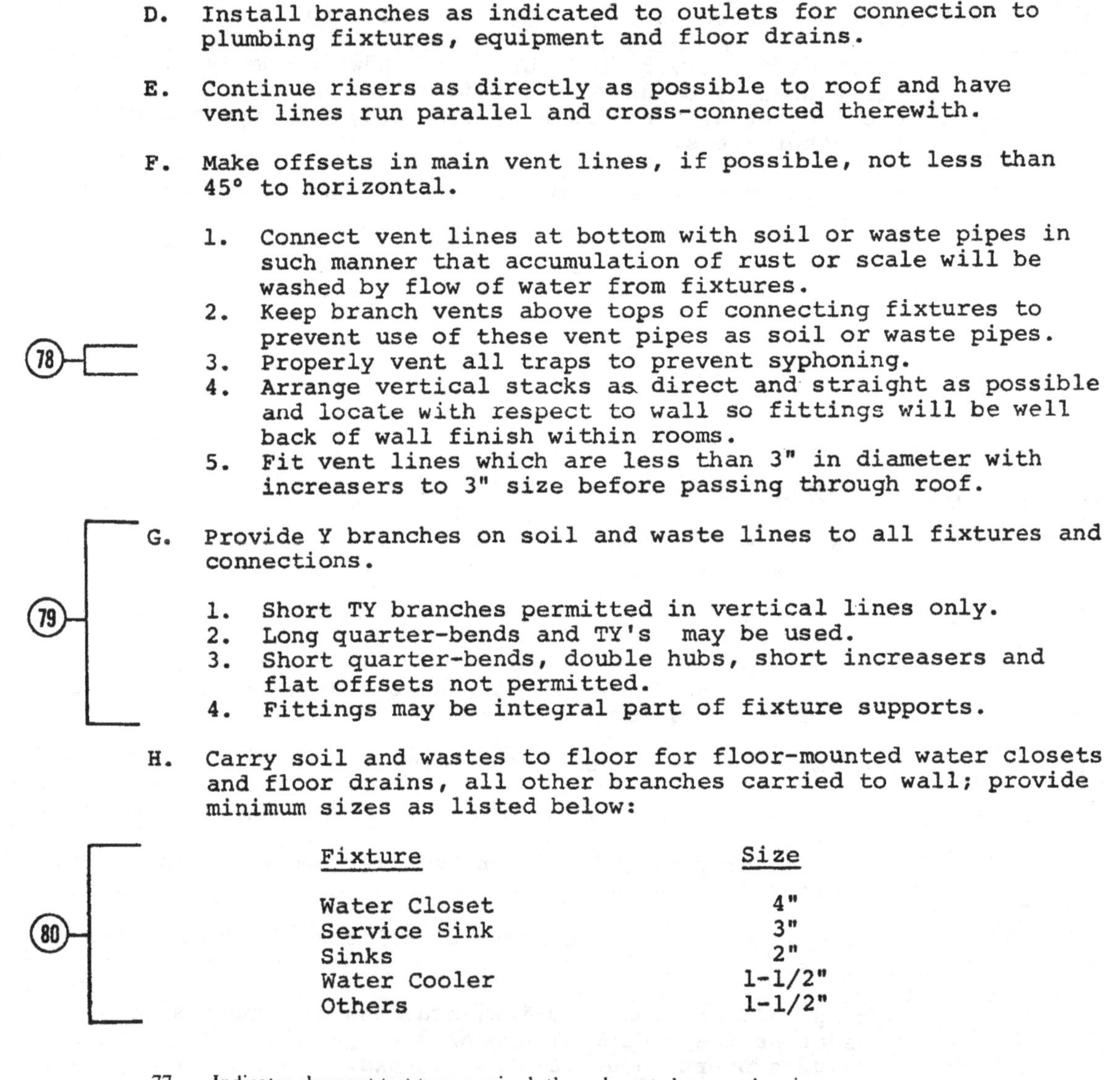

Three-Story Service Center PLUMBING 3-4

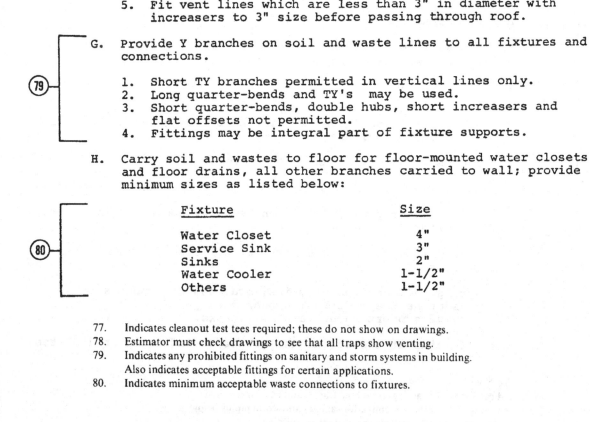

1. Minimum grade of horizontal lines in building 1/4" per ft.
2. Make changes in direction and connection of drain, soil and waste lines with proper fittings.
3. Provide offsets and cleanouts as indicated or required for testing or cleaning. (77)

D. Install branches as indicated to outlets for connection to plumbing fixtures, equipment and floor drains.

E. Continue risers as directly as possible to roof and have vent lines run parallel and cross-connected therewith.

F. Make offsets in main vent lines, if possible, not less than 45° to horizontal.

1. Connect vent lines at bottom with soil or waste pipes in such manner that accumulation of rust or scale will be washed by flow of water from fixtures.
2. Keep branch vents above tops of connecting fixtures to prevent use of these vent pipes as soil or waste pipes.
3. Properly vent all traps to prevent syphoning. (78)
4. Arrange vertical stacks as direct and straight as possible and locate with respect to wall so fittings will be well back of wall finish within rooms.
5. Fit vent lines which are less than 3" in diameter with increasers to 3" size before passing through roof.

(79)

G. Provide Y branches on soil and waste lines to all fixtures and connections.

1. Short TY branches permitted in vertical lines only.
2. Long quarter-bends and TY's may be used.
3. Short quarter-bends, double hubs, short increasers and flat offsets not permitted.
4. Fittings may be integral part of fixture supports.

H. Carry soil and wastes to floor for floor-mounted water closets and floor drains, all other branches carried to wall; provide minimum sizes as listed below:

(80)

Fixture	Size
Water Closet	4"
Service Sink	3"
Sinks	2"
Water Cooler	1-1/2"
Others	1-1/2"

77. Indicates cleanout test tees required; these do not show on drawings.
78. Estimator must check drawings to see that all traps show venting.
79. Indicates any prohibited fittings on sanitary and storm systems in building. Also indicates acceptable fittings for certain applications.
80. Indicates minimum acceptable waste connections to fixtures.

Figure 4.51

(81) I. Install test cleanouts at base of all soil and waste lines.

J. Connect runouts to sanitary sewer 10' outside building.

1. Runouts extended with cast-iron pipe.
2. Make connection by inserting cast-iron soil pipe inside hub of terra cotta pipe not less than 6" or as far as possible in bend to form suitable bearing.

3.02 STORM WATER DRAINAGE

(82) A. Install interior rainwater conductors with hubless cast iron pipe and fitting.

1. Install storm drainage piping below finished first floor with hubless cast iron pipe and fittings.
2. Piping installed generally as specified for sanitary building drainage piping.

(83) B. Install test tee and cleanout at base of each riser.

(84) C. Install roof drains on rainwater conductor, fit with lead flashing and make tight with roofing.

D. Extend runouts with cast-iron pipe to approximately 10' outside of building and connect to storm sewer as specified for sanitary runouts.

3.03 SEWER CONSTRUCTION

A. Construct storm sewer to receive runouts from rainwater conductors, roof drains and catch basin and connect existing sewer as indicated.

B. Construct sanitary sewer to receiving building runouts from soil and waste piping and connect to City of Albany sanitary sewer.

C. Provide sanitary and storm sewer piping of types indicated or listed below.

(85)
1. Sewers more than 10' outside of building of terra cotta pipe for sizes 12" and smaller.
2. Provide concrete pipe for sizes 15" and larger.
3. Provide cast-iron pipe for sizes 12" and smaller under roadways.

(86) D. Construct catch basins of materials as specified herein.

1. Set basins in cement and locate at proper elevation to drain to finish grade.
2. Extend one length of cast-iron pipe from each catch to provide runout.

81. See Note 77.
82. Indicates piping material required for storm system (above grade) in building.
83. See Note 81.

Figure 4.52

3. Fit pipe with cast-iron T set bullheaded in vertical position within basin as indicated.

E. Install sanitary and storm sewer piping with minimum of 36" cover, except as indicated.

1. Commencing at lowest point in system lay sewer piping with bell upgrade.
2. Test pipe for soundness and clean interior and joint surfaces before lowering the pipe into trench.
3. Carefully check invert elevations of sewers to which connections are made.
4. Lay pipe in straight lines and on uniform grades between points where changes in alignment or grade are indicated.
5. Check line and invert grade of each pipe from top line carried on batter boards, not over 25' apart.
6. Fit pipes to form smooth uniform invert.
7. Keep stopper in pipe mouth when pipe laying is not in progress.

F. In jointing sewer pipe, comply fully with directions of manufacturer or pre-formed joint pipe.

1. Two lengths of pipe may be jointed vertically on bank provided they are lowered and laid without injury to joint.
2. In making cement joints in tongue and groove pipe, wet joint surfaces thoroughly.

   a. Spread mortar in lower half of groove end, over upper half of tongue.
   b. After pipe has been shoved home, fill remainder of joint space, inside and outside, with mortar and finish outside with bead or ring around pipe.

3. As soon as joint has set, start backfill operation as specified herein.

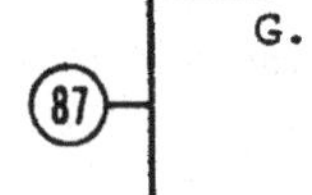

G. Where indicated, provide end runs of branch sewers with cleanout having cast-iron pipe set vertically and equipped with cast-iron cleanout plug, lock type, as specified herein, set flush with finished grade.

84. Indicates type of flashing required for roof drains.
85. Indicates type of piping required for site drainage.
86. Indicates construction method for catch basins.
87. Indicates where cleanouts are required on site drainage system.

Figure 4.53

Three-Story Service Center — Index

DIVISION 15 - PLUMBING

SECTION 4

FIRE PROTECTION SYSTEM

Figure 4.54

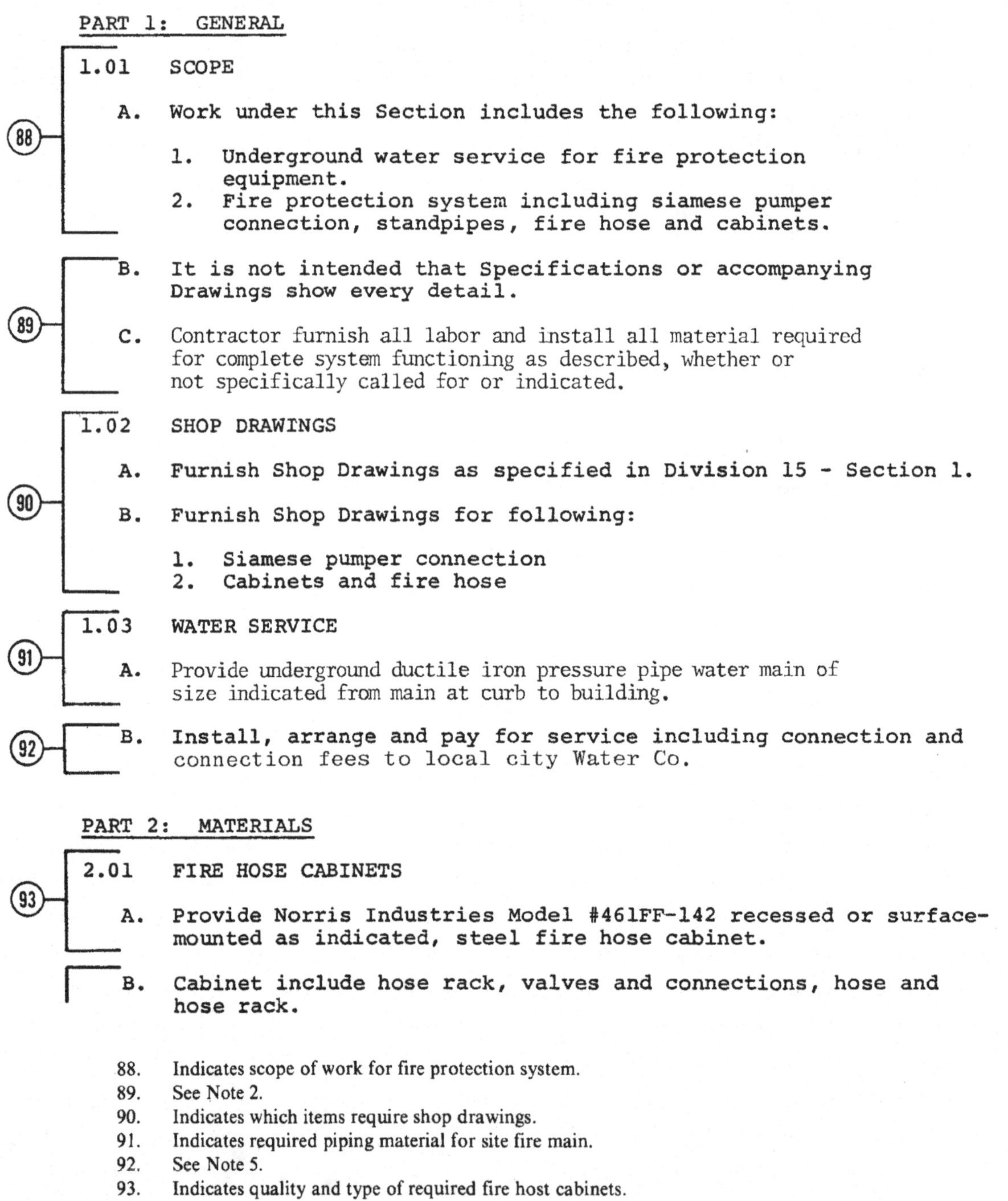

Three-Story Service Center PLUMBING 4-1

DIVISION 15 - PLUMBING

SECTION 4

FIRE PROTECTION SYSTEM

PART 1: GENERAL

1.01 SCOPE

(88) A. Work under this Section includes the following:

1. Underground water service for fire protection equipment.
2. Fire protection system including siamese pumper connection, standpipes, fire hose and cabinets.

(89) B. It is not intended that Specifications or accompanying Drawings show every detail.

C. Contractor furnish all labor and install all material required for complete system functioning as described, whether or not specifically called for or indicated.

(90) 1.02 SHOP DRAWINGS

A. Furnish Shop Drawings as specified in Division 15 - Section 1.

B. Furnish Shop Drawings for following:

1. Siamese pumper connection
2. Cabinets and fire hose

(91) 1.03 WATER SERVICE

A. Provide underground ductile iron pressure pipe water main of size indicated from main at curb to building.

(92) B. Install, arrange and pay for service including connection and connection fees to local city Water Co.

PART 2: MATERIALS

(93) 2.01 FIRE HOSE CABINETS

A. Provide Norris Industries Model #461FF-142 recessed or surface-mounted as indicated, steel fire hose cabinet.

B. Cabinet include hose rack, valves and connections, hose and hose rack.

88. Indicates scope of work for fire protection system.
89. See Note 2.
90. Indicates which items require shop drawings.
91. Indicates required piping material for site fire main.
92. See Note 5.
93. Indicates quality and type of required fire host cabinets.

Figure 4.55

1. Provide two (2) valved hose connections in cabinet.

   a. Provide 2-1/2" bottom connection and 1-1/2" top connection.
   b. Fire hose connection conform to local fire company.

2. Provide one piece, swing type hose rack with steel pins and 100' of 1-1/2" single jacket, polyester rubber lined hose.

94

   a. Provide hose with adjustable hose rack and fog nozzle, all FM approved.

C. Construct cabinet of not less than 18-gauge steel.

1. Cabinet have continuous hinged steel door with glass front and chrome lever.
2. Finish cabinet inside with white enamel; provide rust resistant primer on door and trim.

D. Provide fire hose cabinets as manufactured by Norris Industries, Seco Manufacturing Inc., Elkhart Brass Co., or Fyr-Fyter Co.

2.02 SIAMESE PUMPER CONNECTION

A. Provide Seco Manufacturing Inc. No. 251, wall type Siamese pumper connection.

1. Unit constructed of highly polished brass.
2. Siamese pumper connection equipped with double clapper valves.
3. Provide "Standpipe" engraved on escutcheon plate.
4. Unit complete with brass plugs and chains of polished brass.

95

B. Provide 2-1/2" fire department inlets with threads to match local fire department.

C. Provide outlet size as indicated.

D. Provide Siamese pumper connection as manufactured by Seco Manufacturing, Inc., Elkhart Brass Co., or Fyr-Fyter Co.

PART 3: EXECUTION

3.01 STANDPIPE SYSTEM

A. Provide standpipe system as indicated connected from service entrance to fire hose cabinets.

94. See Note 93.
95. Indicates quality and type of required Siamese connection.

Figure 4.56

Three-Story Service Center PLUMBING 4- 3

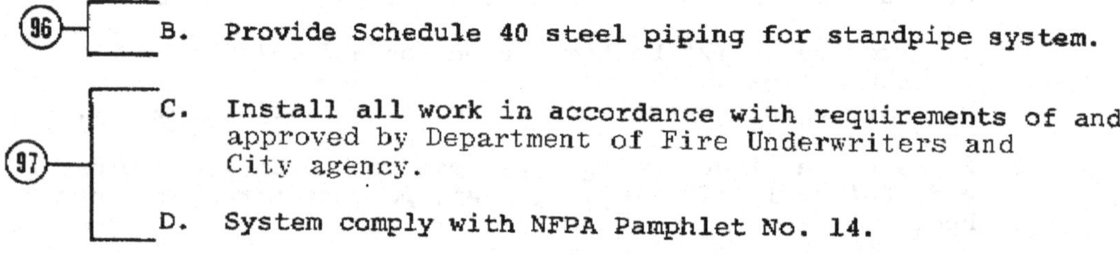

(96) B. Provide Schedule 40 steel piping for standpipe system.

(97) C. Install all work in accordance with requirements of and approved by Department of Fire Underwriters and City agency.

D. System comply with NFPA Pamphlet No. 14.

96. Indicates required piping material for fire standpipe system in building.
97. Indicates codes system may comply to.

Figure 4.57

After the specifications, you will find the actual quantity takeoff for the plumbing work on the Three-Story Service Center (Figures 4.58 through 4.84). As with the specifications, there will be accompanying "takeoff notes" corresponding to numbered items flagged on the takeoff sheet.

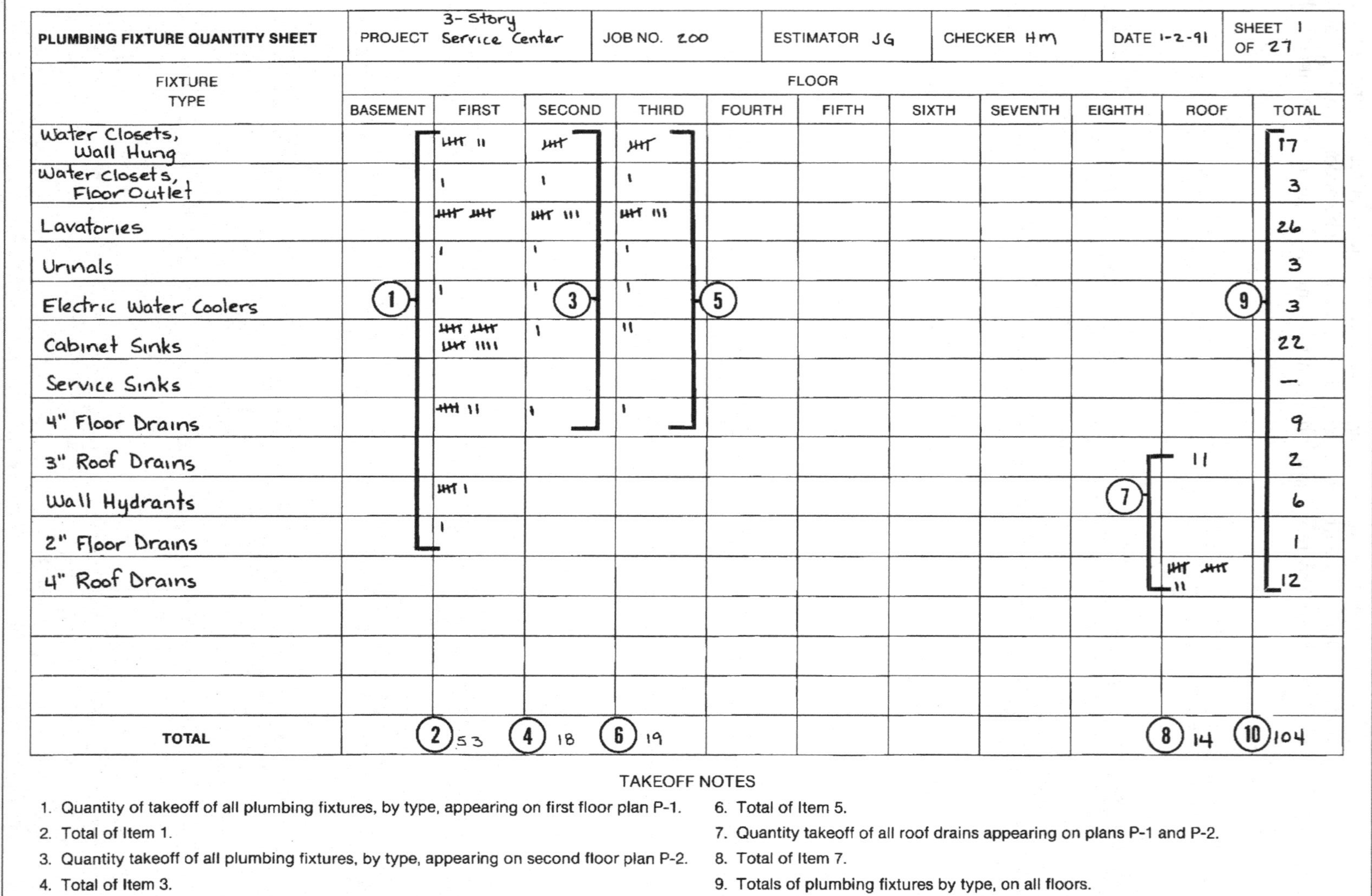

PLUMBING FIXTURE QUANTITY SHEET	PROJECT 3-Story Service Center		JOB NO. 200		ESTIMATOR JG		CHECKER HM		DATE 1-2-91		SHEET 1 OF 27
FIXTURE TYPE	FLOOR										
	BASEMENT	FIRST	SECOND	THIRD	FOURTH	FIFTH	SIXTH	SEVENTH	EIGHTH	ROOF	TOTAL
Water Closets, Wall Hung		卌 ‖	卌	卌							17
Water Closets, Floor Outlet		ǀ	ǀ	ǀ							3
Lavatories		卌 卌	卌 ‖ǀ	卌 ‖ǀ							26
Urinals		ǀ	ǀ	ǀ							3
Electric Water Coolers	(1)	ǀ	ǀ (3)	ǀ	(5)					(9)	3
Cabinet Sinks		卌 卌 卌 ‖‖	ǀ	‖							22
Service Sinks											—
4" Floor Drains		卌 ‖	ǀ	ǀ							9
3" Roof Drains										‖	2
Wall Hydrants		卌 ǀ							(7)		6
2" Floor Drains		ǀ									1
4" Roof Drains										卌 卌 ‖	12
TOTAL		(2) 53	(4) 18	(6) 19						(8) 14	(10) 104

TAKEOFF NOTES

1. Quantity of takeoff of all plumbing fixtures, by type, appearing on first floor plan P-1.
2. Total of Item 1.
3. Quantity takeoff of all plumbing fixtures, by type, appearing on second floor plan P-2.
4. Total of Item 3.
5. Quantity takeoff of all plumbing fixtures, by type, appearing on third floor plan P-3.
6. Total of Item 5.
7. Quantity takeoff of all roof drains appearing on plans P-1 and P-2.
8. Total of Item 7.
9. Totals of plumbing fixtures by type, on all floors.
10. Total of all plumbing fixtures on job.

Figure 4.58

EQUIPMENT QUANTITY SHEET	PROJECT 3-Story Service Center	JOB NO. 200	ESTIMATOR JG	CHECKER HM	DATE 1-2-91	SHEET 2 OF 27

ITEM	QTY.	SYSTEM	CAPACITIES, DESCRIPTION AND OTHER INFORMATION
3" Domestic Water Meter	1 (1)	Hot and Cold Water	
Electric Domestic Hot Water Heater - H-3	1 (2)	Hot and Cold Water (5)	73.8 G.P.H. @ 100° Rise Recovery, 120 G.P.H. (3) Element, 18 KW, 480 V, 3-Phase Jackson 'H' Series # GRE 120-T-H (6)
Hot Water Circulation Pump - P-2	1 (3)	Hot and Cold Water	In-line, Direct Connection, 1750 RPM, 1/6 H.P. 115 V, 1-Phase, Bell + Gossett # 7-PR (7)
Elevator Sump Pump	1 (4)	Sanitary	9 G.P.M., 1/3 H.P., Simplex, 24 Ft. Head, 1750 RPM, 120 V, 1-Phase, Weil # 55-805-ND. (8)

TAKEOFF NOTES

1. 3" domestic water meter taken off mechanical equipment room plan P-3.
2. Electric hot water heater taken off mechanical equipment room plan P-3.
3. Hot water circulation pump taken off domestic hot water heater piping schematic plan P-3.
4. Elevator sump pump taken off first floor plan P-1.
5. Denotes applicable system for each item of equipment.
6. Capacity of hot water heater found in specification page 2-2.
7. Capacity of hot water circulation pump taken off domestic hot water heater piping schematic plan P-3.
8. Capacity for elevator sump pump was not given in drawings or specification. Estimator had to verify capacity with design engineer.

Figure 4.59

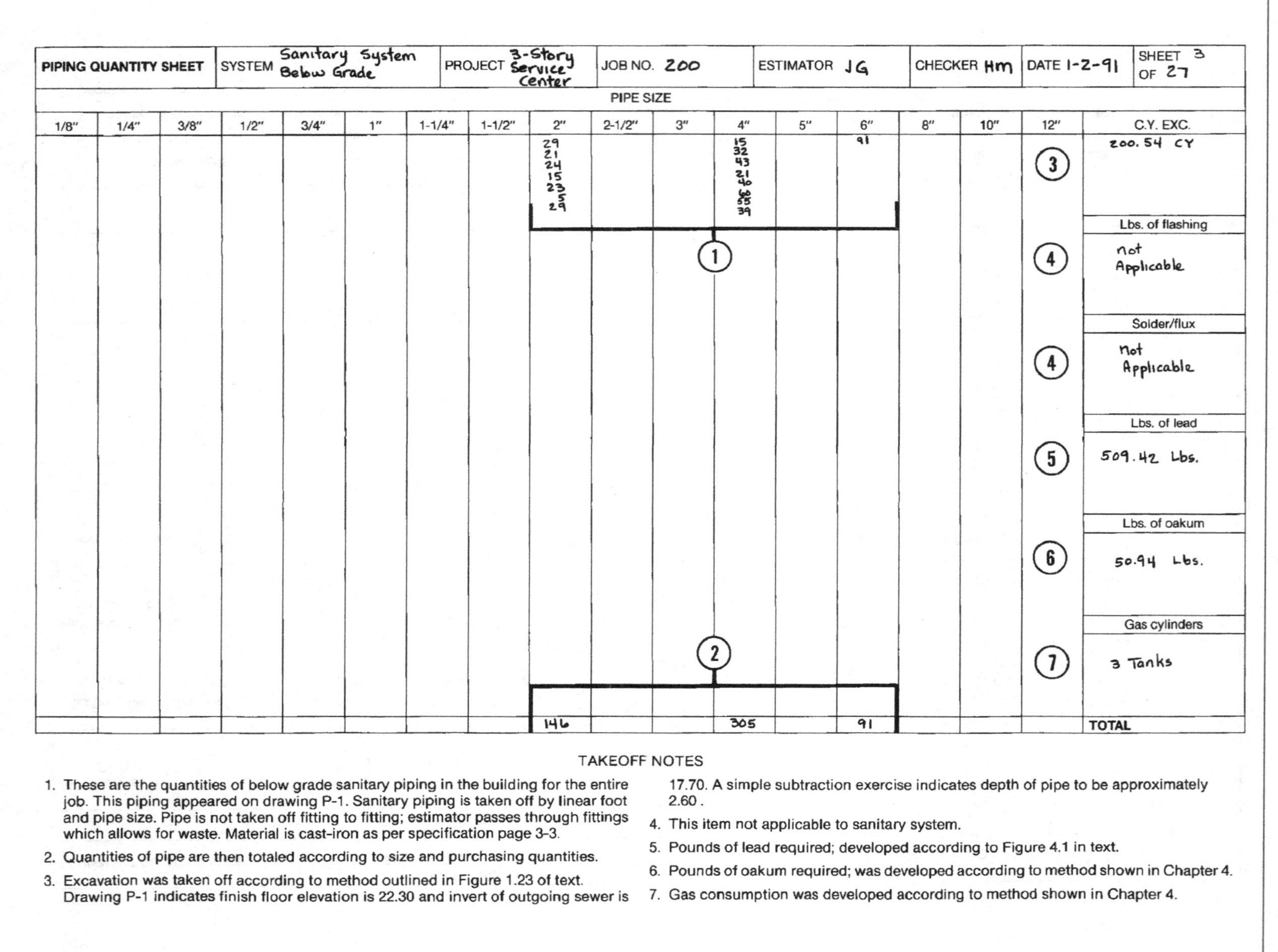

PIPING QUANTITY SHEET	SYSTEM Sanitary System Below Grade	PROJECT 3-Story Service Center	JOB NO. 200	ESTIMATOR JG	CHECKER HM	DATE 1-2-91	SHEET 3 OF 27

PIPE SIZE

1/8"	1/4"	3/8"	1/2"	3/4"	1"	1-1/4"	1-1/2"	2"	2-1/2"	3"	4"	5"	6"	8"	10"	12"	
								29 21 24 15 23 5 29			15 32 43 21 40 60 55 39		91			(3)	C.Y. EXC. 200.54 CY
											(1)					(4)	Lbs. of flashing not Applicable
																(4)	Solder/flux Not Applicable
																(5)	Lbs. of lead 509.42 Lbs.
																(6)	Lbs. of oakum 50.94 Lbs.
											(2)					(7)	Gas cylinders 3 Tanks
								146			305		91				TOTAL

TAKEOFF NOTES

1. These are the quantities of below grade sanitary piping in the building for the entire job. This piping appeared on drawing P-1. Sanitary piping is taken off by linear foot and pipe size. Pipe is not taken off fitting to fitting; estimator passes through fittings which allows for waste. Material is cast-iron as per specification page 3-3.
2. Quantities of pipe are then totaled according to size and purchasing quantities.
3. Excavation was taken off according to method outlined in Figure 1.23 of text. Drawing P-1 indicates finish floor elevation is 22.30 and invert of outgoing sewer is 17.70. A simple subtraction exercise indicates depth of pipe to be approximately 2.60 .
4. This item not applicable to sanitary system.
5. Pounds of lead required; developed according to Figure 4.1 in text.
6. Pounds of oakum required; was developed according to method shown in Chapter 4.
7. Gas consumption was developed according to method shown in Chapter 4.

Figure 4.60

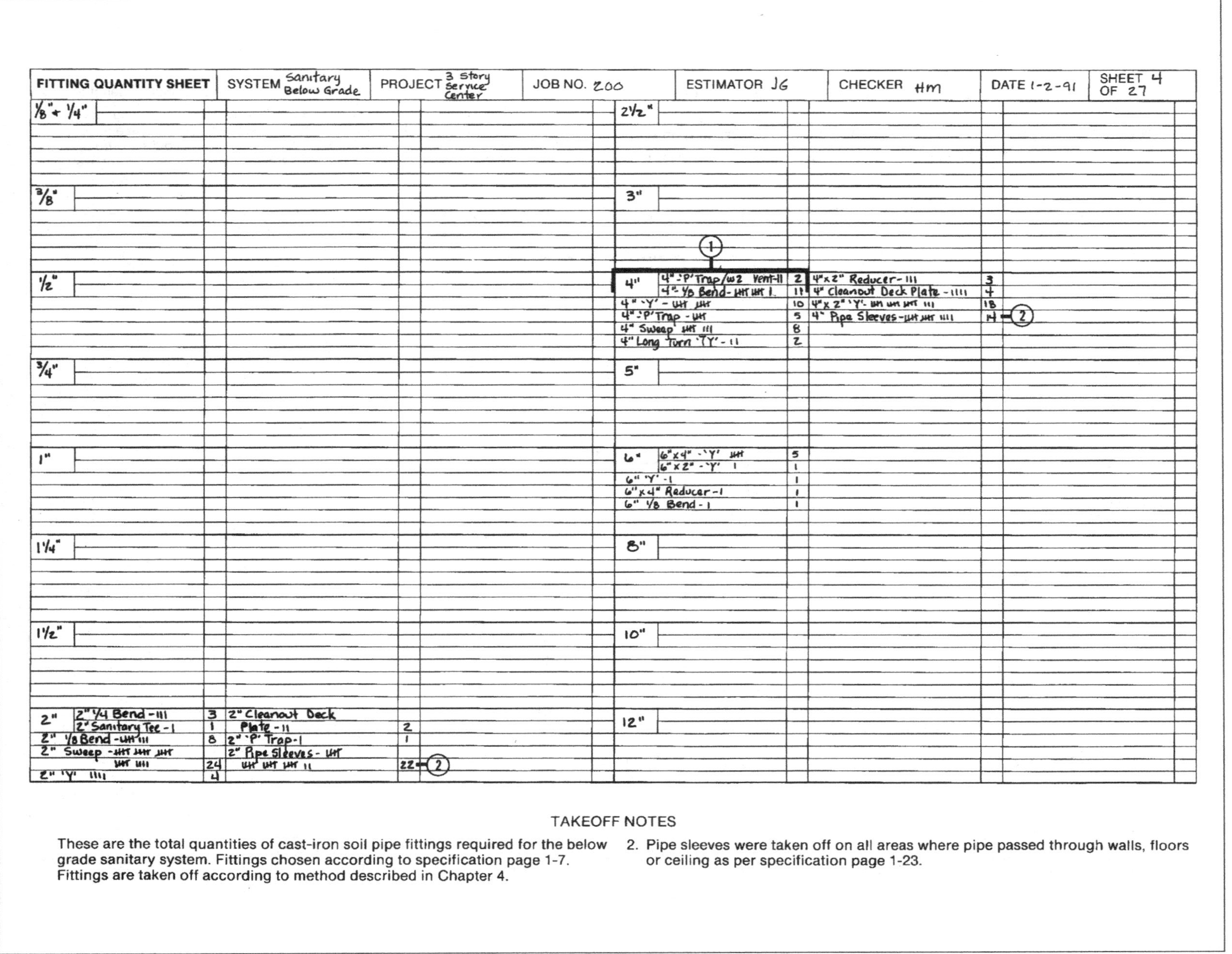

FITTING QUANTITY SHEET	SYSTEM Sanitary Below Grade	PROJECT 3 Story Service Center	JOB NO. 200	ESTIMATOR JG	CHECKER HM	DATE 1-2-91	SHEET 4 OF 27

Size	Fitting	Qty	Fitting	Qty
1/8" + 1/4"				
3/8"				
1/2"				
3/4"				
1"				
1 1/4"				
1 1/2"				
2"	2" 1/4 Bend - III	3	2" Cleanout Deck Plate - II	2
	2" Sanitary Tee - I	1	2" 'P' Trap - I	1
	2" 1/8 Bend - HHT III	8	2" Pipe Sleeves - HHT HHT HHT HHT II	22 (2)
	2" Sweep - HHT HHT HHT HHT IIII	24		
	2" 'Y' IIII	4		
2 1/2"				
3"				
4" (1)	4" 'P' Trap/w2 Vent - II	2	4"x2" Reducer - III	3
	4" 1/8 Bend - HHT HHT I	11	4" Cleanout Deck Plate - IIII	4
	4" 'Y' - HHT HHT	10	4"x2" 'Y' - HHT HHT HHT III	18
	4" 'P' Trap - HHT	5	4" Pipe Sleeves - HHT HHT IIII	14 (2)
	4" Sweep HHT III	8		
	4" Long Turn 'Y' - II	2		
5"				
6"	6"x4" 'Y' HHT	5		
	6"x2" 'Y' I	1		
	6" 'Y' - I	1		
	6"x4" Reducer - I	1		
	6" 1/8 Bend - I	1		
8"				
10"				
12"				

TAKEOFF NOTES

These are the total quantities of cast-iron soil pipe fittings required for the below grade sanitary system. Fittings chosen according to specification page 1-7. Fittings are taken off according to method described in Chapter 4.

2. Pipe sleeves were taken off on all areas where pipe passed through walls, floors or ceiling as per specification page 1-23.

Figure 4.61

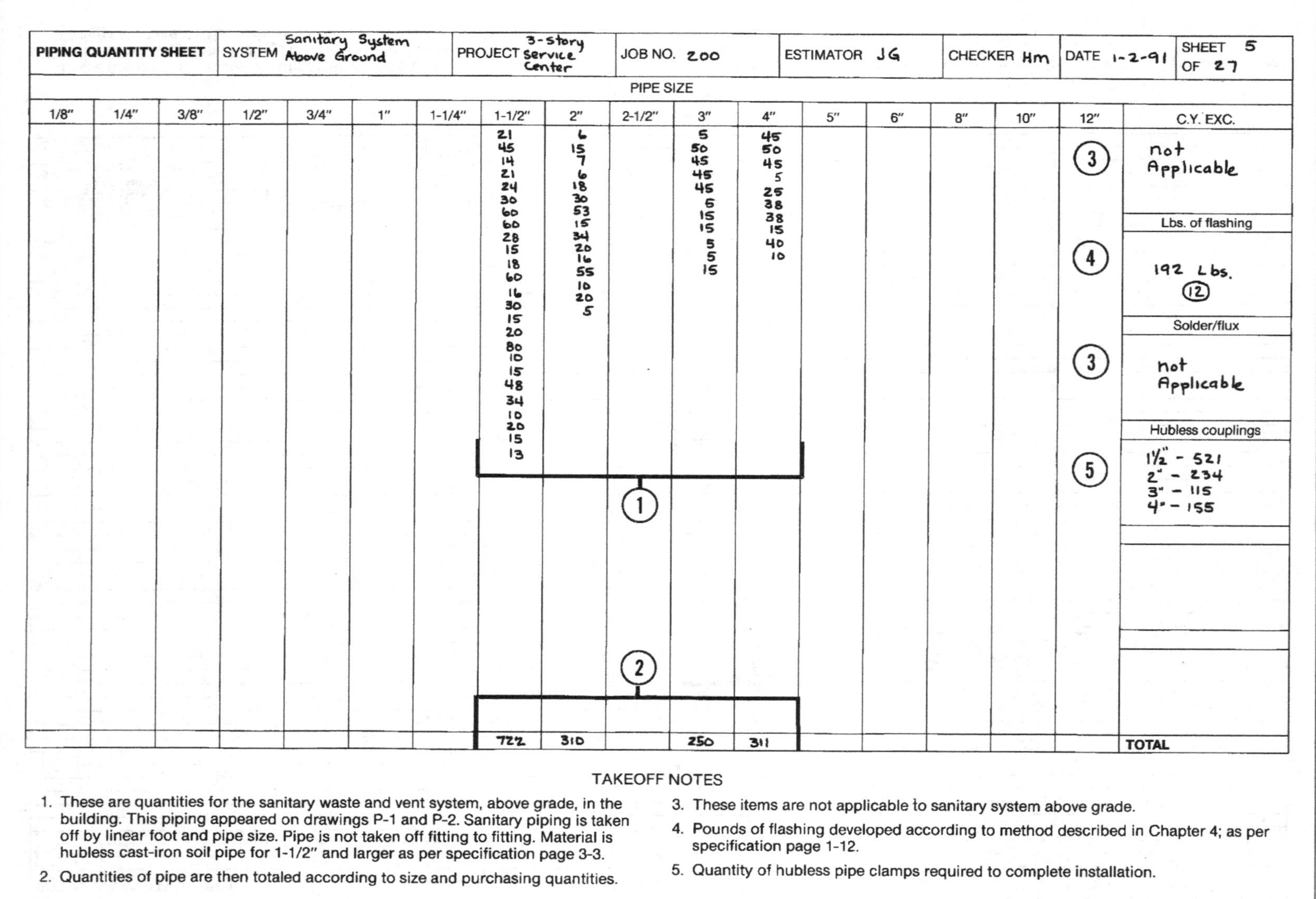

PIPING QUANTITY SHEET	SYSTEM Sanitary System Above Ground	PROJECT 3-Story Service Center	JOB NO. 200	ESTIMATOR JG	CHECKER Hm	DATE 1-2-91	SHEET 5 OF 27

PIPE SIZE

1/8"	1/4"	3/8"	1/2"	3/4"	1"	1-1/4"	1-1/2"	2"	2-1/2"	3"	4"	5"	6"	8"	10"	12"	C.Y. EXC.
							21 45 14 21 24 30 60 60 28 15 18 60 16 30 15 20 80 10 15 48 34 10 20 15 13	6 15 7 6 18 30 53 15 34 20 16 55 10 20 5	(1)	5 50 45 45 45 6 15 15 5 5 15	45 50 45 5 25 38 38 15 40 10					(3) (4) (3) (5)	not Applicable Lbs. of flashing 192 Lbs. (12) Solder/flux not Applicable Hubless couplings 1½" - 521 2" - 234 3" - 115 4" - 155
									(2)								
							722	310		250	311						TOTAL

TAKEOFF NOTES

1. These are quantities for the sanitary waste and vent system, above grade, in the building. This piping appeared on drawings P-1 and P-2. Sanitary piping is taken off by linear foot and pipe size. Pipe is not taken off fitting to fitting. Material is hubless cast-iron soil pipe for 1-1/2" and larger as per specification page 3-3.
2. Quantities of pipe are then totaled according to size and purchasing quantities.
3. These items are not applicable to sanitary system above grade.
4. Pounds of flashing developed according to method described in Chapter 4; as per specification page 1-12.
5. Quantity of hubless pipe clamps required to complete installation.

Figure 4.62

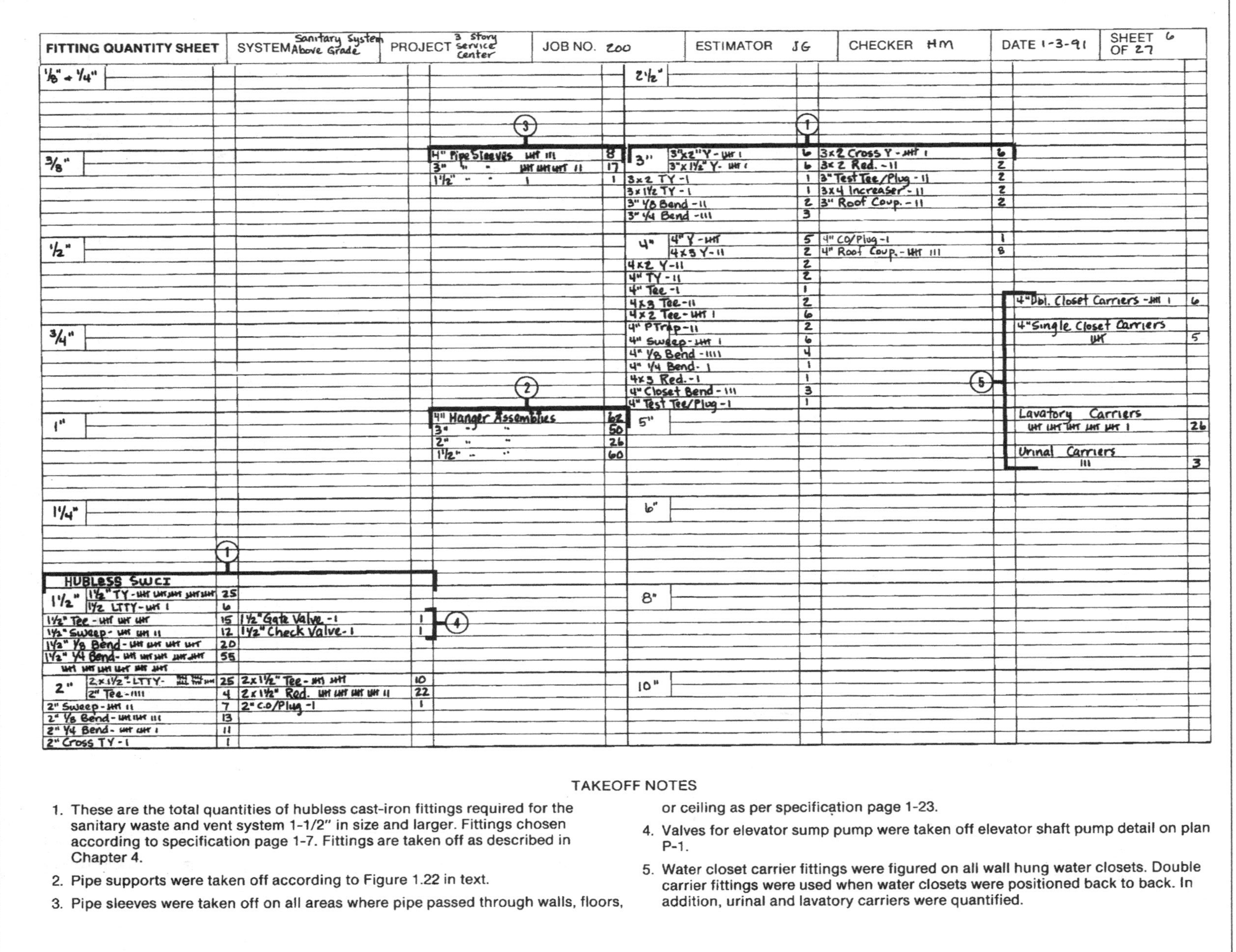

FITTING QUANTITY SHEET	SYSTEM Sanitary System Above Grade	PROJECT 3 Story Service Center	JOB NO. 200	ESTIMATOR JG	CHECKER HM	DATE 1-3-91	SHEET 6 OF 27

Size	Item	Qty
1/8" + 1/4"		
3/8"		
1/2"		
3/4"		
1"		
1 1/4"		
HUBLESS SWCI (1)		
1 1/2"	1 1/2" TY	25
	1 1/2 LTTY	6
	1 1/2" Tee	15
	1 1/2" Sweep	12
	1 1/2" 1/8 Bend	20
	1 1/2" 1/4 Bend	55
2"	2 x 1 1/2" LTTY	25
	2" Tee	4
	2" Sweep	7
	2" 1/8 Bend	13
	2" 1/4 Bend	11
	2" Cross TY	1
	2 x 1 1/2" Tee	10
	2 x 1 1/2" Red.	22
	2" C.O/Plug	1
(4)	1 1/2" Gate Valve	1
	1 1/2" Check Valve	1
(3)	4" Pipe Sleeves	8
	3" Pipe Sleeves	17
	1 1/2" Pipe Sleeves	1
(2)	4" Hanger Assemblies	62
	3" Hanger Assemblies	50
	2" Hanger Assemblies	26
	1 1/2" Hanger Assemblies	60
2 1/2"		
3" (1)	3"x2" Y	6
	3"x1 1/2" Y	6
	3x2 TY	1
	3x1 1/2 TY	1
	3" 1/8 Bend	2
	3" 1/4 Bend	3
	3x2 Cross Y	6
	3x2 Red.	2
	3" Test Tee/Plug	2
	3x4 Increaser	2
	3" Roof Coup.	2
4"	4" Y	5
	4x3 Y	2
	4x2 Y	2
	4" TY	2
	4" Tee	1
	4x3 Tee	2
	4x2 Tee	6
	4" P Trap	2
	4" Sweep	6
	4" 1/8 Bend	4
	4" 1/4 Bend	1
	4x3 Red.	1
	4" Closet Bend	3
	4" Test Tee/Plug	1
	4" CO/Plug	1
	4" Roof Coup.	8
5"		
6"		
8"		
10"		
(5)	4" Dbl. Closet Carriers	6
	4" Single Closet Carriers	5
	Lavatory Carriers	26
	Urinal Carriers	3

TAKEOFF NOTES

1. These are the total quantities of hubless cast-iron fittings required for the sanitary waste and vent system 1-1/2" in size and larger. Fittings chosen according to specification page 1-7. Fittings are taken off as described in Chapter 4.
2. Pipe supports were taken off according to Figure 1.22 in text.
3. Pipe sleeves were taken off on all areas where pipe passed through walls, floors, or ceiling as per specification page 1-23.
4. Valves for elevator sump pump were taken off elevator shaft pump detail on plan P-1.
5. Water closet carrier fittings were figured on all wall hung water closets. Double carrier fittings were used when water closets were positioned back to back. In addition, urinal and lavatory carriers were quantified.

Figure 4.63

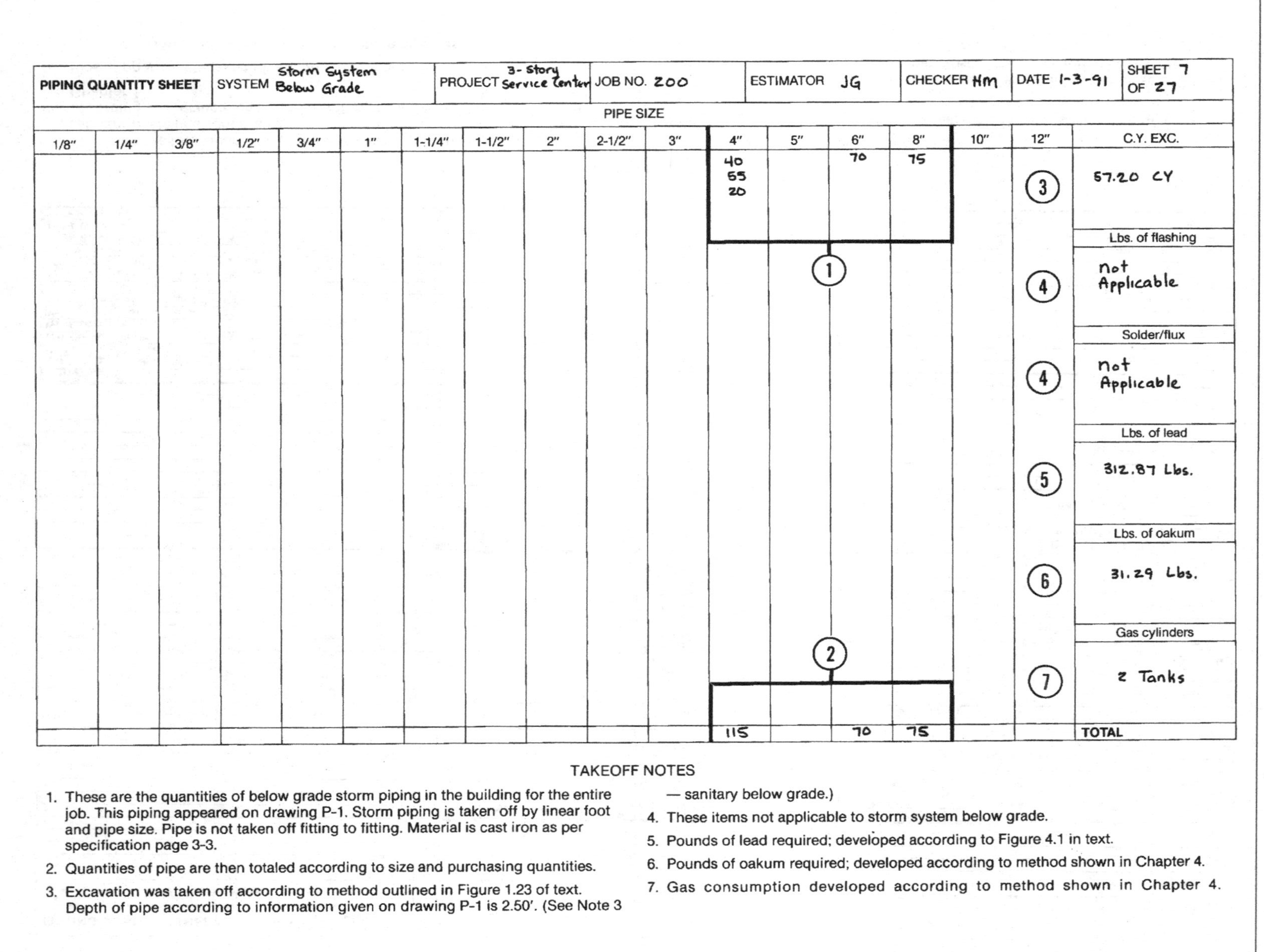

PIPING QUANTITY SHEET	SYSTEM Storm System Below Grade	PROJECT 3-Story Service Center	JOB NO. 200	ESTIMATOR JG	CHECKER HM	DATE 1-3-91	SHEET 7 OF 27

PIPE SIZE

4"	5"	6"	8"	10"	12"	
40 55 20		70	75			C.Y. EXC.: 57.20 CY
						Lbs. of flashing: Not Applicable
						Solder/flux: Not Applicable
						Lbs. of lead: 312.87 Lbs.
						Lbs. of oakum: 31.29 Lbs.
						Gas cylinders: 2 Tanks
115		70	75			TOTAL

Item	Note
Pipe entries (4", 6", 8")	(1)
Pipe totals (4", 6", 8")	(2)
C.Y. EXC.	(3)
Lbs. of flashing	(4)
Solder/flux	(4)
Lbs. of lead	(5)
Lbs. of oakum	(6)
Gas cylinders	(7)

TAKEOFF NOTES

1. These are the quantities of below grade storm piping in the building for the entire job. This piping appeared on drawing P-1. Storm piping is taken off by linear foot and pipe size. Pipe is not taken off fitting to fitting. Material is cast iron as per specification page 3-3.
2. Quantities of pipe are then totaled according to size and purchasing quantities.
3. Excavation was taken off according to method outlined in Figure 1.23 of text. Depth of pipe according to information given on drawing P-1 is 2.50′. (See Note 3 — sanitary below grade.)
4. These items not applicable to storm system below grade.
5. Pounds of lead required; developed according to Figure 4.1 in text.
6. Pounds of oakum required; developed according to method shown in Chapter 4.
7. Gas consumption developed according to method shown in Chapter 4.

Figure 4.64

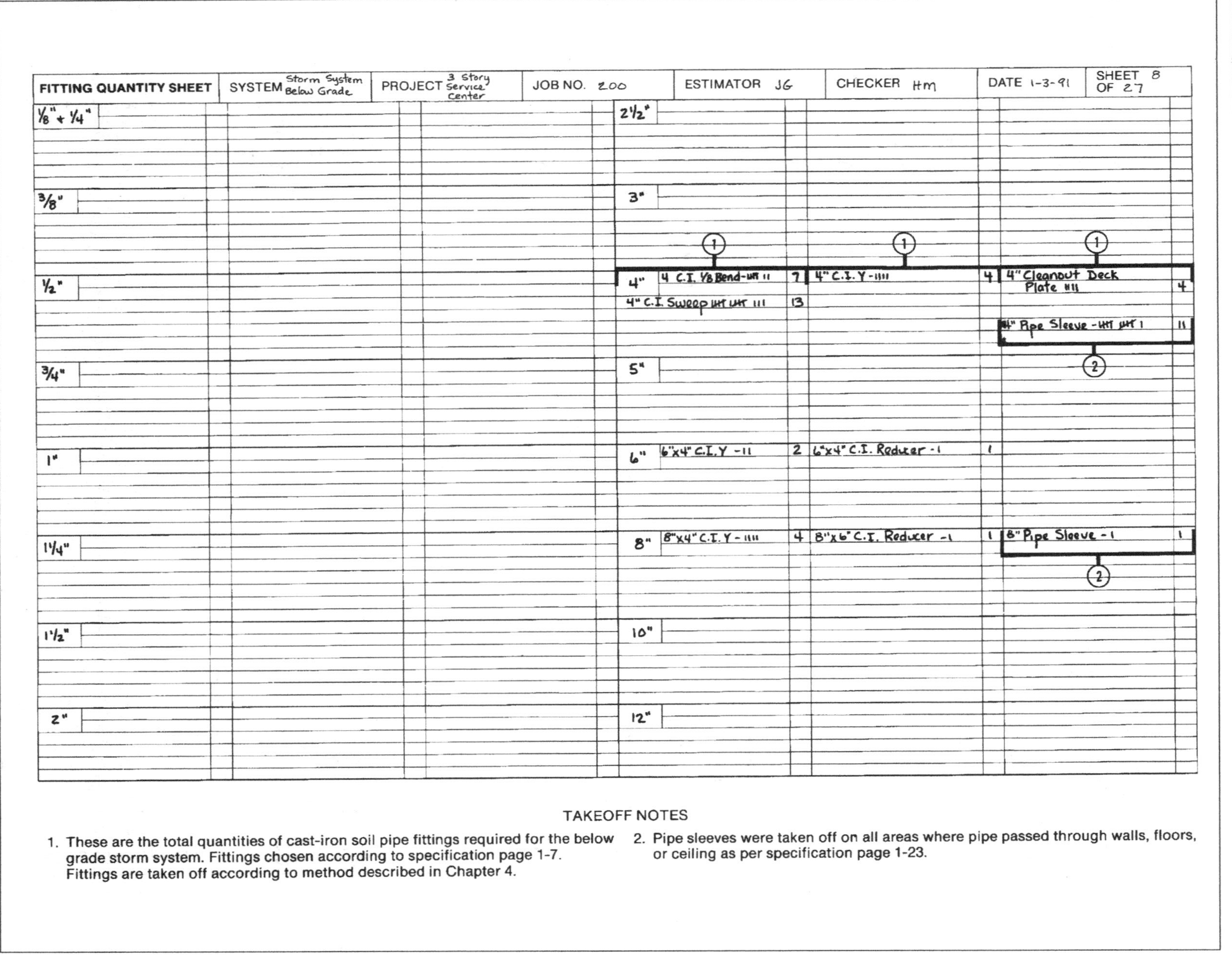

FITTING QUANTITY SHEET	SYSTEM Storm System Below Grade	PROJECT 3 Story Service Center	JOB NO. 200	ESTIMATOR JG	CHECKER HM	DATE 1-3-91	SHEET 8 OF 27

Size	Item	Qty	Item	Qty	Item	Qty
1/8" + 1/4"						
3/8"						
1/2"						
3/4"						
1"						
1 1/4"						
1 1/2"						
2"						
2 1/2"						
3"						
4"	4 C.I. 1/8 Bend-卌 11 (1)	7	4" C.I. Y-1111 (1)	4	4" Cleanout Deck Plate 111 (1)	4
	4" C.I. Sweep 卌 卌 111	13			4" Pipe Sleeve-卌 卌 1 (2)	11
5"						
6"	6"x4" C.I. Y-11	2	6"x4" C.I. Reducer-1	1		
8"	8"x4" C.I. Y-1111	4	8"x6" C.I. Reducer-1	1	8" Pipe Sleeve-1 (2)	1
10"						
12"						

TAKEOFF NOTES

1. These are the total quantities of cast-iron soil pipe fittings required for the below grade storm system. Fittings chosen according to specification page 1-7. Fittings are taken off according to method described in Chapter 4.
2. Pipe sleeves were taken off on all areas where pipe passed through walls, floors, or ceiling as per specification page 1-23.

Figure 4.65

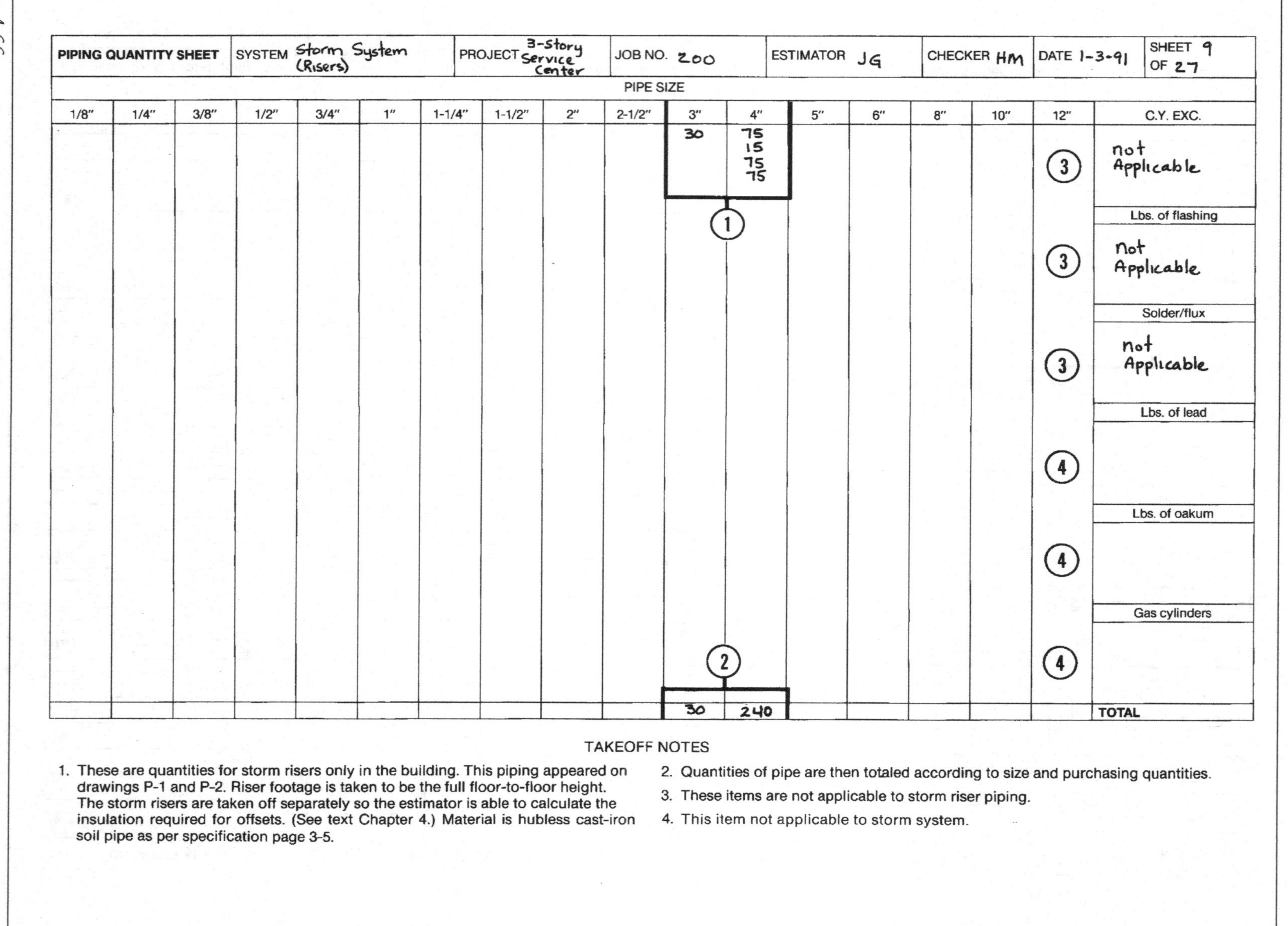

PIPING QUANTITY SHEET	SYSTEM Storm System (Risers)	PROJECT 3-Story Service Center	JOB NO. 200	ESTIMATOR JG	CHECKER HM	DATE 1-3-91	SHEET 9 OF 27

PIPE SIZE																	
1/8"	1/4"	3/8"	1/2"	3/4"	1"	1-1/4"	1-1/2"	2"	2-1/2"	3"	4"	5"	6"	8"	10"	12"	C.Y. EXC.
										30	75 15 75 75					(3)	not Applicable
										(1)							Lbs. of flashing
																(3)	not Applicable
																	Solder/flux
																(3)	not Applicable
																	Lbs. of lead
																(4)	
																	Lbs. of oakum
																(4)	
																	Gas cylinders
										(2)						(4)	
										30	240						TOTAL

TAKEOFF NOTES

1. These are quantities for storm risers only in the building. This piping appeared on drawings P-1 and P-2. Riser footage is taken to be the full floor-to-floor height. The storm risers are taken off separately so the estimator is able to calculate the insulation required for offsets. (See text Chapter 4.) Material is hubless cast-iron soil pipe as per specification page 3-5.
2. Quantities of pipe are then totaled according to size and purchasing quantities.
3. These items are not applicable to storm riser piping.
4. This item not applicable to storm system.

Figure 4.66

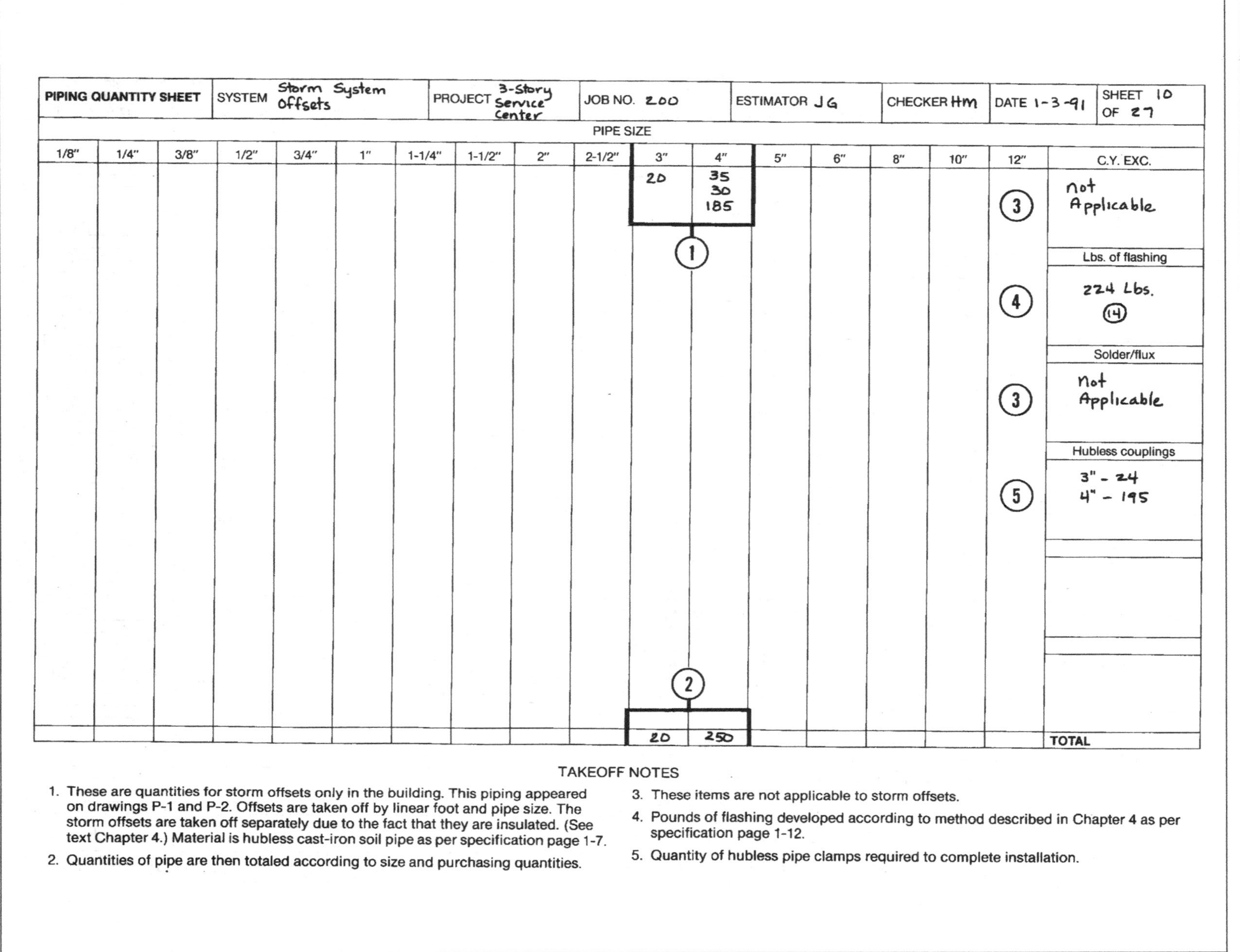

PIPING QUANTITY SHEET	SYSTEM Storm System Offsets	PROJECT 3-Story Service Center	JOB NO. 200	ESTIMATOR JG	CHECKER HM	DATE 1-3-91	SHEET 10 OF 27

PIPE SIZE

1/8"	1/4"	3/8"	1/2"	3/4"	1"	1-1/4"	1-1/2"	2"	2-1/2"	3"	4"	5"	6"	8"	10"	12"	C.Y. EXC.
										20	35 30 185 (1)					(3)	Not Applicable
																	Lbs. of flashing
																(4)	224 Lbs. (14)
																	Solder/flux
																(3)	Not Applicable
																	Hubless couplings
																(5)	3" - 24 4" - 195
										20	250 (2)						TOTAL

TAKEOFF NOTES

1. These are quantities for storm offsets only in the building. This piping appeared on drawings P-1 and P-2. Offsets are taken off by linear foot and pipe size. The storm offsets are taken off separately due to the fact that they are insulated. (See text Chapter 4.) Material is hubless cast-iron soil pipe as per specification page 1-7.
2. Quantities of pipe are then totaled according to size and purchasing quantities.
3. These items are not applicable to storm offsets.
4. Pounds of flashing developed according to method described in Chapter 4 as per specification page 1-12.
5. Quantity of hubless pipe clamps required to complete installation.

Figure 4.67

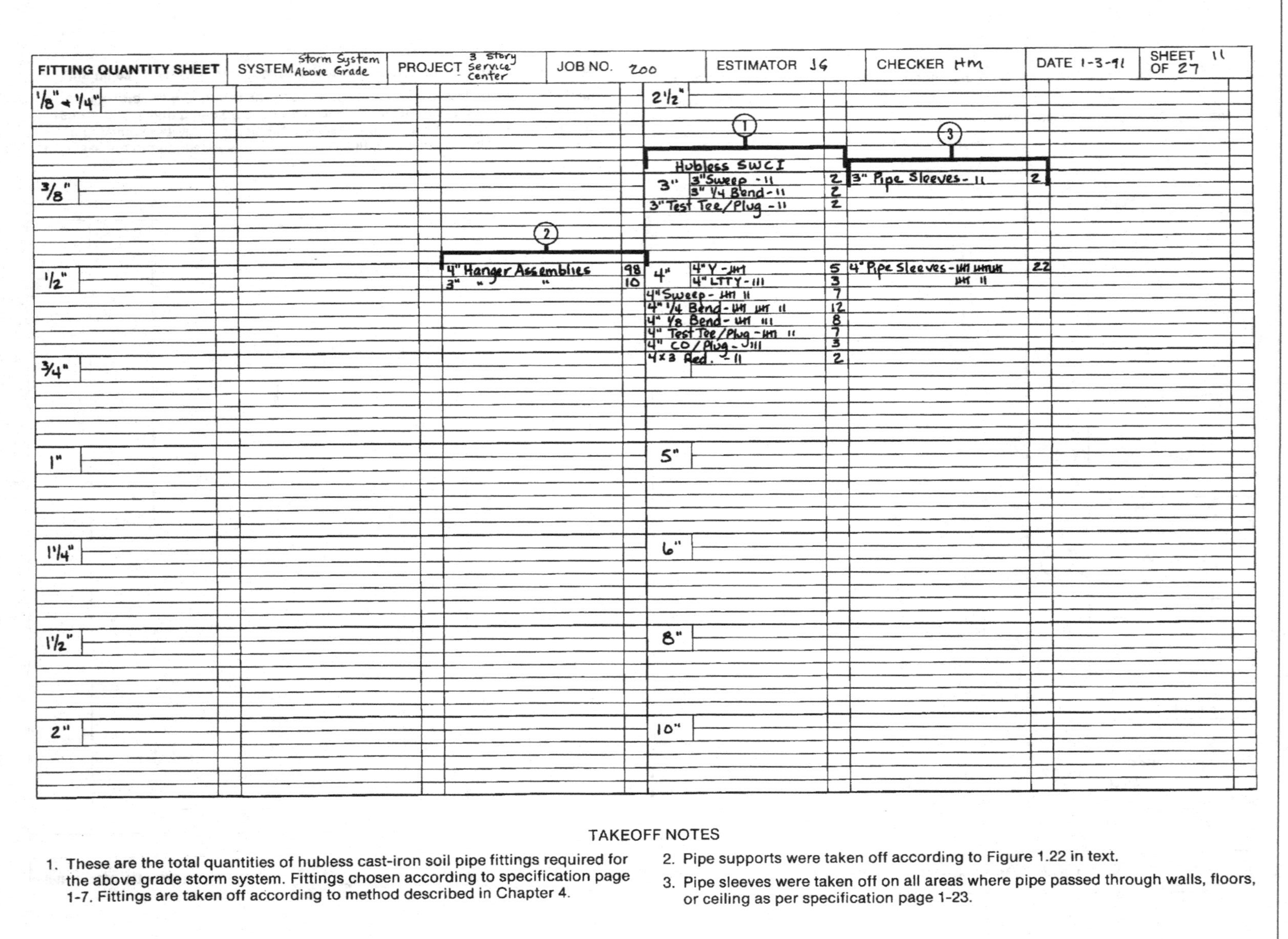

FITTING QUANTITY SHEET	SYSTEM Storm System Above Grade	PROJECT 3 Story Service Center	JOB NO. 200	ESTIMATOR JG	CHECKER HM	DATE 1-3-91	SHEET 11 OF 27

Size	Item	Qty	Size	Item	Qty	Item	Qty
1/8" + 1/4"			2 1/2"				
				(1) Hubless SWCI		(3)	
3/8"			3"	3" Sweep - 11	2	3" Pipe Sleeves - 11	2
				3" 1/4 Bend - 11	2		
				3" Test Tee/Plug - 11	2		
	(2)						
1/2"	4" Hanger Assemblies	98	4"	4" Y	5	4" Pipe Sleeves	22
	3" " "	10		4" LTTY - 111	3		
				4" Sweep	7		
				4" 1/4 Bend	12		
				4" 1/8 Bend	8		
				4" Test Tee/Plug	7		
				4" CO/Plug	3		
				4x3 Red. - 11	2		
3/4"							
1"			5"				
1 1/4"			6"				
1 1/2"			8"				
2"			10"				

TAKEOFF NOTES

1. These are the total quantities of hubless cast-iron soil pipe fittings required for the above grade storm system. Fittings chosen according to specification page 1-7. Fittings are taken off according to method described in Chapter 4.
2. Pipe supports were taken off according to Figure 1.22 in text.
3. Pipe sleeves were taken off on all areas where pipe passed through walls, floors, or ceiling as per specification page 1-23.

Figure 4.68

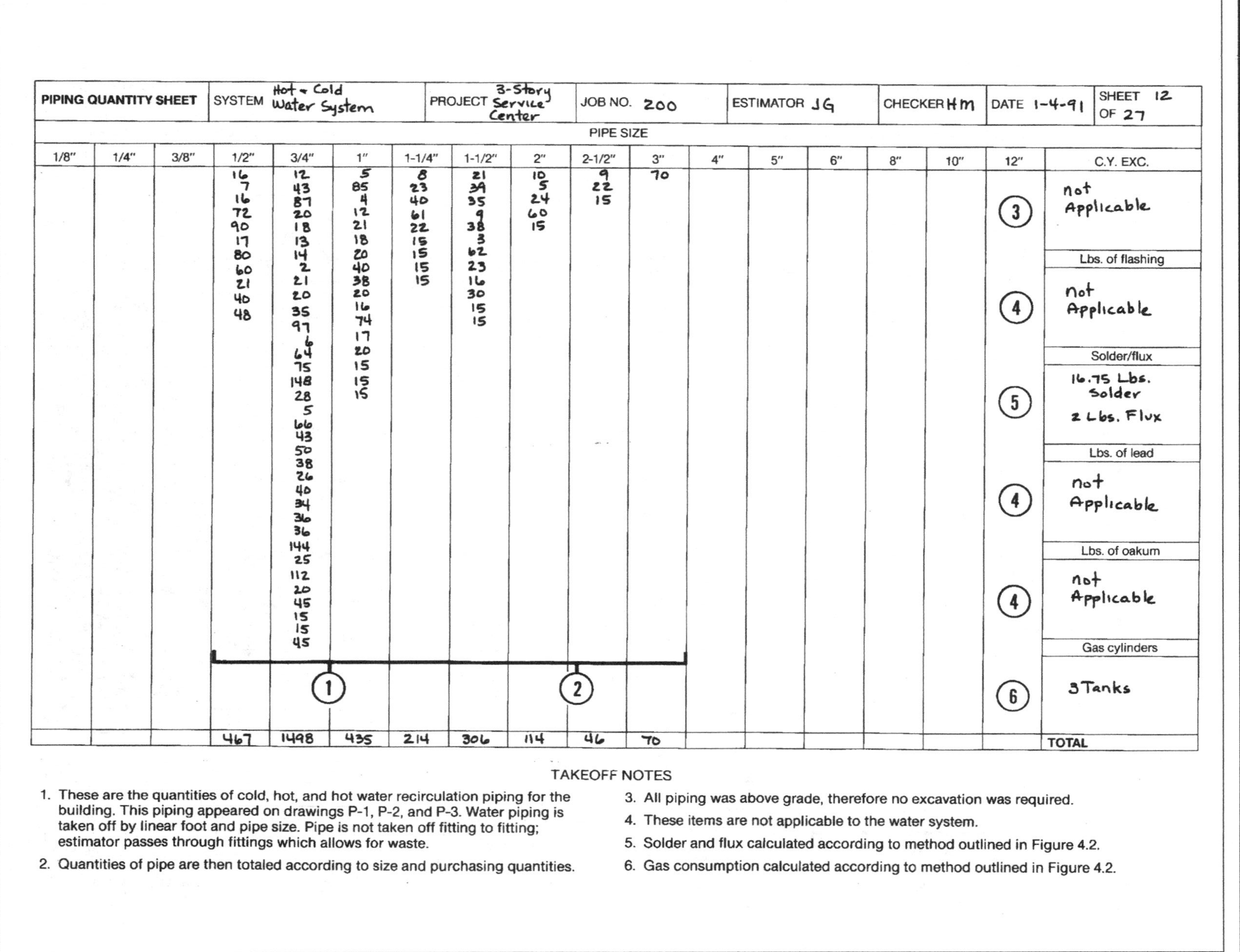

PIPING QUANTITY SHEET	SYSTEM Hot + Cold Water System	PROJECT 3-Story Service Center	JOB NO. 200	ESTIMATOR JG	CHECKER HM	DATE 1-4-91	SHEET 12 OF 27

PIPE SIZE															
1/8"	1/4"	3/8"	1/2"	3/4"	1"	1-1/4"	1-1/2"	2"	2-1/2"	3"	4"	5"	6"	8"	10"
			16 7 16 72 90 17 80 60 21 40 48	12 43 87 20 18 13 14 2 21 20 35 97 6 64 75 148 28 5 66 43 50 38 26 40 34 36 36 144 25 112 20 45 15 15 45	5 85 4 12 21 18 20 40 38 20 16 74 17 20 15 15 15	8 23 40 61 22 15 15 15 15	21 39 35 9 38 3 62 23 16 30 15 15	10 5 24 60 15	9 22 15	70					
				(1)					(2)						
			467	1498	435	214	306	114	46	70					

12"	
	C.Y. EXC.
(3)	not Applicable
	Lbs. of flashing
(4)	not Applicable
	Solder/flux
(5)	16.75 Lbs. Solder 2 Lbs. Flux
	Lbs. of lead
(4)	not Applicable
	Lbs. of oakum
(4)	not Applicable
	Gas cylinders
(6)	3 Tanks
	TOTAL

TAKEOFF NOTES

1. These are the quantities of cold, hot, and hot water recirculation piping for the building. This piping appeared on drawings P-1, P-2, and P-3. Water piping is taken off by linear foot and pipe size. Pipe is not taken off fitting to fitting; estimator passes through fittings which allows for waste.
2. Quantities of pipe are then totaled according to size and purchasing quantities.
3. All piping was above grade, therefore no excavation was required.
4. These items are not applicable to the water system.
5. Solder and flux calculated according to method outlined in Figure 4.2.
6. Gas consumption calculated according to method outlined in Figure 4.2.

Figure 4.69

FITTING QUANTITY SHEET	SYSTEM Hot + Cold water System	PROJECT 3 Story Service Center	JOB NO. 200	ESTIMATOR JG	CHECKER HM	DATE 1-4-91	SHEET 13 OF 27

Size	Fitting	Qty
1/8" + 1/4"	Misc. Nipples	60
3/8"		
1/2"	1/2" Wrought Ell	143
	1/2" Wrought 45	29
	1/2" Wrought Tees	2
	1/2" Copper x Female Adapters	54
	1/2" Wrought Cap	2
	1/2" Wrought Couplings	10
3/4"	3/4" Wrought Ells	58
	3/4" Wrought Tee	8
	3/4" Wrought Tee, Red.	53
	3/4" Wrought Cap	6
	3/4" Copper x Female Adapter	9
	3/4" Wrought 45	14
	3/4" x Red. Wrought Ells	2
	3/4" Wrought Couplings	29
1"	1" x Red Coupling	1
	1" Wrought Ell	35
	1" Red. Wrought Tee	11
	1" Wrought Tee	1
	1" Wrought Cap	1
	1" Copper x Female Adapter	18
	1" Wrought 45	2
	1" Wrought Coupling	12
1 1/4"	1 1/4 x Red. Wrought Tee	10
	1 1/4" Wrought Tee	1
	1 1/4" Wrought Ells	6
	1 1/4" Wrought Cap	1
	1 1/4" Wrought Coupling	7
	1 1/4" Wrought 45	1
1 1/2"	1 1/2" Wrought Ell	23
	1 1/2" x Red. Wrought Ell	1
	1 1/2" x Red. Wrought Tee	30
	1 1/2" Wrought 45	4
	1 1/2" Wrought Coupling	11
	1 1/2" Wrought Cap	1
	1 1/2" Pipe Sleeves	25
2	2" Wrought Ells	3
	2" Red. Wrought Tee	4
	2" Wrought Coupling	5
	2" Pipe Sleeve	1
2 1/2"	2 1/2" Red. Wrought Tee	3
	2 1/2" Wrought Ell	1
	2 1/2" Wrought Couplings	2
	2 1/2" Pipe Sleeve	1
3"	3" Wrought Ell	7
	3" Red. Wrought Tee	5
	3" Wrought Couplings	2
	3" Pipe Sleeve	2
4"		
5"		
6"		
8"		
10"		
12"		

Pipe Supports	Qty
3" Pipe Supports	7
2 1/2" " "	5
2" " "	12
1 1/2" " "	31
1 1/4" " "	36
1" " "	13
3/4" " "	250
1/2" " "	78

Note references: (1) 1/2" through 1 1/2" fittings and 3" and 4" fittings; (2) Misc. Nipples; (3) Pipe Sleeves and couplings; (4) Pipe Supports.

TAKEOFF NOTES

1. These are the total quantities of wrought copper solder joint fittings required for the hot and cold water system in the building. Fittings chosen according to specification page 1-7. Fittings are taken off according to method described in Chapter 4.
2. Pipe nipples were taken off all areas requiring threaded connections such as around roughing of fixtures and equipment. Size and length of nipples were not taken off since they will be priced on an average basis.
3. Pipe sleeves were taken off on all areas where pipe passed through walls, floors, or ceiling as per specification page 1-23.
4. Pipe supports were taken off according to Figure 1.22 in text.

Figure 4.70

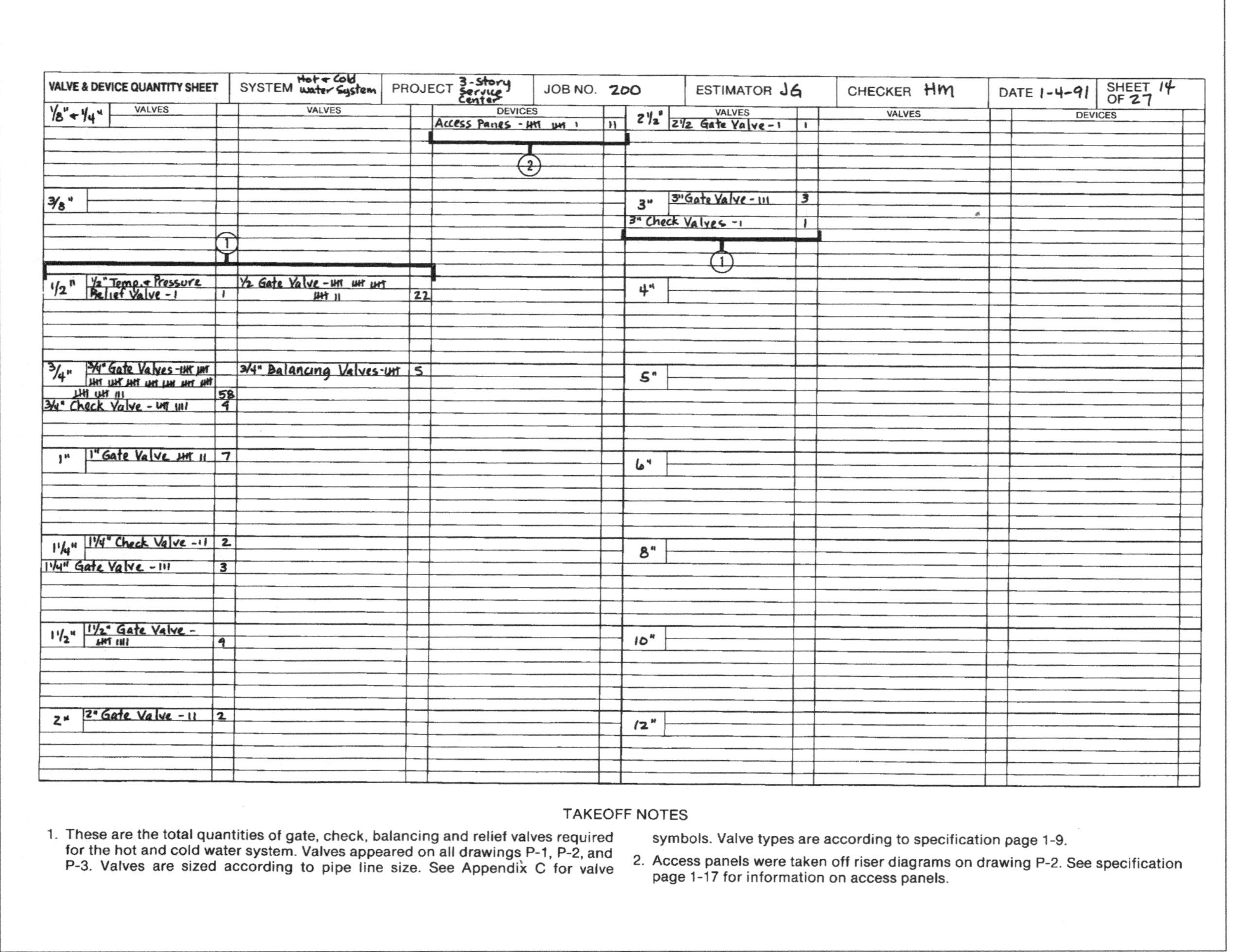

VALVE & DEVICE QUANTITY SHEET	SYSTEM Hot + Cold water System	PROJECT 3-Story Service Center	JOB NO. 200	ESTIMATOR JG	CHECKER HM	DATE 1-4-91	SHEET 14 OF 27

Size	Valves	Qty	Valves	Qty	Devices	Qty
1/8" + 1/4"					Access Panes	11
3/8"						
1/2"	1/2" Temp. + Pressure Relief Valve	1	1/2 Gate Valve	22		
3/4"	3/4" Gate Valves	58	3/4" Balancing Valves	5		
	3/4" Check Valve	9				
1"	1" Gate Valve	7				
1 1/4"	1 1/4" Check Valve	2				
	1 1/4" Gate Valve	3				
1 1/2"	1 1/2" Gate Valve	9				
2"	2" Gate Valve	2				

Size	Valves	Qty	Valves	Qty	Devices	Qty
2 1/2"	2 1/2 Gate Valve	1				
3"	3" Gate Valve	3				
	3" Check Valves	1				
4"						
5"						
6"						
8"						
10"						
12"						

TAKEOFF NOTES

1. These are the total quantities of gate, check, balancing and relief valves required for the hot and cold water system. Valves appeared on all drawings P-1, P-2, and P-3. Valves are sized according to pipe line size. See Appendix C for valve symbols. Valve types are according to specification page 1-9.
2. Access panels were taken off riser diagrams on drawing P-2. See specification page 1-17 for information on access panels.

Figure 4.71

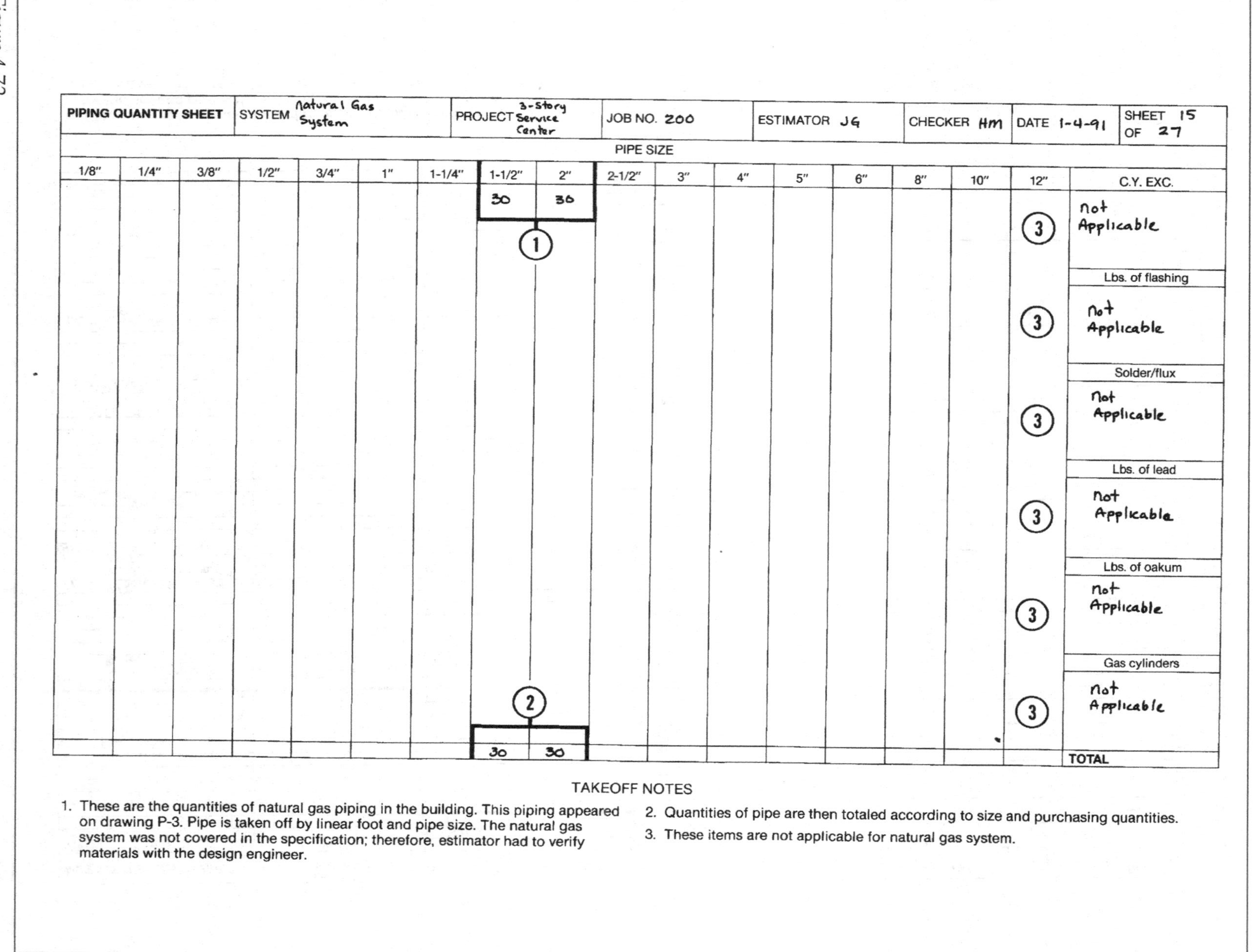

PIPING QUANTITY SHEET	SYSTEM Natural Gas System	PROJECT 3-Story Service Center	JOB NO. 200	ESTIMATOR JG	CHECKER HM	DATE 1-4-91	SHEET 15 OF 27

PIPE SIZE	1/8"	1/4"	3/8"	1/2"	3/4"	1"	1-1/4"	1-1/2"	2"	2-1/2"	3"	4"	5"	6"	8"	10"	12"	C.Y. EXC.
								30	30 (1)								(3)	Not Applicable
Lbs. of flashing																	(3)	Not Applicable
Solder/flux																	(3)	Not Applicable
Lbs. of lead																	(3)	Not Applicable
Lbs. of oakum																	(3)	Not Applicable
Gas cylinders																	(3)	Not Applicable
TOTAL								30	30 (2)									

TAKEOFF NOTES

1. These are the quantities of natural gas piping in the building. This piping appeared on drawing P-3. Pipe is taken off by linear foot and pipe size. The natural gas system was not covered in the specification; therefore, estimator had to verify materials with the design engineer.
2. Quantities of pipe are then totaled according to size and purchasing quantities.
3. These items are not applicable for natural gas system.

Figure 4.72

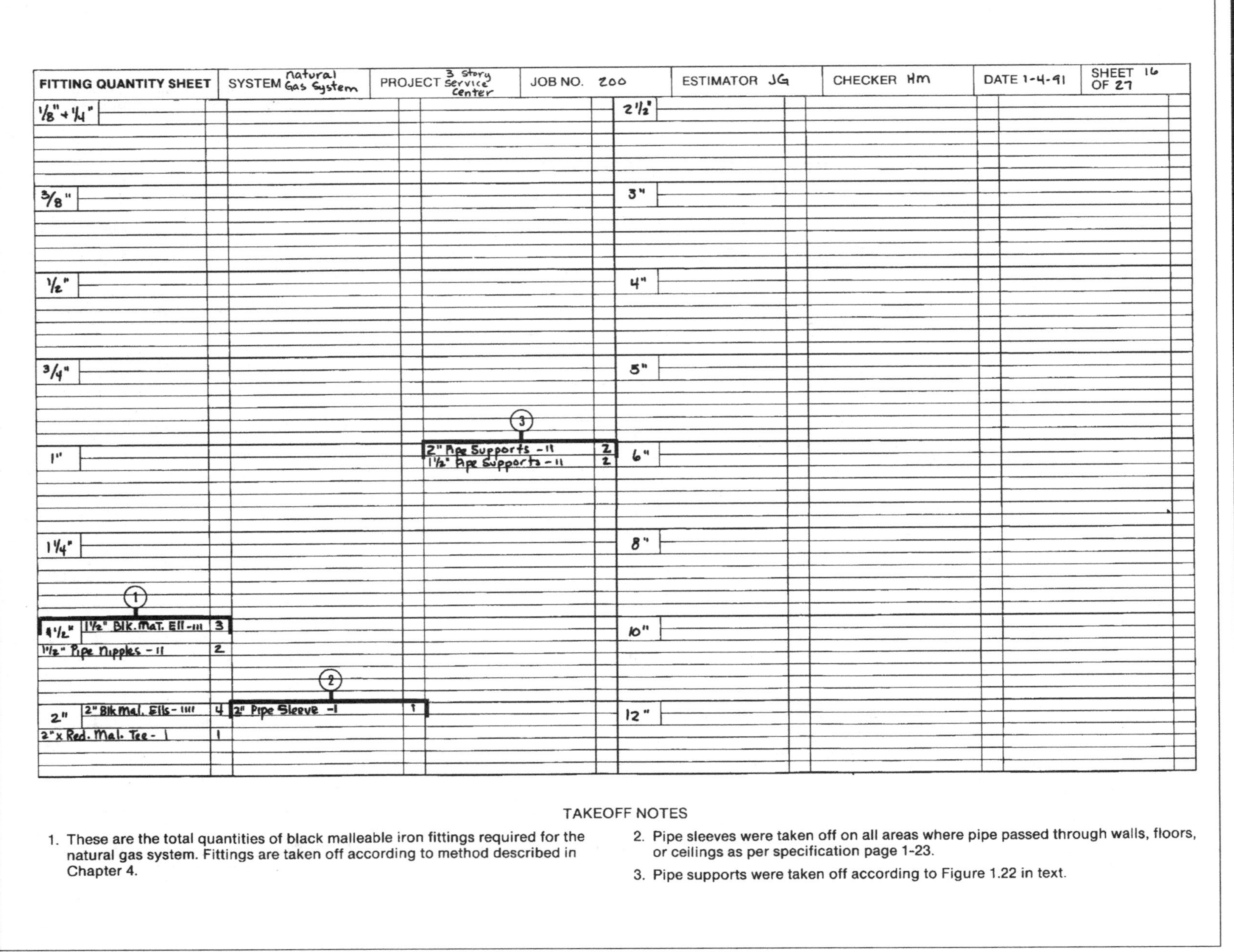

FITTING QUANTITY SHEET | SYSTEM Natural Gas System | PROJECT 3 Story Service Center | JOB NO. 200 | ESTIMATOR JG | CHECKER HM | DATE 1-4-91 | SHEET 16 OF 27

Size	Entries	Qty
1/8" + 1/4"		
3/8"		
1/2"		
3/4"		
1"		
1 1/4"		
1 1/2"	① 1 1/2" Blk. Mal. Ell - III	3
	1 1/2" Pipe Nipples - II	2
2"	2" Blk Mal. Ells - IIII	4
	2" x Red. Mal. Tee - I	1
	② 2" Pipe Sleeve - I	1
2 1/2"		
3"		
4"		
5"		
6"	③ 2" Pipe Supports - II	2
	1 1/2" Pipe Supports - II	2
8"		
10"		
12"		

TAKEOFF NOTES

1. These are the total quantities of black malleable iron fittings required for the natural gas system. Fittings are taken off according to method described in Chapter 4.
2. Pipe sleeves were taken off on all areas where pipe passed through walls, floors, or ceilings as per specification page 1-23.
3. Pipe supports were taken off according to Figure 1.22 in text.

Figure 4.73

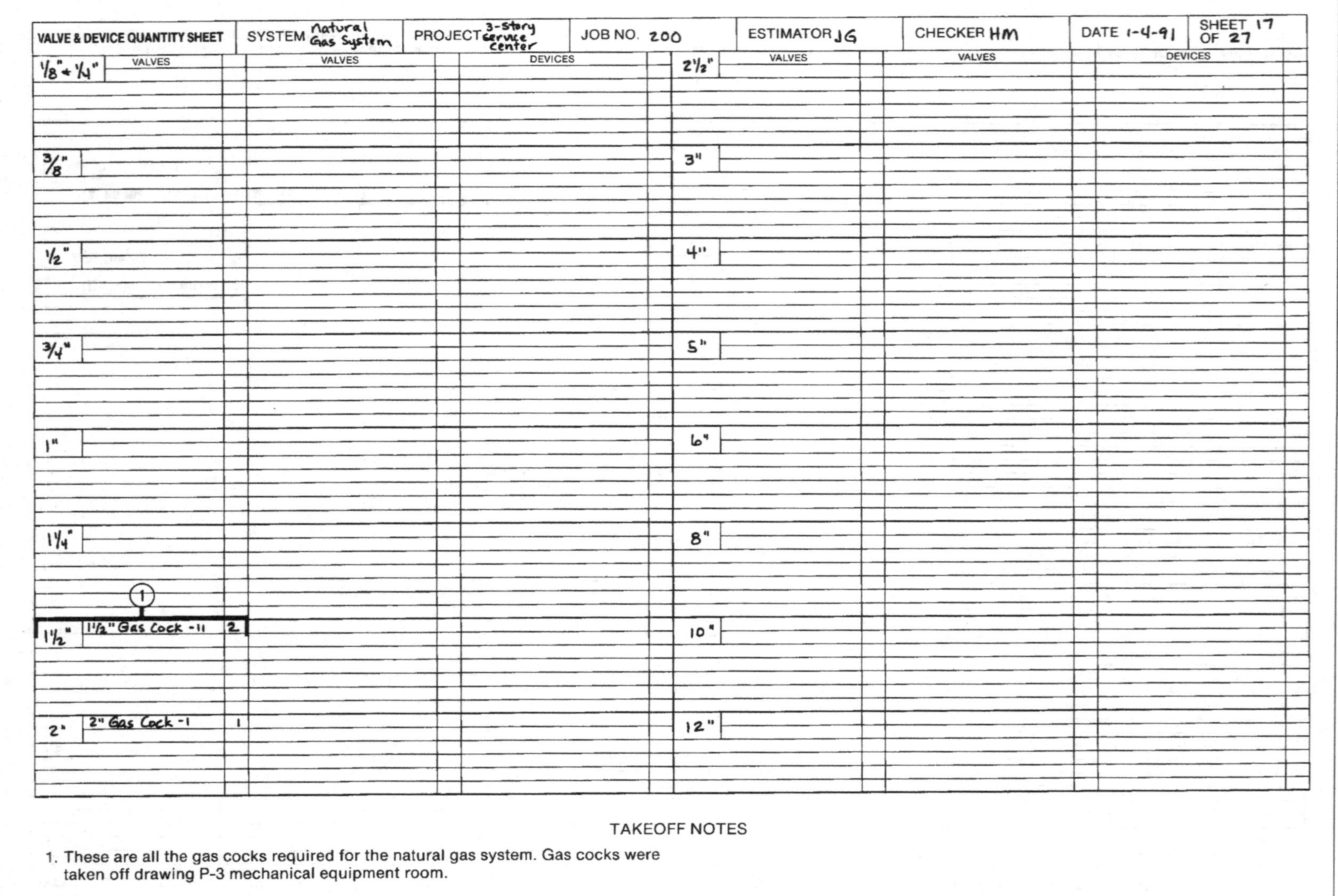

VALVE & DEVICE QUANTITY SHEET	SYSTEM Natural Gas System	PROJECT 3-Story Service Center	JOB NO. 200	ESTIMATOR JG	CHECKER HM	DATE 1-4-91	SHEET 17 OF 27

Size	VALVES		VALVES		DEVICES		Size	VALVES		VALVES		DEVICES	
1/8" + 1/4"							2½"						
3/8"							3"						
½"							4"						
¾"							5"						
1"							6"						
1¼"							8"						
1½"	1½" Gas Cock -11 (1)	2					10"						
2"	2" Gas Cock -1	1					12"						

TAKEOFF NOTES

1. These are all the gas cocks required for the natural gas system. Gas cocks were taken off drawing P-3 mechanical equipment room.

Figure 4.74

PIPING QUANTITY SHEET		SYSTEM Fire Standpipe System			PROJECT 3-Story Service Center		JOB NO. 200			ESTIMATOR JG			CHECKER HM		DATE 1-7-91		SHEET 18 OF 27
								PIPE SIZE									
1/8"	1/4"	3/8"	1/2"	3/4"	1"	1-1/4"	1-1/2"	2"	2-1/2"	3"	4"	5"	6"	8"	10"	12"	C.Y. EXC.
					10		12		5	70						(3)	Not Applicable
									12	5							
									5	27							
										25							
																	Lbs. of flashing
																(3)	Not Applicable
																	Solder/flux
																(3)	Not Applicable
																	Lbs. of lead
																(3)	Not Applicable
																	Lbs. of oakum
																(3)	Not Applicable
																	Gas cylinders
																(3)	Not Applicable
					10		12		22	127							TOTAL

TAKEOFF NOTES

1. These are the quantities of fire standpipe piping in the building. This piping appeared on drawings P-1, P-2 and P-3. Pipe is taken off by linear foot and pipe size. Pipe is not taken off fitting to fitting. Pipe material is black steel schedule 40 as per specification page 4-2.
2. Quantities of pipe are then totaled according to size and purchasing quantities.
3. These items are not applicable for fire standpipe system.

Figure 4.75

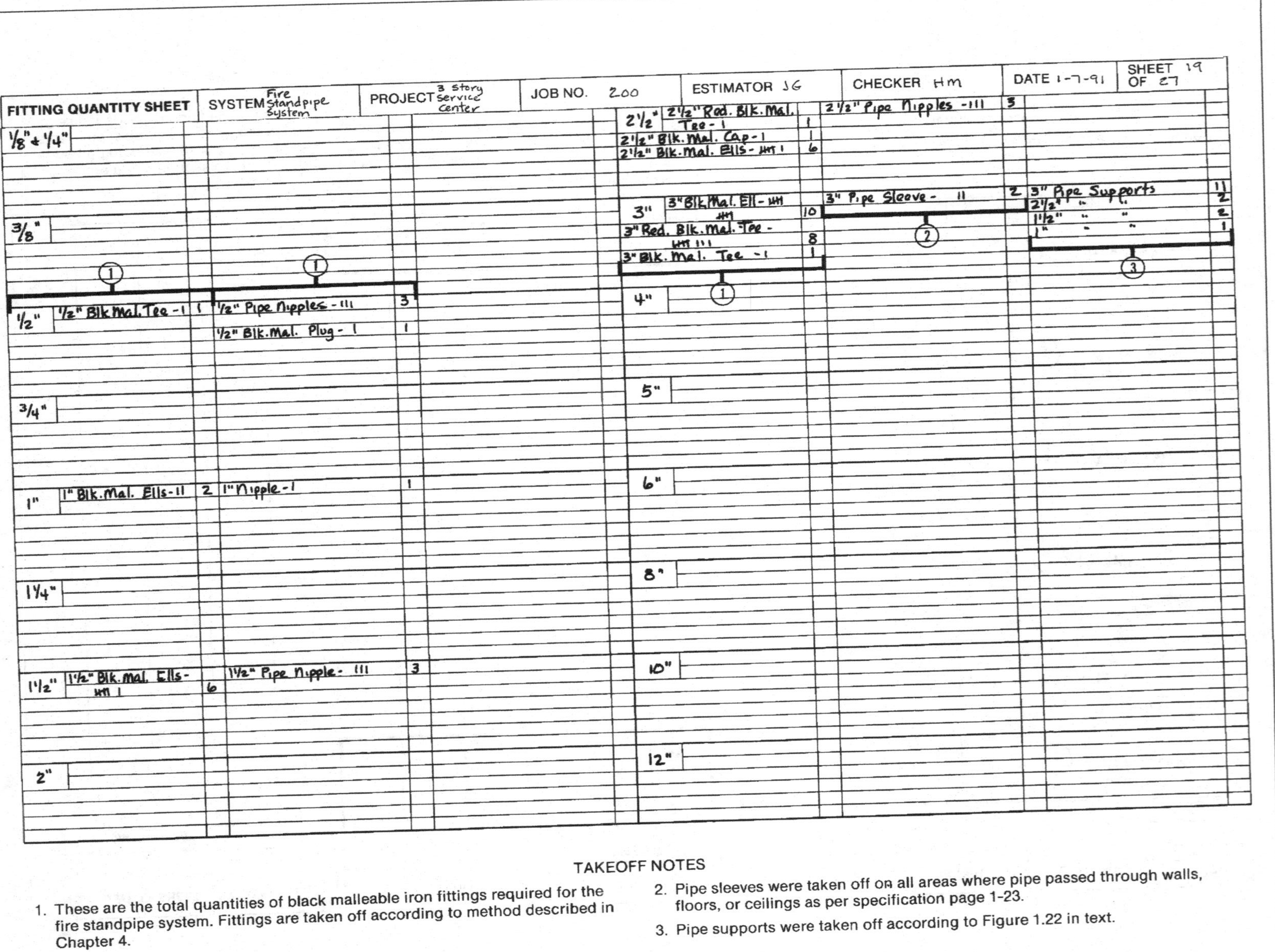

FITTING QUANTITY SHEET	SYSTEM Fire Standpipe System	PROJECT 3 Story Service Center	JOB NO. 200	ESTIMATOR JG	CHECKER HM	DATE 1-7-91	SHEET 19 OF 27

Size	Fitting	Qty	Fitting	Qty
1/8" & 1/4"				
3/8"				
1/2"	1/2" Blk. Mal. Tee - I	1	1/2" Pipe Nipples - III	3
			1/2" Blk. Mal. Plug - I	1
3/4"				
1"	1" Blk. Mal. Ells - II	2	1" Nipple - I	1
1 1/4"				
1 1/2"	1 1/2" Blk. Mal. Ells - IIII I	6	1 1/2" Pipe Nipple - III	3
2"				

Size	Fitting	Qty	Fitting	Qty	Fitting	Qty
2 1/2"	2 1/2" Red. Blk. Mal. Tee - I	1	2 1/2" Pipe Nipples - III	3		
	2 1/2" Blk. Mal. Cap - I	1				
	2 1/2" Blk. Mal. Ells - IIII I	6				
3"	3" Blk. Mal. Ell - IIII IIII	10	3" Pipe Sleeve - II	2	3" Pipe Supports	11
	3" Red. Blk. Mal. Tee - IIII III	8			2 1/2" " "	2
	3" Blk. Mal. Tee - I	1			1 1/2" " "	2
					1" " "	1
4"						
5"						
6"						
8"						
10"						
12"						

TAKEOFF NOTES

1. These are the total quantities of black malleable iron fittings required for the fire standpipe system. Fittings are taken off according to method described in Chapter 4.
2. Pipe sleeves were taken off on all areas where pipe passed through walls, floors, or ceilings as per specification page 1-23.
3. Pipe supports were taken off according to Figure 1.22 in text.

Figure 4.76

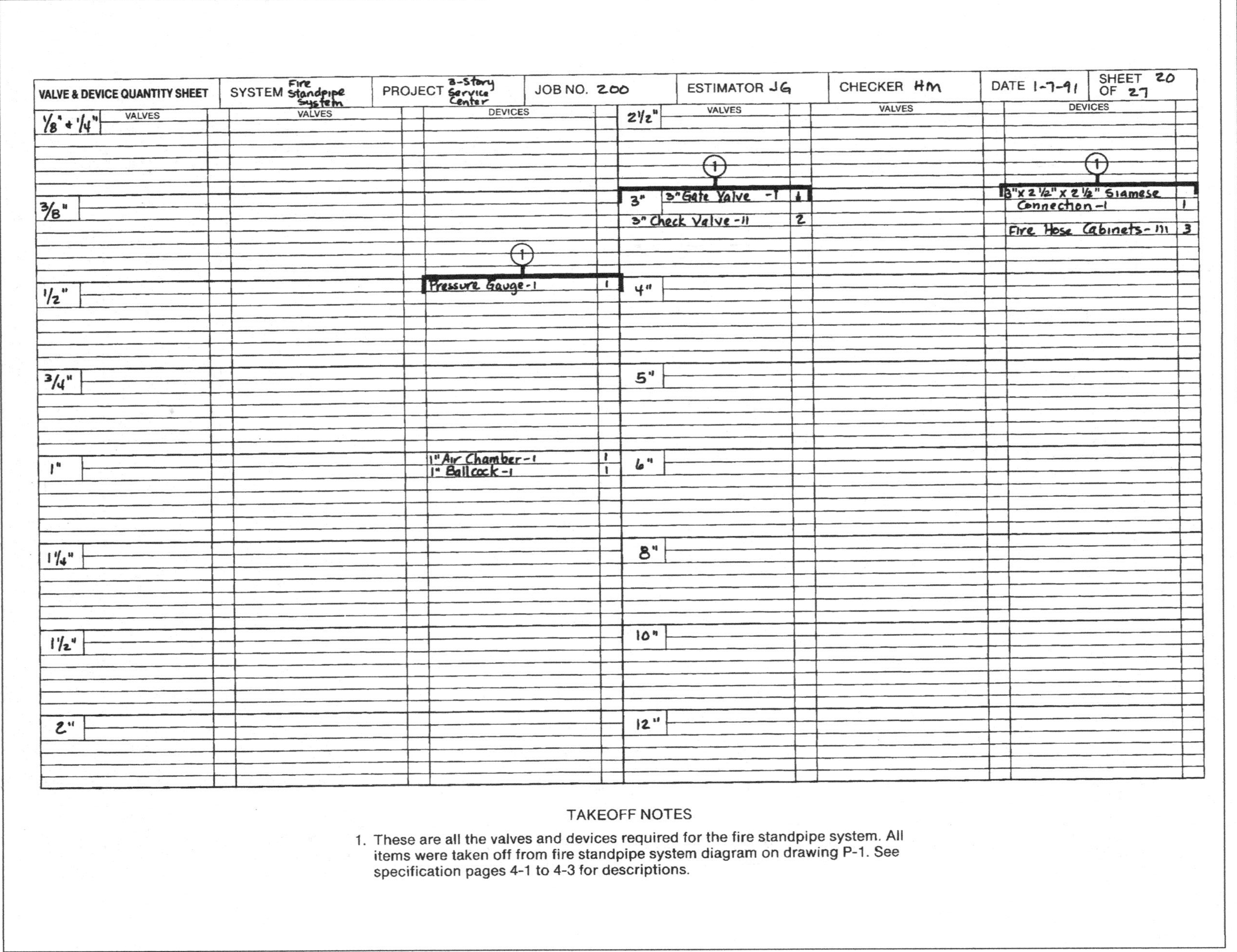
VALVE & DEVICE QUANTITY SHEET
SYSTEM Fire Standpipe System
PROJECT 3-Story Service Center
JOB NO. 200
ESTIMATOR JG
CHECKER HM
DATE 1-7-91
SHEET 20 OF 27
VALVES
DEVICES
1/8" + 1/4"
3/8"
1/2"
3/4"
1"
1 1/4"
1 1/2"
2"
2 1/2"
3"
4"
5"
6"
8"
10"
12"
Pressure Gauge-1
1
1" Air Chamber-1
1" Ballcock-1
1
1
3" Gate Valve -1
1
3" Check Valve-11
2
3" x 2 1/2" x 2 1/2" Siamese Connection-1
1
Fire Hose Cabinets-111
3
TAKEOFF NOTES
1. These are all the valves and devices required for the fire standpipe system. All items were taken off from fire standpipe system diagram on drawing P-1. See specification pages 4-1 to 4-3 for descriptions.

Figure 4.77

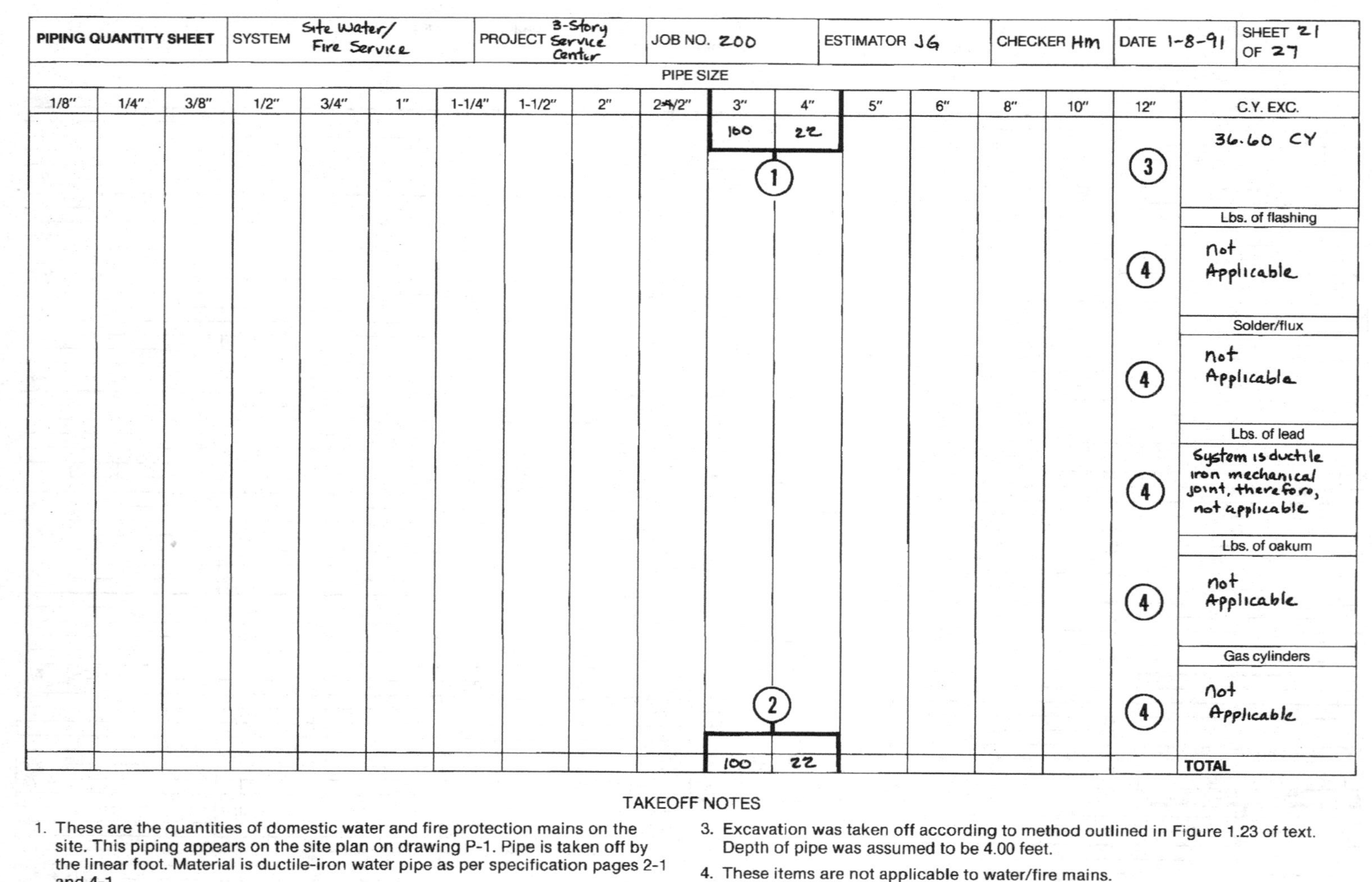

PIPING QUANTITY SHEET	SYSTEM Site Water/ Fire Service			PROJECT 3-Story Service Center			JOB NO. 200			ESTIMATOR JG			CHECKER HM		DATE 1-8-91		SHEET 21 OF 27
PIPE SIZE																	
1/8"	1/4"	3/8"	1/2"	3/4"	1"	1-1/4"	1-1/2"	2"	2-1/2"	3"	4"	5"	6"	8"	10"	12"	C.Y. EXC.
										100	22					(3)	36.60 CY
										(1)							
																	Lbs. of flashing
																(4)	not Applicable
																	Solder/flux
																(4)	not Applicable
																	Lbs. of lead
																(4)	System is ductile iron mechanical joint, therefore, not applicable
																	Lbs. of oakum
																(4)	not Applicable
																	Gas cylinders
										(2)						(4)	not Applicable
										100	22						TOTAL

TAKEOFF NOTES

1. These are the quantities of domestic water and fire protection mains on the site. This piping appears on the site plan on drawing P-1. Pipe is taken off by the linear foot. Material is ductile-iron water pipe as per specification pages 2-1 and 4-1.
2. Quantities of pipe are then totaled according to size and purchasing quantities.
3. Excavation was taken off according to method outlined in Figure 1.23 of text. Depth of pipe was assumed to be 4.00 feet.
4. These items are not applicable to water/fire mains.

Figure 4.78

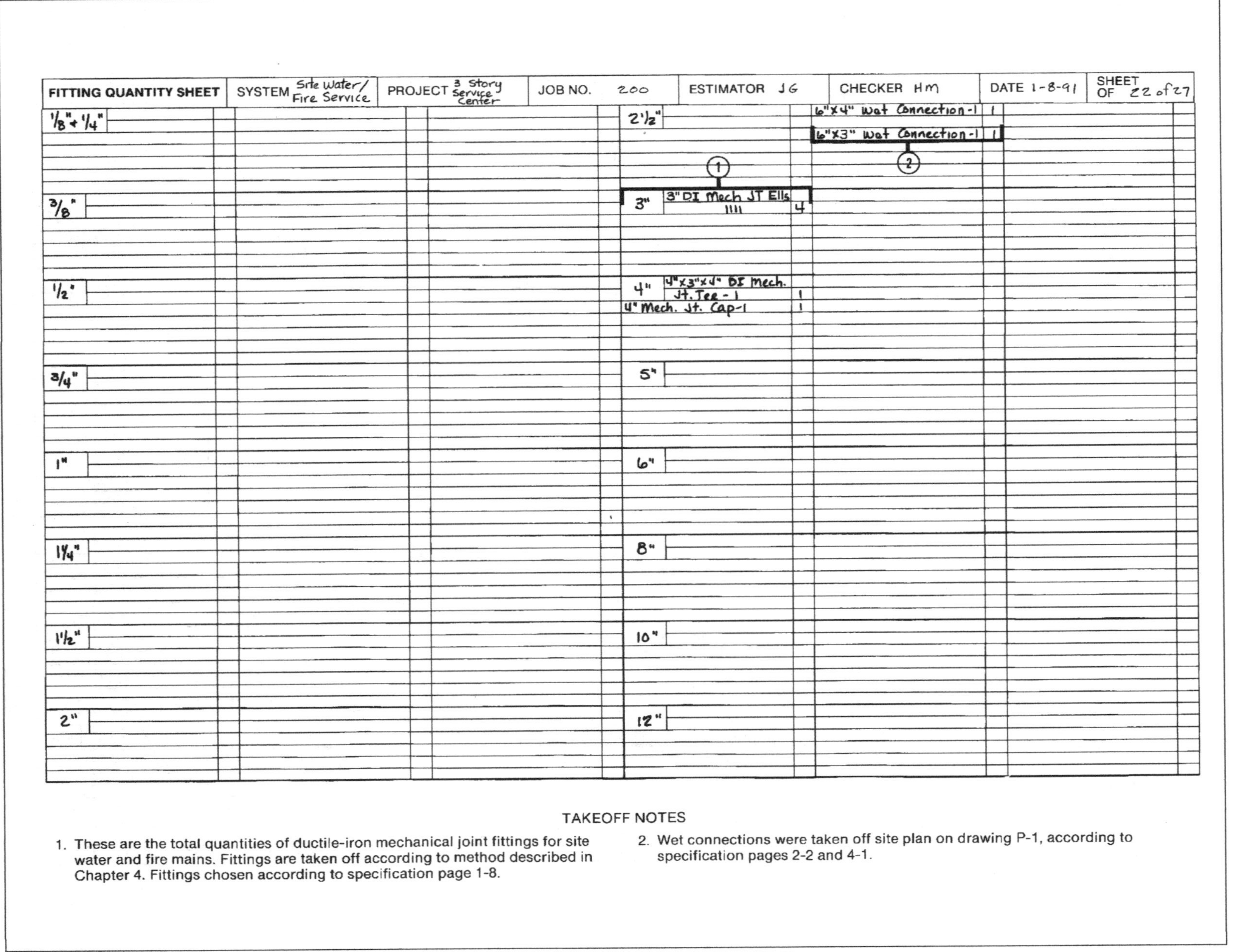

FITTING QUANTITY SHEET	SYSTEM Site Water/ Fire Service	PROJECT 3 Story Service Center	JOB NO. 200	ESTIMATOR JG	CHECKER HM	DATE 1-8-91	SHEET 22 OF 27

Size	Fittings	Quantity
1/8" + 1/4"		
3/8"		
1/2"		
3/4"		
1"		
1¼"		
1½"		
2"		
2½"	6"x4" Wet Connection-1	1
	6"x3" Wet Connection-1 (2)	1
3"	3" DI Mech JT Ells (1) ‖‖	4
4"	4"x3"x4" DI Mech. Jt. Tee-1	1
	4" Mech. Jt. Cap-1	1
5"		
6"		
8"		
10"		
12"		

TAKEOFF NOTES

1. These are the total quantities of ductile-iron mechanical joint fittings for site water and fire mains. Fittings are taken off according to method described in Chapter 4. Fittings chosen according to specification page 1-8.
2. Wet connections were taken off site plan on drawing P-1, according to specification pages 2-2 and 4-1.

Figure 4.79

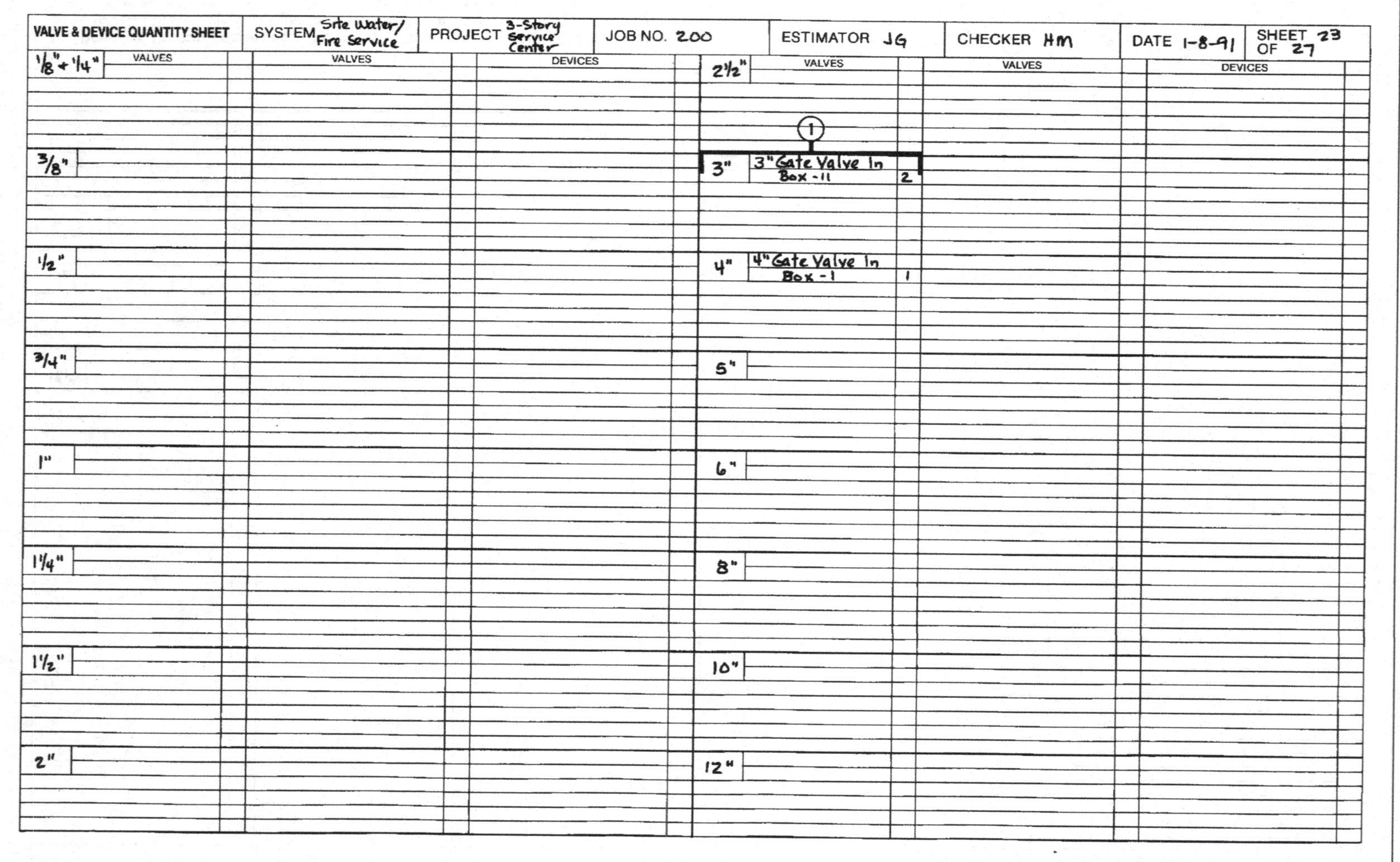

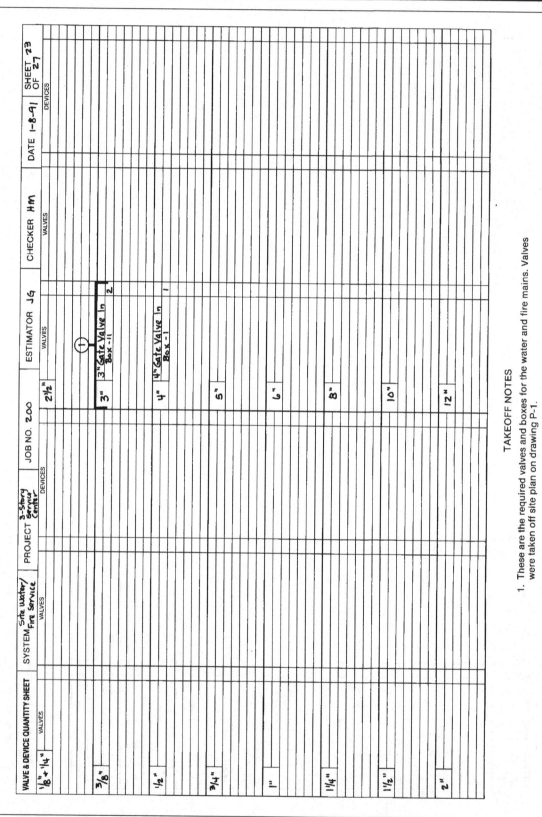

VALVE & DEVICE QUANTITY SHEET

SYSTEM Site Water/Fire Service | PROJECT 3-Story Service Center | JOB NO. 200 | ESTIMATOR JG | CHECKER HM | DATE 1-8-91 | SHEET 23 OF 27

Size	VALVES	VALVES	DEVICES	Size	VALVES	Qty	VALVES	DEVICES
1/8" + 1/4"				2 1/2"				
3/8"				3"	① 3" Gate Valve In Box-11	2		
1/2"				4"	4" Gate Valve In Box-1	1		
3/4"				5"				
1"				6"				
1 1/4"				8"				
1 1/2"				10"				
2"				12"				

TAKEOFF NOTES

1. These are the required valves and boxes for the water and fire mains. Valves were taken off site plan on drawing P-1.

Figure 4.80

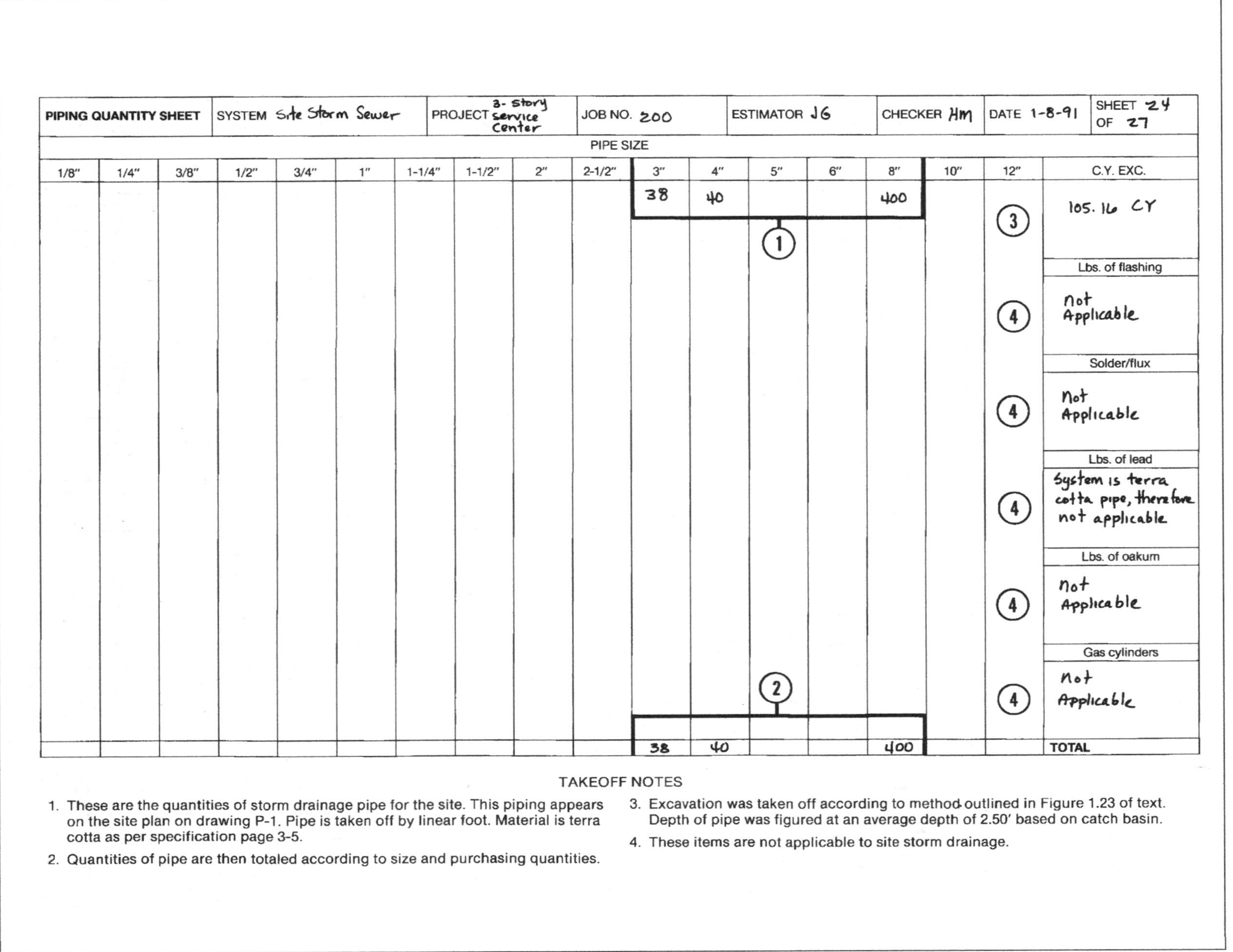

PIPING QUANTITY SHEET	SYSTEM Site Storm Sewer	PROJECT 3-story service Center	JOB NO. 200	ESTIMATOR JG	CHECKER HM	DATE 1-8-91	SHEET 24 OF 27

PIPE SIZE																	
1/8"	1/4"	3/8"	1/2"	3/4"	1"	1-1/4"	1-1/2"	2"	2-1/2"	3"	4"	5"	6"	8"	10"	12"	C.Y. EXC.
										38	40	(1)		400		(3)	105.16 CY
																	Lbs. of flashing
																(4)	not Applicable
																	Solder/flux
																(4)	Not Applicable
																	Lbs. of lead
																(4)	System is terra cotta pipe, therefore not applicable
																	Lbs. of oakum
																(4)	not Applicable
																	Gas cylinders
												(2)				(4)	Not Applicable
										38	40			400			TOTAL

TAKEOFF NOTES

1. These are the quantities of storm drainage pipe for the site. This piping appears on the site plan on drawing P-1. Pipe is taken off by linear foot. Material is terra cotta as per specification page 3-5.
2. Quantities of pipe are then totaled according to size and purchasing quantities.
3. Excavation was taken off according to method outlined in Figure 1.23 of text. Depth of pipe was figured at an average depth of 2.50′ based on catch basin.
4. These items are not applicable to site storm drainage.

Figure 4.81

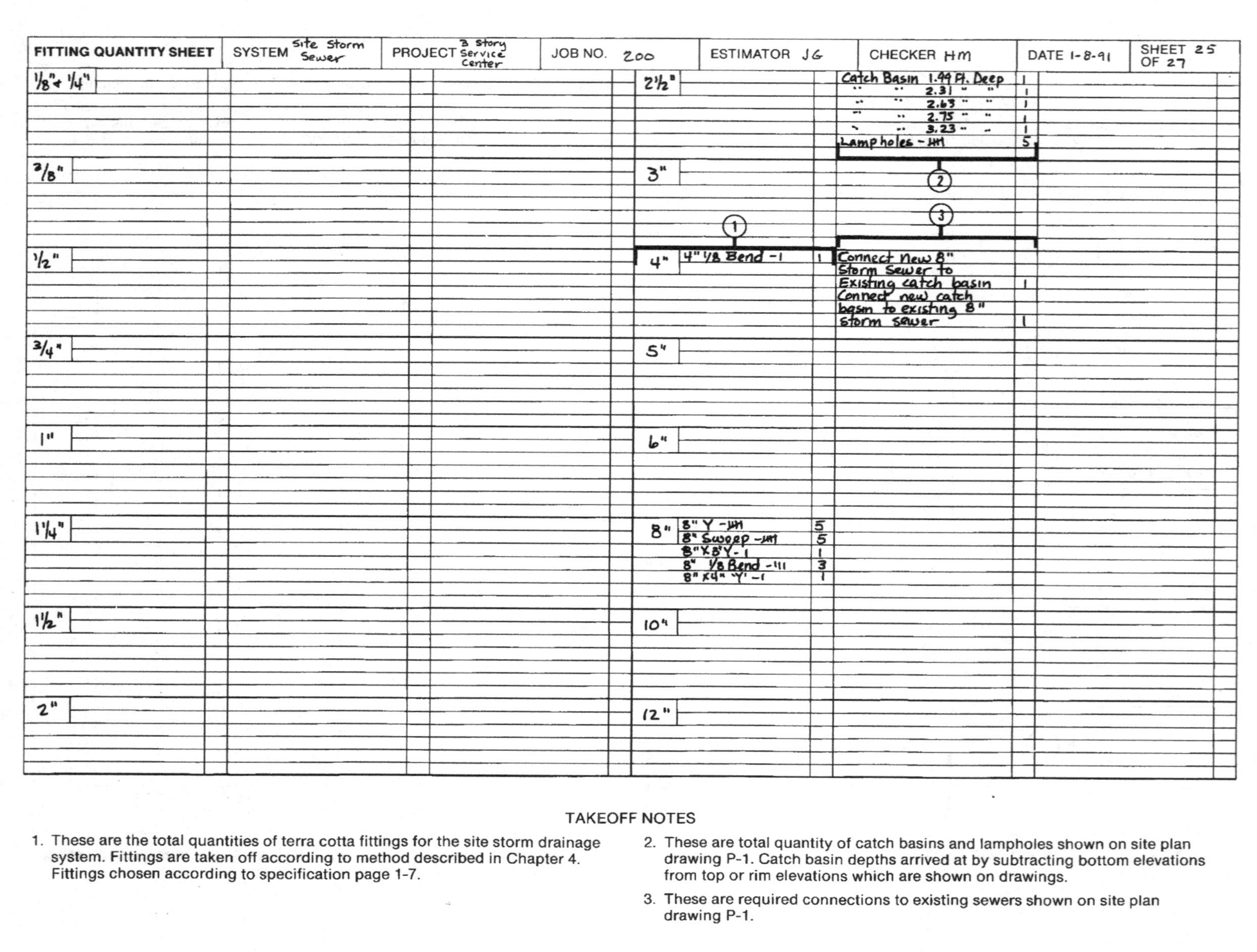

FITTING QUANTITY SHEET	SYSTEM Site Storm Sewer	PROJECT 3 Story Service Center	JOB NO. 200	ESTIMATOR JG	CHECKER HM	DATE 1-8-91	SHEET 25 OF 27

Size	Fitting	Qty	Size	Fitting	Qty	Item	Qty
1/8" + 1/4"			2½"			Catch Basin 1.49 Ft. Deep	1
						" " 2.31 " "	1
						" " 2.63 " "	1
						" " 2.75 " "	1
						" " 3.23 " "	1
						Lampholes – 𝍸	5
						②	
3/8"			3"				
				①		③	
1/2"			4"	4" 1/8 Bend – 1	1	Connect new 8" Storm Sewer to Existing catch basin	1
						Connect new catch basin to existing 8" storm sewer	1
3/4"			5"				
1"			6"				
1¼"			8"	8" Y – 𝍸	5		
				8" Sweep – 𝍸	5		
				8"×8" Y – 1	1		
				8" 1/8 Bend – |||	3		
				8"×4" "Y" – 1	1		
1½"			10"				
2"			12"				

TAKEOFF NOTES

1. These are the total quantities of terra cotta fittings for the site storm drainage system. Fittings are taken off according to method described in Chapter 4. Fittings chosen according to specification page 1-7.
2. These are total quantity of catch basins and lampholes shown on site plan drawing P-1. Catch basin depths arrived at by subtracting bottom elevations from top or rim elevations which are shown on drawings.
3. These are required connections to existing sewers shown on site plan drawing P-1.

Figure 4.82

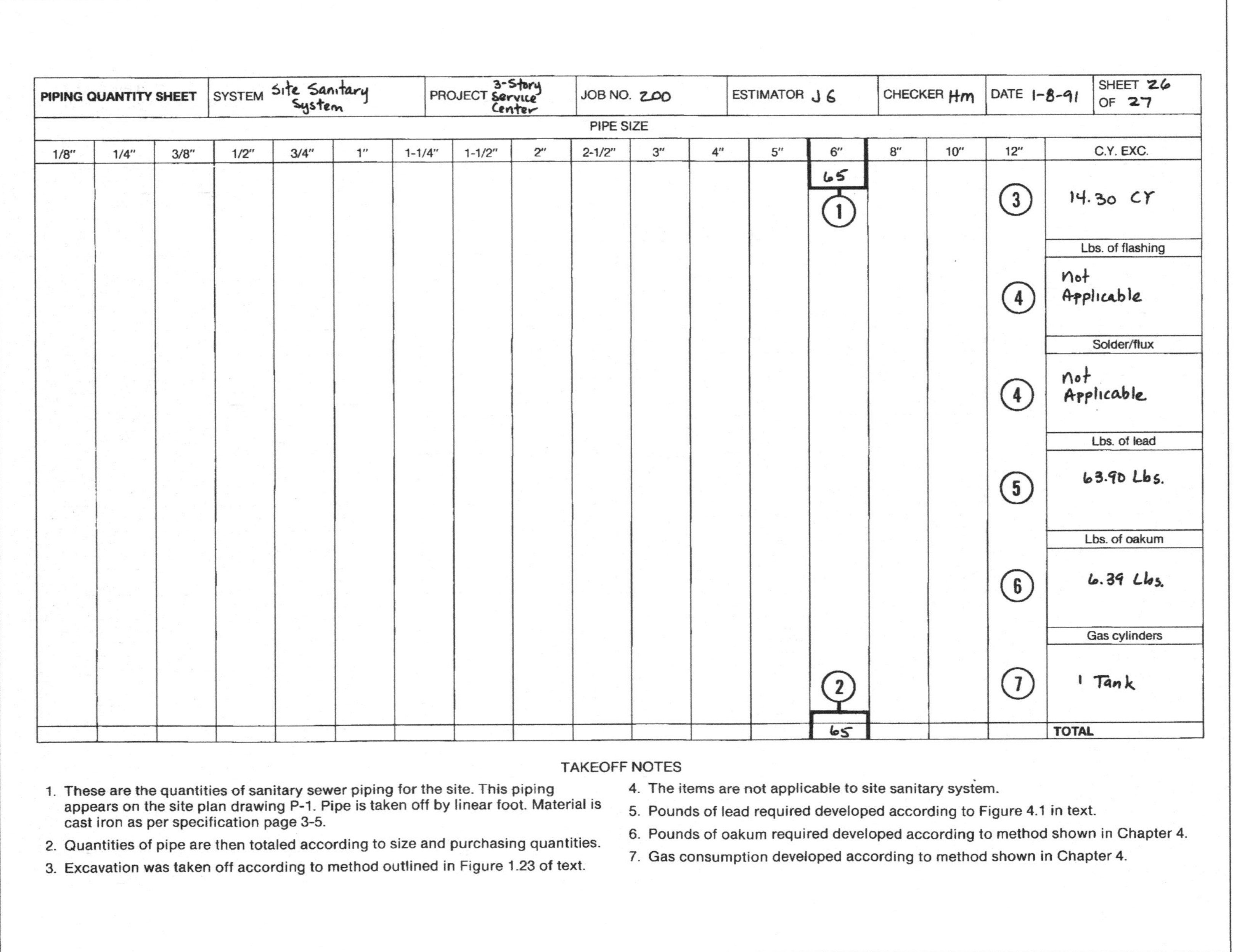

PIPING QUANTITY SHEET	SYSTEM Site Sanitary System	PROJECT 3-Story Service Center	JOB NO. 200	ESTIMATOR J G	CHECKER Hm	DATE 1-8-91	SHEET 26 OF 27

PIPE SIZE

1/8"	1/4"	3/8"	1/2"	3/4"	1"	1-1/4"	1-1/2"	2"	2-1/2"	3"	4"	5"	6"	8"	10"	12"	C.Y. EXC.
													65 (1)			(3)	14.30 CY
																	Lbs. of flashing
																(4)	Not Applicable
																	Solder/flux
																(4)	Not Applicable
																	Lbs. of lead
																(5)	63.90 Lbs.
																	Lbs. of oakum
																(6)	6.39 Lbs.
																	Gas cylinders
													(2)			(7)	1 Tank
													65				TOTAL

TAKEOFF NOTES

1. These are the quantities of sanitary sewer piping for the site. This piping appears on the site plan drawing P-1. Pipe is taken off by linear foot. Material is cast iron as per specification page 3-5.
2. Quantities of pipe are then totaled according to size and purchasing quantities.
3. Excavation was taken off according to method outlined in Figure 1.23 of text.
4. The items are not applicable to site sanitary system.
5. Pounds of lead required developed according to Figure 4.1 in text.
6. Pounds of oakum required developed according to method shown in Chapter 4.
7. Gas consumption developed according to method shown in Chapter 4.

Figure 4.83

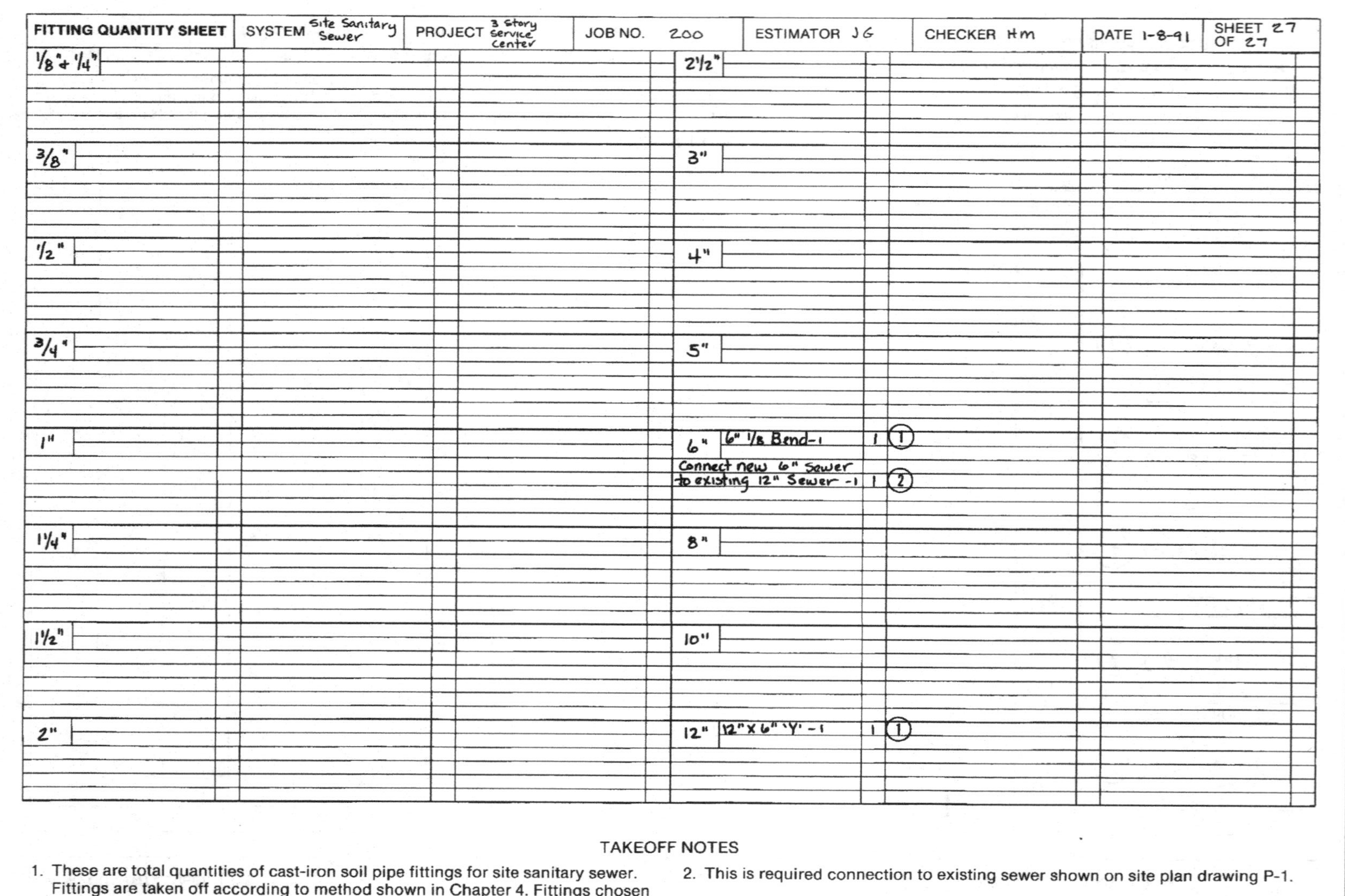

FITTING QUANTITY SHEET | SYSTEM Site Sanitary Sewer | PROJECT 3 Story Service Center | JOB NO. 200 | ESTIMATOR JG | CHECKER HM | DATE 1-8-91 | SHEET 27 OF 27

Size	Fitting	Qty	Note
1/8" & 1/4"			
3/8"			
1/2"			
3/4"			
1"			
1 1/4"			
1 1/2"			
2"			
2 1/2"			
3"			
4"			
5"			
6"	6" 1/8 Bend-1	1	(1)
	Connect new 6" Sewer to existing 12" Sewer -1	1	(2)
8"			
10"			
12"	12" x 6" 'Y' -1	1	(1)

TAKEOFF NOTES

1. These are total quantities of cast-iron soil pipe fittings for site sanitary sewer. Fittings are taken off according to method shown in Chapter 4. Fittings chosen according to specification page 1-7.
2. This is required connection to existing sewer shown on site plan drawing P-1.

Figure 4.84

Chapter 5

Writing Up and Pricing the Estimate

With the takeoff complete and all quantities rechecked, the estimator is now ready to write up and price the estimate. This chapter will cover material pricing, including discounts and labor. The second part of the chapter is a written estimate for the sample job.

Writing Up the Estimate

The estimate is written up on estimate forms similar to the one shown in Chapter 3, Figure 3.7. In addition to the basic information on the top of the form, the following columns are included:

1. Item: used for numbering entries.
2. Classification of Work: used to enter description of item to be priced.
3. Quantity: used to enter quantity and unit of measure. For example: each, linear foot, pound allowance, percentage, etc.
4. Material: used to enter material unit cost per item and then material total cost per item.
5. Labor: used to enter labor-hour production unit per item, and total labor-hours per item.
6. Labor Cost: used to enter the prevailing labor-hour labor rate per item and the total labor cost per item.
7. Total Cost: used to enter total cost (material and labor).

The order in which the estimate is written up corresponds to the systems breakdown on the project summary sheet. (See summary sheet, Chapter 3, Figure 3.8.) To begin the write-up, the estimator transfers all quantities on the takeoff forms to the estimate forms, and draws a red line through the totals on the quantity takeoff as a check-off system to be sure that all items are accounted for.

The estimator must refer to the job specifications while writing up the estimate. This is necessary to account for items that are not specified on the takeoff forms, but must be priced. Items that fall into this category are:

- Plumbing fixture trim
- Painting (if required)
- Tags/charts/identification
- Pipe insulation

- Rigging
- Tests and adjustments

A good habit to acquire is numbering all estimate forms as shown on the sample estimate. This also holds true for all takeoff sheets. In this way, one can put the sheets in order quickly, and be sure that all pages are there.

The Art of Pricing

Although pricing the estimate usually takes less than half the time required for the takeoff, it is more demanding. A takeoff of the same job, performed by two different careful, competent estimators using the same plans, will probably differ only slightly. This is not the case with pricing. Many factors enter into the picture when an estimator prices a job. These are:

- Material costs and availability
- Size and location of job to be bid
- Volume of construction activity in area of job to be bid (if area is inactive, bidding may be very competitive)
- Other jobs currently under contract (workload)
- Labor availability in area of job to be bid
- Union or open shop area
- Labor contract expiration dates

The estimator pricing the job should have all of the above information before assigning any unit prices to the estimate. Pricing a job is more than just taking a material price and labor rate, entering them on the estimate form, and arriving at a bid price. For this reason, job pricing should be reserved for the senior estimator in a contracting office.

Material Pricing and Discounts

As previously mentioned, material costs are in a constant state of change. Therefore, it would be of little value to place any material cost data in this book other than that used for our sample estimate. However, we can discuss how materials are priced and how discounts are taken.

Materials should, of course, always be priced from the latest trade price sheets and quotes issued by a manufacturer. The estimator, however, must realize that the duration of the project might be two or three years. If awarded the job, purchase orders for materials should be issued immediately to ensure that the goods will be received at the current (bid) price. In addition to price sheets, published cost data, such as the current issue of *Means Plumbing Cost Data*, offers the estimator another material price source.

Plumbing contractors are often afforded what is called a contractor's discount on materials. Discounts are percentages deducted from list prices, and the amounts of discount vary according to the type and amount of material. For example, fixtures are discounted at a different rate than copper tubing. If a contractor purchases 25 fixtures, his or her rate of discount will be less than a contractor purchasing 250 fixtures.

To avoid confusion or misinterpretation, all prices in our sample estimate are net prices, showing no material discounts. However, it should be noted here that the estimator should apply all discounts afforded at the bottom of the material list on the estimate form. If the estimator has a list price of $25,000 for copper tubing and is granted a 25% discount, the bid or net price should be $18,750.

Composition of the Labor-Hour

The Labor-Hour Production Table in Appendix B is based on certain assumptions. On medium to large projects, work is performed by teams. The number of teams required depends on the size of the project and the stage of work being performed at a particular time. On a plumbing job of average size, a typical crew might be as follows:

1 foreman
4 journeymen plumbers
1 apprentice

On this crew, two workers usually work at a specific task, such as installing a fixture or a length of copper tubing. Depending on the specific task, the two-worker mix may differ. The number of labor-hours shown on the table is a measure of an average team's productivity, including the time for supervision by the foreman. Conditions such as weather, hazards, or job conditions, all of which affect production, have not been considered. The estimator should evaluate the particular project and adjust the production rates up or down accordingly.

Pricing the Sample Takeoff

Note the first item on page 1 of the estimate sheet (Figure 5.1):

"Water closet wall-hung, flush-valve, seat."

We see from the first page of Appendix B that 2.0 labor-hours are required for its installation. This is placed in the labor column. Knowing that 17 units are required, a quick calculation will show that 17 × 2.0 = 34.0 labor-hours total are required to install the wall-hung water closets.

To determine the labor cost, the total labor-hours must be multiplied by the labor rate in dollars per hour. Appendix C contains a chart of union and open shop wage rates for journeymen plumbers in U.S. cities. The base rate (column A) in Appendix C is then multiplied by the city location factor to arrive at the hourly rate for the desired city. Assume for our sample project the hourly rate is computed to be $31.35. Thus,

34.0 labor-hours × 31.35 per hour = $1,065.90,

which is the total labor cost shown for the water closet.

To this is added the material cost. In this case, the estimator found the current price per unit to be $239.00. The total material price is calculated to be:

17 × 239 = $4,063.00

We are now ready to go through a step-by-step calculation of material and labor costs for all the items transferred to the estimating forms from the takeoff forms in Figures 5.1 through 5.15.

ABC PLUMBING COMPANY, INC.
CONTRACTORS
ANYWHERE, USA

SHEET NO. 1 of 15

PROJECT Three Story Service Center
ESTIMATE NO.

LOCATION Anywhere, USA
CLASSIFICATION Plumbing
DATE

ARCHITECT Jarrett, Assoc.
ESTIMATOR JG
PRICED BY SG
CHECKED BY MG

DESCRIPTION	QUANTITY	UNIT	MATERIAL UNIT COST	MATERIAL TOTAL COST	LABOR UNIT LH	LABOR TOTAL LH	LABOR UNIT RATE	LABOR TOTAL COST	TOTAL COST MAT./LABOR
1. Plumbing Fixtures + Trim									
Water Closet - Wall Hung	17	Ea.	239	4063	2.0	34.0	31.35	1065.90	5128.90
Flush Valve w/stop + V.B.									
Toilet Seat									
Water Closet - Floor Outlet	3	Ea.	207	621	2.25	6.75		211.61	832.61
Flush Valve									
Toilet Seat									
Closet Flange w/Bolt, nuts, washers									
Wax Gasket									
Lavatories 20"x 18" Wall Hung	26	Ea.	210	5460	1.75	45.5		1426.43	6886.43
Faucet w/ Pop-up Waste									
1½"x 1¼" 'P' Trap w/Tail Piece + Esc.									
Pr. ⅜" Supplies w/Angle Stops + Esc.									
Urinals - Wall Hung	3	Ea.	295	885	2.0	6.0		188.10	1073.10
Flush Valve w/Stop + V. B.									
Electric Water Coolers	3	Ea.	575	1725	3.0	9.0		282.15	2007.15
1½" x 1¼" 'P' Trap w/Tail Piece + Esc.									
⅜" Supply w/Angle Stop + Esc.									
Cabinet Sinks 19"x 18"	22	Ea.	300	6600	3.0	66.0		2069.10	8669.10
Faucet w/Gooseneck Discharge									
nozzle + Wrist Action Handles									
8" Centerset + C.I. Drain Plug									
1¼" Tail Piece									
2" 'P' Trap w/Esc.									
Pr. ⅜" Supplies w/Angle Stop + Esc.									
Plumbing Fixtures + Trim									
Total Direct Cost				19354				5243.29	24597.29

Figure 5.1 *The costs shown in this sample estimate are used as examples only. The costs should not be used for estimating purposes.*

ABC PLUMBING COMPANY, INC.
CONTRACTORS
ANYWHERE, USA

SHEET NO. 2 of 15

PROJECT — ESTIMATE NO.

LOCATION — CLASSIFICATION — DATE

ARCHITECT — ESTIMATOR — PRICED BY — CHECKED BY

DESCRIPTION	QUANTITY	UNIT	MATERIAL UNIT COST	MATERIAL TOTAL COST	LABOR UNIT LH	LABOR TOTAL LH	LABOR UNIT RATE	LABOR TOTAL COST	TOTAL COST MAT./LABOR
2. Sanitary Waste/Vent System Below Ground									
6" XHCI Pipe B+S	95	LF	8.93	848.35	.21	19.95	31.35	625.43	1474.78
4" " " "	305	↓	5.76	1756.80	.14	42.70	↓	1334.38	3091.18
2" " " "	150	↓	2.69	403.50	.08	12.0		376.20	779.70
6" XHCI Y	1	Ea.	36.62	36.62	2.11	2.11		66.15	102.77
6x4 Y	5		23.32	116.60	2.11	10.55		330.74	447.34
6x2 Y	1		16.95	16.95	2.11	2.11		66.15	83.10
4" Y	10		20.88	208.80	1.38	13.80		432.63	641.43
4x2 Y	18		13.52	243.36	1.38	24.84		778.73	1022.09
2" Y	4		5.77	23.08	.70	2.80		87.78	110.86
4" Long Turn TY	2		20.88	41.76	1.38	2.76		86.53	128.29
2" Tee	1		7.00	7.00	.70	.70		21.95	28.95
6" 1/8 Bend	1		12.98	12.98	1.06	1.06		33.23	46.21
4" 1/8 Bend	11		8.56	94.16	.70	7.70		241.40	335.56
2" 1/8 Bend	8		2.87	22.96	.36	2.88		90.29	113.25
4" Sweep	8		15.85	126.80	.70	5.60		175.59	302.36
2" Sweep	24		5.77	138.48	.36	8.64		270.86	409.34
2" 1/4 Bend	3		3.59	10.77	.36	1.08		33.86	44.63
4" 'P' Trap w/2" Vent	2		31.27	62.54	.70	1.40		43.89	106.43
4" 'P' Trap	5		17.98	89.90	.70	3.50		109.73	199.63
2" 'P' Trap	1		6.70	6.70	.36	.36		11.29	17.99
6x4 Reducer	1		8.72	8.72	1.06	1.06		33.23	41.95
4x2 Reducer	3	↓	5.07	15.21	.70	2.10		65.84	81.05
Caulking Lead	510	Lbs.	1.04	530.40	–	–		–	530.40
Oakum	51	"	.95	48.45	–	–		–	48.45
Gas	3	CYL	12.00	36.00	–	–		–	36.00
4" Pipe Sleeves	14	Ea.	25.00	350.00	.50	7.0		219.45	569.45
2" " "	22	Ea.	12.00	264.00	.32	7.04		220.70	484.70
4" Floor Drains w/Trap Seal, Valve + Vac. Brk.	7	Ea.	100.00	700.00	1.25	8.75		274.31	974.31
2" " " " " " " "	1		89.00	89.00	1.25	1.25		39.19	128.19
4" Cleanout Deck Plate	4		55.00	220.00	.75	3.0		94.05	314.05
2" " " "	2	↓	49.00	98.00	.75	1.50	↓	47.03	145.03
Sanitary Waste/Vent - Below Ground Total Direct Cost				6627.89				6210.58	12838.47

Figure 5.2 *The costs shown in this sample estimate are used as examples only. The costs should not be used for estimating purposes.*

ABC PLUMBING COMPANY, INC.
CONTRACTORS
ANYWHERE, USA

SHEET NO. 3 of 15

PROJECT | ESTIMATE NO.

LOCATION | CLASSIFICATION | DATE

ARCHITECT | ESTIMATOR | PRICED BY | CHECKED BY

DESCRIPTION		QUANTITY	UNIT	MATERIAL UNIT COST	MATERIAL TOTAL COST	LABOR UNIT LH	LABOR TOTAL LH	LABOR UNIT RATE	LABOR TOTAL COST	TOTAL COST MAT./LABOR
3. Sanitary Waste/Vent System Above Ground										
4" SWCI Hubless Pipe		315	LF	2.97	935.55	.14	44.10	31.35	1382.54	2318.09
3" " " "		250	"	2.29	572.50	.10	25.0		783.75	1356.25
2" " " "		310	"	1.66	514.60	.07	21.70		680.30	1194.90
1½" " " "		725	"	1.58	1145.50	.04	29.0		909.15	2054.65
4" SWCI Hubless Y		5	Ea.	6.98	34.90	1.0	5.0		156.75	191.65
4x3	Y	2		5.83	11.66	1.0	2.0		62.70	74.36
4x2	Y	2		4.35	8.10	1.0	2.0		62.70	70.80
3x2	Y	6		3.39	20.34	.88	5.28		165.53	185.87
3x 1½	Y	6		3.39	20.34	.88	5.28		165.53	185.87
4"	TY	2		6.19	12.38	1.0	2.0		62.70	75.08
3"x2"	TY	1		3.67	3.67	.88	.88		27.59	31.26
3 x 1½	TY	1		3.67	3.67	.88	.88		27.59	31.26
1½"	TY	25		2.95	73.75	.62	15.50		485.93	559.68
2x 1½	Long Turn TY	25		3.43	85.75	.75	18.75		587.81	673.56
1½"	" " TY	6		3.15	18.90	.62	3.72		116.62	135.52
4"	Tee	1		6.19	6.19	1.0	1.0		31.35	37.54
4x3	Tee	2		5.62	11.24	1.0	2.0		62.70	73.94
4x2	Tee	6		4.71	28.50	1.0	6.0		188.10	216.60
2"	Tee	4		3.31	13.24	.75	3.0		94.05	107.29
2x Red.	Tee	10		3.31	33.10	.75	7.50		235.13	268.25
1½"	Tee	15		2.95	44.25	.62	9.30		291.56	335.81
4"	P Trap	2		13.45	26.90	.66	1.32		41.38	68.28
4"	Sweep	6		9.18	55.08	.66	3.96		124.15	179.23
2"	Sweep	7		4.90	34.30	.49	3.43		107.53	141.83
1½"	Sweep	12		4.83	57.96	.41	4.92		154.24	212.20
4"	⅛ Bend	4		3.35	13.40	.66	2.64		82.76	96.16
3"	⅛ Bend	2		2.55	5.10	.58	1.16		36.37	41.47
2"	⅛ Bend	13		1.79	23.27	.49	6.37		199.64	222.91
1½"	⅛ Bend	20		1.75	35.00	.41	8.20		257.07	292.07
4"	¼ Bend	1		4.67	4.67	.66	.66		20.69	25.36
3"	¼ Bend	3		3.11	9.33	.58	1.74		54.55	63.88
2"	¼ Bend	11		2.35	25.85	.49	5.39		168.98	194.83
1½"	¼ Bend	55		2.23	122.65	.41	22.55		706.94	829.59
3x2	Cross TY	6		7.06	42.36	1.17	7.02		220.08	262.44
2"	Cross TY	1		4.99	4.99	1.0	1.0		31.35	36.34
4x3	Reducer	1		2.55	2.55	.66	.66		20.69	23.24
3x2	Reducer	2		1.67	3.34	.58	1.16		36.37	39.71
2x 1½	Reducer	22		1.63	35.86	.49	3.92		122.89	158.75
4"	Closet Bends	3		13.85	41.55	1.0	3.0		94.05	135.60
Continued					4142.29				9059.81	13202.10

Figure 5.3 *The costs shown in this sample estimate are used as examples only. The costs should not be used for estimating purposes.*

ABC PLUMBING COMPANY, INC.
CONTRACTORS
ANYWHERE, USA

SHEET NO. 4 of 15

PROJECT ESTIMATE NO.

LOCATION CLASSIFICATION DATE

ARCHITECT ESTIMATOR PRICED BY CHECKED BY

DESCRIPTION	QUANTITY	UNIT	MATERIAL UNIT COST	MATERIAL TOTAL COST	LABOR UNIT LH	LABOR TOTAL LH	LABOR UNIT RATE	LABOR TOTAL COST	TOTAL COST MAT./LABOR
3. Sanitary Waste/Vent - Aboveground - Continued	Brought Fwd.			4142.29				9059.81	13202.10
4" SWCI Hubless Test Tee Plug	1	Ea.	7.91	7.91	.66	.66	31.35	20.69	28.55
3"	2		4.75	9.50	.58	1.16		36.37	45.87
4" Co w/ Plug	1		3.75	3.75	.33	.33		10.35	14.10
2" C.O. w/ Plug	1		2.56	2.56	.25	.25		7.84	10.40
3x4 Increaser	2		4.27	8.54	.58	1.16		36.37	44.91
4" Roof Coupling	8		20.00	160.00	.66	5.28		165.53	325.53
3" Roof Coupling	2		17.00	34.00	.58	1.16		36.37	70.37
4" Hubless Coupling	155	Ea.	2.58	399.90	–	–		–	399.90
3"	115		2.23	256.45	–	–		–	256.45
2"	234		1.84	430.56	–	–		–	430.56
1½"	521		1.84	958.64	–	–		–	958.64
4" Double Water Closet Carriers	6	Ea.	290.	1740.00	2.0	12.0		376.20	2116.20
4" Single " "	5		156.	780.00	2.0	10.0		313.50	1093.50
Lavatory Carriers	26		97.	2522.00	2.0	52.0		1630.20	4152.20
Urinal Carriers	3		103.	309.00	1.5	4.5		141.08	450.08
4" Pipe Sleeves	8	Ea.	25.	200.00	.50	4.0		125.40	325.40
3" " "	17		18.	306.00	.45	7.65		239.83	545.83
1½" " "	1		11.	11.00	.28	.28		8.78	19.78
4" Hanger Assemblies	62	Ea.	7.11	440.82	.36	22.32		699.73	1140.55
3" " "	50		6.43	321.50	.36	18.0		564.30	885.80
2" " "	26		5.70	148.20	.36	9.36		293.44	441.64
1½" " "	60		5.66	339.60	.36	21.60		677.16	1016.76
Sump Pump - Simplex	1	Ea.	175.	175.00	6.0	6.0		188.10	363.10
9 GPM, 1/3 HP, 24' Head, 1750 RPM, 120V, 1Ø									
1½" Gate Valve	1	Ea.	24.	24.00	.66	.66		165.53	189.53
1½" Check Valve	1	Ea.	28.	28.00	.66	.66		165.53	193.53
4" Floor Drains w/Trap Seal Valve + V.B.	2	Ea.	100.	200.00	1.25	2.50		78.38	278.38
Sheetlead Flashing (12)	192	Lbs.	1.10	211.20	1.0	12.0		376.20	587.40
Sanitary Waste/Vent Above Ground Total Direct Cost				14,178.42				15,416.64	29,587.06

Figure 5.4 *The costs shown in this sample estimate are used as examples only. The costs should not be used for estimating purposes.*

ABC PLUMBING COMPANY, INC.
CONTRACTORS
ANYWHERE, USA

SHEET NO. 5 of 15

PROJECT ESTIMATE NO.

LOCATION CLASSIFICATION DATE

ARCHITECT ESTIMATOR PRICED BY CHECKED BY

DESCRIPTION	QUANTITY	UNIT	MATERIAL UNIT COST	MATERIAL TOTAL COST	LABOR UNIT LH	LABOR TOTAL LH	LABOR UNIT RATE	LABOR TOTAL COST	TOTAL COST MAT./LABOR
4. Storm System- Below Ground									
8" XHCI Pipe B & S	75	LF	13.44	1008.	.32	24.0	31.35	752.40	1760.40
6" " " "	70	↓	8.93	625.10	.21	14.70		460.85	1085.95
4" " " "	115	↓	5.76	662.40	.14	16.10		504.74	1167.14
8"x4" XHCI Y	4	Ea.	43.48	173.92	3.13	12.52		392.50	566.42
6 x 4 Y	2		23.32	46.64	2.11	4.22		132.30	178.94
4" Y	4		20.88	83.52	1.38	5.52		173.05	256.57
4" Sweep	13		15.85	206.05	.70	9.10		285.29	491.34
4" 1/8 Bend	7		8.56	59.92	.70	4.90		153.62	213.54
8 x 6 Reducer	1		19.73	19.73	1.57	1.57		49.22	68.95
6 x 4 Reducer	1	↓	8.72	8.72	1.06	1.06		33.23	41.95
4" Cleanout Deck Plate	4	Ea.	55.	220.	.75	3.0		94.05	314.05
8" Pipe Sleeves	1	Ea.	40.	40.	.63	.63		19.75	59.75
4" Pipe Sleeves	11	Ea.	25.	275.	.40	4.40		137.94	412.94
Caulking Lead	315	Lbs.	1.04	327.60	-	-		-	327.60
Oakum	32	Lbs.	.95	30.40	-	-		-	30.40
Gas	2	Cyl.	12.	24.	-	-	↓	-	24.00
Storm System- Below Ground Total Direct Cost				3811.00				3188.94	6999.94

Figure 5.5 *The costs shown in this sample estimate are used as examples only. The costs should not be used for estimating purposes.*

ABC PLUMBING COMPANY, INC.
CONTRACTORS
ANYWHERE, USA

SHEET NO. 6 of 15

PROJECT　　　　ESTIMATE NO.

LOCATION　　　　CLASSIFICATION　　　　DATE

ARCHITECT　　　　ESTIMATOR　　　　PRICED BY　　　　CHECKED BY

DESCRIPTION	QUANTITY	UNIT	MATERIAL UNIT COST	MATERIAL TOTAL COST	LABOR UNIT LH	LABOR TOTAL LH	LABOR UNIT RATE	LABOR TOTAL COST	TOTAL COST MAT./LABOR
5. Storm System- Above Ground									
4" SWCI Hubless Pipe	490	LF	2.97	1455.30	.14	6.86	31.35	215.06	1670.36
3" " "	50	LF	2.29	114.50	.10	5.0	↓	156.75	271.25
4" SWCI Hubless Long Turn TY	3	Ea.	9.70	29.10	1.0	3.0		94.05	123.15
4" Y	5	↓	6.98	34.90	1.0	5.0		156.75	191.65
4" Sweep	7		9.18	64.26	.66	4.62		144.84	209.10
3" Sweep	2		5.55	11.10	.58	1.16		36.37	47.47
4" 1/4 Bend	12		4.67	56.04	.66	7.92		248.29	304.33
3" 1/4 Bend	2		3.11	6.22	.58	1.16		36.37	42.59
4" 1/8 Bend	8		3.35	26.80	.66	5.28		165.53	192.33
4" Test Tee w/ Plug	7		7.91	55.37	.66	4.62		144.84	200.21
3" " " " "	2		4.75	9.50	.58	1.16		36.37	45.87
4" C.O. w/ Plug	3		3.75	11.25	.33	.99		31.04	42.29
4x3 Reducer	2	↓	2.55	5.10	.66	1.32		41.38	46.48
4" Hubless Couplings	195	Ea.	2.58	503.10	-	-		-	503.10
3" " "	24	Ea.	2.23	53.52	-	-		-	53.52
4" Pipe Sleeves	22	Ea.	25.	550.	.40	8.80		275.88	825.88
3" " "	2	Ea.	18.	36.	.32	.64		20.06	56.06
4" Hanger Assemblies	98	Ea.	7.11	696.78	.36	35.28		1106.03	1802.81
3" " "	10	Ea.	6.43	64.30	.36	3.60		112.86	177.16
4" Roof Drains w/Integral Expansion Jt.	12	Ea.	165.	1980.	1.50	18.0		564.30	2544.30
Brass Sleeve, Sediment Cup, Flashing,									
Clamp + Gravel Stop									
3" Roof Drains - Same as 4"	2	Ea.	155.	310.	1.50	3.0		94.05	404.05
Sheet Lead Flashing (14)	224	Lbs.	1.10	246.40	1.0	14.0	↓	438.90	685.30
Storm System- Above Ground Total Direct Cost				6319.54				4119.72	10,439.26

Figure 5.6 *The costs shown in this sample estimate are used as examples only. The costs should not be used for estimating purposes.*

ABC PLUMBING COMPANY, INC.
CONTRACTORS
ANYWHERE, USA

SHEET NO. 7 of 15

PROJECT　　　　ESTIMATE NO.

LOCATION　　　　CLASSIFICATION　　　　DATE

ARCHITECT　　　　ESTIMATOR　　　　PRICED BY　　　　CHECKED BY

DESCRIPTION	QUANTITY	UNIT	MATERIAL UNIT COST	MATERIAL TOTAL COST	LABOR UNIT LH	LABOR TOTAL LH	LABOR UNIT RATE	LABOR TOTAL COST	TOTAL COST MAT./LABOR
6. Hot + Cold Water System									
3" Type "K" Copper	70	LF	7.10	497.	.12	8.40	31.35	263.34	760.34
2½" " "	50	↓	5.14	257.	.09	4.50	↓	141.08	398.08
2" " " "	115		3.55	408.25	.06	6.90		216.32	624.57
1½" Type "L" Copper	305		1.94	591.70	.06	18.30		573.71	1165.41
1¼" " " "	215		1.52	326.80	.05	10.75		337.01	663.81
1" " " "	435		1.12	487.20	.04	17.40		545.49	1032.69
¾" " " "	1500		.80	1200.	.03	45.0		1410.75	2610.75
½" " " "	470		.51	239.70	.03	14.10		442.04	681.74
3" Copper Wrot Ells	7	Ea.	18.15	127.05	.89	6.23		195.31	322.36
2½"	1	↓	12.90	12.90	.70	.70		21.95	34.85
2"	3		6.83	20.49	.57	1.71		53.61	74.10
1½"	23		3.75	86.25	.50	11.50		360.53	446.78
1¼"	6		2.65	15.90	.43	2.58		80.88	96.78
1"	35		1.54	53.90	.39	13.65		427.93	481.83
¾"	58		.65	37.70	.32	18.56		581.86	619.56
½"	143		.29	41.47	.29	41.47		1300.08	1341.55
3 x Red. Tee	5		30.50	152.50	1.34	6.70		210.05	362.55
2½ x Red.	3		27.50	82.50	1.05	3.15		98.75	181.25
2 x Red.	4		9.05	36.20	.86	3.44		107.84	144.04
1½ x Red.	30		6.08	182.40	.75	22.50		705.38	887.78
1¼ x Red.	10		5.35	53.50	.65	6.50		203.78	257.28
1 x Red.	11		3.68	40.48	.59	6.49		203.46	243.94
¾ x Red.	53		1.11	58.83	.48	25.44		797.54	856.37
1¼"	1		5.55	5.55	.65	.65		20.38	25.93
1"	1		3.50	3.50	.59	.59		18.50	22.00
¾"	8		1.71	13.68	.48	3.84		120.38	134.06
½"	2		.49	.98	.43	.86		26.96	27.94
1½" 45	4		3.78	15.12	.50	2.0		62.70	77.82
1¼"	1		3.15	3.15	.43	.43		13.48	16.63
1"	2		2.29	4.58	.39	.78		24.45	29.03
¾"	14		.89	12.46	.32	4.48		140.45	152.91
½"	29		.53	15.37	.29	8.41		263.65	279.02
1½ x Red. Ell	1		10.60	10.60	.50	.50		15.68	26.28
¾ x Red. Ell	2		1.64	3.28	.32	.64		20.06	23.34
1½" Cap	1		1.81	1.81	.25	.25		7.84	9.65
1¼"	1		1.25	1.25	.22	.22		6.90	8.15
1"	1		.87	.87	.20	.20		6.27	7.14
¾"	6		.35	2.10	.16	.96		30.10	32.20
½"	2	↓	.20	.40	.15	.30	↓	9.41	9.81
Continued				5104.42				10,065.90	15,170.32

Figure 5.7 *The costs shown in this sample estimate are used as examples only. The costs should not be used for estimating purposes.*

ABC PLUMBING COMPANY, INC.
CONTRACTORS
ANYWHERE, USA

SHEET NO. 8 of 15

PROJECT | ESTIMATE NO.

LOCATION | CLASSIFICATION | DATE

ARCHITECT | ESTIMATOR | PRICED BY | CHECKED BY

DESCRIPTION	QUANTITY	UNIT	MATERIAL UNIT COST	MATERIAL TOTAL COST	LABOR UNIT LH	LABOR TOTAL LH	LABOR UNIT RATE	LABOR TOTAL COST	TOTAL COST MAT./LABOR
6. Hot & Cold Water - Continued	Brought Fwd.			5104.42				10065.90	15170.32
3" Copper Wrot Coupling	2	Ea.	11.15	22.30	.89	1.78	31.35	55.80	78.10
2½"	2		6.08	12.16	.70	1.40		43.89	56.05
2"	5		3.40	17.00	.57	2.85		89.35	106.35
1½"	11		2.74	30.14	.50	5.50		172.43	202.57
1¼"	7		1.71	11.97	.43	3.01		94.36	106.33
1"	12		.98	11.76	.39	4.68		146.72	158.48
3/4"	29		.48	13.92	.32	9.28		290.93	304.85
1/2"	10		.24	2.40	.29	2.90		90.92	93.32
1 x Red. XFE Adapters	1		1.45	1.45	.39	.39		12.23	13.68
1"	18		2.70	48.60	.20	3.60		112.86	161.46
3/4"	9		1.32	11.88	.16	1.44		45.14	57.02
1/2"	54		.94	50.76	.15	8.10		253.94	304.70
Miscellaneous Brass Nipples	60	Ea.	8.87	532.20	.04	2.40		75.24	607.44
3" Pipe Sleeves	2	Ea.	20.	40.	.45	.90		28.22	68.22
2½"	1		17.	17.	.40	.40		12.54	29.54
2"	1		14.	14.	.32	.32		10.03	24.03
1½"	25		12.	300.	.28	7.0		219.45	519.45
3" Hanger Assemblies	7	Ea.	9.32	65.24	.36	2.52		79.00	144.24
2½"	5		8.59	42.95	.36	1.80		56.43	99.38
2"	12		6.07	72.84	.36	4.32		135.43	208.27
1½"	31		5.63	174.53	.36	11.16		349.87	524.40
1¼"	36		5.53	199.08	.36	12.96		406.30	605.38
1"	13		5.40	70.20	.36	4.68		146.72	216.92
3/4"	250		5.21	1302.50	.36	90.0		2821.50	4124.00
1/2"	78		5.07	395.46	.36	28.08		880.31	1275.77
Solder 95/5	17	Lbs.	11.80	200.60	–	–		–	200.60
Flux	2	"	4.05	8.10	–	–		–	8.10
Gas	3	Cyl.	12.	36.	–	–		–	36.00
Access Panels	11	Ea.	30.	330.	1.0	11.0		344.85	674.85
Continued				9139.46				17040.36	26179.82

Figure 5.8 *The costs shown in this sample estimate are used as examples only. The costs should not be used for estimating purposes.*

ABC PLUMBING COMPANY, INC.
CONTRACTORS
ANYWHERE, USA

SHEET NO. 9 of 15

PROJECT

ESTIMATE NO.

LOCATION

CLASSIFICATION

DATE

ARCHITECT

ESTIMATOR

PRICED BY

CHECKED BY

DESCRIPTION	QUANTITY	UNIT	MATERIAL UNIT COST	MATERIAL TOTAL COST	LABOR UNIT LH	LABOR TOTAL LH	LABOR UNIT RATE	LABOR TOTAL COST	TOTAL COST MAT./LABOR
6. Hot & Cold Water - Continued	Brought Fwd.			9139.46				17040.36	26,179.82
3" Gate Valves	3	Ea.	166.	498.	.98	2.94	31.35	92.17	590.17
2½"	1		119.	119.	.77	.77		24.14	143.14
2"	2		56.	112.	.63	1.26		39.50	151.50
1½"	9		35.	315.	.54	4.86		152.36	467.36
1¼"	3		29.	87.	.47	1.41		44.20	131.20
1"	7		21.	147.	.41	2.87		89.97	236.97
¾"	58		16.	928.	.34	19.72		618.22	1546.22
½"	22		13.	286.	.30	6.60		206.91	492.91
3" Check Valve	1		155.	155.	.98	.98		30.72	185.72
1¼"	2		30.	60.	.47	.94		29.47	89.47
¾"	9		14.	126.	.34	3.06		95.93	221.93
¾" Balancing Valve	5		17.	85.	.34	1.70		53.30	138.30
½" Temp. & Press. Relief Valve	1		9.	9.	.30	.30		9.41	18.41
¾" Non-Freeze Wall Hydrants	6		75.	450.	.50	3.0		94.05	544.05
3" Domestic Compound Water Meter	1	Ea.	750.	750.	2.50	2.50		78.38	828.38
Elec. Hot Water Heater - 73.8 GPH	1	Ea.	650.	650.	6.0	6.0		188.10	838.10
@ 100° Rise Recov. 120 GPH (3) Element									
18 KW, 480 V, 3Φ									
¾" HW Circulator Pump - Inline Direct	1	Ea.	230.	230.	1.25	1.25		39.19	269.19
Conn. 1750 RPM, ⅙ HP, 115 V, 1Φ									
Hot & Cold Water System									
Total Direct Cost				14,146.40				18,926.38	33,072.84

Figure 5.9 *The costs shown in this sample estimate are used as examples only. The costs should not be used for estimating purposes.*

ABC PLUMBING COMPANY, INC.
CONTRACTORS
ANYWHERE, USA

SHEET NO. 10 of 15

PROJECT | ESTIMATE NO.

LOCATION | CLASSIFICATION | DATE

ARCHITECT | ESTIMATOR | PRICED BY | CHECKED BY

DESCRIPTION	QUANTITY	UNIT	MATERIAL UNIT COST	MATERIAL TOTAL COST	LABOR UNIT LH	LABOR TOTAL LH	LABOR UNIT RATE	LABOR TOTAL COST	TOTAL COST MAT./LABOR
7. Natural Gas System									
2" Black Steel Sch. 40 Pipe	30	LF	2.83	84.90	.07	2.10	31.35	65.84	150.74
1½" " " " " "	30	"	2.09	62.70	.06	1.80	↓	56.43	119.13
2" Black Mal. Ells	4	Ea.	3.61	14.44	.67	2.68		84.02	98.46
1½" " " "	3	"	2.46	7.38	.57	1.71		53.61	60.99
2 x Red. " " Tee	1	"	6.03	6.03	1.01	1.01		31.66	37.69
1½" Black Nipples	2	Ea.	3.98	7.96	.03	.06		1.88	9.84
2" Pipe Sleeves	1	Ea.	12.	12.	.32	.32		10.03	22.03
2" Hanger Assemblies	2	Ea.	5.70	11.40	.36	.72		22.57	33.97
1½" " "	2	"	5.66	11.32	.36	.72		22.57	33.89
2" Gas Cocks	1	Ea.	32.	32.	.80	.80		25.08	57.08
1½" " "	2	"	23.	46.	.75	1.50	↓	47.03	93.03
Natural Gas System Total Direct Cost				296.13				420.72	716.85

Figure 5.10 *The costs shown in this sample estimate are used as examples only. The costs should not be used for estimating purposes.*

ABC PLUMBING COMPANY, INC.
CONTRACTORS
ANYWHERE, USA

SHEET NO. 11 of 15

PROJECT | ESTIMATE NO.

LOCATION | CLASSIFICATION | DATE

ARCHITECT | ESTIMATOR | PRICED BY | CHECKED BY

DESCRIPTION	QUANTITY	UNIT	MATERIAL UNIT COST	MATERIAL TOTAL COST	LABOR UNIT LH	LABOR TOTAL LH	LABOR UNIT RATE	LABOR TOTAL COST	TOTAL COST MAT./LABOR
8. Fire Standpipe System									
3" Black Steel Sch. 40 Pipe	130	LF	5.93	770.90	.11	14.30	31.35	448.31	1219.21
2½"	25		4.52	113.00	.10	2.50		78.38	191.38
1½"	15		2.09	31.35	.06	.90		28.22	59.57
1"	10		1.36	13.60	.04	.40		12.54	26.14
3" Black Mal. Ells	10	Ea.	13.69	136.90	1.62	16.20		507.87	644.77
2½"	6		9.17	55.02	1.35	8.10		253.94	308.96
1½"	6		2.46	14.76	.57	3.42		107.22	121.98
1"	2		1.14	2.28	.47	.94		29.47	31.75
3 x Red. Tee	8		19.43	155.44	2.44	19.52		611.95	767.39
3"	1		21.24	21.24	2.44	2.44		76.49	97.73
2½ x Red.	1		14.55	14.55	2.03	2.03		63.64	78.19
½"	1		.75	.75	.58	.58		18.18	18.93
2½" Cap	1		4.71	4.71	.65	.65		20.38	25.09
½" Plug	1		.90	.90	.19	.19		5.96	6.86
2½" Blk. Nipples	3	Ea.	10.49	31.47	.05	.15		4.70	36.17
1½"	3		3.98	11.94	.04	.12		3.76	15.70
1"	1		2.72	2.72	.03	.03		.94	3.66
½"	3		1.51	4.53	.03	.09		2.82	7.35
3" Pipe Sleeves	2	Ea.	18.00	36.00	.45	.90		28.22	64.22
3" Hanger Assemblies	11	Ea.	18.00	198.00	.36	3.96		124.15	322.15
2½"	2		15.00	30.00	.36	.72		22.57	52.57
1½"	2		11.00	22.00	.36	.72		22.57	44.57
1"	1		7.00	7.00	.36	.36		11.29	18.29
3" I.B. OS+Y Valve	1	Ea.	152.	152.00	2.05	2.05		64.27	216.27
3" Check Valve	2		132.	264.00	2.05	4.10		128.54	392.54
3"x 2½" x 2½" Siamese Conn.	1		320.	320.00	3.0	3.0		94.05	414.05
Fire Hose Cabinet w/Rack + Valves	3		350.	1050.00	3.0	9.0		282.15	1332.15
Pressure Gauge	1		75.	75.00	.30	.30		9.41	84.41
1" Air Chamber	1		15.	15.00	.30	.30		9.41	24.41
1" Auto Ball Drip	1		35.	35.00	.30	.30		9.41	44.41
Fire Standpipe System									
Total Direct Cost				3590.06				3080.81	6670.87

Figure 5.11 *The costs shown in this sample estimate are used as examples only. The costs should not be used for estimating purposes.*

ABC PLUMBING COMPANY, INC.
CONTRACTORS
ANYWHERE, USA

SHEET NO. 12 of 15

PROJECT ESTIMATE NO.

LOCATION CLASSIFICATION DATE

ARCHITECT ESTIMATOR PRICED BY CHECKED BY

	DESCRIPTION	QUANTITY	UNIT	MATERIAL		LABOR				TOTAL COST MAT./LABOR
				UNIT COST	TOTAL COST	UNIT LH	TOTAL LH	UNIT RATE	TOTAL COST	
9.	Insulation									
	4" Pipe Insulation ½" Thick (Storm)	250	LF	2.85	712.50	.073	18.25	31.90	582.18	1294.68
	3" ↓ ↓ "	20	↓	2.40	48.00	.070	1.40	↓	44.66	92.66
	3" ↓ ↓ (Water)	70	↓	2.40	168.00	.070	4.90	↓	156.31	324.31
	2½" ↓ ↓ ↓	50	↓	2.10	105.00	.067	3.35	↓	106.87	211.87
	2" ↓ ↓ ↓	115	↓	1.90	218.50	.065	7.48	↓	238.61	457.11
	1½" ↓ ↓ ↓	305	↓	1.70	518.50	.055	16.78	↓	535.28	1053.78
	1¼" ↓ ↓ ↓	215	↓	1.60	344.00	.050	10.75	↓	342.93	686.93
	1" ↓ ↓ ↓	435	↓	1.52	661.20	.050	21.75	↓	693.83	1355.03
	¾" ↓ ↓ ↓	1500	↓	1.46	2190.00	.045	67.50	↓	2153.25	4343.25
	½" ↓ ↓ ↓	470	↓	1.35	634.50	.045	21.15	↓	674.69	1309.19
	Insulation									
	Total Direct Cost				5600.20				5528.61	11128.81

Figure 5.12 *The costs shown in this sample estimate are used as examples only. The costs should not be used for estimating purposes.*

ABC PLUMBING COMPANY, INC.
CONTRACTORS
ANYWHERE, USA

SHEET NO. 13 of 15

PROJECT ESTIMATE NO.

LOCATION CLASSIFICATION DATE

ARCHITECT ESTIMATOR PRICED BY CHECKED BY

DESCRIPTION	QUANTITY	UNIT	MATERIAL		LABOR				TOTAL COST MAT./LABOR
			UNIT COST	TOTAL COST	UNIT LH	TOTAL LH	UNIT RATE	TOTAL COST	
10. Sitework									
10A Water + Fire Service									
4" Ductile Iron Pipe Cem. Lined M.J.	22	LF	8.36	183.92	.13	2.86	31.35	89.66	273.58
3" " " " " "	100	"	7.68	768.00	.11	11.0	↓	344.85	1112.85
4" Ductile Iron M.J. Cem. Lined Cap	1	Ea.	9.95	9.95	Incl.	with Pipe		—	9.95
4x3 " " " " " Tee	1	"	89.80	89.80	"	"	"	—	89.80
3" " " " " " Ell	4	"	29.30	117.20	"	"	"	—	117.20
4" M.J. Gate Valve w/Box	1	Ea.	408.	408.00	3.0	3.0		94.05	502.05
3" " " " "	2	"	353.	706.00	3.0	6.0		188.10	894.10
6x4 Wet Connection	1	Ea.	795.	795.00	3.0	3.0		94.05	889.05
6x3 " "	1	"	795.	795.00	3.0	3.0	↓	94.05	889.05
Water + Fire Service									
Total Direct Cost				3872.87				904.76	4777.63

Figure 5.13 *The costs shown in this sample estimate are used as examples only. The costs should not be used for estimating purposes.*

ABC PLUMBING COMPANY, INC.
CONTRACTORS
ANYWHERE, USA

SHEET NO. 14 of 15

PROJECT | ESTIMATE NO.

LOCATION | CLASSIFICATION | DATE

ARCHITECT | ESTIMATOR | PRICED BY | CHECKED BY

DESCRIPTION	QUANTITY	UNIT	MATERIAL UNIT COST	MATERIAL TOTAL COST	LABOR UNIT LH	LABOR TOTAL LH	LABOR UNIT RATE	LABOR TOTAL COST	TOTAL COST MAT./LABOR
10. Sitework - Continued									
10B Storm Sewer									
8" XH Terracotta Pipe	400	LF	2.67	1068.	.12	48.0	31.35	1504.80	2572.80
4" " " "	40	"	1.03	41.20	.10	4.0		125.40	166.60
3" Bituminous Fibre Pipe	40	"	1.16	46.40	.10	4.0		125.40	171.80
8" Terracotta Y	5	Ea.	29.40	147.	Incl. with pipe			–	147.00
8x4 Y	1		29.40	29.40				–	29.40
8x3 Y	1		29.40	29.40				–	29.40
8" Sweep	5		29.40	147.				–	147.00
8" 1/8 Bend	3		29.40	88.20				–	88.20
4" 1/8 Bend	1		12.10	12.10				–	12.10
Catch Basins 1.99 Ft. Dp.	1	Ea.	575.	575.00	4.0	4.0		125.40	700.40
2.31	1								
2.63	1								
2.75	1								
3.23	1								
Lamp Holes	5	Ea.	50.	250.00	3.0	15.0		470.25	720.25
Connect new 8" storm sewer to existing catch basin	1	Ea.	50.	50.00	6.0	6.0		188.10	238.10
Connect new catch basin to existing 8" storm sewer	1	Ea.	50.	50.00	6.0	6.0		188.10	238.10
Storm Sewer Total Direct Cost				4833.70				3229.05	8062.75

Figure 5.14 *The costs shown in this sample estimate are used as examples only. The costs should not be used for estimating purposes.*

ABC PLUMBING COMPANY, INC.
CONTRACTORS
ANYWHERE, USA

SHEET NO. 15 of 15

PROJECT ESTIMATE NO.

LOCATION CLASSIFICATION DATE

ARCHITECT ESTIMATOR PRICED BY CHECKED BY

DESCRIPTION	QUANTITY	UNIT	MATERIAL		LABOR				TOTAL COST MAT./LABOR
			UNIT COST	TOTAL COST	UNIT LH	TOTAL LH	UNIT RATE	TOTAL COST	
10. Sitework- Continued									
10c Sanitary Sewer									
6" XHCI Soil Pipe B & S	65	LF	8.93	580.45	.21	13.65	31.35	427.93	1008.38
12"X6" XHCI Y	1	Ea.	108.05	108.05	5.46	5.46		171.17	279.22
6" 1/8 Bend	1	"	12.98	12.98	1.06	1.06		33.23	46.21
Connect new 6" Sanitary Sewer to existing 12" Sewer	1	Ea.	75.	75.00	6.0	6.0		188.10	263.10
Caulking Lead	65	Lbs.	1.04	67.60	–	–		–	67.60
Oakum	6	"	.95	5.70	–	–		–	5.70
Gas	1	Cyl.	12.	12.00	–	–		–	12.00
Sanitary Sewer Total Direct Cost				861.78				820.43	1682.21

Figure 5.15 *The costs shown in this sample estimate are used as examples only. The costs should not be used for estimating purposes.*

Chapter 6

Marking Up and Completing the Estimate

The estimator has now arrived at direct costs for all the systems within the estimate, direct cost being the cost of a project prior to the addition of any supplementary expenses, such as overhead and profit. At this point, the estimator transfers all the direct cost totals, both material and labor, onto the project summary sheet to apply the necessary markups. Markups are all the supplementary costs necessary to arrive at a bid price.

Breakdown of Markups

The following is a list of items that comprise the markup of most plumbing estimates:

1. Sales tax (material only): At this writing, some states are contemplating changes that would base sales tax on the total contract bid or selling price rather than on material only.
2. Field overhead
   a. Temporary facilities
   b. Trailer field office
   c. Tools
   d. Engineering/shop drawings
   e. Job supervision
3. Office overhead
4. Permits and fees
5. Insurance
6. Bid bond
7. Profit

Sales Tax

This is the prescribed local tax imposed on all purchased materials. The amount of sales tax to be included in the estimate is determined by the cost of materials and the tax rate in the area of the project. (In some areas there are both city and state sales taxes.)

Field Overhead

These items are the overhead costs attributed to the particular job being bid.

1. **Temporary Facilities**: For the plumbing contractor, temporary facilities are the provision of water and/or sanitary facilities during construction. Depending on the particular project requirements, temporary water and sanitary facilities may mean providing only a fresh water outlet and a rented portable toilet. On the other hand, temporary facilities may mean providing metered water to the construction site with cold water outlets on every floor, in addition to working toilets in a temporary structure on the site or in the building under construction. There is a wide range of possibilities regarding temporary facilities, from relatively inexpensive to very expensive set-ups. Temporary facilities should be estimated on an individual project basis. The estimator should take great care in interpreting the temporary facilities section in the job specifications. When estimated, temporary facilities costs should be entered as a lump sum cost on the summary sheet. (See Figures 6.1 and 6.2.)
2. **Trailer Field Office**: A prime contractor on a project usually includes in the bid provisions for a field office or trailer. The field office is furnished with office equipment, such as plan tables, desks, file cabinets, and telephones, all of which become part of the field office expense. To estimate field office expenses, the estimator must envision the duration of the project to calculate trailer rentals, office equipment rentals, and telephone, light, and power charges.
3. **Tools**: An allowance for small tools should be included in the field overhead expenses. Tools such as pipe wrenches, caulking irons, and pipe threading machines inevitably get worn out, damaged, stolen, or lost. The tool allowance can be estimated by consulting the purchasing records for small tools on previous projects, usually expressed as a percentage of the project labor.
4. **Engineering and Shop Drawings**: In anticipation of being awarded the contract, the estimator should include in the field overhead costs an allowance for engineering and shop drawings. This allowance should cover the expenses incurred by the contractor to develop dimensioned layout drawings necessary to coordinate the project. The allowance should also include the preparation of any record, or "as-built," drawing sets furnished to the owner.
5. **Job Supervision**: This is time charged by foremen and others for supervising productive labor. Depending on the project, the estimator may want to carry this cost at the full salary of the superintendent every week for the duration of the project, at full salary part-time, or at partial salary full-time.

Depending on the project, field overhead can range from 5% to 15% of the total direct cost of the job.

Office Overhead

A contractor incurs project-related expenses in running the main office, shop, and yard. Examples of these expenses are rent; utilities; wages for draftsmen, secretaries, clerks, and estimators; taxes; stationery; and office equipment. The office overhead, depending on the project, can range from 2% to 4% of the sum of the total direct cost and field overhead.

Summary Sheet

ABC PLUMBING COMPANY, INC.
CONTRACTORS
ANYWHERE, USA

SHEET NO. ______

PROJECT ______ ESTIMATE NO. ______

LOCATION ______ CLASSIFICATION ______ DATE ______

ARCHITECT ______ ESTIMATOR ______ PRICED BY ______ CHECKED BY ______

DESCRIPTION	QUAN-TITY	UNIT	MATERIAL UNIT COST	MATERIAL TOTAL COST	LABOR UNIT LH	LABOR TOTAL LH	LABOR UNIT RATE	LABOR TOTAL COST	TOTAL COST MAT./LABOR

Figure 6.1

ABC PLUMBING COMPANY, INC.
CONTRACTORS
ANYWHERE, USA

SHEET NO. 1 of 1

PROJECT 3-Story Service Center | ESTIMATE NO.

LOCATION Anywhere, USA | CLASSIFICATION | DATE

ARCHITECT Jarrett Assoc. | ESTIMATOR JG | PRICED BY SG | CHECKED BY MS

DESCRIPTION	QUANTITY	UNIT	MATERIAL		LABOR				TOTAL COST MAT./LABOR
			UNIT COST	TOTAL COST	UNIT LH	TOTAL LH	UNIT RATE	TOTAL COST	
1. Plumbing Fixtures + Trim				19354.00				5243.29	24597.29
2. Sanitary Waste/Vent - Below Ground				6627.89				6210.58	12838.47
3. Sanitary Waste/Vent - Above Ground				14170.42				15416.64	29587.06
4. Storm System - Below Ground				3811.00				3188.94	6999.94
5. Storm System - Above Ground				6319.54				4119.72	10439.26
6. Hot + Cold Water System				14146.46				18926.38	33072.84
7. Natural Gas System				296.13				420.72	716.85
8. Fire Protection System				3590.06				3080.81	6670.87
9. Insulation				5600.20				5528.61	11128.81
10. Sitework									
10A Domestic Water + Fire Service				3872.87				904.76	4777.63
10B Storm Sewer				4833.70				3229.05	8062.75
10C Sanitary Sewer				861.78				820.43	1682.21
11. Test + Adjustment (Included with Labor)				—				-	-
Tags, Charts, Pipe Identification				500.00				750.00	1250.00
Excavation + Backfill									
Sanitary	201	CY		-			5.00	1005.00	1005.00
Storm	57			-				285.00	285.00
Site Water	37			-				185.00	185.00
Site Storm	105			-				526.00	525.00
Site Sanitary	14	↓		-			↓	70.00	70.00
Sales Tax (material Only) 7%				5878.88				-	5878.88
Direct Job Cost				89862.93				69909.93	159772.86
Field Overhead									
Temporary Facilities									2500.00
Trailer/Field Office									4000.00
Tools									1500.00
Engineering/Shop Drawings									1200.00
Job Supervision									7000.00
Office Overhead 2.5%									4400.00
Subtotal Cost									180372.86
Permits + Fees 74 Fixt. @ $5.00 Ea.									370.00
Insurance 1.5%									2705.00
Bid Bond 2.0%									3807.00
Profit 10%									18725.49
Total Job Cost									205980.35

Figure 6.2 *The costs shown in this sample estimate are used as examples only. The costs should not be used for estimating purposes.*

Permits and Fees

Most local municipalities require the plumbing contractor to obtain permits for installation of plumbing, based on the number of fixtures. Depending on local jurisdiction, permits may range in cost from one to ten dollars per fixture.

Insurance

The insurance the plumbing contractor will have to carry depends on local laws and project requirements. The different types of insurance the plumbing contractor may be required to obtain are:

- Public liability
- Property damage
- Fire insurance

Worker's compensation and liability insurance are usually carried in the hourly wage rate. (See Wage Rate Table, Appendix C.) Insurance can be estimated at 1-1/2% of the subtotal cost on the summary sheet (Figures 6.1 and 6.2).

Bid Bond

When required, a bid bond is furnished by the contractor at the time of bid as proof that he or she will accept the contract for work if chosen to do the work. The cost of the bid bond is usually between 2% and 4% of the subtotal cost on the summary sheet (Figures 6.1 and 6.2).

Profit

This is the amount a contractor hopes to gain for performing the work. Profit is usually a percentage of the total material, labor, and overhead costs. The percentage of profit that the contractor allows usually depends on the size and type of job. Generally, the larger the project, the smaller the profit mark-up. Another factor that determines the contractor's profit percentage is, of course, need. If the shop is rather slow, it may need to bid the job very tight to ensure success. In a recession, the main goal is often to merely keep the workers busy and on the payroll. The percentage of profit mark-up for a plumbing job can range from 5% to 15%.

Contingencies

Often a contractor may be required to bid on a project that has incomplete drawings and specifications or other unknown factors. It is important that the estimator keep both mental and written notes regarding these unknowns, which should be evaluated during the pricing stage. If vague projects come into the office, it is to the contractor's advantage to give them to the most seasoned estimator to scrutinize. An estimator with many years of experience can readily pick up any missing items on the drawings or in the specifications, and assign properly weighted contingency costs to cover these unknown factors. Contingency factors are generally added immediately after the direct costs and prior to any mark-ups.

Completing the Sample Estimate

The summary sheet of our sample estimate (Figure 6.2) is ready to be filled out with the appropriate markups to complete the estimate.

The estimator must now transfer the total material and labor direct costs for each system from the estimating sheets to the summary sheets. Once this is done, sales tax (when applicable) is entered in the Material column of the summary. The estimator then adds the material and labor costs for each system and enters the totals into the Total column. The next calculation is adding the Material, Labor, and Total columns down, with their respective totals entered on the line labeled direct project cost. As a check, the material and labor direct costs, when added together, should equal the total direct project cost as shown in the Total column.

The estimator is now ready to add the appropriate markups, shown on the summary sheet (Figure 6.2). This involves calculating and entering field and overhead costs in the Total column. These figures are then added to the direct job cost and the sum entered as a subtotal cost. Permits, insurance, bid bond, and profit are now calculated and entered in the Total column. The sum of these figures and the subtotal cost is the total project cost (bid price).

Submitting the Estimate

With the estimate complete, the plumbing contractor is now prepared to submit the bid to the respective owner, agency, or general contractor. In submitting a bid, the plumbing contractor will function either as a prime or subcontractor. A prime contractor is generally directly responsible to the owner or agency. A subcontractor usually furnishes an estimate for work to a general contractor, to become part of one bid price. The subcontractor is responsible to the general contractor only for his or her portion of work as called for in the plans, specifications, and general conditions for the total project. Whether one is acting as a prime contractor or subcontractor, there are various ways to submit an estimate, some of which are described below.

Competitive and Negotiated Bids

Competitive bids are responses to formal announcements made by government agencies or owners requesting bids from any contractors who can meet their requirements. A contractor bidding competitively must estimate strictly according to plans and specifications. Bid closing dates and times are strictly adhered to in competitive bidding, and many bids have been refused because the contractor was seconds late to the bid-opening session. Almost all large projects will be bid in the competitive manner.

Negotiated bids are quite different from competitive bids, as demonstrated by the following example. A general contractor or owner has had good results on past jobs using two or three particular plumbing contractors in the area. Not wanting to tamper with success, the general contractor may invite only these contractors to bid on the next job.

Unlike competitive bidding, negotiated bid closing dates and times may not be firm. The plumbing contractor may suggest to the general contractor or owner alternate methods of design that will save the owner money and, at the same time, ensure the lowest possible bid. These contractor-owner communications are often held prior to the bid date.

Cover Letter (List of Exceptions)

Cover letters defining exactly what the contractor has or has not figured into the estimate are commonplace in negotiated bidding. As previously mentioned, competitive bids require the bidder to estimate strictly by plans and specifications, and cover letters with lists of exceptions are not required or permitted. In the informality of negotiated bidding, however, it is good practice for a cover letter to accompany the bid price to explain all inclusions and exclusions.

Value Engineering Analysis

Every estimator should use a value engineering approach while visualizing the project. Unaware of the true meaning of value engineering, many people often misinterpret it as a simple cost-cutting exercise. Alphonse Dell'Isola defined value engineering in his book, *Value Engineering: Practical Applications*, as "a systematic approach to obtaining optimum value for every dollar spent. Through a system of investigation, unnecessary expenditures are avoided, resulting in improved value and economy."

The use of value engineering by plumbing contractors can result in considerable cost savings and reduced installation time. In the process of applying value engineering to a project, the estimator should place all alternate design schemes and material substitutions on an idea-listing sheet. It might also be advantageous for the estimator to conduct a brainstorming session with other estimators and/or engineers in the firm to obtain their ideas. The brainstorming session should be a freethinking exercise, with all ideas entered into the listing. Once all ideas are reported, they should be analyzed further for acceptance or rejection. Ideas that involve code violations should be rejected immediately. Lifecycle costs must also be analyzed in this stage. For example, vitrified clay pipe may initially cost less than cast iron soil pipe, but will it have the same life expectancy as the structure? If not, and a new installation is required after 10 or 15 years, the idea has cost money, not saved it. Once all acceptable ideas have been listed, the estimator should present them to the owner and architect for final approval.

Unfortunately, the contractor has little hope of applying value engineering to jobs bid competitively until after being awarded the contract. Having to bid by plans and specifications, the contractor is locked into figuring the job as designed on the bid documents. In this case, he or she can only hope that any perceived value engineering ideas can be implemented after construction has begun.

Due to its informality, negotiated bidding is very conducive to a value engineering exercise, which can mean the difference in getting or losing the job.

Some expensive specifications that are often noticed while brainstorming a plumbing job are:

1. Fixtures are not designed in back-to-back groups wherever possible.
2. One large capacity hot water generator and hot water recirculation line are designed when small, isolated water heaters will suffice, eliminating the high-cost generator and recirculation pump, piping, and insulation.
3. Cast iron bell and spigot pipe is specified for above grade soil, waste, and storm lines. Consider the use of hubless soil pipe, DWV copper, or DWV PVC ABS drainage pipe and fittings.
4. Galvanized steel pipe is specified for above grade soil, waste, and storm lines. Consider the use of hubless cast iron soil pipe, DWV copper, or PVC drainage pipe and fittings.

Typed or Handwritten Estimates

With the possible exception of the summary sheet, all estimates should remain in handwritten form. Typing is an extra operation, requiring the transfer of figures to other sheets, and allowing for unnecessary errors in transposition. A neatly printed estimate can be read as easily as a typewritten one. Numerical figures should not be typewritten in any circumstances, as a slip of the finger can cause 1,000 L.F. to read 10,000 L.F. Clients are interested in the bottom line figure on an estimate. It is the low bid, not whether the estimate is typewritten or handwritten, that will determine who will be awarded the contract.

Coordination with Other Trades

For the plumbing contractor who is the successful low bidder, close coordination before and during construction with other trades, such as structural, heating, ventilating, air-conditioning, and electrical, will be necessary. Upon winning the contract, the plumbing contractor should request two complete sets of architectural, structural, mechanical, and electrical drawings. When shop drawings are being developed showing exact locations for items such as sleeves, fixtures, piping runs, and equipment, it is imperative that the layout draftsman have these other drawings to ensure that the plumbing installation will not interfere with the other trades. A properly coordinated project will make for fewer delays and change orders, and costs will be kept to a minimum.

Chapter 7

Other Plumbing Estimates

Four types of plumbing estimates are discussed in this chapter.

1. Change order analysis
2. Estimating additions and alterations
3. Budget estimating
4. Computer estimates

Any of the above should be attempted only by an estimator with considerable experience. For those who may be unfamiliar with these situations, this chapter explains the circumstances under which each estimate would be used, and the proper approach to be taken.

Change Order Analysis

A change order is any change in the original bid drawings or specifications. There are various reasons why change orders occur, such as errors or omissions from drawings, or simply the owner finding additional or less work necessary to meet the program requirements.

To estimate a change order, both the original and revised documents are needed. The estimator takes off the changed areas only in both the original and revised drawings and keeps each takeoff separate. In preparing a change order estimate, one must carefully take off only the items that are changed, to avoid listing materials or labor twice for the same work. Once the takeoffs are complete, the estimator prepares two estimates — one according to the original design and the other according to the revised design.

There are factors affecting change order pricing that did not affect the original bid. For example, a plumbing contractor can include delivery or returned-item charges paid on materials that are not usable due to the change. The contractor can also price any added materials at current prices while crediting the original material at the discounted price prevailing at bid time. This assumes that the cost of goods purchased at the time of the change order will be higher than for those at the time of the bid. The plumbing contractor can also charge for labor inefficiencies caused by the disruption of installation due to the change.

The estimator has now prepared two estimates showing direct costs. If the difference between the two estimates results in a *net addition*, the contractor adds the appropriate overhead and profit. If the difference results in a *net credit* (unless otherwise noted in the job specifications or by previous written agreement), the contractor does not deduct any original estimated overhead or profit.

Estimating Additions and Alterations

Because of unknown conditions within an existing structure, additions and alterations represent the most uncertain type of estimating in the construction industry. The first thing an estimator must do is visit the site. By doing so, one can determine many important facts that will influence the project estimate. Some of the items the estimator should look for on a site visit are:

- How the existing structure is built
- Locations of existing pipelines
- Difficulty in gaining access to hidden pipelines
- Condition of the building (poorly or well maintained)
- Any unusual conditions in the building or on the site that might make working more difficult, such as very high ceilings, flooded basements, crawl spaces, occupant use, etc.
- Difficulty in getting new equipment through existing entrance ways or corridors, or up stairwells or elevators

With a full understanding of site conditions, the estimator is ready to estimate the project. It should be noted that although most alteration drawings should show existing services, such representations cannot be accepted as accurate. The design engineer for the new work may have been working with old or poor plans and may have had difficulty in interpreting them.

The estimator must be sure to allow enough labor and material costs to cover any unknowns that might exist.

The plumbing estimator must also determine costs for the removal and relocation of fixtures and piping, in addition to connections to existing services. If piping is very old, the estimator must take care not to underestimate the necessary removal time. On the surface, removals may seem easier than new installations, but if the contractor disrupts functioning lines during the installation, he or she must assume responsibility for any damages. Connections to existing lines are generally expensive in terms of time, and the estimator should take them off based on the size of the existing line being cut.

The labor required to install new items, such as fixtures and piping, on an alteration is generally greater than the cost of installing the same item on a new construction project. The estimator must realize that piping production will drop considerably due to expected interference from the existing structure (including the necessity of handling short measures of pipe rather than full lengths).

Reading the specifications is another important part of estimating alterations work. The estimator should check to determine whether cutting and/or patching of existing walls and floors are part of the plumbing contract. The estimator should also check the specifications and general conditions for clauses stating responsibility for items such as dust control and noise abatement. On alteration work, certain parts of the job often must be done at specified times during the day

COST DATA SHEET

JOB NAME ____________________ JOB NO. ____________

TYPE OF BUILDING ____________________ LOCATION ____________

SF ____________________ COST/SF ____________

FIXTURES ____________________ $/FIXTURE ____________

	$	% OF JOB	$/SF
PLUMBING FIXTURES AND TRIM			
EQUIPMENT			
HOT AND COLD WATER (DOMESTIC)			
SANITARY (ABOVE AND BELOW GRADE)			
GAS PIPING			
VALVES, FLASHING AND MISCELLANEOUS			
HANGERS, SLEEVES AND INSERTS			
INSULATION			
TEST-ADJUSTMENTS AND MISCELLANEOUS			
EXCAVATION AND BACKFILL			
SPRINKLER SYSTEM			
KITCHEN EQUIPMENT			
FIRE STANDPIPE SYSTEM			
LAB AND SPECIAL PIPING SYSTEM			
SITEWORK			
OVERHEAD			
PROFIT			
SALES TAX			
TOTAL COST			

NOTES

Figure 7.1

or night so as not to disrupt the daily functions of the building. If this is so, the estimator must figure any necessary overtime charges for workers performing during off hours. After the estimator has priced the alteration, overhead and profit are added. The percentages here might differ due to the nature of the job.

Budget Estimating

An estimator occasionally is required to calculate a plumbing cost for a project based on nothing more than architectural plans. The two methods of measurement used to budget-estimate plumbing are *cost per square foot of building* and *cost per fixture*. Taking off the plumbing fixtures or square footage from the architectural plan is not a difficult task, but allocating the right costs can be a tremendous challenge.

Accurate budget estimates must be based on historical cost information. Published cost data, such as *Means Plumbing Cost Data*, can be helpful for this information. Another approach is to gather data from one's own past projects to compile a historical cost base. To do so, the contractor should fill out a cost data sheet similar to that in Figure 7.1 after completion of each project. After a period of time, the contractor will have developed a method from which to budget-estimate other similar types of projects.

Filling out a cost data sheet is done as follows:

1. Enter job name, number, and location.
2. Enter type of building, i.e., library, hospital, etc.
3. Enter gross square foot area (s.f.)
4. Enter fixture count. (Total amount of plumbing fixtures on job. Floor and roof drains are valued at one half of a fixture.)
5. Enter from Estimate Summary Sheet dollar values for applicable items in the $ column.
6. Enter the percentage each item represents of the total value of the project in the % of job column. The sum of these items should equal 100%
7. Enter cost per square foot for each item in the $/s.f. column.
8. Enter cost per square foot for whole project in cost/s.f. space. The sum of item 7 should equal cost/s.f.
9. Enter cost per fixture in $/fixture space. Cost per fixture is the total job cost, including mark-ups, divided by the number of fixtures.
10. Note that space is provided in Figure 7.1 for any unusual features a particular job may have, such as alterations, a packaged sewage treatment plant on the site, or kitchen equipment to be furnished or installed.

Plumbing costs vary from building type to building type. Therefore, for data to be valid, comparisons should be made with similar types of building usage and construction.

Computerized Estimating Systems

More and more estimators are starting to harness power of the computer as an aid in preparing estimates. This section surveys some of the systems currently available.

Two important points should be emphasized. First, although computers make estimating easier and faster, their use by a person inexperienced in estimating techniques can lead to inaccurate results. Remember, the guiding principle of all computer applications: garbage in, garbage out. If the information put into the computer is flawed, no amount of computer manipulation will correct it. A thorough understanding of estimating is needed to properly use computerized estimating systems.

Second, while the use of a computer can be faster in many cases, this is not always so. For example, small estimates may be accomplished as quickly by hand. It may also take a while to learn the intricacies of a new computer program, and this learning curve must be taken into account. Different programs will take varying amounts of time to learn.

Most offices using computerized estimating systems utilize small desktop computers. While there are programs available for larger and more powerful types of computers, the principles involved are the same, and desktop computers have adequate power to handle large and complex estimates. Other systems may have a digitizer, which is a "pad" of sensitive material laid out on a desk upon which drawings may be placed and "traced" into the computer with a special stylus.

The four classes of estimating programs surveyed on the following pages range from the ultra-simple adaptation of a spreadsheet program to systems that automate almost all aspects of the process.

Spreadsheet Applications

A simple yet powerful tool for computerizing the estimating process is the computer spreadsheet program. (Some well-known programs are Lotus 1-2-3,™ Quattro,™ and Excel™). A computer spreadsheet is a variation on an accountant's ledger sheet, consisting of a matrix of rows and columns of blank spaces called cells (shown as rows 1 thru 12 and columns A thru J, for example). Each cell can act as a label or be related to another cell mathematically. For example, in Figure 7.2, cell A4 (column A, row 4) contains a label with a description: 4" SWCI hubless pipe. The next cell horizontally, B4, contains a quantity: 10. Continuing horizontally, one encounters a label: LF, and a number: $2.97. The next cell, E4, contains the material total cost which is a formula: QUANTITY times UNIT PRICE or in spreadsheet language, B4 times D4 (the formula is not displayed, only the result). In this manner an estimate spreadsheet can be built up containing labels, numbers, and formulas. Cells J9, J10, and J12 contain formulas which respectively sum the prices for each line item, add markups and total the estimate.

The usefulness of the computer spreadsheet will only come into play after the quantity takeoff and pricing. These will still have to be done manually. Once entered, however, the power of the computer can be brought to bear not only in computing extensions and totals for large estimates, but in making adjustments to individual line items. Once entered, a formula, such as that in cell E4, stays as the mathematical formula B4 times D4. If the value of 134 or D4 changes, the value displayed in E4 changes accordingly. "What-if" analyses can be played on a spreadsheet by varying some numbers and seeing what results. One may vary unit prices to estimate future costs or vary quantities to perform value engineering or least-cost studies.

To use a spreadsheet effectively, the estimator must familiarize himself or herself with the ins and outs of the particular computer program being used. Since most offices have spreadsheets for other purposes such as accounting, the cost involved in set-up for estimating is solely the time required for personnel to learn the program.

Library-Based Applications

The previous example (Figure 7.2) involved a broad-use commercial package, tailored for estimating. There are other programs that are vastly more specialized and exclusively oriented to estimating, such as Means *CostWorks*. Such programs also require an estimator to manually take off quantities, but have libraries of cost information from which cost items needed for a particular estimate may be selected. In the case of Means *CostWorks*, the format follows the layout of the annual cost data books published and can be updated yearly. An estimator enters various pieces of information about the cost item selected, such as quantity, whether or not the item is being subcontracted, in what city the work is taking place, markups, etc. The program will then compute the cost extensions and totals and print out numerous reports ranging from the brief summaries to detailed breakdowns. (See Figure 7.3.)

These library-based applications free an estimator from the burden of researching prices by allowing easy use of a database of national costs adjusted to local areas. Modifications can be made to allow for price or labor-hour adjustment. In this way, in-house information maintained by a contractor can be incorporated into the database, as can costs for high-priced or unusually complicated specialty work. These kinds of programs are generally inexpensive, but since they are specialized, must be purchased specifically for estimating. They are easy to learn, but require some basic familiarity with computers.

Input Device Applications

There are several programs currently on the market that take the library-based applications one step further by allowing direct input from a drawing to the computer. They usually use a digitizer to trace

	A	B	C	D	E	F	G	H	I	J
1				MATERIAL			LABOR			TOTAL COST
2	DESCRIPTION	QTY	UNIT	UNIT COST	TOTAL COST	UNIT LH	TOTAL LH	UNIT RATE	TOTAL	MAT/LAB
3										
4	4" SWCI HUBLESS PIPE	10	LF	$4.16	$41.60	0.14	1.40	$47.03	$65.84	$107.44
5	4" SWCI HUBLESS "Y"	1	EA	$9.77	$ 9.77	1.00	1.00	$47.03	$47.03	$56.80
6	4" SWCI HUBLESS COUPLINGS	4	EA	$3.61	$14.44					$14.44
7	4" HANGAR ASSEMBLY	3	EA	$9.24	$27.72	0.36	1.08	$47.03	$50.79	$78.51
8										
9									SUBTOTAL:	$257.19
10							OVERHEAD & PROFIT @ 21%:			$54.01
11										
12									TOTAL:	$311.20

Figure 7.2

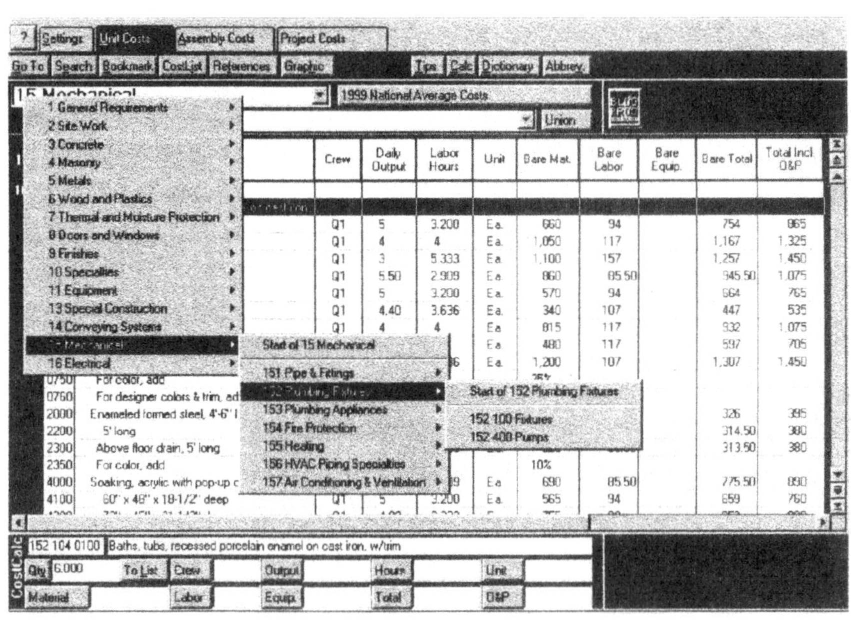

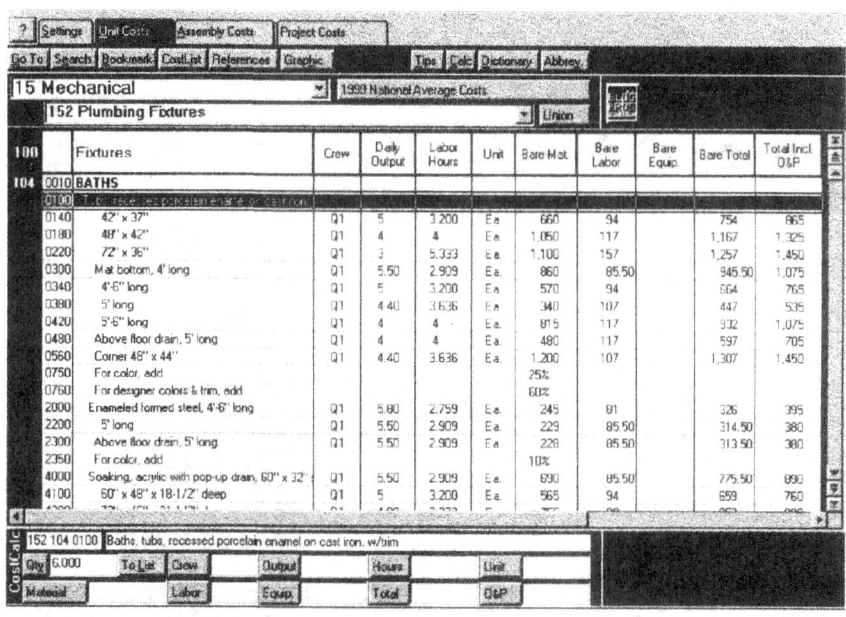

A user has the option of selecting line items from within the standard CSI code divisions, subdivisions, and major classifications for use in estimates. The database has all the standard line items contained in the Means Cost Data books.

Figure 7.3

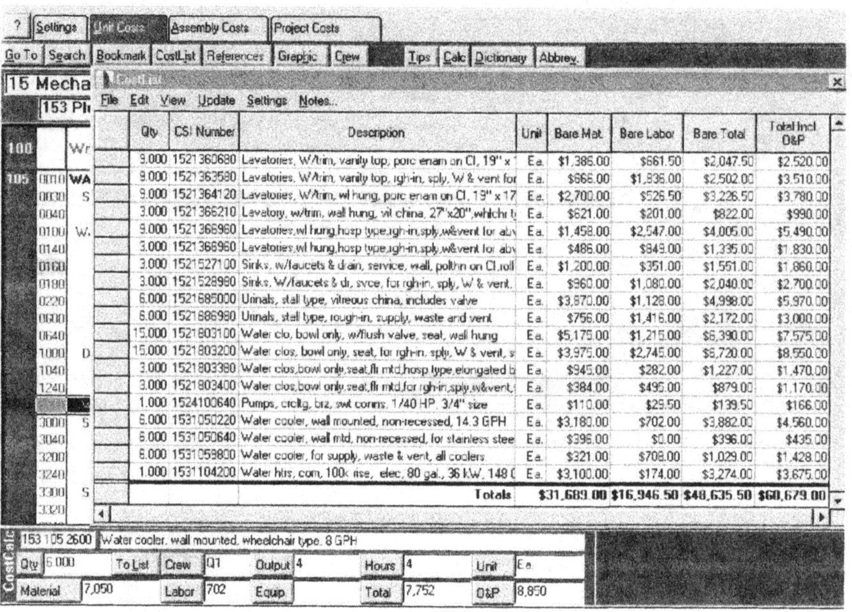

Qty	CSI Number	Description	Unit	Bare Mat.	Bare Labor	Bare Total	Total Incl O&P
9.000	1521360680	Lavatories, W/trim, vanity top, porc enam on CI, 19" x	Ea.	$1,385.00	$661.50	$2,047.50	$2,520.00
9.000	1521363580	Lavatories, W/trim, vanity top, rgh-in, sply, W & vent for	Ea.	$666.00	$1,836.00	$2,502.00	$3,510.00
9.000	1521364120	Lavatories, W/trim, wl hung, porc enam on CI, 19" x 17	Ea.	$2,700.00	$526.50	$3,226.50	$3,780.00
3.000	1521366210	Lavatory, w/trim, wall hung, vit china, 27"x20",whlchr t	Ea.	$621.00	$201.00	$822.00	$990.00
9.000	1521366960	Lavatories,wl hung,hosp type,rgh-in,sply,w&vent for abv	Ea.	$1,458.00	$2,547.00	$4,005.00	$5,490.00
3.000	1521366960	Lavatories,wl hung,hosp type,rgh-in,sply,w&vent for abv	Ea.	$486.00	$849.00	$1,335.00	$1,830.00
3.000	1521527100	Sinks, w/faucets & drain, service, wall, polthn on CI,roll	Ea.	$1,200.00	$351.00	$1,551.00	$1,860.00
3.000	1521528980	Sinks, W/faucets & dr, svce, for rgh-in, sply, W & vent.	Ea.	$960.00	$1,080.00	$2,040.00	$2,700.00
6.000	1521685000	Urinals, stall type, vitreous china, includes valve	Ea.	$3,870.00	$1,128.00	$4,998.00	$5,970.00
6.000	1521686980	Urinals, stall type, rough-in, supply, waste and vent	Ea.	$756.00	$1,416.00	$2,172.00	$3,000.00
15.000	1521803100	Water clo, bowl only, w/flush valve, seat, wall hung	Ea.	$5,175.00	$1,215.00	$6,390.00	$7,575.00
15.000	1521803200	Water clos, bowl only, seat, for rgh-in, sply, W & vent, s	Ea.	$3,975.00	$2,745.00	$6,720.00	$8,550.00
3.000	1521803380	Water clos,bowl only,seat,flr mtd,hosp type,elongated b	Ea.	$945.00	$282.00	$1,227.00	$1,470.00
3.000	1521803400	Water clos,bowl only,seat,flr mtd,for rgh-in,sply,w&vent,	Ea.	$384.00	$495.00	$879.00	$1,170.00
1.000	1524100640	Pumps, crclg, brz, swt conns, 1/40 HP, 3/4" size	Ea.	$110.00	$29.50	$139.50	$166.00
6.000	1531050220	Water cooler, wall mounted, non-recessed, 14.3 GPH	Ea.	$3,180.00	$702.00	$3,882.00	$4,560.00
6.000	1531050640	Water cooler, wall mtd, non-recessed, for stainless stee	Ea.	$396.00	$0.00	$396.00	$435.00
6.000	1531059800	Water cooler, for supply, waste & vent, all coolers	Ea.	$321.00	$708.00	$1,029.00	$1,428.00
1.000	1531104200	Water htrs, com, 100k rise, elec, 80 gal., 36 kW, 148 (	Ea.	$3,100.00	$174.00	$3,274.00	$3,675.00
		Totals		$31,689.00	$16,946.50	$48,635.50	$60,679.00

Each line item, whether selected from among those in the program's database or input by the user, can have a number of attributes (quantity, material cost, labor cost, overhead and profit, etc.) associated with it as shown in the figure above.

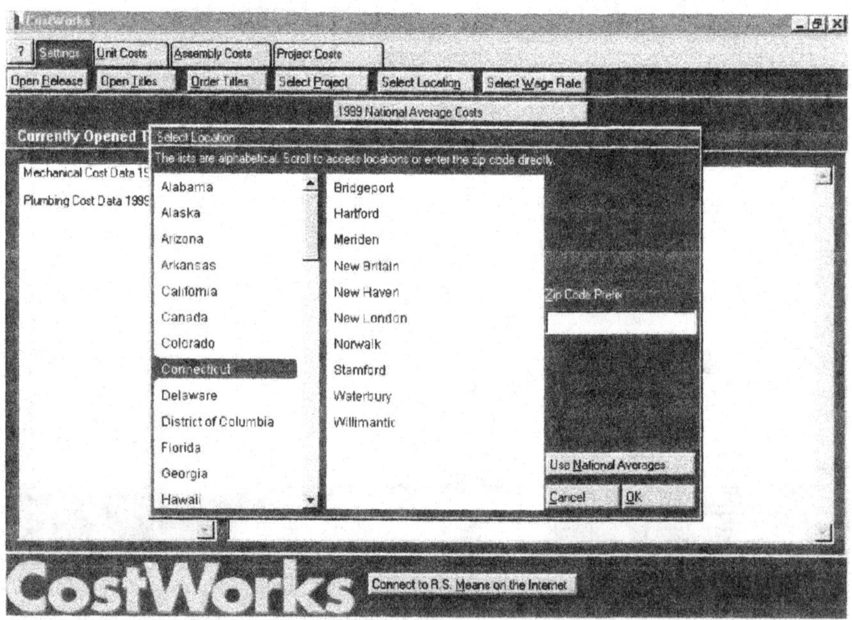

Special cost factors can be used to adjust the estimate to local economic conditions. The figure above shows several cities available in Connecticut. The user selects a city, and the program automatically revises the estimate using the appropriate cost factor.

Figure 7.3 (cont'd.)

drawings directly into the computer. If a building perimeter is traced, wall length, building area, and volume are computed automatically. The system will look up costs, labor-hour, and crew size information from a library database and produce numerous reports.

This type of system eliminates the double work of taking off quantities manually and then entering them into the computer. The takeoff is faster, but care should be exercised since a mistake can throw off the estimate just as easily as in the manual method—and might be harder to find. These systems are highly specialized, and the program itself might cost several times more than the computer hardware it runs on. They also require that time and effort be spent in learning all the features, and usually take much longer than the previous two types of applications.

Another feature commonly found in these more sophisticated programs is the ability to take the data generated by the takeoff and put it in a format that allows the same "what-if" analyses as spreadsheets. Some programs allow the cost and labor-hour data to be used to prepare realistic cost-loaded construction schedules and for budget and payment tracking throughout a project.

Direct Input From CAD

No discussion of computer-aided estimating could be complete without mention of programs that allow direct input from computer-aided design packages (CAD). As long as drawings are still generated by manual draftsmen, the information will have to be manually taken off or entered into the computer with a digitizer. When the drawing is done on a computer screen, however, the takeoff can proceed automatically and simultaneously with the design. Once quantified, the program proceeds in a similar manner to the above-described applications (using database cost information, generating reports, producing schedules, etc.).

These applications are highly specialized, are the most expensive to purchase and set up, and require the user to be somewhat familiar with CAD systems as well as estimating. These are becoming more common in the offices of the larger construction companies.

Chapter 8

Using Means Plumbing Cost Data

There are occasions when a plumbing contractor or estimator is required to prepare an estimate without the benefit of finished plans and specifications. These estimates are usually for budget or preplanning purposes.

The data required for these early estimates may be readily available from the contractor's own records of past projects, especially if the plumbing contractor specializes in a particular type of building, such as hospitals or schools. Analysis of this data may reveal that for a hospital, the plumbing will cost so many dollars per hospital bed, or, in the case of a school, a cost per student or per classroom. The contractor might also maintain historical records based on dollars per square foot, particularly in the case of fire protection. These historical costs must be continually updated in order to be useful and accurate.

Many contractors or designers may not have records of similar, past projects to use as a basis for a budget estimate. They would have to find an outside source for this information, such as the cost data published by RSMeans in both book and electronic format. RSMeans has been compiling and publishing construction cost data on an annual basis since 1942. This data is based on actual construction costs gathered from across the United States and Canada.

Four Levels of Estimates

Means Plumbing Cost Data furnishes information (both historical and current) for four stages of estimating: forecasting, budgeting, planning, and bidding. For the conceptual stage, order of magnitude estimates and data are covered. As the design progresses, the estimator uses the square foot or cubic foot data. When the design has progressed to the stage that occupancy is established and the number of plumbing fixtures or toilet rooms has been determined, an assemblies estimate comes into play. The final completed bid documents containing plans and specifications provide the estimator with the opportunity to perform an accurate unit-by-unit takeoff, estimate, and bid. Unit price information is presented for this level of estimate.

Building construction estimators use these four basic types of estimates, which may be referred to by different names and may not be recognized by all as definitive. Most estimators, however, will agree that each type has its place in the construction estimating process. The following paragraphs briefly describe these estimate types. Further details may be found later in the chapter.

Order of Magnitude Estimates

Order of magnitude costs are defined in relation to the usable units that have been designed for a facility. If, for example, a hospital administrator is planning to enlarge a hospital, he needs to know the projected cost per bed. If an estimator knows the quantity of beds in a proposed hospital (or number of apartments in an apartment building, or tons of air-conditioning in a facility), the cost of the project can be estimated. It is a very quick method, and accuracy may be plus or minus 20%.

Square Foot and Cubic Foot Estimates

This type is most useful when the proposed size and use of a planned building are known, but no further details. This method can be completed within an hour or two. Depending on the source of cost information, an accuracy of plus or minus 15% can be realized.

Assemblies Estimates

An assemblies estimate is best used as a budgetary tool in the planning stages of a project when some parameters have been decided (e.g., occupancy, fixture requirements, etc.). This type of estimate could require as much as one day to complete. Because more specific information is known about the project, a plus or minus of 10% accuracy can be attained from an assemblies level estimate.

Unit Price Estimates

Working drawings and full specifications are required to complete a unit price estimate. It is the most accurate of the four types, but is also the most time-consuming to prepare. Used primarily for bidding purposes, the accuracy of a unit price estimate can be plus or minus 5%. (This means, theoretically, that all bids based on a complete set of bid documents should be within 5% of the average proposal.)

Figure 8.1 demonstrates the relative relationship of required time versus resultant accuracy of a complete estimate for each of these four basic types. It should be recognized that, as an estimator and his company gain repetitive experience on similar or identical projects, the accuracy of all four types of estimates should improve dramatically. In fact, given enough experience, square foot and assemblies estimates may closely approach the accuracy of unit price estimates.

Data Format

The major portion of *Means Plumbing Cost Data* is the unit price section. This is the primary source of unit cost data, and is organized according to the CSI MasterFormat. This index was developed by representatives of all parties concerned with the building construction industry and is has been accepted by the American Institute of Architects (AIA), the Associated General Contractors of America (AGC), and the Construction Specifications Institute (CSI). In *Means Plumbing Cost Data,* relevant parts of other divisions are included along with Division 15, Mechanical. For example, items from most other divisions appear in addition to the Division 15 entries.

CSI MasterFormat Divisions:

Division 1 - General Requirements
Division 2 - Site Construction
Division 3 - Concrete
Division 4 - Masonry
Division 5 - Metals
Division 6 - Wood & Plastics
Division 7 - Thermal & Moisture Protection
Division 8 - Doors & Windows
Division 9 - Finishes
Division 10 - Specialties
Division 11 - Equipment
Division 12 - Furnishings
Division 13 - Special Construction
Division 14 - Conveying Systems
Division 15 - Mechanical
Division 16 - Electrical

In addition to the 16 CSI divisions, Division 17, Square Foot and Cubic Foot costs, presents consolidated data from over 11,200 actual reported construction projects and provides information based on total costs, as well as costs for major components.

The Assemblies Cost Tables contain over 3,000 costs for mechanical and appropriate related assemblies, or systems. Components of the systems are fully detailed and accompanied by illustrations.

The Reference Section contains tables and reference charts. It also provides estimating procedures and explanations of cost development, which support and supplement the unit price and systems cost data. Also in the Reference Section are City Cost Indexes and Location Factors, representing the compilation of construction data for 930 major U.S. and Canadian cities and postal zones. Cost factors are given for each trade, relative to the national average.

Figure 8.1

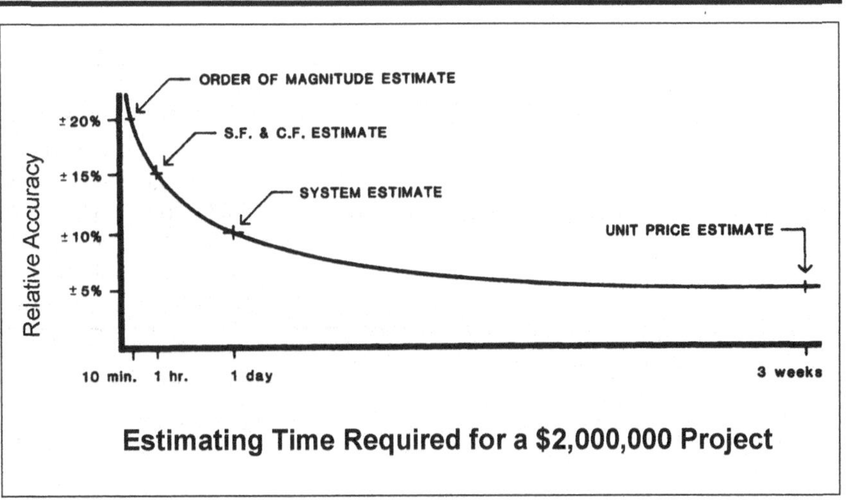

Estimating Time Required for a $2,000,000 Project

The prices presented in *Means Plumbing Cost Data* are national averages. Material and equipment costs are developed through annual contact with manufacturers, dealers, distributors, and contractors throughout the U.S. and Canada. Means' staff of engineers is constantly updating prices and keeping abreast of changes and fluctuations within the industry. Labor rates are the national average of each trade as determined from union agreements from 30 major U.S. cities. Throughout the calendar year, as new agreements are negotiated, labor costs should be factored accordingly.

Following is a list of factors and assumptions on which the costs presented in *Means Plumbing Cost Data* have been based:

- **Quality:** The costs are based on methods, materials, and workmanship in accordance with U.S. government standards and represent good, sound construction practice.
- **Overtime:** The costs, as presented, include no allowance for overtime. If overtime or premium time is anticipated, labor costs must be factored accordingly.
- **Productivity:** The daily output and labor hour figures are based on an eight-hour workday, during daylight hours.
- **Size of Project:** Costs in *Means Plumbing Cost Data* are based on commercial and industrial buildings for which total project costs are $1,000,000 and up. Large residential projects are also included.
- **Local Factors:** Weather conditions, season of the year, local union restrictions, and unusual building code requirements can all have significant impact on construction costs. The availability of a skilled labor force, sufficient materials, and even adequate energy and utilities will also affect costs. These factors vary in impact and are not necessarily dependent on location. They must be reviewed for each project in every area.

In presenting prices in *Means Plumbing Cost Data,* certain rounding rules are employed to make the numbers easy to use without significantly affecting accuracy. The rules are used consistently and are as follows:

Prices From	Rounded to the Nearest
$.01 to $5.00	$.01
$5.01 to $20.00	$.05
$20.01 to $100.00	$.50
$100.01 to $300.00	$1.00
$300.01 to $1,000.00	$5.00
$1,000.01 to $10,000.00	$25.00
$10,000.01 to $50,000.00	$100.00
$50,000.01 and above	$500.00

Unit Price Costs

The Unit Price section of *Means Plumbing Cost Data* contains a great deal of information in addition to the unit cost for each construction components. Figure 8.2 is a typical page showing costs for fire valves. Note that prices are included for several types of valves, each in a wide range of size and capacity ratings. In addition, appropriate crews, workers, and productivity data are indicated. The information and cost data is broken down and itemized in this way to provide the most detailed pricing possible. Both the unit price and the systems sections include detailed illustrations. The reference numbers enclosed in the squares refer the user to an appropriate assemblies section or reference table.

Within each individual line item, there is a description of the construction component, information regarding typical crews designated to perform the work and productivity shown as daily output and labor hours. Costs are presented in two ways: "bare," or unburdened costs, and costs with markups for overhead and profit. Figure 8.3 is a graphic representation of how to use the Unit Price section as presented in *Means Plumbing Cost Data*.

Line Numbers

Every construction item in Means unit price cost data books has a unique line number. The line number acts as an address so that each item can be quickly located. The numbering system is based on the CSI MasterFormat classification by division. In Figure 8.3 note the reverse number 03300, which represents the division number (Levels One and Two of CSI MasterFormat). The number 03310 directly below is CSI Level Three, this subdivision is the most detailed level published in the CSI MasterFormat. The next 3 digits "240" are level four and are assigned by RSMeans as major classifications. This level is predominately alphabetical and is designated by bold type face for both numbers and descriptions. Each item or line is further defined by an individual 4 digit line number. As shown in Figure 8.3 the full line number for each item consists of: a 5 digit (Levels One, Two & Three) CSI MasterFormat number indicating the CSI division and major subdivision, a 3 digit Means classification and the 4 digit unique line number. For example, in Figure 8.3 the line number for strip footings, 18″ × 9″, unreinforced is 03310-240-3900.

Line Description

Each line has a text description of the item for which costs are listed. The description may be self-contained and all inclusive, or, if indented, the complete description for a line is dependent on the information provided above. All indented items are delineations (by size, color, material, etc.) or breakdowns of previously described items. An index is provided in the back of *Means Plumbing Cost Data* to aid in locating particular items.

Crew

For each construction element (each line item), a minimum typical crew is designated as appropriate to perform the work. The crew may include one or more trades, foremen, craftsmen, helpers, and any equipment required for proper installation of the described item. If an individual trade installs the item using only hand tools, the smallest efficient number of tradesmen will be indicated (1 Plum, 2 Spri, etc.). Abbreviations for trades are shown in Figure 8.4. If more than one trade

	13900 | Fire Suppression											
	13910 | Basic Fire Protection Matl/Methd		CREW	DAILY OUTPUT	LABOR-HOURS	UNIT	2004 BARE COSTS MAT.	LABOR	EQUIP.	TOTAL	TOTAL INCL O&P	
400	8020	Three way, projecting, polished brass (R10520-310)										400
	8040	2-1/2"(3) x 4"	Q-12	4.80	3.333	Ea.	605	118		723	845	
	8070	2-1/2" (3) x 6" (R10520-320)	"	4.60	3.478		605	123		728	855	
	8100	For polished chrome, add				↓	12%					
	8200	Four way, square, flush, polished brass,										
	8240	2-1/2"(4) x 6"	Q-12	3.60	4.444	Ea.	1,600	157		1,757	2,025	
	8300	For polished chrome, add				"	10%					
	8550	Wall, vertical, flush, cast brass										
	8600	Two way, 2-1/2" x 2-1/2" x 4"	Q-12	5	3.200	Ea.	650	113		763	885	
	8660	Four way, 2-1/2"(4) x 6"		3.80	4.211		1,450	149		1,599	1,800	
	8680	Six way, 2-1/2"(6) x 6"	↓	3.40	4.706		1,925	166		2,091	2,375	
	8700	For polished chrome, add				↓	10%					
	8800	Sidewalk siamese unit, polished brass, two way										
	8820	2-1/2" x 2-1/2" x 4"	Q-12	2.50	6.400	Ea.	310	226		536	680	
	8850	2-1/2" x 2-1/2" x 6"		2	8		515	283		798	995	
	8860	3" x 3" x 4"		2.50	6.400		495	226		721	885	
	8890	3" x 3" x 6"	↓	2	8	↓	670	283		953	1,175	
	8940	For polished chrome, add					12%					
	9100	Sidewalk siamese unit, polished brass, three way										
	9120	2-1/2" x 2-1/2" x 2-1/2" x 6"	Q-12	2	8	Ea.	740	283		1,023	1,250	
	9160	For polished chrome, add					15%					
	9200	Storage house, hose only, primed steel					465			465	515	
	9220	Aluminum					1,025			1,025	1,125	
	9280	Hose and hydrant house, primed steel					630			630	695	
	9300	Aluminum					915			915	1,000	
	9340	Tools, crowbar and brackets	1 Carp	12	.667		64.50	22		86.50	106	
	9360	Combination hydrant wrench and spanner				↓	17.30			17.30	19	
	9380	Fire axe and brackets										
	9400	6 lb.	1 Carp	12	.667	Ea.	66.50	22		88.50	108	
800	0010	**FIRE VALVES** (D4020-310)										800
	0020	Angle, combination pressure adjust/restricting, rough brass										
	0030	1-1/2" (D4020-330)	1 Spri	12	.667	Ea.	46.50	26		72.50	90.50	
	0040	2-1/2"	"	7	1.143	"	98	45		143	176	
	0042	Nonpressure adjustable/restricting, rough brass										
	0044	1-1/2"	1 Spri	12	.667	Ea.	29.50	26		55.50	72	
	0046	2-1/2" (D4020-410)	"	7	1.143	"	50.50	45		95.50	124	
	0050	For polished brass, add					30%					
	0060	For polished chrome, add (R10520-310)					40%					
	0080	Wheel handle, 300 lb., 1-1/2"	1 Spri	12	.667	Ea.	29.50	26		55.50	72	
	0090	2-1/2" (R10520-320)	"	7	1.143	"	50.50	45		95.50	124	
	0100	For polished brass, add					35%					
	0110	For polished chrome, add					50%					
	1000	Ball drip, automatic, rough brass, 1/2"	1 Spri	20	.400	Ea.	8.45	15.70		24.15	33.50	
	1010	3/4"	"	20	.400	"	9.55	15.70		25.25	34.50	
	1100	Ball, 175 lb., sprinkler system, FM/UL, threaded, bronze										
	1120	Slow close										
	1150	1" size	1 Spri	19	.421	Ea.	82.50	16.55		99.05	116	
	1160	1-1/4" size		15	.533		89	21		110	130	
	1170	1-1/2" size		13	.615		113	24		137	161	
	1180	2" size	↓	11	.727		143	28.50		171.50	200	
	1190	2-1/2" size	Q-12	15	1.067	↓	192	37.50		229.50	268	
	1230	For supervisory switch kit, all sizes										
	1240	One circuit, add	1 Spri	48	.167	Ea.	62.50	6.55		69.05	78.50	
	1280	Quarter turn for trim										
	1300	1/2" size	1 Spri	22	.364	Ea.	10.90	14.25		25.15	33.50	

(Reprinted from Means Plumbing Cost Data 2004)

Figure 8.2

How to Use the Unit Price Pages

The following is a detailed explanation of a sample entry in the Unit Price Section. Next to each bold number below is the item being described with appropriate component of the sample entry following in parenthesis. Some prices are listed as bare costs, others as costs that include overhead and profit of the installing contractor. In most cases, if the work is to be subcontracted, the general contractor will need to add an additional markup (RSMeans suggests using 10%) to the figures in the column "Total Incl. O&P."

Division Number/Title (03300/Cast-In-Place Concrete)

Use the Unit Price Section Table of Contents to locate specific items. The sections are classified according to the CSI MasterFormat (1995 Edition).

Line Numbers (03310 240 3900)

Each unit price line item has been assigned a unique 12-digit code based on the CSI MasterFormat classification.

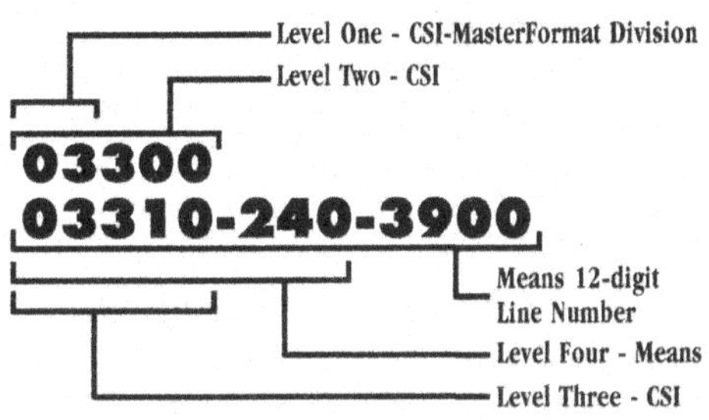

Description (Concrete-In-Place, etc.)

Each line item is described in detail. Sub-items and additional sizes are indented beneath the appropriate line items. The first line or two after the main item (in boldface) may contain descriptive information that pertains to all line items beneath this boldface listing.

Reference Number Information

R03310 -010 You'll see reference numbers shown in bold rectangles at the beginning of some sections. These refer to related items in the Reference Section, visually identified by a vertical gray bar on the edge of pages.

The relation may be: (1) an estimating procedure that should be read before estimating, (2) an alternate pricing method, or (3) technical information.

The "R" designates the Reference Section. The numbers refer to the MasterFormat classification system.

It is strongly recommended that you review all reference numbers that appear within the section in which you are working.

Note: Not all reference numbers appear in all Means publications.

03300 | Cast-In-Place Concrete

	03310 | Structural Concrete		CREW	DAILY OUTPUT	LABOR-HOURS	UNIT	2004 BARE COSTS MAT.	LABOR	EQUIP.	TOTAL	TOTAL INCL O&P	
240	0010	**CONCRETE IN PLACE** Including forms (4 uses), reinforcing	R03310 -010									240
	0050	steel and finishing unless otherwise indicated										
	0300	Beams, 5 kip per L.F., 10' span	R03310 -100	C-14A	15	04	C.Y.	233	425	8	706	975	
	0350	25' span		"	18.55	10.782	↓	214	355	40	609		
	3520	For 7 to 20 stories high		C-8	21,200	.003	S.F.		.08	.03	.11		
	3800	Footings, spread under 1 C.Y.		C-14C	38.07	2.942	C.Y.	105	93	.68	198.68	263	
	3850	Over 5 C.Y.			81.04	1.382		96.50	43.50	.32	140.32	175	
	3900	Footings, strip, 18" x 9", unreinforced			40	2.800		91	88.50	.64	180.14	240	
	3920	18" x 9", reinforced			35	3.200		105	101	.74	206.74	275	
	3925	20" x 10", unreinforced			45	2.489		88	78.50	.57	167.07	221	
	3930	20" x 10", reinforced			40	2.800		99.50	88.50	.64	188.64	249	
	3935	24" x 12", unreinforced			55	2.036		87	64.50	.47	151.97	198	
	3940	24" x 12", reinforced			48	2.333		98.50	74	.54	173.04	226	
	3945	36" x 12", unreinforced			70	1.600		83.50	50.50	.37	134.37	172	
	3950	36" x 12", reinforced			60	1.867		94	59	.43	153.43	196	
	4000	Foundation mat, under 10 C.Y.		↓	38.67	2.896	↓	129	91.50	.67	221.17	287	

(Reprinted from Means Plumbing Cost Data 2004)

Figure 8.3

Crew (C-14C)

The "Crew" column designates the typical trade or crew used to install the item. If an installation can be accomplished by one trade and requires no power equipment, that trade and the number of workers are listed (for example, "2 Carpenters"). If an installation requires a composite crew, a crew code designation is listed (for example, "C-14C"). You'll find full details on all composite crews in the Crew Listings.

- For a complete list of all trades utilized in this book and their abbreviations, see the inside back cover.

Crews

Crew No.	Bare Costs		Incl. Subs O & P		Cost Per Labor-Hour	
Crew C-14C	Hr.	Daily	Hr.	Daily	Bare Costs	Incl. O&P
1 Carpenter Foreman (out)	$35.00	$280.00	$54.55	$436.40	$31.63	$49.56
6 Carpenters	33.00	1584.00	51.40	2467.20		
2 Rodmen (reinf.)	37.10	593.60	61.25	980.00		
4 Laborers	26.00	832.00	40.50	1296.00		
1 Cement Finisher	31.55	252.40	46.45	371.60		
1 Gas Engine Vibrator		26.00		28.60	.23	.26
112 L.H., Daily Totals		$3568.00		$5579.80	$31.86	$49.82

Productivity: Daily Output (40.0)/Labor-Hours (2.80)

The "Daily Output" represents the typical number of units the designated crew will install in a normal 8-hour day. To find out the number of days the given crew would require to complete the installation, divide your quantity by the daily output. For example:

Quantity	÷	Daily Output	=	Duration
100 C.Y.	÷	40.0/ Crew Day	=	2.50 Crew Days

The "Labor-Hours" figure represents the number of labor-hours required to install one unit of work. To find out the number of labor-hours required for your particular task, multiply the quantity of the item times the number of labor-hours shown. For example:

Quantity	x	Productivity Rate	=	Duration
100 C.Y.	x	2.80 Labor-Hours/ C.Y.	=	280 Labor-Hours

Unit (C.Y.)

The abbreviated designation indicates the unit of measure upon which the price, production, and crew are based (C.Y. = Cubic Yard). For a complete listing of abbreviations refer to the Abbreviations Listing in the Reference Section of this book.

Bare Costs:

Mat. (Bare Material Cost) (91)

The unit material cost is the "bare" material cost with no overhead and profit included. *Costs shown reflect national average material prices for January of the current year and include delivery to the job site. No sales taxes are included.*

Labor (88.50)

The unit labor cost is derived by multiplying bare labor-hour costs for Crew C-14C by labor-hour units. The bare labor-hour cost is found in the Crew Section under C-14C. (If a trade is listed, the hourly labor cost—the wage rate—is found on the inside back cover.)

Labor-Hour Cost Crew C-14C	x	Labor-Hour Units	=	Labor
$31.63	x	2.80	=	$88.50

Equip. (Equipment) (.64)

Equipment costs for each crew are listed in the description of each crew. Tools or equipment whose value justifies purchase or ownership by a contractor are considered overhead as shown on the inside back cover. The unit equipment cost is derived by multiplying the bare equipment hourly cost by the labor-hour units.

Equipment Cost Crew C-14C	x	Labor-Hour Units	=	Equip.
.23	x	2.80	=	$.64

Total (180.14)

The total of the bare costs is the arithmetic total of the three previous columns: mat., labor, and equip.

Material	+	Labor	+	Equip.	=	Total
$91	+	$88.50	+	$.64	=	$180.14

Total Costs Including O&P

This figure is the sum of the bare material cost plus 10% for profit; the bare labor cost plus total overhead and profit (per the inside back cover or, if a crew is listed, from the crew listings); and the bare equipment cost plus 10% for profit.

Material is Bare Material Cost + 10% = 91 + 9.10	=	$100.10
Labor for Crew C-14C = Labor-Hour Cost (49.56) x Labor-Hour Units (2.80)	=	$138.77
Equip. is Bare Equip. Cost + 10% = .64 + .06	=	$ .70
Total (Rounded)	=	$240

(Reprinted from Means Plumbing Cost Data 2004)

Figure 8.3 (cont'd.)

is required to install the item and/or if powered equipment is needed, a crew member will be designated (Q-19, Q-8, etc.). A complete listing of crews is presented in the reference pages of *Means Plumbing Cost Data*. (See Figure 8.5.) On these pages, each crew is broken down into the following components:

1. Number and type of workers designated.
2. Number, size, and type of equipment required.
3. Hourly labor costs listed two ways: "bare," with the base rate including fringe benefits only; and "including overhead and profit," the installing contractor's billing rate. (See Figure 8.4 for labor rate information.)
4. Daily equipment costs, based on the weekly equipment rental cost divided by 5, plus the hourly operating cost, times 8 hours. This cost is listed two ways: as a bare cost with a 10% markup to cover handling and management costs.
5. Labor and equipment are broken down further into: costs per labor hour for labor, and cost per labor hour for the equipment.
6. The total daily labor hours for the crew.
7. The total bare costs per day for the crew, including equipment.
8. The total daily cost of the crew, including the installing contractor's overhead and profit.

The total daily cost of the required crew is used to calculate the unit installation cost for each item (for both bare costs and costs including overhead and profit).

The crew designation does not mean that this is the only crew that can perform the work. Crew size and content have been developed and chosen based on practical experience and feedback from contractors. These designations represent a labor and equipment makeup commonly found in the industry. The most appropriate crew for a given task is best determined based on particular project requirements. Unit costs may vary if crew sizes or content are significantly changed.

Figure 8.6 is a page from Division 01 of *Means Plumbing Cost Data*. This page lists the equipment costs used in the presentation and calculation of the crew costs and unit price data. Rental costs are shown as daily, weekly, and monthly rates. The Hourly Operating Cost represents the cost of fuel. Lubrication, and routine maintenance. Equipment costs used in the crews are calculated as follows:

Line Number:	01590-400-7800
Equipment:	Electric Welder, 300Amp.
Rent per week:	\$180.00
Hourly Operating Cost:	\$5.22

$$\frac{\text{Weekly rental}}{\text{5 days}} + (\text{Hourly Oper. Cost} \times \text{8 hrs/day}) = \text{Daily Equipment Costs}$$

$$\frac{180}{5} + (\$5.22 \times 8) = \$77.76 \text{ per day}$$

Installing Contractor's Overhead & Profit

Below are the **average** installing contractor's percentage mark-ups applied to base labor rates to arrive at typical billing rates.

Column A: Labor rates are based on union wages averaged for 30 major U.S. cities. Base rates including fringe benefits are listed hourly and daily. These figures are the sum of the wage rate and employer-paid fringe benefits such as vacation pay, employer-paid health and welfare costs, pension costs, plus appropriate training and industry advancement funds costs.

Column B: Workers' Compensation rates are the national average of state rates established for each trade.

Column C: Column C lists average fixed overhead figures for all trades. Included are Federal and State Unemployment costs set at 6.2%; Social Security Taxes (FICA) set at 7.65%; Builder's Risk Insurance costs set at 0.44%; and Public Liability costs set at 2.02%. All the percentages except those for Social Security Taxes vary from state to state as well as from company to company.

Columns D and E: Percentages in Columns D and E are based on the presumption that the installing contractor has annual billing of $4,000,000 and up. Overhead percentages may increase with smaller annual billing. The overhead percentages for any given contractor may vary greatly and depend on a number of factors, such as the contractor's annual volume, engineering and logistical support costs, and staff requirements. The figures for overhead and profit will also vary depending on the type of job, the job location, and the prevailing economic conditions. All factors should be examined very carefully for each job.

Column F: Column F lists the total of Columns B, C, D, and E.

Column G: Column G is Column A (hourly base labor rate) multiplied by the percentage in Column F (O&P percentage).

Column H: Column H is the total of Column A (hourly base labor rate) plus Column G (Total O&P).

Column I: Column I is Column H multiplied by eight hours.

		A		B	C	D	E	F	G	H	I
		Base Rate Incl. Fringes		Workers' Comp. Ins.	Average Fixed Over-head	Over-head	Profit	Total Overhead & Profit		Rate with O & P	
Abbr.	Trade	Hourly	Daily					%	Amount	Hourly	Daily
Skwk	Skilled Workers Average (35 trades)	$33.65	$269.20	16.2%	16.3%	13.0%	10%	55.5%	$18.70	$52.35	$418.80
	Helpers Average (5 trades)	24.55	196.40	17.9		11.0		55.2	13.55	38.10	304.80
	Foreman Average, Inside ($.50 over trade)	34.15	273.20	16.2		13.0		55.5	18.95	53.10	424.80
	Foreman Average, Outside ($2.00 over trade)	35.65	285.20	16.2		13.0		55.5	19.80	55.45	443.60
Clab	Common Building Laborers	26.00	208.00	18.5		11.0		55.8	14.50	40.50	324.00
Asbe	Asbestos/Insulation Workers/Pipe Coverers	36.00	288.00	15.2		16.0		57.5	20.70	56.70	453.60
Boil	Boilermakers	41.25	330.00	13.1		16.0		55.4	22.85	64.10	512.80
Bric	Bricklayers	34.25	274.00	15.0		11.0		52.3	17.90	52.15	417.20
Brhe	Bricklayer Helpers	25.60	204.80	15.0		11.0		52.3	13.40	39.00	312.00
Carp	Carpenters	33.00	264.00	18.5		11.0		55.8	18.40	51.40	411.20
Cefi	Cement Finishers	31.55	252.40	9.9		11.0		47.2	14.90	46.45	371.60
Elec	Electricians	39.40	315.20	6.4		16.0		48.7	19.20	58.60	468.80
Elev	Elevator Constructors	41.60	332.80	7.2		16.0		49.5	20.60	62.20	497.60
Eqhv	Equipment Operators, Crane or Shovel	34.80	278.40	10.3		14.0		50.6	17.60	52.40	419.20
Eqmd	Equipment Operators, Medium Equipment	33.65	269.20	10.3		14.0		50.6	17.05	50.70	405.60
Eqlt	Equipment Operators, Light Equipment	32.15	257.20	10.3		14.0		50.6	16.25	48.40	387.20
Eqol	Equipment Operators, Oilers	29.20	233.60	10.3		14.0		50.6	14.80	44.00	352.00
Eqmm	Equipment Operators, Master Mechanics	35.20	281.60	10.3		14.0		50.6	17.80	53.00	424.00
Glaz	Glaziers	32.05	256.40	13.8		11.0		51.1	16.40	48.45	387.60
Lath	Lathers	30.55	244.40	10.7		11.0		48.0	14.65	45.20	361.60
Marb	Marble Setters	31.95	255.60	15.0		11.0		52.3	16.70	48.65	389.20
Mill	Millwrights	34.35	274.80	10.3		11.0		47.6	16.35	50.70	405.60
Mstz	Mosaic & Terrazzo Workers	31.60	252.80	9.6		11.0		46.9	14.80	46.40	371.20
Pord	Painters, Ordinary	29.60	236.80	12.9		11.0		50.2	14.85	44.45	355.60
Psst	Painters, Structural Steel	30.05	240.40	50.0		11.0		87.3	26.25	56.30	450.40
Pape	Paper Hangers	29.35	234.80	12.9		11.0		50.2	14.75	44.10	352.80
Pile	Pile Drivers	32.05	256.40	22.9		16.0		65.2	20.90	52.95	423.60
Plas	Plasterers	29.95	239.60	14.6		11.0		51.9	15.55	45.50	364.00
Plah	Plasterer Helpers	25.80	206.40	14.6		11.0		51.9	13.40	39.20	313.60
Plum	Plumbers	39.60	316.80	7.8		16.0		50.1	19.85	59.45	475.60
Rodm	Rodmen (Reinforcing)	37.10	296.80	24.8		14.0		65.1	24.15	61.25	490.00
Rofc	Roofers, Composition	28.45	227.60	31.8		11.0		69.1	19.65	48.10	384.80
Rots	Roofers, Tile & Slate	28.65	229.20	31.8		11.0		69.1	19.80	48.45	387.60
Rohe	Roofers, Helpers (Composition)	20.95	167.60	31.8		11.0		69.1	14.50	35.45	283.60
Shee	Sheet Metal Workers	38.80	310.40	11.1		16.0		53.4	20.70	59.50	476.00
Spri	Sprinkler Installers	39.25	314.00	9.0		16.0		51.3	20.15	59.40	475.20
Stpi	Steamfitters or Pipefitters	39.75	318.00	7.8		16.0		50.1	19.90	59.65	477.20
Ston	Stone Masons	33.85	270.80	15.0		11.0		52.3	17.70	51.55	412.40
Sswk	Structural Steel Workers	37.15	297.20	38.9		14.0		79.2	29.40	66.55	532.40
Tilf	Tile Layers	31.50	252.00	9.6		11.0		46.9	14.75	46.25	370.00
Tilh	Tile Layers Helpers	24.45	195.60	9.6		11.0		46.9	11.45	35.90	287.20
Trlt	Truck Drivers, Light	25.70	205.60	15.1		11.0		52.4	13.45	39.15	313.20
Trhv	Truck Drivers, Heavy	26.45	211.60	15.1		11.0		52.4	13.85	40.30	322.40
Sswl	Welders, Structural Steel	37.15	297.20	38.9		14.0		79.2	29.40	66.55	532.40
Wrck	*Wrecking	26.00	208.00	40.5		11.0		77.8	20.25	46.25	370.00

*Not included in averages

(Reprinted from Means Plumbing Cost Data 2004)

Figure 8.4

Crews

Crew No.	Bare Costs		Incl. Subs O & P		Cost Per Labor-Hour	
Crew L-7	**Hr.**	**Daily**	**Hr.**	**Daily**	**Bare Costs**	**Incl. O&P**
2 Carpenters	$33.00	$528.00	$51.40	$822.40	$31.91	$49.31
1 Building Laborer	26.00	208.00	40.50	324.00		
.5 Electrician	39.40	157.60	58.60	234.40		
28 L.H., Daily Totals		$893.60		$1380.80	$31.91	$49.31
Crew L-8	**Hr.**	**Daily**	**Hr.**	**Daily**	**Bare Costs**	**Incl. O&P**
2 Carpenters	$33.00	$528.00	$51.40	$822.40	$34.32	$53.01
.5 Plumber	39.60	158.40	59.45	237.80		
20 L.H., Daily Totals		$686.40		$1060.20	$34.32	$53.01
Crew L-9	**Hr.**	**Daily**	**Hr.**	**Daily**	**Bare Costs**	**Incl. O&P**
1 Labor Foreman (inside)	$26.50	$212.00	$41.30	$330.40	$30.08	$48.48
2 Building Laborers	26.00	416.00	40.50	648.00		
1 Struc. Steel Worker	37.15	297.20	66.55	532.40		
.5 Electrician	39.40	157.60	58.60	234.40		
36 L.H., Daily Totals		$1082.80		$1745.20	$30.08	$48.48
Crew L-10	**Hr.**	**Daily**	**Hr.**	**Daily**	**Bare Costs**	**Incl. O&P**
1 Structural Steel Foreman	$39.15	$313.20	$70.15	$561.20	$37.03	$63.03
1 Structural Steel Worker	37.15	297.20	66.55	532.40		
1 Equip. Oper. (crane)	34.80	278.40	52.40	419.20		
1 Hyd. Crane, 12 Ton		612.00		673.20	25.50	28.05
24 L.H., Daily Totals		$1500.80		$2186.00	$62.53	$91.08
Crew M-1	**Hr.**	**Daily**	**Hr.**	**Daily**	**Bare Costs**	**Incl. O&P**
3 Elevator Constructors	$41.60	$998.40	$62.20	$1492.80	$39.52	$59.10
1 Elevator Apprentice	33.30	266.40	49.80	398.40		
5 Hand Tools		70.00		77.00	2.19	2.41
32 L.H., Daily Totals		$1334.80		$1968.20	$41.71	$61.51
Crew M-3	**Hr.**	**Daily**	**Hr.**	**Daily**	**Bare Costs**	**Incl. O&P**
1 Electrician Foreman (out)	$41.40	$331.20	$61.55	$492.40	$35.46	$53.35
1 Common Laborer	26.00	208.00	40.50	324.00		
.25 Equipment Operator, Medium	33.65	67.30	50.70	101.40		
1 Elevator Constructor	41.60	332.80	62.20	497.60		
1 Elevator Apprentice	33.30	266.40	49.80	398.40		
.25 Crane, SP, 4 x 4, 20 ton		143.60		157.95	4.22	4.65
34 L.H., Daily Totals		$1349.30		$1971.75	$39.68	$58.00
Crew M-4	**Hr.**	**Daily**	**Hr.**	**Daily**	**Bare Costs**	**Incl. O&P**
1 Electrician Foreman (out)	$41.40	$331.20	$61.55	$492.40	$35.18	$52.92
1 Common Laborer	26.00	208.00	40.50	324.00		
.25 Equipment Operator, Crane	34.80	69.60	52.40	104.80		
.25 Equipment Operator, Oiler	29.20	58.40	44.00	88.00		
1 Elevator Constructor	41.60	332.80	62.20	497.60		
1 Elevator Apprentice	33.30	266.40	49.80	398.40		
.25 Crane, Hyd, SP, 4WD, 40 Ton		215.30		236.85	5.98	6.58
36 L.H., Daily Totals		$1481.70		$2142.05	$41.16	$59.50
Crew Q-1	**Hr.**	**Daily**	**Hr.**	**Daily**	**Bare Costs**	**Incl. O&P**
1 Plumber	$39.60	$316.80	$59.45	$475.60	$35.65	$53.53
1 Plumber Apprentice	31.70	253.60	47.60	380.80		
16 L.H., Daily Totals		$570.40		$856.40	$35.65	$53.53
Crew Q-1C	**Hr.**	**Daily**	**Hr.**	**Daily**	**Bare Costs**	**Incl. O&P**
1 Plumber	$39.60	$316.80	$59.45	$475.60	$34.98	$52.58
1 Plumber Apprentice	31.70	253.60	47.60	380.80		
1 Equip. Oper. (medium)	33.65	269.20	50.70	405.60		
1 Trencher, Chain		1310.00		1441.00	54.58	60.04
24 L.H., Daily Totals		$2149.60		$2703.00	$89.56	$112.62

Crew No.	Bare Costs		Incl. Subs O & P		Cost Per Labor-Hour	
Crew Q-2	**Hr.**	**Daily**	**Hr.**	**Daily**	**Bare Costs**	**Incl. O&P**
2 Plumbers	$39.60	$633.60	$59.45	$951.20	$36.97	$55.50
1 Plumber Apprentice	31.70	253.60	47.60	380.80		
24 L.H., Daily Totals		$887.20		$1332.00	$36.97	$55.50
Crew Q-3	**Hr.**	**Daily**	**Hr.**	**Daily**	**Bare Costs**	**Incl. O&P**
1 Plumber Foreman (inside)	$40.10	$320.80	$60.20	$481.60	$37.75	$56.67
2 Plumbers	39.60	633.60	59.45	951.20		
1 Plumber Apprentice	31.70	253.60	47.60	380.80		
32 L.H., Daily Totals		$1208.00		$1813.60	$37.75	$56.67
Crew Q-4	**Hr.**	**Daily**	**Hr.**	**Daily**	**Bare Costs**	**Incl. O&P**
1 Plumber Foreman (inside)	$40.10	$320.80	$60.20	$481.60	$37.75	$56.67
1 Plumber	39.60	316.80	59.45	475.60		
1 Welder (plumber)	39.60	316.80	59.45	475.60		
1 Plumber Apprentice	31.70	253.60	47.60	380.80		
1 Electric Welding Mach.		77.75		85.55	2.43	2.67
32 L.H., Daily Totals		$1285.75		$1899.15	$40.18	$59.34
Crew Q-5	**Hr.**	**Daily**	**Hr.**	**Daily**	**Bare Costs**	**Incl. O&P**
1 Steamfitter	$39.75	$318.00	$59.65	$477.20	$35.78	$53.70
1 Steamfitter Apprentice	31.80	254.40	47.75	382.00		
16 L.H., Daily Totals		$572.40		$859.20	$35.78	$53.70
Crew Q-6	**Hr.**	**Daily**	**Hr.**	**Daily**	**Bare Costs**	**Incl. O&P**
2 Steamfitters	$39.75	$636.00	$59.65	$954.40	$37.10	$55.68
1 Steamfitter Apprentice	31.80	254.40	47.75	382.00		
24 L.H., Daily Totals		$890.40		$1336.40	$37.10	$55.68
Crew Q-7	**Hr.**	**Daily**	**Hr.**	**Daily**	**Bare Costs**	**Incl. O&P**
1 Steamfitter Foreman (inside)	$40.25	$322.00	$60.40	$483.20	$37.89	$56.86
2 Steamfitters	39.75	636.00	59.65	954.40		
1 Steamfitter Apprentice	31.80	254.40	47.75	382.00		
32 L.H., Daily Totals		$1212.40		$1819.60	$37.89	$56.86
Crew Q-8	**Hr.**	**Daily**	**Hr.**	**Daily**	**Bare Costs**	**Incl. O&P**
1 Steamfitter Foreman (inside)	$40.25	$322.00	$60.40	$483.20	$37.89	$56.86
1 Steamfitter	39.75	318.00	59.65	477.20		
1 Welder (steamfitter)	39.75	318.00	59.65	477.20		
1 Steamfitter Apprentice	31.80	254.40	47.75	382.00		
1 Electric Welding Mach.		77.75		85.55	2.43	2.67
32 L.H., Daily Totals		$1290.15		$1905.15	$40.32	$59.53
Crew Q-9	**Hr.**	**Daily**	**Hr.**	**Daily**	**Bare Costs**	**Incl. O&P**
1 Sheet Metal Worker	$38.80	$310.40	$59.50	$476.00	$34.92	$53.58
1 Sheet Metal Apprentice	31.05	248.40	47.65	381.20		
16 L.H., Daily Totals		$558.80		$857.20	$34.92	$53.58
Crew Q-10	**Hr.**	**Daily**	**Hr.**	**Daily**	**Bare Costs**	**Incl. O&P**
2 Sheet Metal Workers	$38.80	$620.80	$59.50	$952.00	$36.22	$55.55
1 Sheet Metal Apprentice	31.05	248.40	47.65	381.20		
24 L.H., Daily Totals		$869.20		$1333.20	$36.22	$55.55
Crew Q-11	**Hr.**	**Daily**	**Hr.**	**Daily**	**Bare Costs**	**Incl. O&P**
1 Sheet Metal Foreman (inside)	$39.30	$314.40	$60.30	$482.40	$36.99	$56.74
2 Sheet Metal Workers	38.80	620.80	59.50	952.00		
1 Sheet Metal Apprentice	31.05	248.40	47.65	381.20		
32 L.H., Daily Totals		$1183.60		$1815.60	$36.99	$56.74

(Reprinted from Means Plumbing Cost Data 2004)

Figure 8.5

01500 | Temporary Facilities & Controls

	01590 | Equipment Rental		UNIT	HOURLY OPER. COST	RENT PER DAY	RENT PER WEEK	RENT PER MONTH	CREW EQUIPMENT COST/DAY	
400	7690	Large production vacuum loader, 3150 CFM	Ea.	15.63	630	1,890	5,675	503.05	400
	7700	Welder, electric, 200 amp		3.74	57	171	515	64.10	
	7800	300 amp		5.22	60	180	540	77.75	
	7900	Gas engine, 200 amp		5.30	35.50	106	320	63.60	
	8000	300 amp		6.20	42.50	128	385	75.20	
	8100	Wheelbarrow, any size		.06	10.35	31	93	6.70	
	8200	Wrecking ball, 4000 lb.		1.85	68.50	205	615	55.80	
500	0010	**HIGHWAY EQUIPMENT RENTAL**							500
	0050	Asphalt batch plant, portable drum mixer, 100 ton/hr.	Ea.	55.10	1,350	4,020	12,100	1,245	
	0060	200 ton/hr.		61.10	1,400	4,215	12,600	1,332	
	0070	300 ton/hr.		71.20	1,675	4,990	15,000	1,568	
	0100	Backhoe attachment, long stick, up to 185 HP, 10.5' long		.30	19.65	59	177	14.20	
	0140	Up to 250 HP, 12' long		.32	21.50	64	192	15.35	
	0180	Over 250 HP, 15' long		.42	27.50	83	249	19.95	
	0200	Special dipper arm, up to 100 HP, 32' long		.86	57.50	172	515	41.30	
	0240	Over 100 HP, 33' long		1.08	72	216	650	51.85	
	0300	Concrete batch plant, portable, electric, 200 CY/Hr		12.67	645	1,940	5,825	489.35	
	0500	Grader attachment, ripper/scarifier, rear mounted							
	0520	Up to 135 HP	Ea.	2.70	58.50	175	525	56.60	
	0540	Up to 180 HP		3.20	75	225	675	70.60	
	0580	Up to 250 HP		3.55	86.50	260	780	80.40	
	0700	Pvmt. removal bucket, for hyd. excavator, up to 90 HP		1.30	41.50	125	375	35.40	
	0740	Up to 200 HP		1.50	63.50	190	570	50	
	0780	Over 200 HP		1.65	76.50	230	690	59.20	
	0900	Aggregate spreader, self-propelled, 187 HP		37.15	790	2,370	7,100	771.20	
	1000	Chemical spreader, 3 C.Y.		2.20	80.50	242	725	66	
	1900	Hammermill, traveling, 250 HP		38.85	1,675	5,020	15,100	1,315	
	2000	Horizontal borer, 3" diam, 13 HP gas driven		3.85	53.50	160	480	62.80	
	2200	Hydromulchers, gas power, 3000 gal., for truck mounting		10.50	187	560	1,675	196	
	2400	Joint & crack cleaner, walk behind, 25 HP		2.05	46.50	140	420	44.40	
	2500	Filler, trailer mounted, 400 gal., 20 HP		6	182	545	1,625	157	
	3000	Paint striper, self propelled, double line, 30 HP		5.15	155	465	1,400	134.20	
	3200	Post drivers, 6" I-Beam frame, for truck mounting		8.15	415	1,250	3,750	315.20	
	3400	Road sweeper, self propelled, 8' wide, 90 HP		23.50	385	1,160	3,475	420	
	4000	Road mixer, self-propelled, 130 HP		28.35	590	1,770	5,300	580.80	
	4100	310 HP		53.85	1,975	5,895	17,700	1,610	
	4200	Cold mix paver, incl pug mill and bitumen tank,							
	4220	165 HP	Ea.	65.05	1,925	5,770	17,300	1,674	
	4250	Paver, asphalt, wheel or crawler, 130 H.P., diesel		58.05	1,650	4,965	14,900	1,457	
	4300	Paver, road widener, gas 1' to 6', 67 HP		28.70	635	1,900	5,700	609.60	
	4400	Diesel, 2' to 14', 88 HP		38.80	995	2,980	8,950	906.40	
	4600	Slipform pavers, curb and gutter, 2 track, 75 HP		22.95	650	1,950	5,850	573.60	
	4700	4 track, 165 HP		31.05	730	2,195	6,575	687.40	
	4800	Median barrier, 215 HP		31.50	755	2,260	6,775	704	
	4901	Trailer, low bed, 75 ton capacity		7.85	182	545	1,625	171.80	
	5000	Road planer, walk behind, 10" cutting width, 10 HP		2	26	78	234	31.60	
	5100	Self propelled, 12" cutting width, 64 HP		5.30	288	865	2,600	215.40	
	5200	Pavement profiler, 4' to 6' wide, 450 HP		141.40	2,800	8,380	25,100	2,807	
	5300	8' to 10' wide, 750 HP		225.75	4,250	12,725	38,200	4,351	
	5400	Roadway plate, steel, 1"x8'x20'		.06	9.65	29	87	6.30	
	5600	Stabilizer,self-propelled, 150 HP		25.25	560	1,685	5,050	539	
	5700	310 HP		42.50	1,200	3,605	10,800	1,061	
	5800	Striper, thermal, truck mounted 120 gal. paint, 150H.P.		32.40	485	1,450	4,350	549.20	
	6000	Tar kettle, 330 gal., trailer mounted		2.37	36.50	110	330	40.95	
	7000	Tunnel locomotive, diesel, 8 to 12 ton		20.45	560	1,675	5,025	498.60	
	7005	Electric, 10 ton		20.45	635	1,905	5,725	544.60	
	7010	Muck cars, 1/2 C.Y. capacity		1.50	20.50	62	186	24.40	

(Reprinted from Means Plumbing Cost Data 2004)

Figure 8.6

Units

The unit column (see Figure 8.2) defines the component for which the costs have been calculated. It is this "unit" on which unit price estimating is based. The units represent standards estimating and quantity takeoff procedures. However, the estimator should always check to be sure that the units taken off are the same as those priced. A list of standard abbreviations is included at the back of *Means Plumbing Cost Data*.

Bare Costs

The four columns listed under bare costs (material, labor, equipment, and total) represent the actual cost of construction items to the contractor. In other words, bare costs are those that do not include overhead and profit of the installing contractor, whether it is a subcontractor or a general contracting company using its own crews.

Material: Material costs are based on the national average contractor purchase price delivered to the job site. Delivered costs are assumed to be within a 20-mile radius of metropolitan areas. No sales tax is included because of variations from state to state.

The prices are based on quantities that would normally be purchases for complete buildings or projects costing $1,000,000 and up. Prices for small quantities must be adjusted accordingly. If more current costs for materials are available for the appropriate location, it is recommended that adjustments be made for the unit costs to reflect any cost difference.

Labor: Labor costs are calculated by multiplying the "bare labor cost" per labor hour by the number of labor hours, from the "labor hours" column. The bare labor rate is determined by adding the base rate plus fringe benefits. The base rate is the actual hourly wage of a worker used in figuring payroll. It is from this figure that employee deductions are taken (federal withholdings, FICA, state withholdings, and so forth). Fringe benefits include all employer-paid benefits, above and beyond the payroll amount (employer-paid health insurance, vacation pay, pension, profit sharing, and so forth). The "bare labor cost" is, therefore, the actual amount that the contractor must pay directly for construction workers. Figure 8.4 shows labor rates for the 35 construction trades plus skilled worker, helper, and foreman averages. These rates are the average of union wage agreements effective January 1 of the current year from 30 major cities in the U.S. The bare labor cost for each trade is shown in column A as the base rate including fringes. Refer to the "crew" column to determine what rate is used to calculate the bare labor cost for a particular line item.

Equipment: Equipment costs are calculated by multiplying the "bare equipment cost" per labor hour, from the appropriate crew listing, by the labor hours in the "labor hours" column. The calculation of the equipment portion of installation costs is outlined earlier in this chapter.

Total Bare Costs

This column simply represents the arithmetic sum of the bare material, labor, and equipment costs. This total is the average cost to the contractor for the particular item of construction, supplied and installed, or "in place." No overhead and/or profit is included.

Total Including Overhead and Profit

This column represents the total cost of an item, including the installing contractor's overhead and profit. The installing contractor could be either the prime mechanical contractor or a subcontractor. If these costs are used for an item to be installed by a subcontractor, the prime mechanical contractor should include an additional percentage (usually 10% to 20%) to cover the expenses of supervision and management. Consideration must be given also to sub-subcontractors who often appear in mechanical contracting. An example might be an electrical sub to the temperature control subcontractor who, of course, is a sub to the HVAC contractor. Each contractor has his own overhead and profit markup. The costs in the "total including overhead and profit" are the mathematical sum of the following three calculations:

- Bare material costs plus 10%
- Labor costs, including overhead and profit, per labor hour times the number of labor hours
- Equipment costs, including overhead and profit, per labor hour times the number of labor hours

The labor and equipment costs, including overhead and profit, are found in the appropriate crew listings. The overhead and profit percentage factor for labor is obtained from Column F in Figure 8.4. The overhead and profit for equipment is 10% of the "bare" cost.

Labor costs are increased by percentages for overhead and profit, depending on trade, as shown in Figure 8.4. The resulting rates are listed in the right-hand columns of the same figure. Note that the percentage increase for overhead and profit for plumbers is 50.1% of the base rate. The following items are included in the increase for overhead and profit, as shown in Figure 8.4.

Workers' Compensation and Employer's Liability: Workers' Compensation and Employer's Liability insurance rates vary from state to state and are tied into the construction trade safety records of each state. Rates also vary by state according to hazards involved. (See Figure 8.7, average insurance rates as of January 2004.) The proper authorities will most likely keep the contractor well informed of the rates and obligations.

State and Federal Unemployment Insurance: The employer's tax rate is adjusted by a merit-rating system according to the number of former employees applying for benefits. Contractors who offer a maximum of steady employment can enjoy a reduction in the unemployment tax rate.

Employer-Paid Social Security (FICA): The tax rate is adjusted annually by the federal government. It is a percentage of an employee's salary up to a maximum annual contribution.

Builder's Risk and Public Liability: These insurance rates very according to the trades involved and the state in which the work is done.

Overhead: The column listed as "overhead" provides percentages to be added for office or operating overhead. This is the cost of doing business. Percentages are presented as national averages by trade as shown in Figure 8.4. Note that the operating overhead costs are applied to labor only, in *Means Plumbing Cost Data*.

General Requirements	R011	Overhead & Miscellaneous Data

R01100-060 Workers' Compensation Insurance Rates by Trade

The table below tabulates the national averages for Workers' Compensation insurance rates by trade and type of building. The average "Insurance Rate" is multiplied by the "% of Building Cost" for each trade. This produces the "Workers' Compensation Cost" by % of total labor cost, to be added for each trade by building type to determine the weighted average Workers' Compensation rate for the building types analyzed.

Trade	Insurance Rate (% Labor Cost)		% of Building Cost			Workers' Compensation		
	Range	Average	Office Bldgs.	Schools & Apts.	Mfg.	Office Bldgs.	Schools & Apts.	Mfg.
Excavation, Grading, etc.	3.5% to 18.5%	10.3%	4.8%	4.9%	4.5%	.49%	.50%	.46%
Piles & Foundations	7.3 to 76.7	22.9	7.1	5.2	8.7	1.63	1.19	1.99
Concrete	5.7 to 35.2	15.8	5.0	14.8	3.7	.79	2.34	.58
Masonry	5.1 to 31.0	15.0	6.9	7.5	1.9	1.04	1.13	.29
Structural Steel	7.2 to 112.0	38.9	10.7	3.9	17.6	4.16	1.52	6.85
Miscellaneous & Ornamental Metals	5.2 to 25.4	12.6	2.8	4.0	3.6	.35	.50	.45
Carpentry & Millwork	6.7 to 53.2	18.5	3.7	4.0	0.5	.68	.74	.09
Metal or Composition Siding	5.3 to 35.5	16.1	2.3	0.3	4.3	.37	.05	.69
Roofing	7.3 to 77.1	31.8	2.3	2.6	3.1	.73	.83	.99
Doors & Hardware	4.5 to 25.3	10.9	0.9	1.4	0.4	.10	.15	.04
Sash & Glazing	4.9 to 38.0	13.8	3.5	4.0	1.0	.48	.55	.14
Lath & Plaster	4.0 to 45.5	14.6	3.3	6.9	0.8	.48	1.01	.12
Tile, Marble & Floors	3.7 to 23.3	9.6	2.6	3.0	0.5	.25	.29	.05
Acoustical Ceilings	2.5 to 24.5	10.7	2.4	0.2	0.3	.26	.02	.03
Painting	4.3 to 29.6	12.9	1.5	1.6	1.6	.19	.21	.21
Interior Partitions	6.7 to 53.2	18.5	3.9	4.3	4.4	.72	.80	.81
Miscellaneous Items	2.5 to 110.1	17.3	5.2	3.7	9.7	.90	.64	1.68
Elevators	2.5 to 15.3	7.2	2.1	1.1	2.2	.15	.08	.16
Sprinklers	2.8 to 23.1	9.0	0.5	—	2.0	.05	—	.18
Plumbing	3.0 to 12.5	7.8	4.9	7.2	5.2	.38	.56	.41
Heat., Vent., Air Conditioning	4.0 to 28.1	11.1	13.5	11.0	12.9	1.50	1.22	1.43
Electrical	2.7 to 12.5	6.4	10.1	8.4	11.1	.65	.54	.71
Total	2.5% to 110.1%	—	100.0%	100.0%	100.0%	16.35%	14.87%	18.36%
	Overall Weighted Average 16.53%							

Workers' Compensation Insurance Rates by States

The table below lists the weighted average Workers' Compensation base rate for each state with a factor comparing this with the national average of 16.2%.

State	Weighted Average	Factor	State	Weighted Average	Factor	State	Weighted Average	Factor
Alabama	28.0%	173	Kentucky	16.6%	102	North Dakota	12.9%	80
Alaska	17.9	110	Louisiana	28.2	174	Ohio	12.7	78
Arizona	7.1	44	Maine	20.8	128	Oklahoma	22.2	137
Arkansas	14.7	91	Maryland	12.0	74	Oregon	14.3	88
California	18.8	116	Massachusetts	14.6	90	Pennsylvania	16.0	99
Colorado	14.0	86	Michigan	19.0	117	Rhode Island	21.2	131
Connecticut	24.9	154	Minnesota	27.7	171	South Carolina	13.9	86
Delaware	11.9	73	Mississippi	17.0	105	South Dakota	14.5	90
District of Columbia	19.1	118	Missouri	20.7	128	Tennessee	16.4	101
Florida	31.4	194	Montana	20.6	127	Texas	14.6	90
Georgia	23.0	142	Nebraska	20.1	124	Utah	12.3	76
Hawaii	16.9	104	Nevada	16.2	100	Vermont	20.0	123
Idaho	10.4	64	New Hampshire	22.6	140	Virginia	12.5	77
Illinois	18.2	112	New Jersey	10.6	65	Washington	10.6	65
Indiana	6.2	38	New Mexico	15.2	94	West Virginia	12.8	79
Iowa	12.0	74	New York	13.5	83	Wisconsin	15.9	98
Kansas	8.8	54	North Carolina	13.7	85	Wyoming	8.0	49
			Weighted Average for U.S. is 16.5% of payroll = 100%					

Rates in the following table are the base or manual costs per $100 of payroll for Workers' Compensation in each state. Rates are usually applied to straight time wages only and not to premium time wages and bonuses.

The weighted average skilled worker rate for 35 trades is 16.2%. For bidding purposes, apply the full value of Workers' Compensation directly to total labor costs, or if labor is 38%, materials 42% and overhead and profit 20% of total cost, carry 38/80 x 16.2% =7.7% of cost (before overhead and profit) into overhead. Rates vary not only from state to state but also with the experience rating of the contractor.

Rates are the most current available at the time of publication.

(Reprinted from Means Plumbing Cost Data 2004)

Figure 8.7

Profit: This percentage is the fee added by the contractor to offer both a return on investment and an allowance to cover the risk involved in the type of construction being bid. The profit percentage may vary from 4% on large, straightforward projects to as much as 25% on smaller, high-risk jobs. Profit percentages are directly affected by economic conditions, the expected number of bidders, and the estimated risk involved in the project. For estimating purposes, *Means Plumbing Cost Data* assumes 10% (applied to labor) as a reasonable average profit factor.

Means Order of Magnitude and Square Foot Cost Data

A sample page from Division 17 of *Means Plumbing Cost Data* 2004 (Figure 8.8) indicates order of magnitude costs per rental unit, per apartment, and per bed (for nursing homes). Division 17 also gives square foot costs, cubic foot costs, and percentages of total costs for plumbing, HVAC, and electrical. The division prices 59 building types, from apartments to warehouses.

Division 17 has been developed to facilitate the preparation of rapid preliminary budget estimates. The cost figures in this division are derived from more than 11,200 actual building projects contained in the Means data bank of construction costs, and include the contractor's overhead and profit. The prices shown *do not* include architectural fees or land costs. The files are updated each year with costs for new projects. In no case are all subdividions of a project listed.

These projects were located throughout the U.S. and reflect differences in square foot and cubic foot costs due to both the variations in labor and material costs, and the differences in owners' requirements. For instance, a bank in a large city would have different features and costs than one in a rural area. This is true of all the different types of building analyzed. All individual costs were computed and tabulted separately. Thus, the sum of the median figures for plumbing, HVAC, and electrical will not normally add up to the total mechanical and electrical costs arrived at by separate analysis and tabulation of the projects.

The data and prices presented on a Division 17 page (Figure 8.8) are listed both as square foot or cubic foot costs and as a percentage of total costs. Each category tabulates the data in a similar manner. The median, or middle figure, is listed. This means that 50% of all projects had lower costs, and 50% had higher costs than the median figure. Figures in the "1/4" column indicate that 25% of the projects had lower costs and 75% had higher costs. Similarly, figures in the "3/4" column indicate that 75% had lower costs and 25% of the projects had higher costs.

The costs and figures represent all projects and do not take into account project size. As a rule, larger buildings (of the same type and relative location) will cost less to build per square foot than similar buildings of a smaller size. This cost difference is due to economies of scale as well as a lower exterior envelope-to-floor area ratio. A conversion is necessary to adjust project costs based on size relative to the norm. Figure 8.9 from the reference pages of *Plumbing Cost Data* 2004 provides instructions in doing size modification calculations.

There are two stages of project development when square foot cost estimates are most useful. The first is during the conceptual stage when few, if any, details are available. At this time, square foot costs are

17100 | S.F., C.F. and % of Total Costs

	17100 | S.F. & C.F. Costs			UNIT	UNIT COSTS 1/4	UNIT COSTS MEDIAN	UNIT COSTS 3/4	% OF TOTAL 1/4	% OF TOTAL MEDIAN	% OF TOTAL 3/4	
500	0010	**HOUSING** Public (Low Rise)	R17100 -100	S.F.	54.50	76	99				500
	0020	Total project costs		C.F.	4.93	6.10	7.70				
	2720	Plumbing		S.F.	3.95	5.35	6.60	7.15%	9.05%	11.45%	
	2730	Heating, ventilating, air conditioning			1.98	3.85	4.22	4.26%	6.05%	6.45%	
	2900	Electrical			3.27	4.93	6.85	5.50%	6.75%	8.50%	
	3100	Total: Mechanical & Electrical			15.70	20	23.50	14.70%	19.20%	26.50%	
	9000	Per apartment, total cost		Apt.	60,000	68,500	86,000				
	9500	Total: Mechanical & Electrical		"	12,800	15,800	17,500				
510	0010	**ICE SKATING RINKS**	R17100 -100	S.F.	48.50	109	120				510
	0020	Total project costs		C.F.	3.44	3.52	4.06				
	2720	Plumbing		S.F.	1.75	3.28	3.35	3.23%	5.65%	6.75%	
	2900	Electrical			5	7.70	8.15	6.80%	10.15%	15.05%	
	3100	Total: Mechanical & Electrical			8.50	12	15	18.95%	18.95%	18.95%	
520	0010	**JAILS**	R17100 -100	S.F.	144	184	237				520
	0020	Total project costs		C.F.	13.45	17.90	22				
	2720	Plumbing		S.F.	14.50	18.40	24.50	7.05%	8.90%	9.60%	
	2770	Heating, ventilating, air conditioning			12.85	17.15	33	8%	9.45%	12.20%	
	2900	Electrical			15.65	20	24.50	9.80%	11.70%	14.95%	
	3100	Total: Mechanical & Electrical			40	71	84	27.50%	30%	31%	
530	0010	**LIBRARIES**	R17100 -100	S.F.	88.50	114	147				530
	0020	Total project costs		C.F.	6.15	7.45	9.85				
	2720	Plumbing		S.F.	3.32	4.84	6.50	3.38%	4.26%	5.50%	
	2770	Heating, ventilating, air conditioning			7.35	12.45	16.25	7.80%	10.95%	12.80%	
	2900	Electrical			9.05	11.65	14.80	8.40%	10.25%	11.85%	
	3100	Total: Mechanical & Electrical			27	34	42.50	19.65%	23%	26.50%	
540	0010	**LIVING, ASSISTED**	R17100 -100	S.F.	86.50	99	117				540
	0020	Total project costs		C.F.	7.20	8.20	9.65				
	2720	Plumbing		S.F.	7	9.35	9.90	6.05%	8.15%	10.60%	
	2770	Heating, ventilating, air conditioning			8.30	8.65	9.50	7.95%	9.35%	9.70%	
	2900	Electrical			8.20	9.30	10.60	9.05%	10.30%	10.80%	
	3100	Total: Mechanical & Electrical			23.50	28.50	31	25%	30%	32%	
550	0010	**MEDICAL CLINICS**	R17100 -100	S.F.	83	106	135				550
	0020	Total project costs		C.F.	6.20	8.70	11.05				
	2720	Plumbing		S.F.	5.65	7.85	10.95	6%	8.35%	10.35%	
	2770	Heating, ventilating, air conditioning			6.70	8.80	12.95	6.70%	8.85%	11.80%	
	2900	Electrical			7.10	10.35	13.65	8.15%	10%	12.40%	
	3100	Total: Mechanical & Electrical			23	32	44	22.50%	27.50%	33.50%	
570	0010	**MEDICAL OFFICES**	R17100 -100	S.F.	78	97.50	120				570
	0020	Total project costs		C.F.	5.80	8.05	10.90				
	2720	Plumbing		S.F.	4.41	6.80	9.15	5.60%	6.80%	8.50%	
	2770	Heating, ventilating, air conditioning			5.30	7.70	10.15	6.20%	8.25%	9.70%	
	2900	Electrical			6.35	9.25	12.65	7.35%	9.75%	11.10%	
	3100	Total: Mechanical & Electrical			17.10	24	37.50	19.65%	23.50%	30.50%	
590	0010	**MOTELS**	R17100 -100	S.F.	50.50	74	96				590
	0020	Total project costs		C.F.	4.48	6	10.10				
	2720	Plumbing		S.F.	5.10	6.50	7.75	9.45%	10.60%	12.55%	
	2770	Heating, ventilating, air conditioning			3.11	4.63	8.30	5.60%	5.60%	10%	
	2900	Electrical			4.79	6.10	7.90	7.45%	9.20%	10.80%	
	3100	Total: Mechanical & Electrical			16.15	20	34.50	18.50%	21%	25.50%	
	9000	Per rental unit, total cost		Unit	25,600	48,900	52,500				
	9500	Total: Mechanical & Electrical		"	5,000	7,550	8,775				
600	0010	**NURSING HOMES**	R17100 -100	S.F.	78.50	103	122				600
	0020	Total project costs		C.F.	6	7.95	10.80				
	2720	Plumbing		S.F.	7.15	10.25	12.55	9.40%	10.25%	12.20%	
	2770	Heating, ventilating, air conditioning			7.10	10.80	14.35	10.65%	11.80%	12.70%	
	2900	Electrical			7.80	10.05	13.55	9.15%	10.55%	12.55%	
	3100	Total: Mechanical & Electrical			19.20	26	43.50	26%	29.50%	30.50%	

(Reprinted from Means Plumbing Cost Data 2004)

Figure 8.8

S.F. & C.F. Costs	R171	Information

R17100-100 Square Foot Project Size Modifier

One factor that affects the S.F. cost of a particular building is the size. In general, for buildings built to the same specifications in the same locality, the larger building will have the lower S.F. cost. This is due mainly to the decreasing contribution of the exterior walls plus the economy of scale usually achievable in larger buildings. The Area Conversion Scale shown below will give a factor to convert costs for the typical size building to an adjusted cost for the particular project.

The Square Foot Base Size lists the median costs, most typical project size in our accumulated data and the range in size of the projects.

The Size Factor for your project is determined by dividing your project area in S.F. by the typical project size for the particular Building Type. With this factor, enter the Area Conversion Scale at the appropriate Size Factor and determine the appropriate cost multiplier for your building size.

Example: Determine the cost per S.F. for a 100,000 S.F. Mid-rise apartment building.

$$\frac{\text{Proposed building area} = 100{,}000 \text{ S.F.}}{\text{Typical size from below} = 50{,}000 \text{ S.F.}} = 2.00$$

Enter Area Conversion scale at 2.0, intersect curve, read horizontally the appropriate cost multiplier of .94. Size adjusted cost becomes .94 x $77.00 = $72.40 based on national average costs.

Note: For Size Factors less than .50, the Cost Multiplier is 1.1
For Size Factors greater than 3.5, the Cost Multiplier is .90

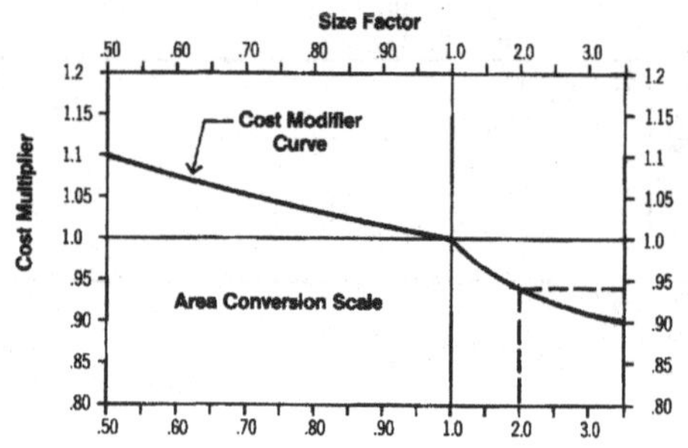

Building Type	Median Cost per S.F.	Typical Size Gross S.F.	Typical Range Gross S.F.	Building Type	Median Cost per S.F.	Typical Size Gross S.F.	Typical Range Gross S.F.
Square Foot Base Size							
Apartments, Low Rise	$ 60.50	21,000	9,700 - 37,200	Jails	$184.00	40,000	5,500 - 145,000
Apartments, Mid Rise	77.00	50,000	32,000 - 100,000	Libraries	114.00	12,000	7,000 - 31,000
Apartments, High Rise	87.00	145,000	95,000 - 600,000	Living, Assisted	99.00	32,300	23,500 - 50,300
Auditoriums	101.00	25,000	7,600 - 39,000	Medical Clinics	106.00	7,200	4,200 - 15,700
Auto Sales	75.50	20,000	10,800 - 28,600	Medical Offices	97.50	6,000	4,000 - 15,000
Banks	135.00	4,200	2,500 - 7,500	Motels	74.00	40,000	15,800 - 120,000
Churches	92.50	17,000	2,000 - 42,000	Nursing Homes	103.00	23,000	15,000 - 37,000
Clubs, Country	95.00	6,500	4,500 - 15,000	Offices, Low Rise	86.00	20,000	5,000 - 80,000
Clubs, Social	90.00	10,000	6,000 - 13,500	Offices, Mid Rise	85.00	120,000	20,000 - 300,000
Clubs, YMCA	106.00	28,300	12,800 - 39,400	Offices, High Rise	109.00	260,000	120,000 - 800,000
Colleges (Class)	118.00	50,000	15,000 - 150,000	Police Stations	136.00	10,500	4,000 - 19,000
Colleges (Science Lab)	173.00	45,600	16,600 - 80,000	Post Offices	101.00	12,400	6,800 - 30,000
College (Student Union)	132.00	33,400	16,000 - 85,000	Power Plants	750.00	7,500	1,000 - 20,000
Community Center	96.00	9,400	5,300 - 16,700	Religious Education	84.00	9,000	6,000 - 12,000
Court Houses	129.00	32,400	17,800 - 106,000	Research	142.00	19,000	6,300 - 45,000
Dept. Stores	56.00	90,000	44,000 - 122,000	Restaurants	122.00	4,400	2,800 - 6,000
Dormitories, Low Rise	100.00	25,000	10,000 - 95,000	Retail Stores	60.00	7,200	4,000 - 17,600
Dormitories, Mid Rise	126.00	85,000	20,000 - 200,000	Schools, Elementary	88.50	41,000	24,500 - 55,000
Factories	54.50	26,400	12,900 - 50,000	Schools, Jr. High	91.50	92,000	52,000 - 119,000
Fire Stations	96.00	5,800	4,000 - 8,700	Schools, Sr. High	98.00	101,000	50,500 - 175,000
Fraternity Houses	93.50	12,500	8,200 - 14,800	Schools, Vocational	89.50	37,000	20,500 - 82,000
Funeral Homes	105.00	10,000	4,000 - 20,000	Sports Arenas	74.00	15,000	5,000 - 40,000
Garages, Commercial	68.00	9,300	5,000 - 13,600	Supermarkets	60.00	44,000	12,000 - 60,000
Garages, Municipal	86.00	8,300	4,500 - 12,600	Swimming Pools	139.00	20,000	10,000 - 32,000
Garages, Parking	36.00	163,000	76,400 - 225,300	Telephone Exchange	163.00	4,500	1,200 - 10,600
Gymnasiums	91.00	19,200	11,600 - 41,000	Theaters	88.50	10,500	8,800 - 17,500
Hospitals	160.00	55,000	27,200 - 125,000	Town Halls	104.00	10,800	4,800 - 23,400
House (Elderly)	82.50	37,000	21,000 - 66,000	Warehouses	44.50	25,000	8,000 - 72,000
Housing (Public)	76.00	36,000	14,400 - 74,400	Warehouse & Office	47.00	25,000	8,000 - 72,000
Ice Rinks	109.00	29,000	27,200 - 33,600				

(Reprinted from Means Plumbing Cost Data 2004)

Figure 8.9

appropriate for ballpark budget purposes. As soon as details become available in the project design, the square foot approach should be discontinued and the project priced more accurately. After the estimate is completed, square foot costs can be used again—this time for verification and as a check against gross errors.

When using the figures in Division 17, it is recommended that the median cost column be consulted for preliminary figures if no additional information is available. When costs have been converted for location (see City Cost Indexes), the median numbers (as shown in Figure 8.8) should provide a fairly accurate base figure. This figure should then be adjusted according to the estimator's experience, local economic conditions, code requirements, and the owner's particular project requirements. There is no need to factor the percentage figures, as these should remain relatively constant from city to city.

Repair and Remodeling

Cost figures in *Means Plumbing Cost Data* are based on new construction utilizing the most effective combination of labor, equipment, and material. The work is scheduled in the proper sequence to allow the various trades to accomplish their tasks in an efficient manner. Figure 8.10 (from Division 01250 of *Means Plumbing Cost Data*) show factors that can be used to adjust figures in other sections of the book for repair and remodeling projects. For expanded coverage, consult *Means Repair & Remodeling Cost Data*.

Assemblies Cost Tables

Means' assemblies data are divided into seven UNIFORMAT II divisions, which organize the components of construction into logical groupings. The systems or assemblies approach was devised to provide quick and easy methods for estimating even when only preliminary design data are available. The groupings, or systems, are presented in such a way so that the estimator can substitute one system for another. This is extremely useful when adapting to budget, design, or other considerations. Figure 8.11, a representative page from the 2004 edition of *Means Plumbing Cost Data,* shows battery or group mounting of lavatories and the system components. The savings realized by side-by-side or back-to-back installation are indicated. All assemblies prices include the installing (plumbing) contractor's overhead and profit. Figure 8.12 shows how the data is presented in the assemblies section. Each system is illustrated and accompanied by a detailed description. The book lists the components and sizes of each system. Each individual component is found in the unit price section.

Quantity

A unit of measure is established for each assembly. For example, sprinkler systems are measured by the square foot of floor area, plumbing fixture systems are measured by "each," and HVAC systems are measured by the square foot of floor area. Within each system, the components are measured by industry standard, using the same units as in the unit price section.

Material

The cost of each component in the "material" column is the "bare material cost" plus 10% handling for the unit and quantity as defined in the "quantity" column.

01200 | Price & Payment Procedures

		01250 | Contract Modification Procedures	CREW	DAILY OUTPUT	LABOR-HOURS	UNIT	2004 BARE COSTS MAT.	LABOR	EQUIP.	TOTAL	TOTAL INCL O&P	
200	0010	**CONTINGENCIES** for estimate at conceptual stage				Project					20%	200
	0050	Schematic stage									15%	
	0100	Preliminary working drawing stage (Design Dev.)									10%	
	0150	Final working drawing stage									3%	
300	0010	**CREWS** For building construction, see How To Use This Book										300
400	0010	**FACTORS** Cost adjustments R01250-010										400
	0100	Add to construction costs for particular job requirements										
	0500	Cut & patch to match existing construction, add, minimum				Costs	2%	3%				
	0550	Maximum					5%	9%				
	0800	Dust protection, add, minimum					1%	2%				
	0850	Maximum					4%	11%				
	1100	Equipment usage curtailment, add, minimum					1%	1%				
	1150	Maximum					3%	10%				
	1400	Material handling & storage limitation, add, minimum					1%	1%				
	1450	Maximum					6%	7%				
	1700	Protection of existing work, add, minimum					2%	2%				
	1750	Maximum					5%	7%				
	2000	Shift work requirements, add, minimum						5%				
	2050	Maximum						30%				
	2300	Temporary shoring and bracing, add, minimum					2%	5%				
	2350	Maximum					5%	12%				
	2400	Work inside prisons and high security areas, add, minimum						30%				
	2450	Maximum						50%				
500	0010	**JOB CONDITIONS** Modifications to total										500
	0020	project cost summaries										
	0100	Economic conditions, favorable, deduct				Project					2%	
	0200	Unfavorable, add									5%	
	0300	Hoisting conditions, favorable, deduct									2%	
	0400	Unfavorable, add									5%	
	0700	Labor availability, surplus, deduct									1%	
	0800	Shortage, add									10%	
	0900	Material storage area, available, deduct									1%	
	1000	Not available, add									2%	
	1100	Subcontractor availability, surplus, deduct									5%	
	1200	Shortage, add									12%	
	1300	Work space, available, deduct									2%	
	1400	Not available, add									5%	
600	0010	**OVERTIME** For early completion of projects or where R01100-110										600
	0020	labor shortages exist, add to usual labor, up to				Costs		100%				
		01255 | Cost Indexes										
200	0010	**CONSTRUCTION COST INDEX** (Reference) over 930 zip code locations in										200
	0020	The U.S. and Canada, total bldg cost, min. (Clarksdale, MS)				%					65.50%	
	0050	Average									100%	
	0100	Maximum (New York, NY)									133.90%	
400	0010	**HISTORICAL COST INDEXES** (Reference) Back to 1954										400
500	0010	**LABOR INDEX** (Reference) For over 930 zip code locations in										500
	0020	the U.S. and Canada, minimum (Clarksdale, MS)				%		31.30%				
	0050	Average						100%				
	0100	Maximum (New York, NY)						163.20%				
600	0011	**MATERIAL INDEX** For over 930 zip code locations in										600
	0020	the U.S. and Canada, minimum (Elizabethtown, KY)				%	92.10%					
	0040	Average					100%					
	0060	Maximum (Ketchikan, AK)					145.20%					

(Reprinted from Means Plumbing Cost Data 2004)

Figure 8.10

D20 Plumbing

D2010 Plumbing Fixtures

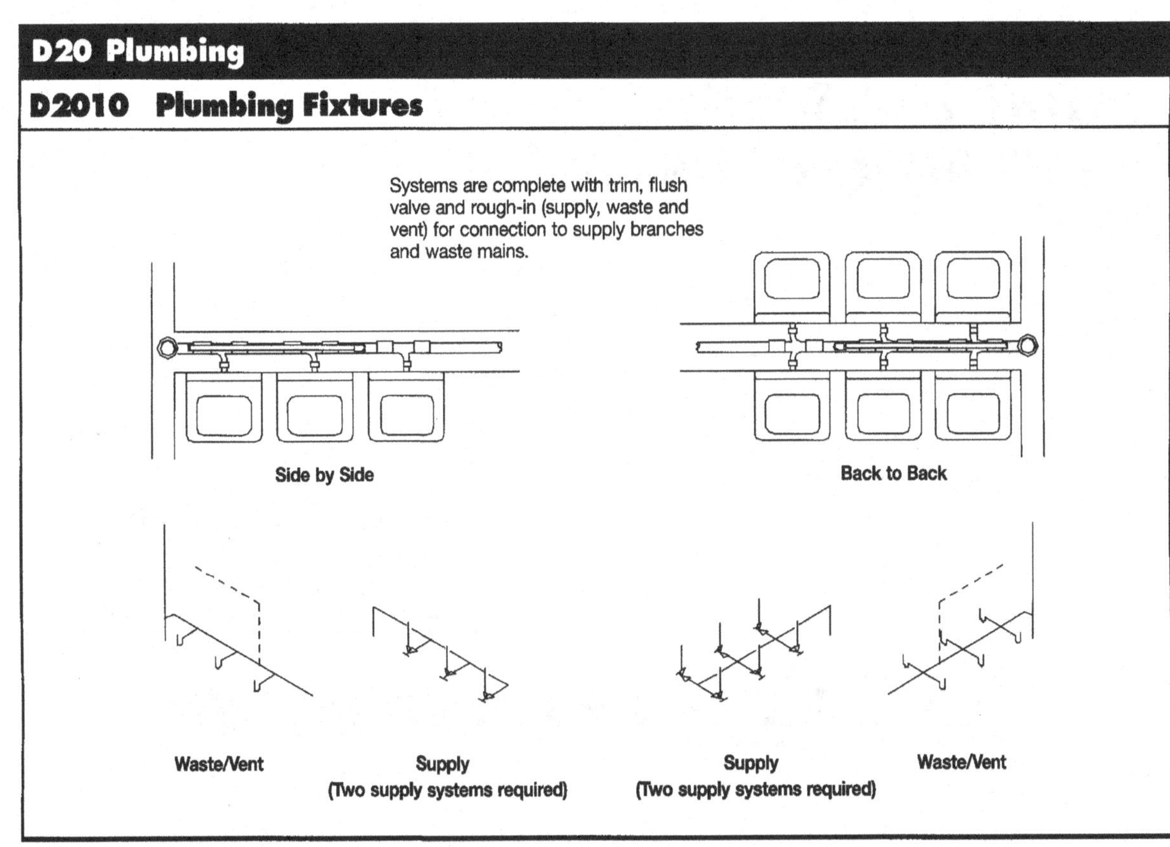

Systems are complete with trim, flush valve and rough-in (supply, waste and vent) for connection to supply branches and waste mains.
Side by Side
Back to Back
Waste/Vent
Supply
(Two supply systems required)
Supply
(Two supply systems required)
Waste/Vent

System Components	QUANTITY	UNIT	COST EACH MAT.	COST EACH INST.	COST EACH TOTAL
SYSTEM D2010 320 1760					
LAVATORIES, BATTERY MOUNT, WALL HUNG, SIDE BY SIDE, FIRST LAVATORY					
Lavatory w/trim wall hung PE on CI 20" x 18"	1.000	Ea.	261	107	368
Stop, chrome, angle supply, 3/8" diameter	2.000	Ea.	10.40	39.60	50
Concealed arm support	1.000	Ea.	186	79.50	265.50
P trap w/cleanout, 20 ga. C.P., 1-1/4" diameter	1.000	Ea.	28.50	26.50	55
Copper tubing, type L, 1/2" diameter	10.000	L.F.	11.10	58.50	69.60
Copper tubing, type DWV, 1-1/4" diameter	4.000	L.F.	9.16	31.80	40.96
Copper 90° elbow, 1/2" diameter	2.000	Ea.	.66	48	48.66
Copper tee, 1/2" diameter	2.000	Ea.	1.12	73	74.12
DWV copper sanitary tee, 1-1/4" diameter	2.000	Ea.	13.30	106	119.30
Galvanized steel pipe, 1-1/4" diameter	4.000	L.F.	12.12	38.40	50.52
Black cast iron 90° elbow, 1-1/4" diameter	1.000	Ea.	4.33	39	43.33
TOTAL			537.69	647.30	1,184.99

D2010 320	Lavatory Systems, Battery Mount	COST EACH MAT.	COST EACH INST.	COST EACH TOTAL
1760	Lavatories, battery mount, side by side, first lavatory	540	645	1,185
1800	Each additional lavatory, add	510	450	960
2000	Back to back, first pair of lavatories	930	1,025	1,955
2100	Each additional pair of lavatories, back to back	910	850	1,760

(Reprinted from Means Plumbing Cost Data 2004)

Figure 8.11

How to Use the Assemblies Cost Tables

The following is a detailed explanation of a sample Assemblies Cost Table. Most Assembly Tables are separated into three parts: 1) an illustration of the system to be estimated; 2) the components and related costs of a typical system; and 3) the costs for similar systems with dimensional and/or size variations. For costs of the components that comprise these systems or "assemblies" refer to the Unit Price Section. Next to each bold number below is the item being described with the appropriate component of the sample entry following in parenthesis. In most cases, if the work is to be subcontracted, the general contractor will need to add an additional markup (RSMeans suggests using 10%) to the "Total" figures.

System/Line Numbers (D3010 510 1760)

Each Assemblies Cost Line has been assigned a unique identification number based on the UniFormat classification sytem.

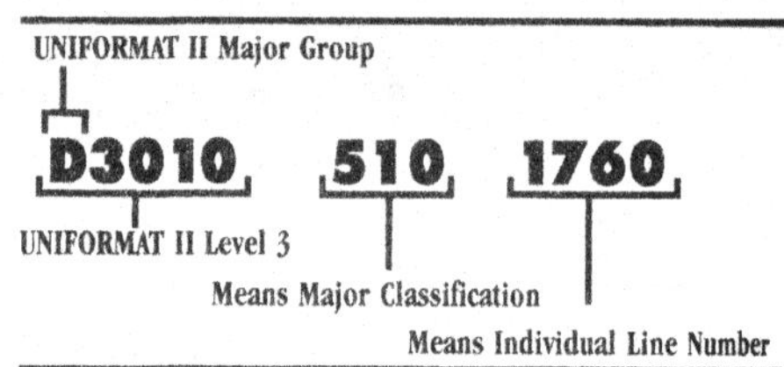

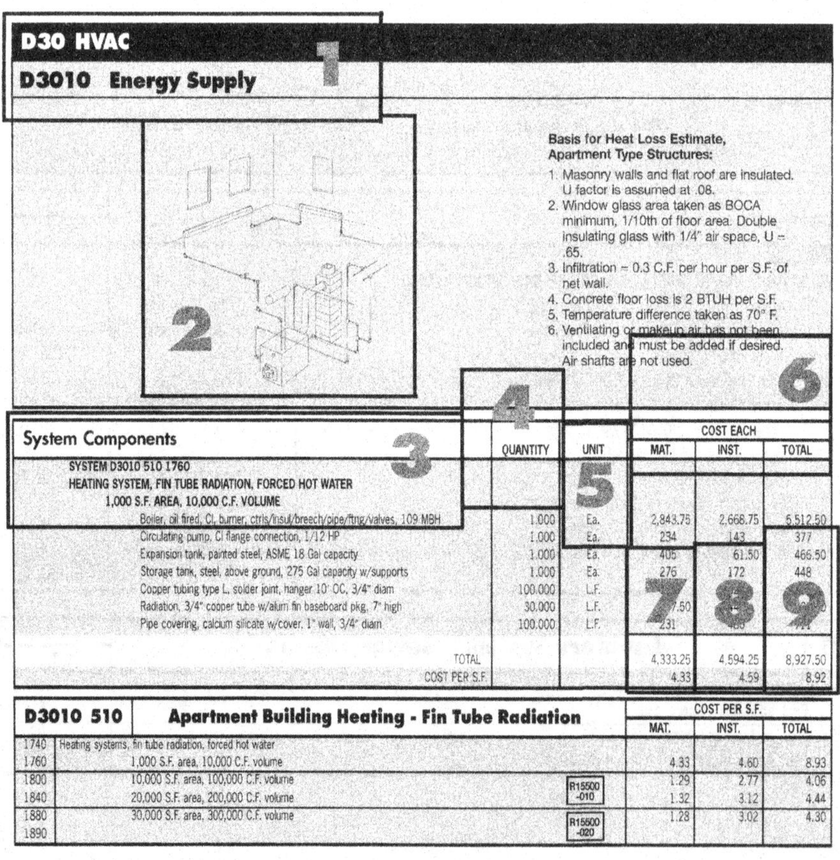

D30 HVAC

D3010 Energy Supply

Basis for Heat Loss Estimate, Apartment Type Structures:

1. Masonry walls and flat roof are insulated. U factor is assumed at .08.
2. Window glass area taken as BOCA minimum, 1/10th of floor area. Double insulating glass with 1/4" air space, U = .65.
3. Infiltration = 0.3 C.F. per hour per S.F. of net wall.
4. Concrete floor loss is 2 BTUH per S.F.
5. Temperature difference taken as 70° F.
6. Ventilating or makeup air has not been included and must be added if desired. Air shafts are not used.

System Components	QUANTITY	UNIT	COST EACH MAT.	COST EACH INST.	COST EACH TOTAL
SYSTEM D3010 510 1760					
HEATING SYSTEM, FIN TUBE RADIATION, FORCED HOT WATER					
1,000 S.F. AREA, 10,000 C.F. VOLUME					
Boiler, oil fired, CI, burner, ctrls/insul/breech/pipe/ftng/valves, 109 MBH	1.000	Ea.	2,843.75	2,668.75	5,512.50
Circulating pump, CI flange connection, 1/12 HP	1.000	Ea.	234	143	377
Expansion tank, painted steel, ASME 18 Gal capacity	1.000	Ea.	405	61.50	466.50
Storage tank, steel, above ground, 275 Gal capacity w/supports	1.000	Ea.	276	172	448
Copper tubing type L, solder joint, hanger 10' OC, 3/4" diam	100.000	L.F.	[illegible]	[illegible]	[illegible]
Radiation, 3/4" copper tube w/alum fin baseboard pkg, 7" high	30.000	L.F.	[illegible].50	[illegible]	[illegible]
Pipe covering, calcium silicate w/cover, 1" wall, 3/4" diam	100.000	L.F.	231	[illegible]	[illegible]
TOTAL			4,333.25	4,594.25	8,927.50
COST PER S.F.			4.33	4.59	8.92

D3010 510	Apartment Building Heating - Fin Tube Radiation		COST PER S.F. MAT.	COST PER S.F. INST.	COST PER S.F. TOTAL
1740	Heating systems, fin tube radiation, forced hot water				
1760	1,000 S.F. area, 10,000 C.F. volume		4.33	4.60	8.93
1800	10,000 S.F. area, 100,000 C.F. volume	R15500 -010	1.29	2.77	4.06
1840	20,000 S.F. area, 200,000 C.F. volume		1.32	3.12	4.44
1880	30,000 S.F. area, 300,000 C.F. volume	R15500 -020	1.28	3.02	4.30
1890					

(Reprinted from Means Plumbing Cost Data 2004)

Figure 8.12

Illustration

At the top of most assembly pages is an illustration, a brief description, and the design criteria used to develop the cost.

System Components

The components of a typical system are listed separately to show what has been included in the development of the total system price. The table below contains prices for other similar systems with dimensional and/or size variations.

Quantity

This is the number of line item units required for one system unit. For example, we assume that it will take 30 linear feet of radiation for the 1000 S.F. area shown.

Unit of Measure for Each Item

The abbreviated designation indicates the unit of measure, as defined by industry standards, upon which the price of the component is based. For example, baseboard radiation is priced by the linear foot. For a complete listing of abbreviations, see the Reference Section.

Unit of Measure for Each System (Each)

Costs shown in the three right hand columns have been adjusted by the component quantity and unit of measure for the entire system. In this example, "Cost Each" is the unit of measure for this system or "assembly."

Materials (4,333.25)

This column contains the Materials Cost of each component. These cost figures are bare costs plus 10% for profit.

Installation (4,594.25)

Installation includes labor and equipment plus the installing contractor's overhead and profit. Equipment costs are the bare rental costs plus 10% for profit. The labor overhead and profit is defined on the inside back cover of this book.

Total (8,927.50)

The figure in this column is the sum of the material and installation costs.

Material Cost	+	Installation Cost	=	Total
$4,333.25	+	$4,594.25	=	$8,927.50

(Reprinted from Means Plumbing Cost Data 2004)

Figure 8.12 (cont'd.)

Installation

Installation costs as listed in the assemblies pages contain both labor and equipment costs. The labor rate includes the "bare labor costs" plus the installing contractor's overhead and profit. These rates are shown in Figure 8.4. The equipment rate is the "bare equipment cost" plus 10%.

Estimating References

Throughout the unit price and assemblies sections are reference numbers highlighted with bold squares. These numbers serve as footnotes, referring to illustrations, charts, and estimating tables in the reference pages, as well as to related information in the assemblies section. Figure 8.13 shows example reference numbers for plumbing fixtures as they appear on a unit price page. Figure 8.14 shows a corresponding reference page from the assemblies pages. The development of unit costs for many items is explained in the reference tables. Design criteria for many types of mechanical systems are also included to aid the designer/estimator in making appropriate choices. See Figure 8.9 for another example of a reference table.

City Cost Indexes

The unit prices in *Means Plumbing Cost Data* are national averages. When they are to be applied to a particular location, these prices must be adjusted to local conditions. Means has developed the City Cost Indexes and Location Factors for just that purpose. *Means Plumbing Cost Data* contains tables of indexes for 930 U.S. and Canadian cities based on a 30 major city average of 100. The figures are broken down into material and installation for all trades, as shown in Figure 8.15. Please note that for each city, there is a weighted average based on total project costs. This average is based on the relative contribution of each division to the construction process as a whole.

In addition to adjusting the figures in *Means Plumbing Cost Data* for particular locations, the City Cost Index can also be used to adjust costs from one city to another. For example, the price of the mechanical work for a particular building type is known for City A. In order to budget the costs of the same building type in City B, the following calculation can be made:

$$\frac{\text{City B Index}}{\text{City A Index}} \times \text{City A Cost} = \text{City B Cost}$$

When City Cost Indexes provide a means to adjust prices for location, the Historical Cost Index (also included in *Means Plumbing Cost Data* and shown in Figure 8.16) provides a means to adjust for time. Using the same principle as above, a time-adjusted factor can be calculated:

$$\frac{\text{Index for Year X}}{\text{Index for Year Y}} \times \text{Time-Adjustment Factor}$$

This time-adjustment factor can be used to determine the budget costs for a particular building type in Year X, based on costs for a similar building type from Year Y. Used together, the two indexes allow for cost adjustments from one city during a given year to another city in another year (the present or otherwise). For example, an office building built in San Francisco in 1974 originally cost $1,000,000. How much

15400 | Plumbing Fixtures & Equipment

	15411 | Commercial/Indust Fixtures		Crew	Daily Output	Labor-Hours	Unit	2004 Bare Costs Mat.	Labor	Equip.	Total	Total Incl O&P
700	5000	Stall type, vitreous china, includes valve	Q-1	2.50	6.400	Ea.	460	228		688	850
	5100	3" seam cover, add		12	1.333		125	47.50		172.50	210
	5200	6" seam cover, add (R15100-410)		12	1.333		174	47.50		221.50	263
	6980	Rough-in, supply, waste and vent		1.99	8.040		119	287		406	560
	8000	Waterless (no flush) urinal									
	8010	Wall hung									
	8020	Standard unit	Q-1	21.30	.751	Ea.	385	27		412	460
	8030	ADA compliant unit	"	21.30	.751		400	27		427	480
	8070	For solid color, add					60			60	66
	8080	For 2" brass flange, (new const.), add	Q-1	96	.167		19.20	5.95		25.15	30
	8090	Rough-in, supply, waste & vent	"	3.86	4.145		82	148		230	310
	8100	Trap liquid									
	8110	1 quart				Ea.	12.80			12.80	14.10
	8120	1 gallon				"	51			51	56.50
800	0010	**WASH CENTER** Prefabricated,stainless steel, semirecessed									
	0050	Lavatory, storage cabinet, mirror, light & switch, electric									
	0060	outlet, towel dispenser, waste receptacle & trim									
	0100	Foot water valve, cup & soap dispenser,16" W x 54-3/4" H	Q-1	8	2	Ea.	2,025	71.50		2,096.50	2,325
	0200	Handicap, wrist blade handles, 17" W x 66-1/2" H		8	2		1,050	71.50		1,121.50	1,250
	0220	20" W x 67-3/8" H		8	2		2,500	71.50		2,571.50	2,850
	0300	Push button metering & thermostatic mixing valves									
	0320	Handicap 17" W x 27-1/2" H	Q-1	8	2	Ea.	1,250	71.50		1,321.50	1,475
	0400	Rough-in, supply, waste and vent	"	2.10	7.619	"	48	272		320	465
840	0010	**WASH FOUNTAINS** (D2010 610)									
	1900	Group, foot control									
	2000	Precast terrazzo, circular, 36" diam., 5 or 6 persons	Q-2	3	8	Ea.	2,650	296		2,946	3,350
	2100	54" diameter for 8 or 10 persons		2.50	9.600		3,200	355		3,555	4,050
	2400	Semi-circular, 36" diam. for 3 persons		3	8		2,350	296		2,646	3,050
	2500	54" diam. for 4 or 5 persons		2.50	9.600		2,925	355		3,280	3,725
	2700	Quarter circle (corner), 54" for 3 persons		3.50	6.857		2,925	254		3,179	3,575
	3000	Stainless steel, circular, 36" diameter		3.50	6.857		2,900	254		3,154	3,575
	3100	54" diameter		2.80	8.571		3,825	315		4,140	4,700
	3400	Semi-circular, 36" diameter		3.50	6.857		2,525	254		2,779	3,150
	3500	54" diameter		2.80	8.571		3,300	315		3,615	4,100
	5000	Thermoplastic, pre-assembled, circular, 36" diameter		6	4		2,300	148		2,448	2,750
	5100	54" diameter		4	6		2,700	222		2,922	3,300
	5400	Semi-circular, 36" diameter		6	4		2,150	148		2,298	2,600
	5600	54" diameter		4	6		2,575	222		2,797	3,150
	5610	Group, infrared control, barrier free									
	5614	Precast terrazzo									
	5620	Semi-circular 36" diam. for 3 persons	Q-2	3	8	Ea.	3,475	296		3,771	4,275
	5630	46" diam. for 4 persons		2.80	8.571		3,900	315		4,215	4,750
	5640	Circular, 54" diam. for 8 persons, button control		2.50	9.600		5,675	355		6,030	6,775
	5700	Rough-in, supply, waste and vent for above wash fountains	Q-1	1.82	8.791		90.50	315		405.50	570
	6200	Duo for small washrooms, stainless steel		2	8		1,725	285		2,010	2,325
	6400	Bowl with backsplash		2	8		1,450	285		1,735	2,025
	6500	Rough-in, supply, waste & vent for duo fountains		2.02	7.921		52	282		334	480
900	0010	**WATER CHILLERS REMOTE** 80° F inlet									
	0100	Air cooled, 50° F outlet, 115V, 4.1 GPH	1 Plum	6	1.333	Ea.	555	53		608	690
	0200	5.7 GPH		5.50	1.455		780	57.50		837.50	940
	0300	8.0 GPH		5	1.600		575	63.50		638.50	730
	0400	10.0 GPH		4.50	1.778		705	70.50		775.50	880
	0500	13.4 GPH		4	2		1,075	79		1,154	1,325
	0700	29 GPH	Q-1	5	3.200		1,325	114		1,439	1,625
	1000	230V, 32 GPH	"	5	3.200		1,500	114		1,614	1,825

(Reprinted from Means Plumbing Cost Data 2004)

Figure 8.13

D20 Plumbing

D2010 Plumbing Fixtures

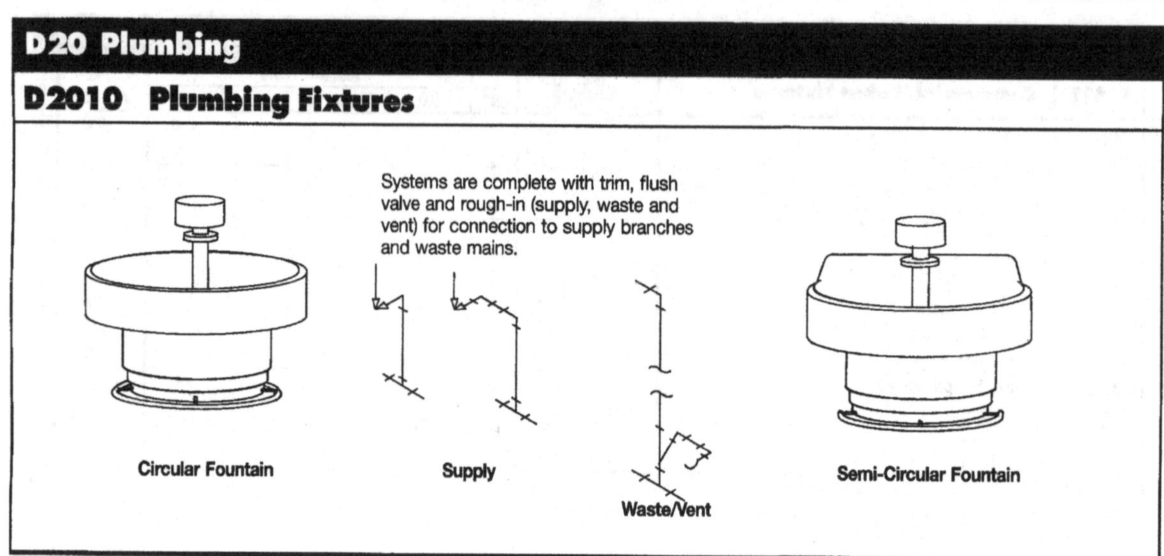

Systems are complete with trim, flush valve and rough-in (supply, waste and vent) for connection to supply branches and waste mains.
Circular Fountain
Supply
Waste/Vent
Semi-Circular Fountain

System Components	QUANTITY	UNIT	COST EACH MAT.	COST EACH INST.	COST EACH TOTAL
SYSTEM D2010 610 1760					
GROUP WASH FOUNTAIN, PRECAST TERRAZZO					
CIRCULAR, 36" DIAMETER					
Wash fountain, group, precast terrazzo, foot control 36" diam	1.000	Ea.	2,900	445	3,345
Copper tubing type DWV, solder joint, hanger 10'OC, 2" diam	10.000	L.F.	36.20	108	144.20
P trap, standard, copper, 2" diam	1.000	Ea.	29.50	31.50	61
Wrought copper, Tee, sanitary, 2" diam	1.000	Ea.	9.65	68	77.65
Copper tubing type L, solder joint, hanger 10' OC 1/2" diam	20.000	L.F.	22.20	117	139.20
Wrought copper 90° elbow for solder joints 1/2" diam	3.000	Ea.	.99	72	72.99
Wrought copper Tee for solder joints, 1/2" diam	2.000	Ea.	1.12	73	74.12
TOTAL			2,999.66	914.50	3,914.16

D2010 610	Group Wash Fountain Systems		COST EACH MAT.	COST EACH INST.	COST EACH TOTAL
1740	Group wash fountain, precast terrazzo				
1760	Circular, 36" diameter		3,000	915	3,915
1800	54" diameter	R15100 -410	3,625	1,000	4,625
1840	Semi-circular, 36" diameter		2,700	915	3,615
1880	54" diameter		3,300	1,000	4,300
1960	Stainless steel, circular, 36" diameter		3,300	850	4,150
2000	54" diameter		4,325	945	5,270
2040	Semi-circular, 36" diameter		2,875	850	3,725
2080	54" diameter		3,725	945	4,670
2160	Thermoplastic, circular, 36" diameter		2,625	690	3,315
2200	54" diameter		3,075	805	3,880
2240	Semi-circular, 36" diameter		2,475	690	3,165
2280	54" diameter		2,925	805	3,730

(Reprinted from Means Plumbing Cost Data 2004)

Figure 8.14

City Cost Indexes

	DIVISION	UNITED STATES			ALABAMA														
		30 CITY AVERAGE			BIRMINGHAM			HUNTSVILLE			MOBILE			MONTGOMERY			TUSCALOOSA		
		MAT.	INST.	TOTAL	MAT.	INST.	TOTAL	MAT.	INST.	TOTAL	MAT.	INST.	TOTAL	MAT.	INST.	TOTAL	MAT.	INST.	TOTAL
01590	EQUIPMENT RENTAL	.0	100.0	100.0	.0	101.4	101.4	.0	101.3	101.3	.0	97.8	97.8	.0	97.8	97.8	.0	101.3	101.3
02	SITE CONSTRUCTION	100.0	100.0	100.0	85.9	93.3	91.5	83.9	92.8	90.6	95.3	86.3	88.5	95.7	86.8	89.0	84.4	92.1	90.2
03100	CONCRETE FORMS & ACCESSORIES	100.0	100.0	100.0	91.9	76.1	78.1	93.7	70.5	73.3	93.7	54.8	59.6	92.2	49.6	54.9	93.6	40.9	47.4
03200	CONCRETE REINFORCEMENT	100.0	100.0	100.0	92.4	87.1	89.2	92.4	79.2	84.4	95.4	55.2	71.1	95.4	85.6	89.5	92.4	85.9	88.5
03300	CAST-IN-PLACE CONCRETE	100.0	100.0	100.0	93.5	69.8	83.6	88.3	67.4	79.6	93.3	56.3	77.8	95.0	50.7	76.5	92.0	48.4	73.8
03	CONCRETE	100.0	100.0	100.0	91.0	77.1	83.9	88.6	72.3	80.2	91.4	57.0	73.8	92.1	58.6	74.9	90.4	54.0	71.7
04	MASONRY	100.0	100.0	100.0	85.7	78.1	80.9	85.6	66.6	73.7	86.2	54.0	66.0	86.8	37.5	55.9	85.9	39.8	57.1
05	METALS	100.0	100.0	100.0	97.7	95.8	97.0	97.3	91.4	95.1	95.9	80.9	90.3	95.4	92.5	94.3	96.4	93.1	95.1
06	WOOD & PLASTICS	100.0	100.0	100.0	93.1	76.5	84.2	92.4	71.2	81.0	92.4	54.9	72.2	90.7	50.3	68.9	92.4	39.9	64.1
07	THERMAL & MOISTURE PROTECTION	100.0	100.0	100.0	95.7	81.5	88.8	95.4	75.7	85.8	95.4	70.2	83.1	95.1	63.4	79.7	95.3	62.2	79.2
08	DOORS & WINDOWS	100.0	100.0	100.0	98.5	79.1	93.6	98.5	67.8	90.7	98.5	55.1	87.5	98.5	59.4	88.6	98.5	59.3	88.5
09200	PLASTER & GYPSUM BOARD	100.0	100.0	100.0	102.1	76.4	84.6	98.7	71.0	79.8	98.7	54.2	68.4	98.7	49.4	65.2	98.7	38.7	57.9
095,098	CEILINGS & ACOUSTICAL TREATMENT	100.0	100.0	100.0	95.0	76.4	82.4	95.0	71.0	78.7	95.0	54.2	67.4	95.0	49.4	64.2	95.0	38.7	56.9
09600	FLOORING	100.0	100.0	100.0	104.3	52.5	91.0	104.3	45.4	89.1	114.7	56.9	99.8	112.9	29.4	91.4	104.3	43.2	88.6
097,099	WALL FINISHES, PAINTS & COATINGS	100.0	100.0	100.0	95.3	72.0	81.4	95.3	66.2	77.9	100.2	58.7	75.3	95.3	57.7	72.8	95.3	49.7	68.0
09	FINISHES	100.0	100.0	100.0	99.0	71.1	83.9	98.4	65.1	80.4	103.5	55.2	77.4	102.6	45.6	71.8	98.4	40.7	67.2
10 - 14	TOTAL DIV. 10000 - 14000	100.0	100.0	100.0	100.0	80.8	95.9	100.0	78.9	95.5	100.0	73.3	94.3	100.0	71.2	93.8	100.0	68.5	93.2
15	MECHANICAL	100.0	100.0	100.0	99.8	67.8	85.1	99.8	64.9	83.8	99.8	63.2	83.0	99.8	43.1	73.7	99.8	37.9	71.3
16	ELECTRICAL	100.0	100.0	100.0	97.6	65.6	78.9	97.8	69.2	81.1	97.8	53.3	71.8	97.6	66.8	79.6	97.8	65.6	79.0
01 - 16	WEIGHTED AVERAGE	100.0	100.0	100.0	96.6	76.5	86.9	96.2	72.6	84.8	97.2	62.8	80.6	97.1	59.0	78.7	96.3	57.1	77.4

	DIVISION	ALASKA									ARIZONA								
		ANCHORAGE			FAIRBANKS			JUNEAU			FLAGSTAFF			MESA/TEMPE			PHOENIX		
		MAT.	INST.	TOTAL	MAT.	INST.	TOTAL	MAT.	INST.	TOTAL	MAT.	INST.	TOTAL	MAT.	INST.	TOTAL	MAT.	INST.	TOTAL
01590	EQUIPMENT RENTAL	.0	118.4	118.4	.0	118.4	118.4	.0	118.4	118.4	.0	94.7	94.7	.0	97.5	97.5	.0	98.1	98.1
02	SITE CONSTRUCTION	143.9	134.0	136.5	127.3	134.0	132.3	139.5	134.0	135.4	82.2	101.2	96.4	85.5	104.0	99.3	85.9	104.9	100.1
03100	CONCRETE FORMS & ACCESSORIES	133.3	118.0	119.9	135.2	124.0	125.4	134.9	118.0	120.1	103.1	63.4	68.3	100.7	65.3	69.6	101.9	70.6	74.4
03200	CONCRETE REINFORCEMENT	137.9	108.1	119.9	115.9	108.1	111.2	102.6	108.1	105.9	101.6	75.2	85.7	100.0	73.4	83.9	98.1	75.9	84.7
03300	CAST-IN-PLACE CONCRETE	193.1	117.7	161.6	160.9	118.2	143.1	193.9	117.7	162.0	94.7	79.8	88.5	101.4	70.6	88.5	101.5	79.9	92.5
03	CONCRETE	151.3	115.4	132.9	127.9	118.3	123.0	147.7	115.4	131.2	119.0	71.2	94.5	100.2	68.5	84.0	99.8	74.6	86.9
04	MASONRY	220.7	124.7	160.6	214.1	124.7	158.2	223.4	124.7	161.6	102.5	57.0	74.0	107.2	54.4	74.1	95.1	68.1	78.2
05	METALS	129.4	103.2	119.5	129.5	103.5	119.7	129.7	103.2	119.7	97.8	69.1	86.9	98.6	69.1	87.5	100.0	71.8	89.4
06	WOOD & PLASTICS	115.0	116.0	115.5	115.2	123.8	119.8	115.0	116.0	115.5	108.6	62.5	83.8	103.7	71.4	86.3	104.8	71.5	86.9
07	THERMAL & MOISTURE PROTECTION	199.5	119.0	160.3	195.5	121.7	159.5	196.2	119.0	158.6	107.4	67.8	88.1	105.8	65.5	86.2	105.7	71.0	88.8
08	DOORS & WINDOWS	127.0	111.8	123.1	124.1	116.0	122.0	124.1	111.8	121.0	102.2	66.4	93.1	99.1	67.1	91.0	100.2	71.3	92.9
09200	PLASTER & GYPSUM BOARD	136.6	116.3	122.8	136.6	124.3	128.2	136.6	116.3	122.8	95.2	61.5	72.2	99.6	70.5	79.8	100.1	70.7	80.1
095,098	CEILINGS & ACOUSTICAL TREATMENT	126.4	116.3	119.5	126.4	124.3	124.9	126.4	116.3	119.5	107.2	61.5	76.2	108.0	70.5	82.6	108.0	70.7	82.7
09600	FLOORING	164.6	132.1	156.2	164.6	132.1	156.2	164.6	132.1	156.2	95.7	53.9	84.9	97.8	67.2	89.9	98.1	70.2	90.9
097,099	WALL FINISHES, PAINTS & COATINGS	167.8	110.4	133.4	167.8	124.7	141.9	167.8	110.4	133.4	94.8	50.4	68.2	105.2	54.8	75.0	105.2	62.3	79.5
09	FINISHES	157.0	119.9	137.0	154.5	126.1	139.1	155.4	119.9	136.2	98.1	59.5	77.2	100.4	64.3	80.9	100.6	69.2	83.7
10 - 14	TOTAL DIV. 10000 - 14000	100.0	119.7	104.2	100.0	120.8	104.5	100.0	119.7	104.2	100.0	76.5	95.0	100.0	73.0	94.2	100.0	78.1	95.3
15	MECHANICAL	100.4	108.8	104.3	100.4	114.7	107.0	100.4	105.9	102.9	100.2	77.2	89.6	100.2	69.1	85.9	100.2	77.3	89.6
16	ELECTRICAL	153.3	116.7	131.9	155.8	116.7	133.0	155.8	116.7	133.0	100.6	49.9	70.9	96.4	62.6	76.7	104.2	65.0	81.3
01 - 16	WEIGHTED AVERAGE	135.1	116.5	126.1	131.0	119.2	125.3	134.4	115.9	125.5	102.0	68.9	86.1	99.7	69.2	85.0	100.0	74.8	87.9

	DIVISION	ARIZONA						ARKANSAS											
		PRESCOTT			TUCSON			FORT SMITH			JONESBORO			LITTLE ROCK			PINE BLUFF		
		MAT.	INST.	TOTAL	MAT.	INST.	TOTAL	MAT.	INST.	TOTAL	MAT.	INST.	TOTAL	MAT.	INST.	TOTAL	MAT.	INST.	TOTAL
01590	EQUIPMENT RENTAL	.0	94.7	94.7	.0	97.5	97.5	.0	85.9	85.9	.0	107.7	107.7	.0	85.9	85.9	.0	85.9	85.9
02	SITE CONSTRUCTION	70.0	100.6	92.9	82.3	104.6	99.0	77.4	84.0	82.3	101.0	99.5	99.9	77.2	84.0	82.3	79.5	84.0	82.9
03100	CONCRETE FORMS & ACCESSORIES	98.4	58.0	63.0	101.4	70.1	73.9	99.8	43.5	50.4	85.7	49.6	54.0	93.6	61.4	65.3	77.8	61.2	63.2
03200	CONCRETE REINFORCEMENT	101.6	73.0	84.3	97.1	75.2	83.9	97.6	76.4	84.8	93.3	50.8	67.6	97.8	73.9	83.4	97.7	73.9	83.3
03300	CAST-IN-PLACE CONCRETE	94.6	65.8	82.6	104.3	79.8	94.1	89.9	68.7	81.0	85.6	58.7	74.3	89.9	68.8	81.1	82.4	68.7	76.7
03	CONCRETE	103.7	63.6	83.2	101.1	74.2	87.3	87.0	59.2	72.7	83.7	54.7	68.8	86.6	66.7	76.4	84.6	66.5	75.3
04	MASONRY	103.0	60.6	76.5	97.0	56.9	71.9	96.8	55.0	70.6	91.9	48.3	64.6	95.1	55.0	70.0	117.3	55.0	78.3
05	METALS	97.8	67.8	86.5	98.6	69.8	87.7	97.2	72.6	87.9	91.4	77.3	86.1	96.8	72.1	87.5	95.9	71.9	86.8
06	WOOD & PLASTICS	104.2	56.7	78.6	104.0	71.5	86.5	104.0	42.2	70.7	90.0	50.8	68.8	100.9	65.8	82.0	82.7	65.8	73.6
07	THERMAL & MOISTURE PROTECTION	105.6	63.6	85.2	107.0	64.6	86.4	99.3	50.0	75.3	109.1	53.4	82.0	98.0	52.5	75.8	98.0	52.5	75.8
08	DOORS & WINDOWS	102.2	60.3	91.6	96.1	71.3	89.8	97.1	47.4	84.5	98.7	49.6	86.2	97.1	60.4	87.8	92.5	60.4	84.3
09200	PLASTER & GYPSUM BOARD	93.1	55.5	67.6	100.4	70.7	80.2	86.1	41.3	55.6	93.5	49.7	63.7	86.1	65.6	72.1	80.1	65.6	70.2
095,098	CEILINGS & ACOUSTICAL TREATMENT	107.2	55.5	72.2	109.4	70.7	83.1	98.1	41.3	59.6	98.7	49.7	65.5	98.1	65.6	76.1	94.2	65.6	74.8
09600	FLOORING	94.0	53.7	83.6	97.2	53.9	86.1	116.5	72.9	105.2	77.8	47.4	70.0	117.7	72.9	106.2	105.4	72.9	97.0
097,099	WALL FINISHES, PAINTS & COATINGS	94.8	50.4	68.2	102.8	50.4	71.3	96.6	60.2	74.7	85.0	54.1	66.5	96.6	61.7	75.7	96.6	61.7	75.7
09	FINISHES	96.1	55.9	74.3	100.3	64.8	81.1	99.0	50.2	72.6	89.7	49.7	68.1	99.3	64.2	80.4	94.0	64.2	77.9
10 - 14	TOTAL DIV. 10000 - 14000	100.0	75.4	94.7	100.0	78.1	95.3	100.0	67.3	93.0	100.0	58.2	91.0	100.0	70.4	93.6	100.0	70.4	93.6
15	MECHANICAL	100.2	75.0	88.6	100.1	71.1	86.8	100.1	44.7	74.6	100.3	44.6	74.6	100.1	58.6	81.0	100.1	47.2	75.8
16	ELECTRICAL	100.2	48.0	69.7	99.2	63.3	78.2	95.9	67.0	79.0	101.8	49.6	71.3	96.5	72.8	82.7	95.0	72.8	82.0
01 - 16	WEIGHTED AVERAGE	99.6	66.4	83.6	99.0	71.2	85.6	96.4	58.6	78.2	95.6	56.7	76.9	96.3	65.9	81.6	95.8	63.6	80.3

(Reprinted from Means Plumbing Cost Data 2004)

Figure 8.15

will a similar building cost in Phoenix in 2004? Adjustment factors are developed as shown above using the data from Figures 8.15 and 8.16.

$$\frac{\text{Phoenix Index}}{\text{San Francisco Index}} = \frac{87.9}{123.6} = 0.71$$

$$\frac{\text{2004 Index}}{\text{1974 Index}} = \frac{133.0}{41.4} = 3.21$$

Original cost × location adjustment x time adjustment = proposed new cost.

$$\$1{,}000{,}000 \times 0.71 \times 3.21 = \$2{,}279{,}100$$

Square Foot and Systems Estimating Examples

Often the contractor is faced with the need to develop a preliminary estimate for a project. The estimate may be for the entire project or for only a portion of the work, which is often the case for the mechanical or plumbing contractor. This chapter contains two complete sample estimates for the plumbing portion of an office building: a *square foot estimate* and a *systems estimate*. In these step-by-step examples, forms are filled out and calculations made. All cost and reference tables are from the annual *Means Plumbing Cost Data*.

Historical Cost Indexes

The table below lists both the Means Historical Cost Index based on Jan. 1, 1993 = 100 as well as the computed value of an index based on Jan. 1, 2004 costs. Since the Jan. 1, 2004 figure is estimated, space is left to write in the actual index figures as they become available through either the quarterly "Means Construction Cost Indexes" or as printed in the "Engineering News-Record." To compute the actual index based on Jan. 1, 2004 = 100, divide the Historical Cost Index for a particular year by the actual Jan. 1, 2004 Construction Cost Index. Space has been left to advance the index figures as the year progresses.

Year	Historical Cost Index Jan. 1, 1993 = 100		Current Index Based on Jan. 1, 2004 = 100		Year	Historical Cost Index Jan. 1, 1993 = 100	Current Index Based on Jan. 1, 2004 = 100		Year	Historical Cost Index Jan. 1, 1993 = 100	Current Index Based on Jan. 1, 2004 = 100	
	Est.	Actual	Est.	Actual		Actual	Est.	Actual		Actual	Est.	Actual
Oct 2004					July 1989	92.1	69.3		July 1971	32.1	24.1	
July 2004					1988	89.9	67.6		1970	28.7	21.6	
April 2004					1987	87.7	65.9		1969	26.9	20.2	
Jan 2004	133.0		100.0	100.0	1986	84.2	63.3		1968	24.9	18.7	
July 2003		132.0	99.2		1985	82.6	62.1		1967	23.5	17.7	
2002		128.7	96.8		1984	82.0	61.6		1966	22.7	17.1	
2001		125.1	94.1		1983	80.2	60.3		1965	21.7	16.3	
2000		120.9	90.9		1982	76.1	57.3		1964	21.2	15.9	
1999		117.6	88.4		1981	70.0	52.6		1963	20.7	15.6	
1998		115.1	86.5		1980	62.9	47.3		1962	20.2	15.2	
1997		112.8	84.8		1979	57.8	43.5		1961	19.8	14.9	
1996		110.2	82.9		1978	53.5	40.2		1960	19.7	14.8	
1995		107.6	80.9		1977	49.5	37.2		1959	19.3	14.5	
1994		104.4	78.5		1976	46.9	35.3		1958	18.8	14.1	
1993		101.7	76.5		1975	44.8	33.7		1957	18.4	13.8	
1992		99.4	74.8		1974	41.4	31.1		1956	17.6	13.2	
1991		96.8	72.8		1973	37.7	28.3		1955	16.6	12.5	
1990		94.3	70.9		1972	34.8	26.2		1954	16.0	12.0	

(Reprinted from Means Plumbing Cost Data 2004)

Figure 8.16

Project Description

A plumbing contractor has been invited by a familiar general contractor to submit a budget estimate on an office building. The project is a three-story office building with a penthouse and garage. Area calculations for this building are shown in Figure 8.17.

For purposes of illustration, a budget estimate will first be developed as a square foot estimate based on the bare minimum of information supplied. A sketch of the building concept with overall dimensions is shown in Figure 8.17.

Square Foot Estimating Example

The dimensions of the project (based on Figure 8.17) and the mathematical procedures to obtain the areas of the office building and the parking garage are shown in Figure 8.18.

Once the area is of the proposed building is known, size modification calculations can be made as shown in Figure 8.19 using data from *Means Plumbing Cost Data*. The appropriate building type is located and its typical size provides the denominator for the modifier equation. Dividing through produces the factor of 2.94 for the office and .12 for the garage. It will be seen that 2.94 is near the end of the horizontal scale to the right, and .12 is off the scale to the left. In both cases, use the value indicated on the vertical scale. NOTE: the modifier for any size factor less than .5 will be 1.1 and for any size factor greater than 3.5 will be .90, as the example problem illustrates.

Square foot costs are then found in Division 17 of *Means Plumbing Cost Data*, a portion of which appears in Figure 8.20, or can be taken from your own cost records. These square foot costs are multiplied by the appropriate modified areas, and costs for plumbing, heating, ventilating, and air conditioning are determined. The total mechanical cost is then modified for a geographic area.

Systems Estimating Example

The next phase of the estimating process, after the preliminary square foot estimate, may be to complete a systems estimate. However, the systems estimate, like the square foot estimate, is not a substitute for a *unit price estimate*. A systems estimate is most often prepared during the conceptual stage of a project when certain parameters are known, but the building plans are not yet completed. This preliminary estimate may help the designer keep the project within the owner's budget. The sketch and dimensions of the proposed building would be identical to that used for the square foot estimate. However, in addition to the areas, the occupied volume is calculated. (See Figure 8.21).

Once the physical area and occupied volume have been defined, the number of occupants needs to be determined. Occupancy is based on the net floor area. Figures 8.22 and 8.23 can be used to determine the occupancy requirements. The floor area ratio table in Figure 8.22 indicates a net to gross ratio for offices as 75% on a per occupied floor basis.

Therefore, 18,900 S.F. (area of each office floor as shown in Figure 8.21) is reduced to 14,175 S.F. net area. This net or reduced area has excluded the space occupied by stairwells, corridors, mechanical rooms, etc. In a commercial building, the net area might be referred to as the "leasable area." The occupancy determinations table in Figure 8.23

indicates the occupancy recommendations of four major building code authorities for several building classifications. All agree that for office buildings, the maximum density is one person per 100 S.F. Based on this calculation, the net floor area of 14,175 S.F. allows a maximum of 142 persons per floor. Without any data to the contrary, a male/female ratio of 50:50 should be assumed. With the basic requirements of the building established, the next step is to develop costs using the systems estimating method.

In a systems estimate, individual components are grouped together into systems that reflect the way buildings are constructed. The grouping of many components into a single system allows the estimator to compare various systems and select the one best suited to accommodate costs, usage, compatibility, and any special requirements for the particular project.

Prior to beginning the estimate, all pertinent data must be assembled and analyzed. A typical preprinted form to record the available data is shown in Figure 8.24

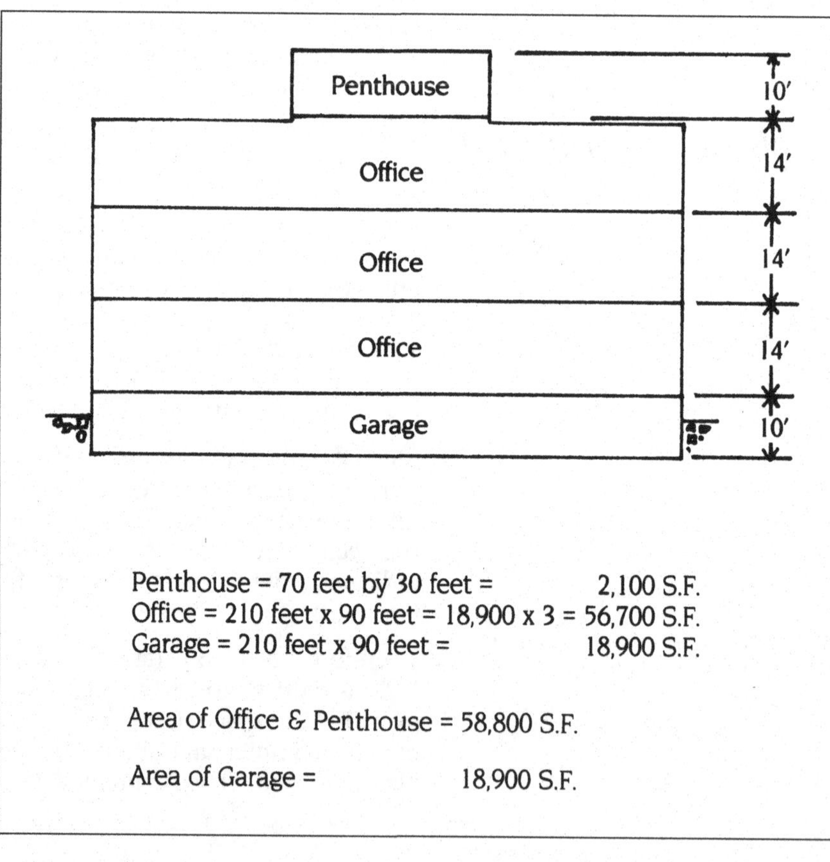

Figure 8.17

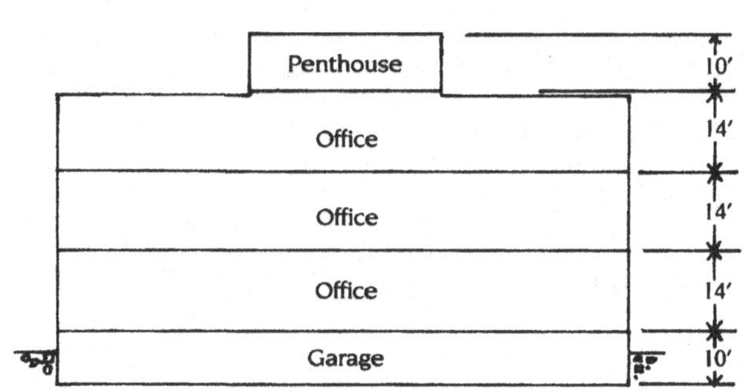

Penthouse	= 70 feet x 30 feet =		2,100 S.F.
Office	= 210 feet x 90 feet =	18,900 x 3 =	56,700 S.F.
Garage	= 210 feet x 90 feet =		18,900 S.F.

Area of Office & Penthouse = A_O = 58,800 S.F.
Area of Garage = A_G = 18,900 S.F.

Size Modification

Office $\frac{58,800}{20,000}$ = (2.94) modifier = (.91) multiplier (M_O)

Garage $\frac{18,900}{163,000}$ = (.12)modifier = (1 .1) multiplier (M_G)

Cost

Office				
17100-610-2720 Plumbing	\$ 3.69 x	(M_O) .91 x	(A_O) 58,800 S.F.	= \$ 197,445
-2770 HVAC	\$ 7.40 x	(M_O) .91 x	(A_O) 58,800 S.F.	= \$ 395,959
-2900 Electrical	\$ 7.80 x	(M_O) .91 x	(A_O) 58,800 S.F.	= \$ 417,362
Garage				
17100-410-2720 Plumbing	\$ 1.09 x	(M_G) 1.1 x	(A_G) 18,900 S.F.	= \$ 22,661
-2900 Electrical	\$ 1.61 x	(M_G) 1.1 x	(A_G) 18,900 S.F.	= \$ 33,472

Total Plumbing	\$ 220,106		
Total HVAC	\$ 395,959		
Total Mechanical	\$ 616,065		
(Geographic correction)	\$ 616,065	x 1.111	= \$ 684,448
Total Electrical	\$ 450,834		
(Geographic correction)	\$ 450,834	x 1.119	= \$ 504,483

(Reprinted from Means Mechanical/Electrical Seminar Workbook 2004)

Figure 8.18

S.F. & C.F. Costs | **R171** | **Information**

R17100-100 Square Foot Project Size Modifier

One factor that affects the S.F. cost of a particular building is the size. In general, for buildings built to the same specifications in the same locality, the larger building will have the lower S.F. cost. This is due mainly to the decreasing contribution of the exterior walls plus the economy of scale usually achievable in larger buildings. The Area Conversion Scale shown below will give a factor to convert costs for the typical size building to an adjusted cost for the particular project.

The Square Foot Base Size lists the median costs, most typical project size in our accumulated data and the range in size of the projects.

The Size Factor for your project is determined by dividing your project area in S.F. by the typical project size for the particular Building Type. With this factor, enter the Area Conversion Scale at the appropriate Size Factor and determine the appropriate cost multiplier for your building size.

Example: Determine the cost per S.F. for a 100,000 S.F. Mid-rise apartment building.

$$\frac{\text{Proposed building area} = 100{,}000 \text{ S.F.}}{\text{Typical size from below} = 50{,}000 \text{ S.F.}} = 2.00$$

Enter Area Conversion scale at 2.0, intersect curve, read horizontally the appropriate cost multiplier of .94. Size adjusted cost becomes .94 x $77.00 = $72.40 based on national average costs.

Note: For Size Factors less than .50, the Cost Multiplier is 1.1
For Size Factors greater than 3.5, the Cost Multiplier is .90

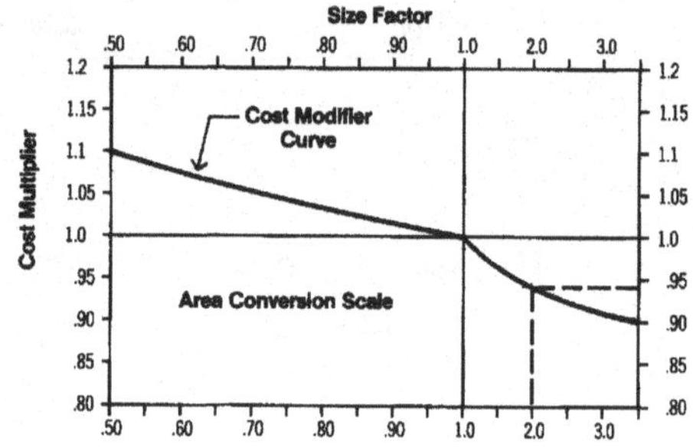

Square Foot Base Size							
Building Type	Median Cost per S.F.	Typical Size Gross S.F.	Typical Range Gross S.F.	Building Type	Median Cost per S.F.	Typical Size Gross S.F.	Typical Range Gross S.F.
Apartments, Low Rise	$ 60.50	21,000	9,700 - 37,200	Jails	$184.00	40,000	5,500 - 145,000
Apartments, Mid Rise	77.00	50,000	32,000 - 100,000	Libraries	114.00	12,000	7,000 - 31,000
Apartments, High Rise	87.00	145,000	95,000 - 600,000	Living, Assisted	99.00	32,300	23,500 - 50,300
Auditoriums	101.00	25,000	7,600 - 39,000	Medical Clinics	106.00	7,200	4,200 - 15,700
Auto Sales	75.50	20,000	10,800 - 28,600	Medical Offices	97.50	6,000	4,000 - 15,000
Banks	135.00	4,200	2,500 - 7,500	Motels	74.00	40,000	15,800 - 120,000
Churches	92.50	17,000	2,000 - 42,000	Nursing Homes	103.00	23,000	15,000 - 37,000
Clubs, Country	95.00	6,500	4,500 - 15,000	Offices, Low Rise	86.00	20,000	5,000 - 80,000
Clubs, Social	90.00	10,000	6,000 - 13,500	Offices, Mid Rise	85.00	120,000	20,000 - 300,000
Clubs, YMCA	106.00	28,300	12,800 - 39,400	Offices, High Rise	109.00	260,000	120,000 - 800,000
Colleges (Class)	118.00	50,000	15,000 - 150,000	Police Stations	136.00	10,500	4,000 - 19,000
Colleges (Science Lab)	173.00	45,600	16,600 - 80,000	Post Offices	101.00	12,400	6,800 - 30,000
College (Student Union)	132.00	33,400	16,000 - 85,000	Power Plants	750.00	7,500	1,000 - 20,000
Community Center	96.00	9,400	5,300 - 16,700	Religious Education	84.00	9,000	6,000 - 12,000
Court Houses	129.00	32,400	17,800 - 106,000	Research	142.00	19,000	6,300 - 45,000
Dept. Stores	56.00	90,000	44,000 - 122,000	Restaurants	122.00	4,400	2,800 - 6,000
Dormitories, Low Rise	100.00	25,000	10,000 - 95,000	Retail Stores	60.00	7,200	4,000 - 17,600
Dormitories, Mid Rise	126.00	85,000	20,000 - 200,000	Schools, Elementary	88.50	41,000	24,500 - 55,000
Factories	54.50	26,400	12,900 - 50,000	Schools, Jr. High	91.50	92,000	52,000 - 119,000
Fire Stations	96.00	5,800	4,000 - 8,700	Schools, Sr. High	98.00	101,000	50,500 - 175,000
Fraternity Houses	93.50	12,500	8,200 - 14,800	Schools, Vocational	89.50	37,000	20,500 - 82,000
Funeral Homes	105.00	10,000	4,000 - 20,000	Sports Arenas	74.00	15,000	5,000 - 40,000
Garages, Commercial	68.00	9,300	5,000 - 13,600	Supermarkets	60.00	44,000	12,000 - 60,000
Garages, Municipal	86.00	8,300	4,500 - 12,600	Swimming Pools	139.00	20,000	10,000 - 32,000
Garages, Parking	36.00	163,000	76,400 - 225,300	Telephone Exchange	163.00	4,500	1,200 - 10,600
Gymnasiums	91.00	19,200	11,600 - 41,000	Theaters	88.50	10,500	8,800 - 17,500
Hospitals	160.00	55,000	27,200 - 125,000	Town Halls	104.00	10,800	4,800 - 23,400
House (Elderly)	82.50	37,000	21,000 - 66,000	Warehouses	44.50	25,000	8,000 - 72,000
Housing (Public)	76.00	36,000	14,400 - 74,400	Warehouse & Office	47.00	25,000	8,000 - 72,000
Ice Rinks	109.00	29,000	27,200 - 33,600				

(Reprinted from Means Plumbing Cost Data 2004)

Figure 8.19

17100 | S.F., C.F. and % of Total Costs

	17100 | S.F. & C.F. Costs			UNIT	UNIT COSTS			% OF TOTAL			
					1/4	MEDIAN	3/4	1/4	MEDIAN	3/4	
340	2900	Electrical		S.F.	4.55	7.30	11.20	8.55%	10.50%	14.20%	340
	3100	Total: Mechanical & Electrical		↓	10.80	17.35	26.50	22%	29%	35.50%	
360	0010	**FIRE STATIONS**	R17100 -100	S.F.	70	96	130				360
	0020	Total project costs		C.F.	4.07	5.90	7.75				
	2720	Plumbing		S.F.	4.60	6.80	9.95	6.20%	7.70%	9.70%	
	2770	Heating, ventilating, air conditioning			3.96	6.50	9.85	5.15%	7.50%	9.25%	
	2900	Electrical			5.15	8.90	11.70	6.80%	8.50%	10.95%	
	3100	Total: Mechanical & Electrical		↓	25.50	30	35.50	19.60%	23%	27%	
370	0010	**FRATERNITY HOUSES** and Sorority Houses	R17100 -100	S.F.	72.50	93.50	132				370
	0020	Total project costs		C.F.	6.95	7.55	9.65				
	2720	Plumbing		S.F.	5.45	6.30	11.50	7.55%	8%	10.90%	
	2900	Electrical			4.79	10.35	12.70	6.60%	9.90%	10.70%	
	3100	Total: Mechanical & Electrical		↓	13.30	18.80	22.50	14.60%	19.90%	20.70%	
380	0010	**FUNERAL HOMES**	R17100 -100	S.F.	76.50	105	190				380
	0020	Total project costs		C.F.	7.80	8.70	16.70				
	2900	Electrical		S.F.	3.39	6.20	7.30	4.44%	5.95%	11.05%	
	3100	Total: Mechanical & Electrical		"	12.30	18.20	24	12.90%	17.80%	18.80%	
390	0010	**GARAGES, COMMERCIAL** (Service)	R17100 -100	S.F.	43	68	92.50				390
	0020	Total project costs		C.F.	2.92	4.22	6.15				
	2720	Plumbing		S.F.	3	4.63	8.45	4.70%	7.40%	10.45%	
	2730	Heating & ventilating			3.95	5.40	7.45	5.25%	6.85%	9.55%	
	2900	Electrical			4.18	6.40	9.15	7.05%	9.40%	10.65%	
	3100	Total: Mechanical & Electrical		↓	9.35	17.35	26	13.60%	17.40%	27%	
400	0010	**GARAGES, MUNICIPAL** (Repair)	R17100 -100	S.F.	64	86	120				400
	0020	Total project costs		C.F.	3.98	5.05	8.25				
	2720	Plumbing		S.F.	2.86	5.50	10.35	3.59%	6.70%	7.95%	
	2730	Heating & ventilating			4.89	7.10	13.65	6.15%	7.45%	13.50%	
	2900	Electrical			4.71	7.40	10.75	6.90%	9.25%	11.15%	
	3100	Total: Mechanical & Electrical		↓	15.15	24	44	21.50%	25.50%	28.50%	
410	0010	**GARAGES, PARKING**	R17100 -100	S.F.	24.50	36	62				410
	0020	Total project costs		C.F.	2.33	3.17	4.61				
	2720	Plumbing		S.F.	.64	1.09	1.68	1.72%	3.05%	3.85%	
	2900	Electrical			1.30	1.61	2.57	4.27%	4.98%	6.10%	
	3100	Total: Mechanical & Electrical		↓	2.17	3.97	5.10	6.90%	8.80%	11.05%	
	9000	Per car, total cost		Car	10,700	13,200	16,800				
610	0010	**OFFICES** Low Rise (1 to 4 story)	R17100 -100	S.F.	66.50	86	111				610
	0020	Total project costs		C.F.	4.62	6.70	8.80				
	2720	Plumbing		S.F.	2.43	3.69	5.65	3.66%	4.50%	6.10%	
	2770	Heating, ventilating, air conditioning			5.30	7.40	10.80	7.25%	10.40%	11.80%	
	2900	Electrical			5.50	7.80	11	7.50%	9.60%	11.40%	
	3100	Total: Mechanical & Electrical		↓	14	19.75	29	18%	22.50%	26.50%	
620	0010	**OFFICES** Mid Rise (5 to 10 story)	R17100 -100	S.F.	70	85	116				620
	0020	Total project costs		C.F.	4.98	6.25	9				
	2720	Plumbing		S.F.	2.11	3.43	4.85	2.76%	4.34%	4.50%	
	2770	Heating, ventilating, air conditioning			5.35	7.65	12.20	7.65%	9.40%	11%	
	2900	Electrical			5.20	6.70	8.85	6.20%	7.80%	10%	
	3100	Total: Mechanical & Electrical		↓	13.15	16.85	34	19.15%	21%	24.50%	

(Reprinted from Means Plumbing Cost Data 2004)

Figure 8.20

In a systems estimate the UNIFORMAT II grouping of 7 construction divisions is used, instead of the 16 Construction Specifications Institute MasterFormat divisions used for unit price estimates. The plumbing data is found in Division D of the UNIFORMAT II data.

It is desirable that the building dimensions, the owner's/occupants' requirements, local building codes, existing utilities, local labor sources, and anticipated economic conditions be known. With this information and the inclusion of any contingencies, a reasonable estimate may be arrived at, which can be updated as more design parameters are developed. For this example, the proposed building is a regional office building of good quality for an insurance company located in the Boston area.

A systems estimate for the mechanical trades is accomplished by obtaining all the available data, including physical area, function, occupancy, etc., and selecting the appropriate mechanical systems to cover all the requirements for such a building. The figures and other

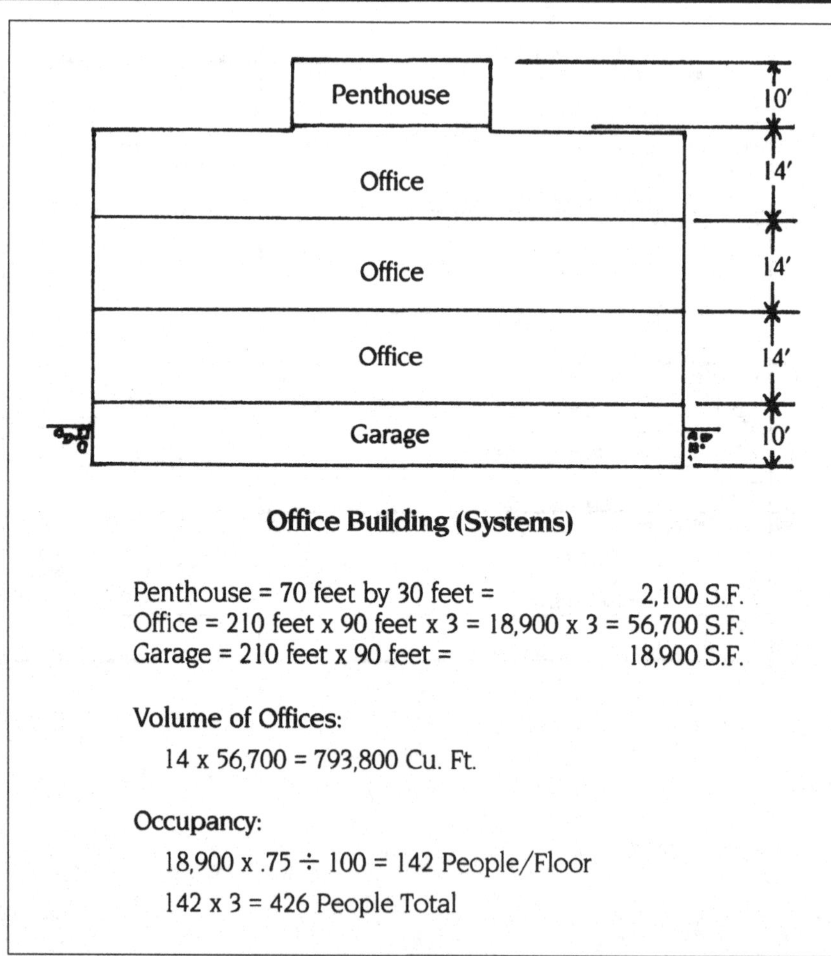

Penthouse
Office
Office
Office
Garage
10'
14'
14'
14'
10'
Office Building (Systems)
Penthouse = 70 feet by 30 feet = 2,100 S.F.
Office = 210 feet x 90 feet x 3 = 18,900 x 3 = 56,700 S.F.
Garage = 210 feet x 90 feet = 18,900 S.F.
Volume of Offices:
14 x 56,700 = 793,800 Cu. Ft.
Occupancy:
18,900 x .75 ÷ 100 = 142 People/Floor
142 x 3 = 426 People Total

Figure 8.21

pertinent information to compile systems and their costs should come from the estimator's own experience and records, or from a reliable cost data source such as *Means Plumbing Cost Data*. Some of these systems are priced by the square foot area of a building (e.g., sprinkler systems), and others are priced by the required number of units (e.g., bathrooms, water heaters, etc.). All systems costs include an allowance for overhead and profit.

Plumbing

The table shown in Figure 8.25, "Minimum Plumbing Fixture Requirements," from *Means Plumbing Cost Data* can be used as the basis for determining the number of plumbing fixtures required depending on building use and the number of occupants or preferably use your own local plumbing code. It has been determined (See Figure 8.23) that this office building will house 142 persons per floor, with a 50:50 male/female ratio.

Based on this determined occupancy of 142 persons per floor, the minimum plumbing fixture requirements based on a local code mandate the following fixture count for business offices:

Fixtures	Men	Women	Handicapped	Misc	Total Per Floor
water closets	2	3	1	–	6
lavatories	3	3	1	–	7
urinals	2	–	–	–	2
water coolers	–	–	1	2	3
service sinks	–	–	–	1	1

The local code allows 1/3 of the water closets to be replaced with urinals. This would change the required three men's water closets to two with one urinal. A second urinal has been arbitrarily added based on an estimate of what an occupant would require. Three handicapped fixtures per floor have also been included in the estimate to conform with local and national codes.

In addition to the fixtures and appliances (water coolers), the plumbing estimate would not be complete without water heaters and roof drains. The roof area is 18,900 S.F. and the building height is 52 feet (see Figure 8.26). An additional two feet is allowed to bury the storm drain. With these measurements, the roof drain system can be sized using the chart shown in Figure 8.26 from *Means Plumbing Cost Data*. Six 4" drains are capable of handling up to 20,760 S.F.

Hot water consumption rates are determined by using the table shown in Figure 8.27 from *Means Plumbing Cost Data*. This table shows that an office building has a minimum hourly demand of 0.4 gallons per person. The office building in this estimating example, having an occupancy of 142 persons per floor or 426 total, requires a 170 gallon minimum hourly demand (gallons per hour-GPH) water heater or series of heaters.

All the plumbing fixtures and appliances should be listed by categories for pricing on a plumbing estimate worksheet, as shown in Figure 8.28. Using the assemblies cost tables shown in Figure 8.29 through 8.35 from *Means Plumbing Cost Data*, a systems table number is assigned for each fixture. The line number for pricing is also recorded. The lavatories, for example, can be priced using Figure 8.29. Vanity top oval lavatories,

R01100-720 Floor Area Ratios

Table below lists commonly used gross to net area and net to gross area ratios expressed in % for various building types.

Building Type	Gross to Net Ratio	Net to Gross Ratio	Building Type	Gross to Net Ratio	Net to Gross Ratio
Apartment	156	64	School Buildings (campus type)		
Bank	140	72	Administrative	150	67
Church	142	70	Auditorium	142	70
Courthouse	162	61	Biology	161	62
Department Store	123	81	Chemistry	170	59
Garage	118	85	Classroom	152	66
Hospital	183	55	Dining Hall	138	72
Hotel	158	63	Dormitory	154	65
Laboratory	171	58	Engineering	164	61
Library	132	76	Fraternity	160	63
Office	135	75	Gymnasium	142	70
Restaurant	141	70	Science	167	60
Warehouse	108	93	Service	120	83
			Student Union	172	59

The gross area of a building is the total floor area based on outside dimensions.

The net area of a building is the usable floor area for the function intended and excludes such items as stairways, corridors and mechanical rooms. In the case of a commercial building, it might be considered as the "leasable area."

(Reprinted from Means Plumbing Cost Data 2004)

Figure 8.22

R01100-730 Occupancy Determinations

Description		S.F. Required per Person		
		BOCA	SBC	UBC
Assembly Areas	Fixed Seats	**	6	7
	Movable Seats		15	15
	Concentrated	7		
	Unconcentrated	15		
	Standing Space	3		
Educational	Unclassified			
	Classrooms	20	40	20
	Shop Areas	50	100	50
Institutional	Unclassified		125	
	In-Patient Areas	240		
	Sleeping Areas	120		
Mercantile	Basement	30	30	20
	Ground Floor	30	30	30
	Upper Floors	60	60	50
Office		100	100	100

BOCA=Building Officials & Code Administrators' National Building Code
SBC=Standard Building Code
UBC=Uniform Building Code

** The occupancy load for assembly area with fixed seats shall be determined by the number of fixed seats installed.

(Reprinted from Means Plumbing Cost Data 2004)

Figure 8.23

ASSEMBLY NUMBER	DESCRIPTION	QTY	UNIT	TOTAL COST		COST PER S.F.
				UNIT	TOTAL	
	Mechanical System					
	Electrical					

Figure 8.24

R15100-410 Minimum Plumbing Fixture Requirements

Minimum Plumbing Fixture Requirements

Type of Building or Occupancy (2)	Water Closets (14) (Fixtures per Person) Male	Water Closets (14) (Fixtures per Person) Female	Urinals (5,10) (Fixtures per Person) Male	Urinals (5,10) (Fixtures per Person) Female	Lavatories (Fixtures per Person) Male	Lavatories (Fixtures per Person) Female	Bathtubs or Showers (Fixtures per Person)	Drinking Fountains (Fixtures per Person) (3, 13)
Assembly Places-Theatres, Auditoriums, Convention Halls, etc.-for permanent employee use	1: 1 - 15 2: 16 - 35 3: 36 - 55 Over 55, add 1 fixture for each additional 40 persons	1: 1 - 15 2: 16 - 35 3: 36 - 55	0: 1 - 9 1: 10 - 50 Add one fixture for each additional 50 males		1 per 40	1 per 40		
Assembly Places-Theatres, Auditoriums, Convention Halls, etc. - for public use	1: 1 - 100 2: 101 - 200 3: 201 - 400 Over 400, add 1 fixture for each additional 500 males and 1 for each additional 125 females	3: 1 - 50 4: 51 - 100 8: 101 - 200 11: 201 - 400	1: 1 - 100 2: 101 - 200 3: 201 - 400 4: 401 - 600 Over 600, add 1 fixture for each additional 300 males		1: 1 - 200 2: 201 - 400 3: 401 - 750 Over 750, add 1 fixture for each additional 500 persons	1: 1 - 200 2: 201 - 400 3: 401 - 750		1: 1 - 150 2: 151 - 400 3: 401 - 750 Over 750, add one fixture for each additional 500 persons
Dormitories (9) School or Labor	1 per 10 Add 1 fixture for each additional 25 males (over 10) and 1 for each additional 20 females (over 8)	1 per 8	1 per 25 Over 150, add 1 fixture for each additional 50 males		1 per 12 Over 12 add 1 fixture for each additional 20 males and 1 for each 15 additional females	1 per 12	1 per 8 For females add 1 bathtub per 30. Over 150, add 1 per 20	1 per 150 (12)
Dormitories-for Staff Use	1: 1 - 15 2: 16 - 35 3: 36 - 55 Over 55, add 1 fixture for each additional 40 persons	1: 1 - 15 3: 16 - 35 4: 36 - 55	1 per 50		1 per 40	1 per 40	1 per 8	
Dwellings: Single Dwelling Multiple Dwelling or Apartment House	1 per dwelling 1 per dwelling or apartment unit				1 per dwelling 1 per dwelling or apartment unit		1 per dwelling 1 per dwelling or apartment unit	
Hospital Waiting rooms	1 per room				1 per room			1 per 150 (12)
Hospitals-for employee use	1: 1 - 15 2: 16 - 35 3: 36 - 55 Over 55, add 1 fixture for each additional 40 persons	1: 1 - 15 3: 16 - 35 4: 36 - 55	0: 1 - 9 1: 10 - 50 Add 1 fixture for each additional 50 males		1 per 40	1 per 40		
Hospitals: Individual Room Ward Room	1 per room 1 per 8 patients				1 per room 1 per 10 patients		1 per room 1 per 20 patients	1 per 150 (12)
Industrial (6) Warehouses Workshops, Foundries and similar establishments-for employee use	1: 1 -10 2: 11 - 25 3: 26 - 50 4: 51 - 75 5: 76 - 100 Over 100, add 1 fixture for each additional 30 persons	1: 1 -10 2: 11 - 25 3: 26 - 50 4: 51 - 75 5: 76 - 100			Up to 100, per 10 persons Over 100, 1 per 15 persons (7, 8)		1 shower for each 15 persons exposed to excessive heat or to skin contamination with poisonous, infectious or irritating material	1 per 150 (12)
Institutional - Other than Hospitals or Penal Institutions (on each occupied floor)	1 per 25	1 per 20	0: 1 - 9 1: 10 - 50 Add 1 fixture for each additional 50 males		1 per 10	1 per 10	1 per 8	1 per 150 (12)
Institutional - Other than Hospitals or Penal Institutions (on each occupied floor)-for employee use	1: 1 - 15 2: 16 - 35 3: 36 - 55 Over 55, add 1 fixture for each additional 40 persons	1: 1 - 15 3: 16 - 35 4: 36 - 55	0: 1 - 9 1: 10 - 50 Add 1 fixture for each additional 50 males		1 per 40	1 per 40	1 per 8	1 per 150 (12)
Office or Public Buildings	1: 1 - 100 2: 101 - 200 3: 201 - 400 Over 400, add 1 fixture for each additional 500 males and 1 for each additional 150 females	3: 1 - 50 4: 51 - 100 8: 101 - 200 11: 201 - 400	1: 1 - 100 2: 101 - 200 3: 201 - 400 4: 401 - 600 Over 600, add 1 fixture for each additional 300 males		1: 1 - 200 2: 201 - 400 3: 401 - 750 Over 750, add 1 fixture for each additional 500 persons	1: 1 - 200 2: 201 - 400 3: 401 - 750		1 per 150 (12)
Office or Public Buildings - for employee use	1: 1 - 15 2: 16 - 35 3: 36 - 55 Over 55, add 1 fixture for each additional 40 persons	1: 1 - 15 3: 16 - 35 4: 36 - 55	0: 1 - 9 1: 10 - 50 Add 1 fixture for each additional 50 males		1 per 40	1 per 40		

(Reprinted from Means Plumbing Cost Data 2004)

Figure 8.25

19″ x 16″, are selected and priced according to line 1600. The same procedure is followed for the remaining fixtures and appliances. The complete plumbing systems estimate is shown in Figure 8.36.

To compensate for materials not included in the systems groupings, certain percentages must be added to the plumbing system. These percentages can be found in table R15100-040 from *Means Plumbing Cost Data* (reproduced in Figure 8.37). The first of these is for "water control" and covers water meter, backflow preventer, shock absorbers, and vacuum breakers. An additional 10-15% should be added to the subtotal for these items.

Pipe and fittings is the next category and includes the interconnecting piping (mains) between the several systems. For pipe and fittings, 30% to 60% should be added to the subtotal. This percentage depends on the building design. For example, a building similar in size to the office building in this example, but having toilet rooms at each of the far ends of the building, would have a much higher percentage than the 30% indicated for this plumbing estimate. In extreme cases, this percentage might reach 100%.

Quality complexity is the last additive. It should be applied to fire protection, HVAC, and plumbing estimates to compensate for quality/complexity over and above the basic level assumed in the Means assemblies prices. Percentages for quality/complexity are 0-5% for economy installation, 5-15% for good quality, medium complexity installation, and 15-25% for above average quality and complexity. Construction and estimating experience is necessary to determine the proper percentages to be added. Keeping cost records of completed projects will contribute to this experience factor.

Roof Drain Systems

Pipe Diameter	Max. S.F. Roof Area	Gallons per Min.
2″	544	23
3″	1,610	67
4″	3,460	144
5″	6,280	261
6″	10,200	424
8″	22,000	913

Design Assumptions: Vertical conductor size is based on a maximum rate of rainfall of 4″ per hour. To convert roof area to other rates multiply "Max S.F. Roof Area" shown by four and divide the result by desired rate. The answer is the local roof area that may be handled by the indicated pipe diameter.

Basic cost is for roof drain, 10′ of vertical leader and 10′ of horizontal, plus connection to the main.

Figure 8.26

R15100-110 Hot Water Consumption Rates

Type of Building	Size Factor	Maximum Hourly Demand	Average Day Demand
Apartment Dwellings	No. of Apartments:		
	Up to 20	12.0 Gal. per apt.	42.0 Gal. per apt.
	21 to 50	10.0 Gal. per apt.	40.0 Gal. per apt.
	51 to 75	8.5 Gal. per apt.	38.0 Gal. per apt.
	76 to 100	7.0 Gal. per apt.	37.0 Gal. per apt.
	101 to 200	6.0 Gal. per apt.	36.0 Gal. per apt.
	201 up	5.0 Gal. per apt.	35.0 Gal. per apt.
Dormitories	Men	3.8 Gal. per man	13.1 Gal. per man
	Women	5.0 Gal. per woman	12.3 Gal. per woman
Hospitals	Per bed	23.0 Gal. per patient	90.0 Gal. per patient
Hotels	Single room with bath	17.0 Gal. per unit	50.0 Gal. per unit
	Double room with bath	27.0 Gal. per unit	80.0 Gal. per unit
Motels	No. of units:		
	Up to 20	6.0 Gal. per unit	20.0 Gal. per unit
	21 to 100	5.0 Gal. per unit	14.0 Gal. per unit
	101 Up	4.0 Gal. per unit	10.0 Gal. per unit
Nursing Homes		4.5 Gal. per bed	18.4 Gal. per bed
Office buildings		0.4 Gal. per person	1.0 Gal. per person
Restaurants	Full meal type	1.5 Gal./max. meals/hr.	2.4 Gal. per meal
	Drive-in snack type	0.7 Gal./max. meals/hr.	0.7 Gal. per meal
Schools	Elementary	0.6 Gal. per student	0.6 Gal. per student
	Secondary & High	1.0 Gal. per student	1.8 Gal. per student

For evaluation purposes, recovery rate and storage capacity are inversely proportional. Water heaters should be sized so that the maximum hourly demand anticipated can be met in addition to allowance for the heat loss from the pipes and storage tank.

R15100-120 Fixture Demands in Gallons Per Fixture Per Hour

Table below is based on 140°F final temperature except for dishwashers in public places (*) where 180°F water is mandatory.

Fixture	Apartment House	Club	Gym	Hospital	Hotel	Indust. Plant	Office	Private Home	School
Bathtubs	20	20	30	20	20			20	
Dishwashers, automatic	15	50-150*		50-150*	50-200*	20-100*		15	20-100*
Kitchen sink	10	20		20	30	20	20	10	20
Laundry, stationary tubs	20	28		28	28			20	
Laundry, automatic wash	75	75		100	150			75	
Private lavatory	2	2	2	2	2	2	2	2	2
Public lavatory	4	6	8	6	8	12	6		15
Showers	30	150	225	75	75	225	30	30	225
Service sink	20	20		20	30	20	20	15	20
Demand factor	0.30	0.30	0.40	0.25	0.25	0.40	0.30	0.30	0.40
Storage capacity factor	1.25	0.90	1.00	0.60	0.80	1.00	2.00	0.70	1.00

To obtain the probable maximum demand multiply the total demands for the fixtures (gal./fixture/hour) by the demand factor. The heater should have a heating capacity in gallons per hour equal to this maximum. The storage tank should have a capacity in gallons equal to the probable maximum demand multiplied by the storage capacity factor.

(Reprinted from Means Plumbing Cost Data 2004)

Figure 8.27

Fire Protection

Fire protection systems are usually composed of sprinkler systems, hose standpipes, or combinations thereof. System classifications according to hazard and design have been developed by the National Fire Protection Association (NFPA). Based on these criteria, a firehose standpipe and sprinkler system for the example office building can be estimated.

Depending on the system's design, the firehose standpipe could be taken off and priced as part of the sprinkler system, or the two could be taken off and priced separately. For this estimate, the firehose standpipe system and the sprinkler system are taken off and priced separately.

Sprinkler System: The system classification for an office building is "Light Hazard Occupancy." A conventional "wet type" distribution is selected for this heated building, and the designer has chosen Schedule #40 black steel pipe and threaded fittings as the piping material.

Using the previously determined area data (18,900 S.F. per floor and 14 L.F. floor heights), and the cost data from *Means Plumbing Cost Data,* shown in Figures 8.38a through 8.38c, the mathematical procedures shown in Figure 8.39 can be carried out. Line number D4010 400 0620 (from Figure 8.38a) indicates a cost per square foot for a light hazard, steel pipe installation for the first floor. This price is for a 10,000 S.F. floor, which is closer to the 18,900 S.F. than the next line number which prices a 50,000 S.F. floor area. For the second and third floors, which do not require the alarm valve, water motor, and other sprinkler systems specialties, a square foot cost is selected. Line number D4010 410 0740 indicates $1.57 per square foot. These prices for the first and upper floors are transferred to the fire protection worksheet shown in Figure 8.39.

Fire Standpipe System: To estimate the cost of a fire standpipe system for the proposed office building, the system classification and type should first be determined. Table R10520-320 from *Means Plumbing Cost Data* (shown in Figure 8.40) indicates the several choices based on the NFPA 14 basic standpipe design. The building is Class III type usage, which allows a system to be operated by either occupants or by fire department personnel. A wet pipe system, a typical system for an office building, is selected. Two standpipes at the building stairwells will adequately cover the areas to be protected. The pipe size is 4″, because the building height is less than 100′.

Figures 8.41a and 8.41b, from *Means Plumbing Cost Data,* illustrate a dry standpipe and list its components, material, and labor costs for the three classifications. Line number D4020 330 1540 (shown in Figure 8.41b) indicates a cost of $3,375 each for the first floor standpipes, including all the system components listed. Note that these components include 20′ of pipe, a 10′ riser and a 10′ allowance for horizontal feed.

Three floors at 14′ + 10′ garage + 2′ roof + 2′ burial = 56′. 56′ - 10′ original = 46′ × 2 standpipes = 92 additional feet required. Line number D4020 330 1560 (Figure 8.41b) indicates $1,045 for one 10-foot floor, or $104.50 per foot. The 92 additional feet totals $9,614. This additional footage includes a valved hose connection at each floor. The totals for two standpipes should be entered on the fire protection worksheet shown in Figure 8.39.

PLUMBING ESTIMATE WORKSHEET

Fixture	Table Number	Qty	UOM	Unit Cost	Total Cost	Notes
Bathtubs						
Drinking fountain						
Kitchen sink						
Laundry sink						
Lavatory	D2010 310	21	Ea			3 floors x 7 lavs/flr
Service sink	D2010 440	3	Ea			3 floors x 1 sink/flr
Shower						
Urinal	D2010 210	6	Ea			3 floors x 2 urinals/flr
Water cooler	D2010 820	6	Ea			3 floors x 2 coolers/flr
Water cooler, handicap	D2010 820	3	Ea			3 x 1 handicap coolers/flr
Water closet	D2010 110	18	Ea			3 floors x W = 3 M = 2, H = 1, Total = 3 x 6 =18
W/C group						
Wash fountain group						
Lavatory group						
Urinals group						
Bathrooms						
Water heater	D2020 250	1	Ea			0.4 Gal. x 142 people x 3 floors = 170 GPH
Roof drains	D2040 210	6	Ea			18,900 / 3460 = 6 drains @ 4" dia.
Roof drains (additional length)	D2040 210	264	LF			14' x 3 Floors = 42' + 10' garage = 52' + 2' for burial = 54' -10' for basic pkg. = 44' x 6 = 264'
SUBTOTAL						
Water Control	R15100-040	10%				% of Subtotal
Pipe & Fittings	R15100-040	30%				% of Subtotal
Quality & Complexity	R15100-040	5%				% of Subtotal
Other						
TOTAL						

(Reprinted from Means Mechanical/Electrical Seminar Workbook 2004)

Figure 8.28

D20 Plumbing

D2010 Plumbing Fixtures

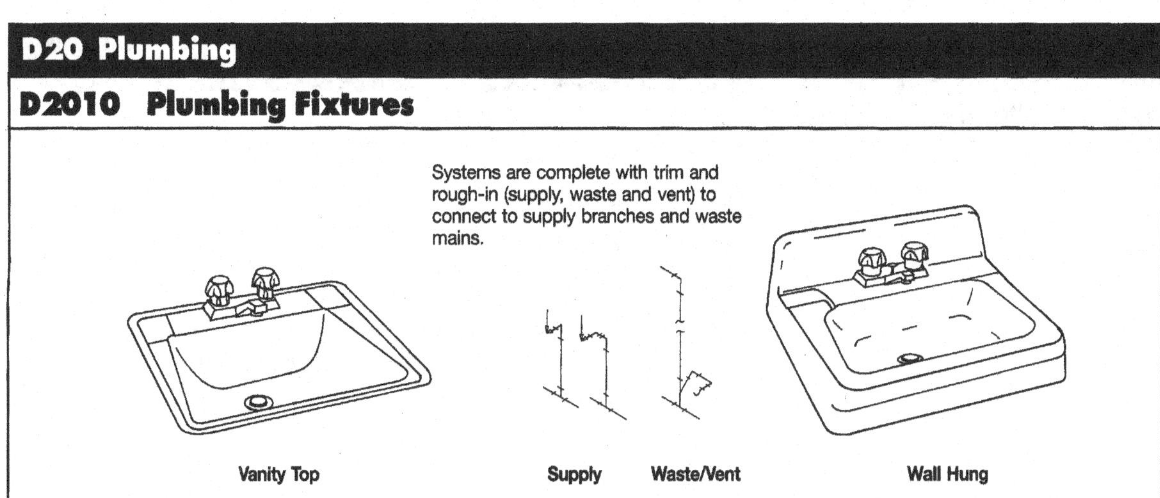
Systems are complete with trim and rough-in (supply, waste and vent) to connect to supply branches and waste mains.
Vanity Top
Supply
Waste/Vent
Wall Hung

System Components	QUANTITY	UNIT	COST EACH MAT.	COST EACH INST.	COST EACH TOTAL
SYSTEM D2010 310 1560					
LAVATORY W/TRIM, VANITY TOP, P.E. ON C.I., 20" X 18"					
Lavatory w/trim, PE on CI, white, vanity top, 20" x 18" oval	1.000	Ea.	227	134	361
Pipe, steel, galvanized, schedule 40, threaded, 1-1/4" diam	4.000	L.F.	12.12	38.40	50.52
Copper tubing type DWV, solder joint, hanger 10'OC 1-1/4" diam	4.000	L.F.	9.16	31.80	40.96
Wrought copper DWV, Tee, sanitary, 1-1/4" diam	1.000	Ea.	6.65	53	59.65
P trap w/cleanout, 20 ga, 1-1/4" diam	1.000	Ea.	18.75	26.50	45.25
Copper tubing type L, solder joint, hanger 10' OC 1/2" diam	10.000	L.F.	11.10	58.50	69.60
Wrought copper 90° elbow for solder joints 1/2" diam	2.000	Ea.	.66	48	48.66
Wrought copper Tee for solder joints, 1/2" diam	2.000	Ea.	1.12	73	74.12
Stop, chrome, angle supply, 1/2" diam	2.000	Ea.	10.40	43	53.40
TOTAL			296.96	506.20	803.16

D2010 310	Lavatory Systems		COST EACH MAT.	COST EACH INST.	COST EACH TOTAL
1560	Lavatory w/trim, vanity top, PE on CI, 20" x 18", Vanity top by others.		297	505	802
1600	19" x 16" oval		212	505	717
1640	18" round		305	505	810
1680	Cultured marble, 19" x 17"		246	505	751
1720	25" x 19"		246	505	751
1760	Stainless, self-rimming, 25" x 22"		315	505	820
1800	17" x 22"	R15100 -410	315	505	820
1840	Steel enameled, 20" x 17"		212	520	732
1880	19" round		210	520	730
1920	Vitreous china, 20" x 16"		315	530	845
1960	19" x 16"		305	530	835
2000	22" x 13"		360	530	890
2040	Wall hung, PE on CI, 18" x 15"		535	560	1,095
2080	19" x 17"		535	560	1,095
2120	20" x 18"		505	560	1,065
2160	Vitreous china, 18" x 15"		445	575	1,020
2200	19" x 17"		425	575	1,000
2240	24" x 20"		695	575	1,270

(Reprinted from Means Plumbing Cost Data 2004)

Figure 8.29

D20 Plumbing

D2010 Plumbing Fixtures

Corrosion resistant service sink systems are complete with trim and rough–in (supply, waste and vent) to connect to supply branches and waste mains.

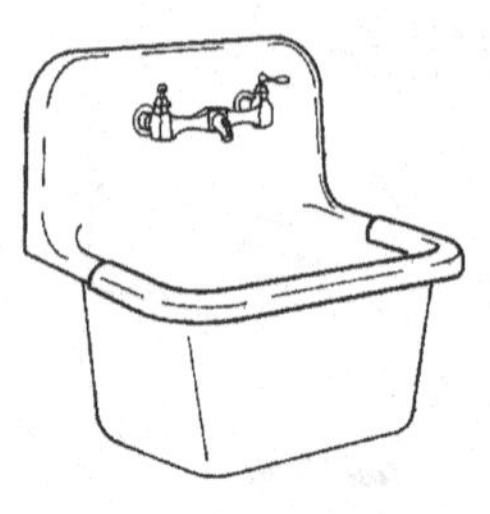

Wall Hung

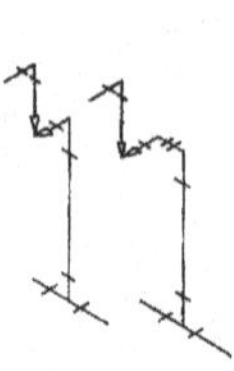

Supply

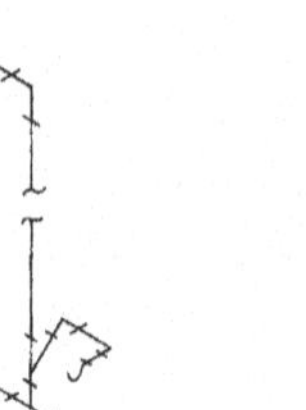

Waste/Vent

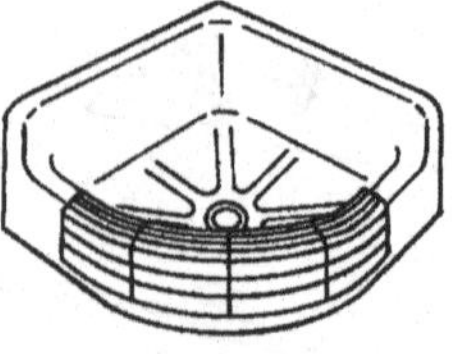

Corner, Floor

System Components	QUANTITY	UNIT	COST EACH		
			MAT.	INST.	TOTAL
SYSTEM D2010 440 4260					
SERVICE SINK, PE ON CI, CORNER FLOOR, 28"X28", W/RIM GUARD & TRIM					
Service sink, corner floor, PE on CI, 28" x 28", w/rim guard & trim	1.000	Ea.	540	195	735
Copper tubing type DWV, solder joint, hanger 10'OC 3" diam	6.000	L.F.	39	88.50	127.50
Copper tubing type DWV, solder joint, hanger 10'OC 2" diam	4.000	L.F.	14.48	43.20	57.68
Wrought copper DWV, Tee, sanitary, 3" diam	1.000	Ea.	35	122	157
P trap with cleanout & slip joint, copper 3" diam	1.000	Ea.	70.50	43	113.50
Copper tubing, type L, solder joints, hangers 10' OC, 1/2" diam	10.000	L.F.	11.10	58.50	69.60
Wrought copper 90° elbow for solder joints 1/2" diam	2.000	Ea.	.66	48	48.66
Wrought copper Tee for solder joints, 1/2" diam	2.000	Ea.	1.12	73	74.12
Stop, angle supply, chrome, 1/2" diam	2.000	Ea.	10.40	43	53.40
TOTAL			722.26	714.20	1,436.46

D2010 440	Service Sink Systems		COST EACH		
			MAT.	INST.	TOTAL
4260	Service sink w/trim, PE on CI, corner floor, 28" x 28", w/rim guard	R15100 -410	720	715	1,435
4300	Wall hung w/rim guard, 22" x 18"		790	830	1,620
4340	24" x 20"		835	830	1,665
4380	Vitreous china, wall hung 22" x 20"		755	830	1,585

(Reprinted from Means Plumbing Cost Data 2004)

Figure 8.30

D20 Plumbing

D2010 Plumbing Fixtures

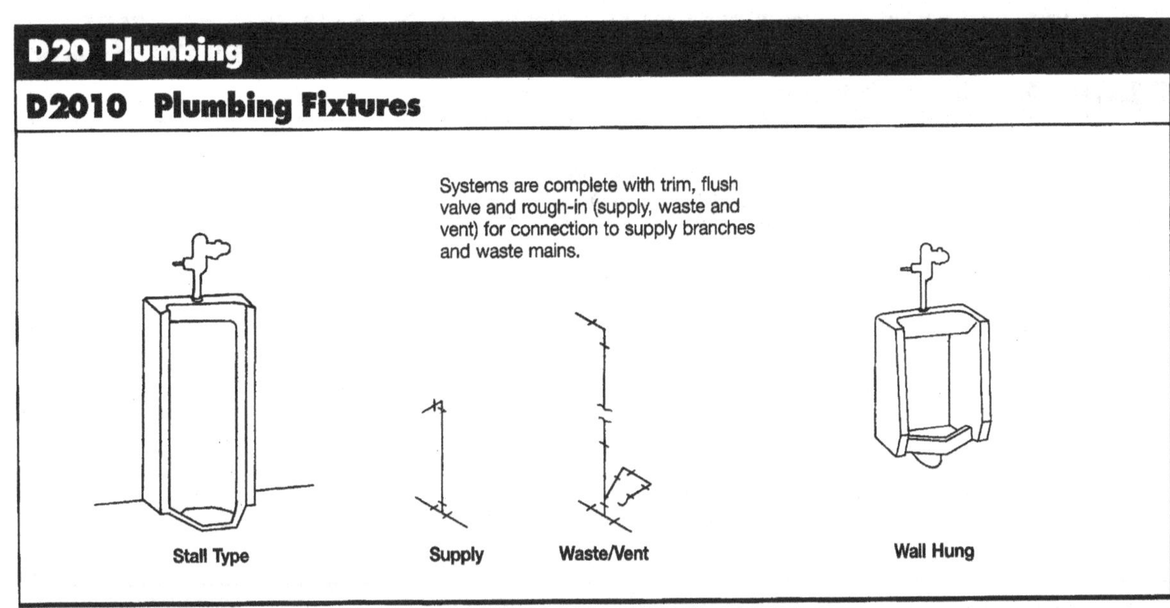
Systems are complete with trim, flush valve and rough-in (supply, waste and vent) for connection to supply branches and waste mains.
Stall Type
Supply
Waste/Vent
Wall Hung

System Components	QUANTITY	UNIT	COST EACH MAT.	COST EACH INST.	COST EACH TOTAL
SYSTEM D2010 210 2000					
URINAL, VITREOUS CHINA, WALL HUNG					
Urinal, wall hung, vitreous china, incl. hanger	1.000	Ea.	320	285	605
Pipe, steel, galvanized, schedule 40, threaded, 1-1/2" diam	5.000	L.F.	16.85	53.50	70.35
Copper tubing type DWV, solder joint, hangers 10'OC, 2" diam	3.000	L.F.	10.86	32.40	43.26
Combination Y & 1/8 bend for CI soil pipe, no hub, 3" diam	1.000	Ea.	10.25		10.25
Pipe, CI, no hub, cplg 10' OC, hanger 5' OC, 3" diam	4.000	L.F.	32.20	53.60	85.80
Pipe coupling standard, CI soil, no hub, 3" diam	3.000	Ea.	12.70	45	57.70
Copper tubing type L, solder joint, hanger 10' OC 3/4" diam	5.000	L.F.	7.80	31.25	39.05
Wrought copper 90° elbow for solder joints 3/4" diam	1.000	Ea.	.74	25	25.74
Wrought copper Tee for solder joints, 3/4" diam	1.000	Ea.	1.36	39.50	40.86
TOTAL			412.76	565.25	978.01

D2010 210	Urinal Systems		COST EACH MAT.	COST EACH INST.	COST EACH TOTAL
2000	Urinal, vitreous china, wall hung	R15100 -410	415	565	980
2040	Stall type		635	680	1,315

(Reprinted from Means Plumbing Cost Data 2004)

Figure 8.31

D20 Plumbing

D2010 Plumbing Fixtures

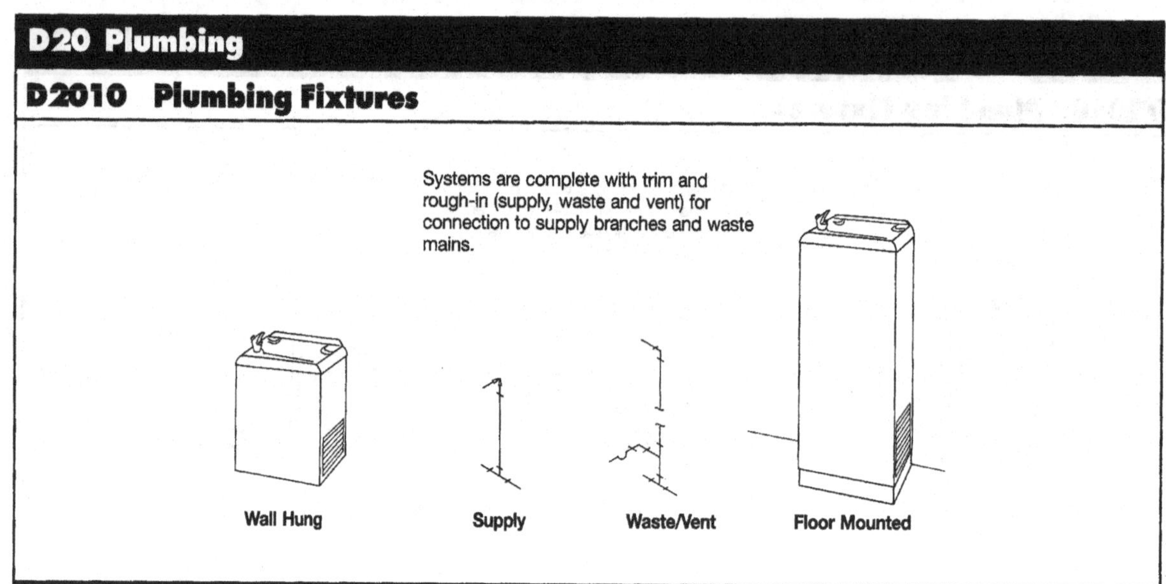

Systems are complete with trim and rough-in (supply, waste and vent) for connection to supply branches and waste mains.
Wall Hung
Supply
Waste/Vent
Floor Mounted

System Components	QUANTITY	UNIT	COST EACH MAT.	COST EACH INST.	COST EACH TOTAL
SYSTEM D2010 820 1840					
WATER COOLER, ELECTRIC, SELF CONTAINED, WALL HUNG, 8.2 GPH					
Water cooler, wall mounted, 8.2 GPH	1.000	Ea.	620	214	834
Copper tubing type DWV, solder joint, hanger 10'OC 1-1/4" diam	4.000	L.F.	9.16	31.80	40.96
Wrought copper DWV, Tee, sanitary 1-1/4" diam	1.000	Ea.	6.65	53	59.65
P trap, copper drainage, 1-1/4" diam	1.000	Ea.	18.75	26.50	45.25
Copper tubing type L, solder joint, hanger 10' OC 3/8" diam	5.000	L.F.	4.95	28.25	33.20
Wrought copper 90° elbow for solder joints 3/8" diam	1.000	Ea.	1.03	21.50	22.53
Wrought copper Tee for solder joints, 3/8" diam	1.000	Ea.	1.77	34	35.77
Stop and waste, straightway, bronze, solder, 3/8" diam	1.000	Ea.	3.95	19.80	23.75
TOTAL			666.26	428.85	1,095.11

D2010 820	Water Cooler Systems		COST EACH MAT.	COST EACH INST.	COST EACH TOTAL
1840	Water cooler, electric, wall hung, 8.2 GPH	R15100 -410	665	430	1,095
1880	Dual height, 14.3 GPH		985	440	1,425
1920	Wheelchair type, 7.5 G.P.H.	♿	1,625	430	2,055
1960	Semi recessed, 8.1 G.P.H.		965	430	1,395
2000	Full recessed, 8 G.P.H.	R15100 -430	1,450	460	1,910
2040	Floor mounted, 14.3 G.P.H.		895	375	1,270
2080	Dual height, 14.3 G.P.H.		1,050	455	1,505
2120	Refrigerated compartment type, 1.5 G.P.H.		1,275	375	1,650

Figure 8.32

D20 Plumbing

D2010 Plumbing Fixtures

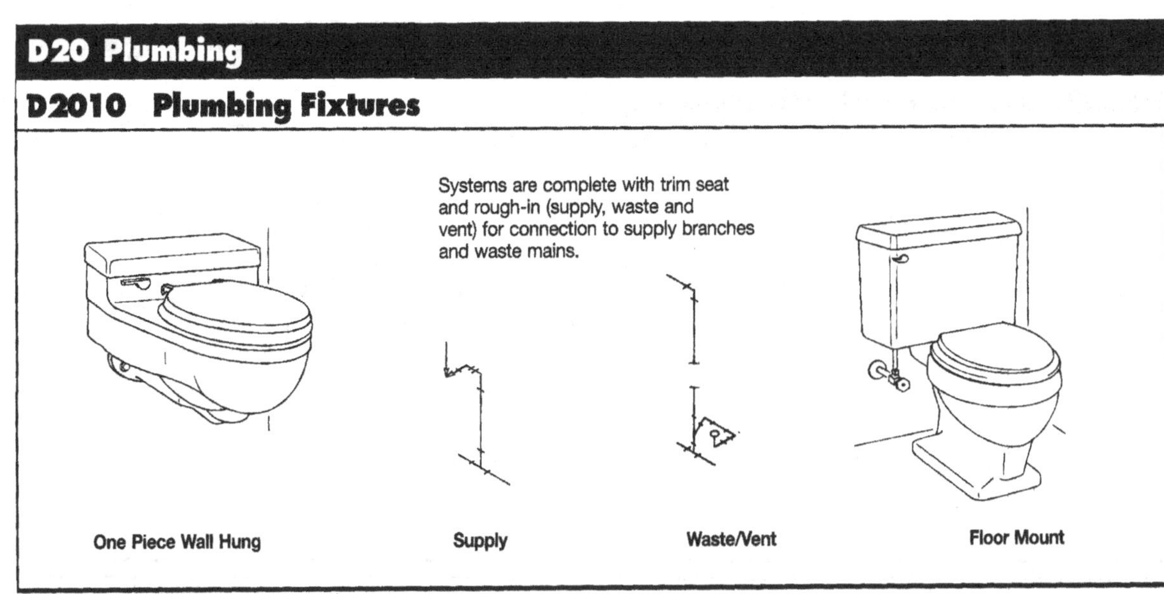
Systems are complete with trim seat and rough-in (supply, waste and vent) for connection to supply branches and waste mains.
One Piece Wall Hung
Supply
Waste/Vent
Floor Mount

System Components	QUANTITY	UNIT	COST EACH MAT.	COST EACH INST.	COST EACH TOTAL
SYSTEM D2010 110 1840					
WATER CLOSET, VITREOUS CHINA, ELONGATED					
TANK TYPE, WALL HUNG, ONE PIECE					
Wtr closet tank type vit china wall hung 1 pc w/seat supply & stop	1.000	Ea.	430	162	592
Pipe steel galvanized, schedule 40, threaded, 2" diam	4.000	L.F.	17.88	53.60	71.48
Pipe, CI soil, no hub, cplg 10′ OC, hanger 5′ OC, 4" diam	2.000	L.F.	20.40	29.50	49.90
Pipe, coupling, standard coupling, CI soil, no hub, 4" diam	2.000	Ea.	15	52	67
Copper tubing type L, solder joint, hanger 10′OC, 1/2" diam	6.000	L.F.	6.66	35.10	41.76
Wrought copper 90° elbow for solder joints 1/2" diam	2.000	Ea.	.66	48	48.66
Wrought copper Tee for solder joints 1/2" diam	1.000	Ea.	.56	36.50	37.06
Support/carrier, for water closet, siphon jet, horiz, single, 4" waste	1.000	Ea.	231	79.50	310.50
TOTAL			722.16	496.20	1,218.36

D2010 110	Water Closet Systems		COST EACH MAT.	COST EACH INST.	COST EACH TOTAL
1800	Water closet, vitreous china, elongated				
1840	Tank type, wall hung, one piece		720	495	1,215
1880	Close coupled two piece	R15100 -410	805	495	1,300
1920	Floor mount, one piece		700	540	1,240
1960	One piece low profile	R15100 -420	595	540	1,135
2000	Two piece close coupled		340	540	880
2040	Bowl only with flush valve				
2080	Wall hung		690	565	1,255
2120	Floor mount		510	545	1,055

Figure 8.33

D20 Plumbing

D2020 Domestic Water Distribution

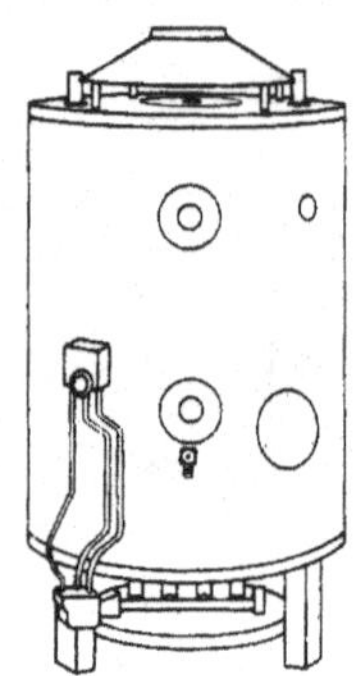

Units may be installed in multiples for increased capacity.

Included below is the heater with self-energizing gas controls, safety pilots, insulated jacket, hi-limit aquastat and pressure relief valve.

Installation includes piping and fittings within 10′ of heater. Gas heaters require vent piping (not included in these prices).

System Components	QUANTITY	UNIT	COST EACH MAT.	COST EACH INST.	COST EACH TOTAL
SYSTEM D2020 250 1780					
GAS FIRED WATER HEATER, COMMERCIAL, 100° F RISE					
75.5 MBH INPUT, 63 GPH					
Water heater, commercial, gas, 75.5 MBH, 63 GPH	1.000	Ea.	1,525	340	1,865
Copper tubing, type L, solder joint, hanger 10′ OC, 1-1/4″ diam	30.000	L.F.	80.70	246	326.70
Wrought copper 90° elbow for solder joints 1-1/4″ diam	4.000	Ea.	11	126	137
Wrought copper Tee for solder joints, 1-1/4″ diam	2.000	Ea.	13.40	106	119.40
Wrought copper union for soldered joints, 1-1/4″ diam	2.000	Ea.	18.80	68	86.80
Valve, gate, bronze, 125 lb, NRS, soldered 1-1/4″ diam	2.000	Ea.	87	63	150
Relief valve, bronze, press & temp, self-close, 3/4″ IPS	1.000	Ea.	81	17	98
Copper tubing, type L, solder joints, 3/4″ diam	8.000	L.F.	12.48	50	62.48
Wrought copper 90° elbow for solder joints 3/4″ diam	1.000	Ea.	.74	25	25.74
Wrought copper, adapter, CTS to MPT, 3/4″ IPS	1.000	Ea.	1.14	28	29.14
Pipe steel black, schedule 40, threaded, 3/4″ diam	10.000	L.F.	15.70	78	93.70
Pipe, 90° elbow, malleable iron black, 150 lb threaded, 3/4″ diam	2.000	Ea.	3.46	68	71.46
Pipe, union with brass seat, malleable iron black, 3/4″ diam	1.000	Ea.	6.90	36.50	43.40
Valve, gas stop w/o check, brass, 3/4″ IPS	1.000	Ea.	11.90	21.50	33.40
TOTAL			1,869.22	1,273	3,142.22

D2020 250	Gas Fired Water Heaters - Commercial Systems		COST EACH MAT.	COST EACH INST.	COST EACH TOTAL
1760	Gas fired water heater, commercial, 100° F rise				
1780	75.5 MBH input, 63 GPH		1,875	1,275	3,150
1860	100 MBH input, 91 GPH	R15100 -110	3,250	1,325	4,575
1980	155 MBH input, 150 GPH		5,025	1,550	6,575
2060	200 MBH input, 192 GPH	R15100 -120	5,225	1,875	7,100
2140	300 MBH input, 278 GPH		7,125	2,325	9,450
2180	390 MBH input, 374 GPH		7,150	2,350	9,500
2220	500 MBH input, 480 GPH		9,775	2,525	12,300
2260	600 MBH input, 576 GPH		11,200	2,725	13,925

(Reprinted from Means Plumbing Cost Data 2004)

Figure 8.34

D20 Plumbing

D2040 Rain Water Drainage

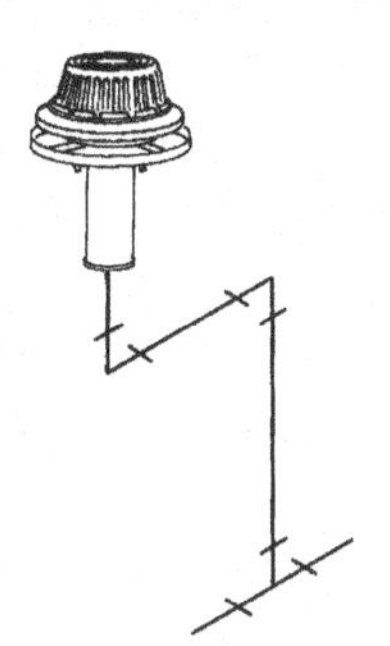

Design Assumptions: Vertical conductor size is based on a maximum rate of rainfall of 4" per hour. To convert roof area to other rates multiply "Max. S.F. Roof Area" shown by four and divide the result by desired local rate. The answer is the local roof area that may be handled by the indicated pipe diameter.

Basic cost is for roof drain, 10' of vertical leader and 10' of horizontal, plus connection to the main.

Pipe Dia.	Max. S.F. Roof Area	Gallons per Min.
2"	544	23
3"	1610	67
4"	3460	144
5"	6280	261
6"	10,200	424
8"	22,000	913

System Components	QUANTITY	UNIT	COST EACH MAT.	COST EACH INST.	COST EACH TOTAL
SYSTEM D2040 210 1880					
ROOF DRAIN, DWV PVC PIPE, 2" DIAM., 10' HIGH					
Drain, roof, main, PVC, dome type 2" pipe size	1.000	Ea.	97.50	61	158.50
Clamp, roof drain, underdeck	1.000	Ea.	20.50	35.50	56
Pipe, Tee, PVC DWV, schedule 40, 2" pipe size	1.000	Ea.	2.09	43	45.09
Pipe, PVC, DWV, schedule 40, 2" diam.	20.000	L.F.	41.80	290	331.80
Pipe, elbow, PVC schedule 40, 2" diam.	2.000	Ea.	3.70	47	50.70
TOTAL			165.59	476.50	642.09

D2040 210	Roof Drain Systems	COST PER SYSTEM MAT.	COST PER SYSTEM INST.	COST PER SYSTEM TOTAL
1880	Roof drain, DWV PVC, 2" diam., piping, 10' high	166	475	641
1920	For each additional foot add	2.09	14.50	16.59
1960	3" diam., 10' high	206	555	761
2000	For each additional foot add	3.50	16.15	19.65
2040	4" diam., 10' high	239	625	864
2080	For each additional foot add	4.44	17.85	22.29
2120	5" diam., 10' high	450	725	1,175
2160	For each additional foot add	6.80	19.90	26.70
2200	6" diam., 10' high	585	800	1,385
2240	For each additional foot add	7.85	22	29.85
2280	8" diam., 10' high	1,350	1,350	2,700
2320	For each additional foot add	14	28	42
3940	C.I., soil, single hub, service wt., 2" diam. piping, 10' high	256	520	776
3980	For each additional foot add	5.15	13.60	18.75
4120	3" diam., 10' high	350	565	915
4160	For each additional foot add	7.15	14.25	21.40
4200	4" diam., 10' high	420	615	1,035
4240	For each additional foot add	9.10	15.55	24.65
4280	5" diam., 10' high	555	680	1,235
4320	For each additional foot add	12.45	17.55	30
4360	6" diam., 10' high	735	730	1,465
4400	For each additional foot add	15.25	18.25	33.50
4440	8" diam., 10' high	1,575	1,475	3,050
4480	For each additional foot add	23.50	30.50	54
6040	Steel galv. sch 40 threaded, 2" diam. piping, 10' high	277	505	782
6080	For each additional foot add	4.47	13.40	17.87
6120	3" diam., 10' high	540	725	1,265
6160	For each additional foot add	9.25	19.90	29.15

(Reprinted from Means Plumbing Cost Data 2004)

Figure 8.35

PLUMBING ESTIMATE WORKSHEET

Fixture	Table Number	Qty	UOM	Unit Cost	Total Cost	Notes
Bathtubs						
Drinking fountain						
Kitchen sink						
Laundry sink						
Lavatory	D2010 310 1600	21	Ea	717.00	$15,057	3 floors x 7 lavs/flr
Service sink	D2010 440 4300	3	Ea	1,620.00	$4,860	3 floors x 1 sink/flr
Shower						
Urinal	D2010 210 2000	6	Ea	980.00	$5,880	3 floors x 2 urinals/flr
Water cooler	D2010 820 1840	6	Ea	1,095.00	$6,570	3 floors x 2 coolers/flr
Water cooler, handicap	D2010 820 1920	3	Ea	2,055.00	$6,165	3 x 1 handicap coolers/flr
Water closet	D2010 110 2080	18	Ea	1,255.00	$22,590	3 floors x W = 3 M = 2, H = 1, Total [3 x 6 = 18]
W/C group						
Wash fountain group						
Lavatory group						
Urinals group						
Bathrooms						
Water heater	D2020 250 2060	1	Ea	7,100.00	$7,100	0.4 Gal. x 142 people x 3 floors = 170 GPH
Roof drains	D2040 210 4200	6	Ea	1,035.00	$6,210	18,900 / 3460 = 6 drains @ 4" dia.
Roof drains (additional length)	D2040 210 4240	264	LF	24.65	$6,508	14' x 3 Floors = 42' + 10' garage = 52' + 2' for burial = 54' -10' for basic pkg. = 44' x 6 = 264'
SUBTOTAL					$80,940	
Water Control	R15100-040	10%			$8,094	% of Subtotal
Pipe & Fittings	R15100-040	30%			$24,282	% of Subtotal
Quality & Complexity	R15100-040	5%			$4,047	% of Subtotal
Other						
TOTAL					$117,362	

(Reprinted from Means Mechanical/Electrical Seminar Workbook 2004)

Figure 8.36

Mechanical | R150 | Basic Materials & Methods

R15050-710 Subcontractors

On the unit cost pages of the R.S. Means Cost Data books, the last column is entitled "Total Incl. O&P". This is normally the cost of the installing contractor. In Division 15, this is the cost of the mechanical contractor. If the particular work being estimated is to be performed by a sub to the mechanical contractor, the mechanical's profit and handling charge (usually 10%) is added to the total of the last column.

R15050-720 Demolition (Selective vs. Removal for Replacement)

Demolition can be divided into two basic categories.

One type of demolition involves the removal of material with no concern for its replacement. The labor-hours to estimate this work are found in Div. 15055 under "Selective Demolition". It is selective in that individual items or all the material installed as a system or trade grouping such as plumbing or heating systems are removed. This may be accomplished by the easiest way possible, such as sawing, torch cutting, or sledge hammer as well as simple unbolting.

The second type of demolition is the removal of some item for repair or replacement. This removal may involve careful draining, opening of unions, disconnecting and tagging of electrical connections, capping of pipes/ducts to prevent entry of debris or leakage of the material contained as well as transport of the item away from its in-place location to a truck/dumpster. An approximation of the time required to accomplish this type of demolition is to use half of the time indicated as necessary to install a new unit. For example; installation of a new pump might be listed as requiring 6 labor-hours so if we had to estimate the removal of the old pump we would allow an additional 3 hours for a total of 9 hours. That is, the complete replacement of a defective pump with a new pump would be estimated to take 9 labor-hours.

Mechanical | R151 | Plumbing

R15100-040 Plumbing Approximations for Quick Estimating

Water Control

Water Meter; Backflow Preventer, Shock Absorbers; Vacuum Breakers; Mixer. .... 10 to 15% of Fixtures

Pipe And Fittings .... 30 to 60% of Fixtures

Note: Lower percentage for compact buildings or larger buildings with plumbing in one area.
Larger percentage for large buildings with plumbing spread out.
In extreme cases pipe may be more than 100% of fixtures.
Percentages **do not** include special purpose or process piping.

Plumbing Labor

1 & 2 Story Residential .... Rough-in Labor = 80% of Materials
Apartment Buildings .... Rough-in Labor = 90 to 100% of Materials
Labor for handling and placing fixtures is approximately 25 to 30% of fixtures

Quality/Complexity Multiplier (for all installations)

Economy installation, add .... 0 to 5%
Good quality, medium complexity, add .... 5 to 15%
Above average quality and complexity, add .... 15 to 25%

R15100-050 Pipe Material Considerations

1. Malleable fittings should be used for gas service.
2. Malleable fittings are used where there are stresses/strains due to expansion and vibration.
3. Cast fittings may be broken as an aid to disassembling of heating lines frozen by long use, temperature and minerals.
4. Cast iron pipe is extensively used for underground and submerged service.
5. Type M (light wall) copper tubing is available in hard temper only and is used for nonpressure and less severe applications than K and L.
6. Type L (medium wall) copper tubing, available hard or soft for interior service.
7. Type K (heavy wall) copper tubing, available in hard or soft temper for use where conditions are severe. For underground and interior service.
8. Hard drawn tubing requires fewer hangers or supports but should not be bent. Silver brazed fittings are recommended, however soft solder is normally used.
9. Type DMV (very light wall) copper tubing designed for drainage, waste and vent plus other non-critical pressure services.

Domestic/Imported Pipe and Fittings Cost

The prices shown in this publication for steel/cast iron pipe and steel, cast iron, malleable iron fittings are based on domestic production sold at the normal trade discounts. The above listed items of foreign manufacture may be available at prices of 1/3 to 1/2 those shown. Some imported items after minor machining or finishing operations are being sold as domestic to further complicate the system.

Caution: Most pipe prices in this book also include a coupling and pipe hangers which for the larger sizes can add significantly to the per foot cost and should be taken into account when comparing "book cost" with quoted supplier's cost.

(Reprinted from Means Plumbing Cost Data 2004)

Figure 8.37

D40 Fire Protection

D4010 Sprinklers

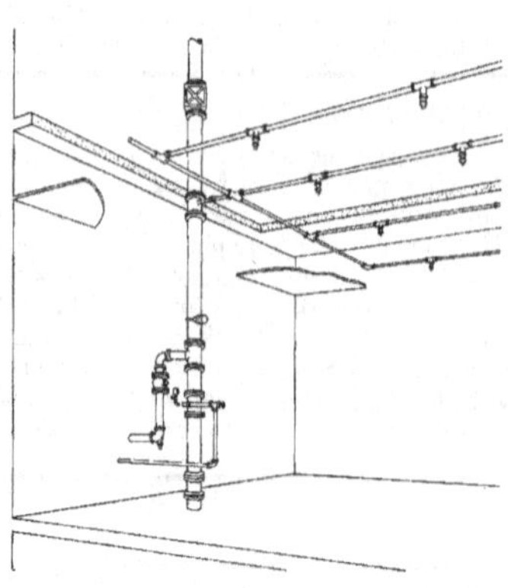

Wet Pipe System. A system employing automatic sprinklers attached to a piping system containing water and connected to a water supply so that water discharges immediately from sprinklers opened by heat from a fire.

All areas are assumed to be open.

System Components	QUANTITY	UNIT	COST EACH MAT.	COST EACH INST.	COST EACH TOTAL
SYSTEM D4010 410 0580					
WET PIPE SPRINKLER, STEEL, BLACK, SCH. 40 PIPE					
LIGHT HAZARD, ONE FLOOR, 2000 S.F.					
Valve, gate, iron body, 125 lb, OS&Y, flanged, 4" diam	1.000	Ea.	307.50	213.75	521.25
Valve, swing check, bronze, 125 lb, regrinding disc, 2-1/2" pipe size	1.000	Ea.	165	42.75	207.75
Valve, angle, bronze, 150 lb, rising stem, threaded, 2" diam	1.000	Ea.	203.25	32.25	235.50
*Alarm valve, 2-1/2" pipe size	1.000	Ea.	686.25	213.75	900
Alarm, water motor, complete with gong	1.000	Ea.	152.25	89.25	241.50
Valve, swing check, w/balldrip CI with brass trim 4" pipe size	1.000	Ea.	132	213.75	345.75
Pipe, steel, black, schedule 40, 4" diam	10.000	L.F.	85.13	189.83	274.96
*Flow control valve, trim & gauges, 4" pipe size	1.000	Set	1,312.50	483.75	1,796.25
Fire alarm horn, electric	1.000	Ea.	30	52.50	82.50
Pipe, steel, black, schedule 40, threaded, cplg & hngr 10'OC, 2-1/2" diam	20.000	L.F.	102.75	257.25	360
Pipe, steel, black, schedule 40, threaded, cplg & hngr 10'OC, 2" diam	12.500	L.F.	37.59	125.63	163.22
Pipe, steel, black, schedule 40, threaded, cplg & hngr 10'OC, 1-1/4" diam	37.500	L.F.	76.78	270	346.78
Pipe steel, black, schedule 40, threaded cplg & hngr 10'OC, 1" diam	112.000	L.F.	188.16	751.80	939.96
Pipe Tee, malleable iron black, 150 lb threaded, 4" pipe size	2.000	Ea.	186	321	507
Pipe Tee, malleable iron black, 150 lb threaded, 2-1/2" pipe size	2.000	Ea.	52.50	142.50	195
Pipe Tee, malleable iron black, 150 lb threaded, 2" pipe size	1.000	Ea.	12.11	58.50	70.61
Pipe Tee, malleable iron black, 150 lb threaded, 1-1/4" pipe size	5.000	Ea.	28.50	228.75	257.25
Pipe Tee, malleable iron black, 150 lb threaded, 1" pipe size	4.000	Ea.	12.60	178.50	191.10
Pipe 90° elbow, malleable iron black, 150 lb threaded, 1" pipe size	6.000	Ea.	13.50	164.25	177.75
Sprinkler head, standard spray, brass 135°-286°F 1/2" NPT, 3/8" orifice	12.000	Ea.	87	354	441
Valve, gate, bronze, NRS, class 150, threaded, 1" pipe size	1.000	Ea.	29.63	18.75	48.38
*Standpipe connection, wall, single, flush w/plug & chain 2-1/2"x2-1/2"	1.000	Ea.	71.63	128.25	199.88
TOTAL			3,972.63	4,530.76	8,503.39
COST PER S.F.			1.99	2.27	4.26
*Not included in systems under 2000 S.F.					

D4010 410	Wet Pipe Sprinkler Systems		COST PER S.F. MAT.	COST PER S.F. INST.	COST PER S.F. TOTAL
0520	Wet pipe sprinkler systems, steel, black, sch. 40 pipe				
0530	Light hazard, one floor, 500 S.F.		1.17	2.17	3.34
0560	1000 S.F.	R10520-110	1.88	2.25	4.13
0580	2000 S.F.		1.99	2.28	4.27
0600	5000 S.F.	R10520-120	.96	1.58	2.54
0620	10,000 S.F.		.62	1.33	1.95

(Reprinted from Means Plumbing Cost Data 2004)

Figure 8.38a

D40 Fire Protection

D4010 Sprinklers

D4010 410	Wet Pipe Sprinkler Systems	COST PER S.F. MAT.	COST PER S.F. INST.	COST PER S.F. TOTAL
0640	50,000 S.F. R10520-130	.44	1.23	1.67
0660	Each additional floor, 500 S.F.	.54	1.85	2.39
0680	1000 S.F.	.57	1.73	2.30
0700	2000 S.F.	.54	1.54	2.08
0720	5000 S.F.	.39	1.31	1.70
0740	10,000 S.F.	.35	1.22	1.57
0760	50,000 S.F.	.30	.97	1.27
1000	Ordinary hazard, one floor, 500 S.F.	1.28	2.33	3.61
1020	1000 S.F.	1.86	2.19	4.05
1040	2000 S.F.	2.04	2.39	4.43
1060	5000 S.F.	1.07	1.70	2.77
1080	10,000 S.F.	.78	1.75	2.53
1100	50,000 S.F.	.67	1.73	2.40
1140	Each additional floor, 500 S.F.	.68	2.09	2.77
1160	1000 S.F.	.56	1.68	2.24
1180	2000 S.F.	.61	1.71	2.32
1200	5000 S.F.	.61	1.61	2.22
1220	10,000 S.F.	.51	1.64	2.15
1240	50,000 S.F.	.51	1.53	2.04
1500	Extra hazard, one floor, 500 S.F.	3.53	3.59	7.12
1520	1000 S.F.	2.29	3.13	5.42
1540	2000 S.F.	2.16	3.22	5.38
1560	5000 S.F.	1.42	2.80	4.22
1580	10,000 S.F.	1.28	2.63	3.91
1600	50,000 S.F.	1.35	2.55	3.90
1660	Each additional floor, 500 S.F.	.95	2.57	3.52
1680	1000 S.F.	.92	2.45	3.37
1700	2000 S.F.	.80	2.48	3.28
1720	5000 S.F.	.70	2.17	2.87
1740	10,000 S.F.	.81	2.01	2.82
1760	50,000 S.F.	.79	1.91	2.70
2020	Grooved steel, black sch. 40 pipe, light hazard, one floor, 2000 S.F.	1.87	1.93	3.80
2060	10,000 S.F.	.67	1.20	1.87
2100	Each additional floor, 2000 S.F.	.44	1.25	1.69
2150	10,000 S.F.	.27	1.04	1.31
2200	Ordinary hazard, one floor, 2000 S.F.	1.90	2.06	3.96
2250	10,000 S.F.	.64	1.48	2.12
2300	Each additional floor, 2000 S.F.	.47	1.38	1.85
2350	10,000 S.F.	.37	1.37	1.74
2400	Extra hazard, one floor, 2000 S.F.	2.01	2.63	4.64
2450	10,000 S.F.	.88	1.92	2.80
2500	Each additional floor, 2000 S.F.	.67	2.02	2.69
2550	10,000 S.F.	.56	1.72	2.28
3050	Grooved steel black sch. 10 pipe, light hazard, one floor, 2000 S.F.	1.85	1.92	3.77
3100	10,000 S.F.	.53	1.13	1.66
3150	Each additional floor, 2000 S.F.	.42	1.23	1.65
3200	10,000 S.F.	.26	1.02	1.28
3250	Ordinary hazard, one floor, 2000 S.F.	1.87	2.04	3.91
3300	10,000 S.F.	.63	1.45	2.08
3350	Each additional floor, 2000 S.F.	.44	1.36	1.80
3400	10,000 S.F.	.36	1.34	1.70
3450	Extra hazard, one floor, 2000 S.F.	1.99	2.62	4.61
3500	10,000 S.F.	.81	1.89	2.70
3550	Each additional floor, 2000 S.F.	.65	2.01	2.66
3600	10,000 S.F.	.52	1.70	2.22
4050	Copper tubing, type M, light hazard, one floor, 2000 S.F.	1.91	1.91	3.82
4100	10,000 S.F.	.62	1.15	1.77
4150	Each additional floor, 2000 S.F.	.49	1.26	1.75

(Reprinted from Means Plumbing Cost Data 2004)

Figure 8.38b

D40 Fire Protection

D4010 Sprinklers

D4010 410	Wet Pipe Sprinkler Systems	COST PER S.F.		
		MAT.	INST.	TOTAL
4200	10,000 S.F.	.35	1.05	1.40
4250	Ordinary hazard, one floor, 2000 S.F.	1.94	2.14	4.08
4300	10,000 S.F.	.69	1.37	2.06
4350	Each additional floor, 2000 S.F.	.52	1.39	1.91
4400	10,000 S.F.	.41	1.24	1.65
4450	Extra hazard, one floor, 2000 S.F.	2.12	2.65	4.77
4500	10,000 S.F.	1.25	2.07	3.32
4550	Each additional floor, 2000 S.F.	.78	2.04	2.82
4600	10,000 S.F.	.78	1.85	2.63
5050	Copper tubing, type M, T-drill system, light hazard, one floor			
5060	2000 S.F.	1.88	1.76	3.64
5100	10,000 S.F.	.58	.94	1.52
5150	Each additional floor, 2000 S.F.	.46	1.11	1.57
5200	10,000 S.F.	.31	.84	1.15
5250	Ordinary hazard, one floor, 2000 S.F.	1.87	1.80	3.67
5300	10,000 S.F.	.66	1.20	1.86
5350	Each additional floor, 2000 S.F.	.44	1.12	1.56
5400	10,000 S.F.	.39	1.09	1.48
5450	Extra hazard, one floor, 2000 S.F.	1.97	2.16	4.13
5500	10,000 S.F.	1.01	1.50	2.51
5550	Each additional floor, 2000 S.F.	.68	1.58	2.26
5600	10,000 S.F.	.54	1.28	1.82

(Reprinted from Means Plumbing Cost Data 2004)

Figure 8.38c

FIRE PROTECTION ESTIMATE WORKSHEET

Component	Table Number	Qty	UOM	Unit Cost	Total Cost	Notes
From table...	R10520-120					
Sprinkler						
Office, light	D4010 410 0620	18,900	SF	1.95	$36,855	First floor
Office, light	D4010 410 0740	37,800	SF	1.57	$59,346	2 Remaining floors
Garage, ordinary	D4010 310 1080	18,900	SF	2.82	$53,298	Garage area
Standpipe						
Class III	D4020 330 1540	2	EA	3,375.00	$6,750	2 standpipes (Dry standpipes are used, assuming stairwell at ends of building are subject to freezing).
	D4020 330 1560	92	LF	104.50	$9,614	14' x 3 = 42', 10' x 1 = 10' Roof + 2'=54' Total +2' for burial = 56' -10' orig. flr = 46' ea.46' x 2 = 92'
Fire Suppression						
Cabinets & Components	D4020 410 8400	8	EA	734.00	$5,872	4 Floors x 2 = 8 cab.
Other						
SUBTOTAL					$171,735	
Quality/Complexity	R10520-140	5%			$8,587	
TOTAL					$180,322	

(Reprinted from Means Mechanical/Electrical Seminar Workbook 2004)

Figure 8.39

Six firehose and cabinet assemblies are priced from *Means Plumbing Cost Data* (shown in Figure 8.42), line number D4020 410 8400, and also entered on the worksheet.

The sprinkler and standpipe costs are totaled, and a quality complexity percentage is added using the parameters from table number R10520-140, shown in Figure 8.40.

These systems may be modified to match conditions other than those shown in this example. For instance, if these two standpipes were fed from one common siamese inlet connection with interconnecting piping, the system cost could be modified by deducting two 4″ × 2-1/2″ × 2-1/2″ standpipe connections (found in the unit price section). One 6″ × 2-1/2″ × 2-1/2″ standpipe connection must then be added with the appropriate markup percentages. Finally, 6″ interconnecting piping for the fire department connection to the two standpipe risers must be added. The pricing information for the additional piping and connections can be found in the unit price section of *Means Plumbing Cost Data*.

The Estimate Summary

In the event that the owner or designer has not informed the estimator of the type or quality of the system to be priced, the estimator should, at this time, qualify his or her proposal by stating the quality and the type of systems being priced.

These rounded-off prices all include subcontractors' overhead and profit. At this time, a location factor adjustment should be added.

Any sales taxes and other adjustments (extra overhead, contingencies, etc.) that are required must also be applied. When these factors have been included, a budget price for the plumbing work is complete.

R10520-140 Adjustment for Sprinkler/Standpipe Installations

Quality/Complexity Multiplier (For all installations)

Economy installation, add . . . . . 0 to 5%
Good quality, medium complexity, add . . . . . 5 to 15%
Above average quality and complexity, add . . . . . 15 to 25%

R10520-310 Standpipe Systems

The basis for standpipe system design is National Fire Protection Association NFPA 14, however, the authority having jurisdiction should be consulted for special conditions, local requirements and approval.

Standpipe systems, properly designed and maintained, are an effective and valuable time saving aid for extinquishing fires, especially in the upper stories of tall buildings, the interior of large commercial or industrial malls, or other areas where construction features or access make the laying of temporary hose lines time consuming and/or hazardous. Standpipes are frequently installed with automatic sprinkler systems for maximum protection.

There are three general classes of service for standpipe systems:
Class I for use by fire departments and personnel with special training for heavy streams (2-1/2″ hose connections).
Class II for use by building occupants until the arrival of the fire department (1-1/2″ hose connector with hose).
Class III for use by either fire departments and trained personnel or by the building occupants (both 2-1/2″ and 1-1/2″ hose connections or one 2-1/2″ hose valve with an easily removable 2-1/2″ by 1-1/2″ adapter).

Standpipe systems are also classified by the way water is supplied to the system. The four basic types are:
Type 1: Wet standpipe system having supply valve open and water pressure maintained at all times.
Type 2: Standpipe system so arranged through the use of approved devices as to admit water to the system automatically by opening a hose valve.
Type 3: Standpipe system arranged to admit water to the system through manual operation of approved remote control devices located at each hose station.
Type 4: Dry standpipe having no permanent water supply.

R10520-320 NFPA 14 Basic Standpipe Design

Class	Design-Use	Pipe Size Minimums	Water Supply Minimums
Class I	2 1/2″ hose connection on each floor All areas within 150′ of an exit in every exit stairway Fire Department Trained Personnel	Height to 100′, 4″ dia. Heights above 100′, 6″ dia. (275′ max. except with pressure regulators 400′ max.)	For each standpipe riser 500 GPM flow For common supply pipe allow 500 GPM for first standpipe plus 250 GPM for each additional standpipe (2500 GPM max. total) 30 min. duration 65 PSI at 500 GPM
Class II	1 1/2″ hose connection with hose on each floor All areas within 130′ of hose connection measured along path of hose travel Occupant personnel	Height to 50′, 2″ dia. Height above 50′, 2 1/2″ dia.	For each standpipe riser 100 GPM flow For multiple riser common supply pipe 100 GPM 300 min. duration, 65 PSI at 100 GPM
Class III	Both of above. Class I valved connections will meet Class III with additional 2 1/2″ by 1 1/2″ adapter and 1 1/2″ hose.	Same as Class I	Same as Class I

*Note: Where 2 or more standpipes are installed in the same building or section of building they shall be interconnected at the bottom.

Combined Systems

Combined systems are systems where the risers supply both automatic sprinklers and 2-1/2″ hose connection outlets for fire department use. In such a system the sprinkler spacing pattern shall be in accordance with NFPA 13 while the risers and supply piping will be sized in accordance with NFPA 14. When the building is completely sprinklered the risers may be sized by hydraulic calculation.The minimum size riser for buildings not completely sprinklered is 6″.
The minimum water supply of a completely sprinklered, light hazard, high-rise occupancy building will be 500 GPM while the supply required for other types of completely sprinklered high-rise buildings is 1000 GPM.

General System Requirements

1. Approved valves will be provided at the riser for controlling branch lines to hose outlets.
2. A hose valve will be provided at each outlet for attachment of hose.
3. Where pressure at any standpipe outlet exceeds 100 PSI a pressure reducer must be installed to limit the pressure to 100 PSI. Note that the pressure head due to gravity in 100′ of riser is 43.4 PSI. This must be overcome by city pressure, fire pumps, or gravity tanks to provide adequate pressure at the top of the riser.
4. Each hose valve on a wet system having linen hose shall have an automatic drip connection to prevent valve leakage from entering the hose.
5. Each riser will have a valve to isolate it from the rest of the system.
6. One or more fire department connections as an auxiliary supply shall be provided for each Class I or Class III standpipe system. In buildings having two or more zones, a connection will be provided for each zone.
7. There will be no shutoff valve in the fire department connection, but a check valve will be located in the line before it joins the system.
8. All hose connections street side will be identified on a cast plate or fitting as to purpose.

(Reprinted from Means Plumbing Cost Data 2004)

Figure 8.40

D40 Fire Protection

D4020 Standpipes

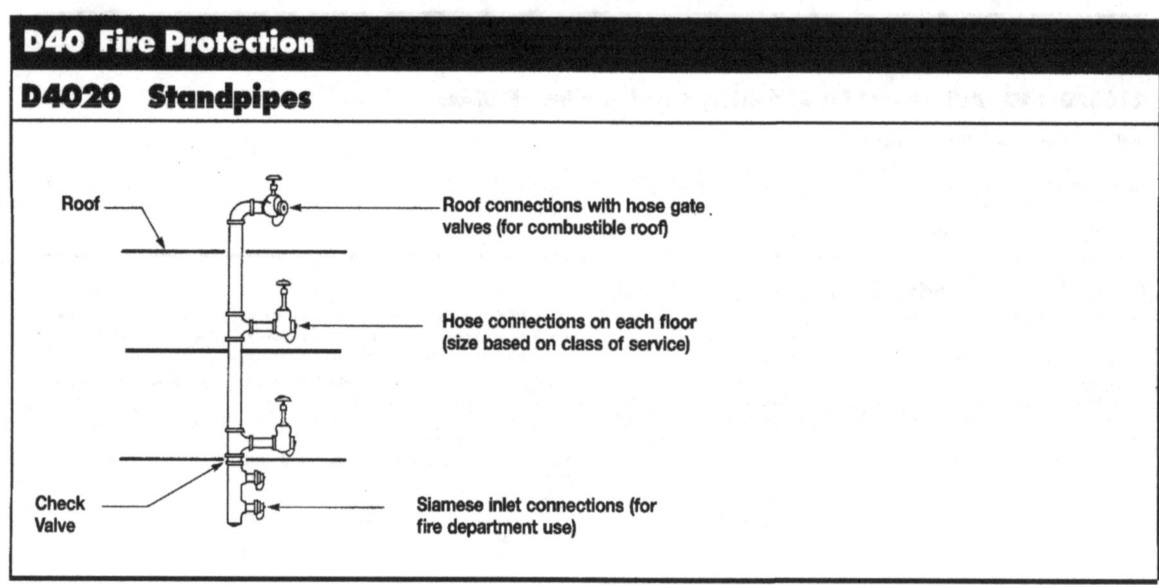

System Components	QUANTITY	UNIT	COST PER FLOOR MAT.	INST.	TOTAL
SYSTEM D4020 330 0540					
DRY STANDPIPE RISER, CLASS I, PIPE, STEEL, BLACK, SCH 40, 10' HEIGHT					
4" DIAMETER PIPE, ONE FLOOR					
Pipe, steel, black, schedule 40, threaded, 4" diam	20.000	L.F.	277	480	757
Pipe, Tee, malleable iron, black, 150 lb threaded, 4" pipe size	2.000	Ea.	248	428	676
Pipe, 90° elbow, malleable iron, black, 150 lb threaded 4" pipe size	1.000	Ea.	78.50	143	221.50
Pipe, nipple, steel, black, schedule 40, 2-1/2" pipe size x 3" long	2.000	Ea.	13.30	107	120.30
Fire valve gate NRS 300 lb, brass w/handwheel, 2-1/2" pipe size	1.000	Ea.	110	68	178
Fire valve, pressure restricting, adj, rgh brs, 2-1/2" pipe size	1.000	Ea.	107	68	175
Standpipe conn wall dble flush brs w/plugs & chains 2-1/2"x2-1/2"x4"	1.000	Ea.	320	171	491
Valve swing check w/ball drip CI w/brs ftngs, 4"pipe size	1.000	Ea.	176	285	461
Roof manifold, fire, w/valves & caps, horiz/vert brs 2-1/2"x2-1/2"x4"	1.000	Ea.	111	178	289
TOTAL			1,440.80	1,928	3,368.80

D4020 330	Dry Standpipe Risers, Class I		COST PER FLOOR MAT.	INST.	TOTAL
0530	Dry standpipe riser, Class I, steel black sch. 40, 10' height				
0540	4" diameter pipe, one floor		1,450	1,925	3,375
0560	Additional floors	D4020 -410	495	695	1,190
0580	6" diameter pipe, one floor		2,725	3,350	6,075
0600	Additional floors	R10520 -310	850	1,125	1,975
0620	8" diameter pipe, one floor		4,025	4,050	8,075
0640	Additional floors	R10520 -320	1,175	1,400	2,575
0660					

D4020 330	Dry Standpipe Risers, Class II	COST PER FLOOR MAT.	INST.	TOTAL
1030	Dry standpipe risers, Class II, steel black sch. 40, 10' height			
1040	2" diameter pipe, one floor	755	890	1,645
1060	Additional floor	254	291	545
1080	2-1/2" diameter pipe, one floor	920	1,050	1,970
1100	Additional floors	294	345	639
1120				

(Reprinted from Means Plumbing Cost Data 2004)

Figure 8.41a

D40 Fire Protection

D4020 Standpipes

D4020 330	Dry Standpipe Risers, Class III	COST PER FLOOR		
		MAT.	INST.	TOTAL
1530	Dry standpipe risers, Class III, steel black sch. 40, 10' height			
1540	4" diameter pipe, one floor	1,475	1,900	3,375
1560	Additional floors	415	630	1,045
1580	6" diameter pipe, one floor	2,750	3,350	6,100
1600	Additional floors	880	1,125	2,005
1620	8" diameter pipe, one floor	4,050	4,050	8,100
1640	Additional floor	1,200	1,400	2,600

(Reprinted from Means Plumbing Cost Data 2004)

Figure 8.41b

D40 Fire Protection

D4020 Standpipes

D4020 410	Fire Hose Equipment	COST EACH		
		MAT.	INST.	TOTAL
0100	Adapters, reducing, 1 piece, FxM, hexagon, cast brass, 2-1/2" x 1-1/2"	34		34
0200	Pin lug, 1-1/2" x 1"	10.30		10.30
0250	3" x 2-1/2"	42		42
0300	For polished chrome, add 75% mat.			
0400	Cabinets, D.S. glass in door, recessed, steel box, not equipped			
0500	Single extinguisher, steel door & frame	65	107	172
0550	Stainless steel door & frame	133	107	240
0600	Valve, 2-1/2" angle, steel door & frame	75	71.50	146.50
0650	Aluminum door & frame	99	71.50	170.50
0700	Stainless steel door & frame	131	71.50	202.50
0750	Hose rack assy, 2-1/2" x 1-1/2" valve & 100' hose, steel door & frame	152	143	295
0800	Aluminum door & frame	225	143	368
0850	Stainless steel door & frame	294	143	437
0900	Hose rack assy,& extinguisher,2-1/2"x1-1/2" valve & hose,steel door & frame	161	171	332
0950	Aluminum	286	171	457
1000	Stainless steel	305	171	476
1550	Compressor, air, dry pipe system, automatic, 200 gal., 1/3 H.P.	790	365	1,155
1600	520 gal., 1 H.P.	935	365	1,300
1650	Alarm, electric pressure switch (circuit closer)	160	18.30	178.30
2500	Couplings, hose, rocker lug, cast brass, 1-1/2"	31.50		31.50
2550	2-1/2"	41.50		41.50
3000	Escutcheon plate, for angle valves, polished brass, 1-1/2"	11.80		11.80
3050	2-1/2"	28		28
3500	Fire pump, electric, w/controller, fittings, relief valve			
3550	4" pump, 30 H.P., 500 G.P.M.	15,100	2,675	17,775
3600	5" pump, 40 H.P., 1000 G.P.M.	22,100	3,025	25,125
3650	5" pump, 100 H.P., 1000 G.P.M.	24,800	3,350	28,150
3700	For jockey pump system, add	2,725	430	3,155
5000	Hose, per linear foot, synthetic jacket, lined,			
5100	300 lb. test, 1-1/2" diameter	1.75	.33	2.08
5150	2-1/2" diameter	2.92	.39	3.31
5200	500 lb. test, 1-1/2" diameter	1.80	.33	2.13
5250	2-1/2" diameter	3.14	.39	3.53
5500	Nozzle, plain stream, polished brass, 1-1/2" x 10"	28.50		28.50
5550	2-1/2" x 15" x 13/16" or 1-1/2"	52		52
5600	Heavy duty combination adjustable fog and straight stream w/handle 1-1/2"	273		273
5650	2-1/2" direct connection	415		415
6000	Rack, for 1-1/2" diameter hose 100 ft. long, steel	35.50	43	78.50
6050	Brass	55	43	98
6500	Reel, steel, for 50 ft. long 1-1/2" diameter hose	81.50	61	142.50
6550	For 75 ft. long 2-1/2" diameter hose	128	61	189
7050	Siamese, w/plugs & chains, polished brass, sidewalk, 4" x 2-1/2" x 2-1/2"	340	340	680
7100	6" x 2-1/2" x 2-1/2"	565	430	995
7200	Wall type, flush, 4" x 2-1/2" x 2-1/2"	320	171	491
7250	6" x 2-1/2" x 2-1/2"	435	186	621
7300	Projecting, 4" x 2-1/2" x 2-1/2"	294	171	465
7350	6" x 2-1/2" x 2-1/2"	485	186	671
7400	For chrome plate, add 15% mat.			
8000	Valves, angle, wheel handle, 300 Lb., rough brass, 1-1/2"	32.50	39.50	72
8050	2-1/2"	55.50	68	123.50
8100	Combination pressure restricting, 1-1/2"	51	39.50	90.50
8150	2-1/2"	108	68	176
8200	Pressure restricting, adjustable, satin brass, 1-1/2"	76	39.50	115.50
8250	2-1/2"	107	68	175
8300	Hydrolator, vent and drain, rough brass, 1-1/2"	45	39.50	84.50
8350	2-1/2"	45	39.50	84.50
8400	Cabinet assy, incls. adapter, rack, hose, and nozzle	475	259	734

(Reprinted from Means Plumbing Cost Data 2004)

Figure 8.42

Chapter 9

Unit Price Estimating Example

This chapter contains a sample plumbing estimate for a three-story office building. To prepare the Unit Price Estimate, procedures and forms are utilized, and calculations made. The estimate that follows is prepared from the perspective of a prime plumbing contractor who has been invited by a familiar general contractor to bid on a regional office building for a nationwide firm. All costs and reference tables are from the annual *Means Plumbing Cost Data.* All forms are from *Means Forms for Building Construction Professionals.*

Project Description

The sample project is a three-story office building with an open parking garage beneath the building. The structure is steel frame with curtain wall enclosure. Plans for the mechanical work are shown in Figures 9.1 through 9.5. These drawings have been reduced in size and are not to be scaled. The plans are for illustration only, and, while fairly representative, do not show all of the detail one would expect from a project of this magnitude. A full set of specifications would normally accompany a set of plans, spelling out the scope of work and level of quality of materials and equipment.

The quantities given on the following takeoff, summary, and cost analysis forms are representative of an actual takeoff of the various systems made from the original plans.

Some liberties have been taken in material selection, etc., to provide the reader with a variety of estimating procedures. For example, in the piping estimates, steel pipe, cast iron, copper tube, and polyvinyl chloride (PVC) have all been used. The estimate also includes a wide variety of joining methods, including threaded, butt weld, flanged, plumbing joint, lead and oakum, no hub, solvent weld, and (no lead) soft solder. This practice has been continued throughout the entire plumbing estimate to provide a diversified estimate while staying within the confines of practicality.

In keeping with accepted practices of the trade in many areas of this country, all building service piping, gas, water, storm, and sanitary systems either stop or begin at the appropriate meter or 5' outside the building foundation.

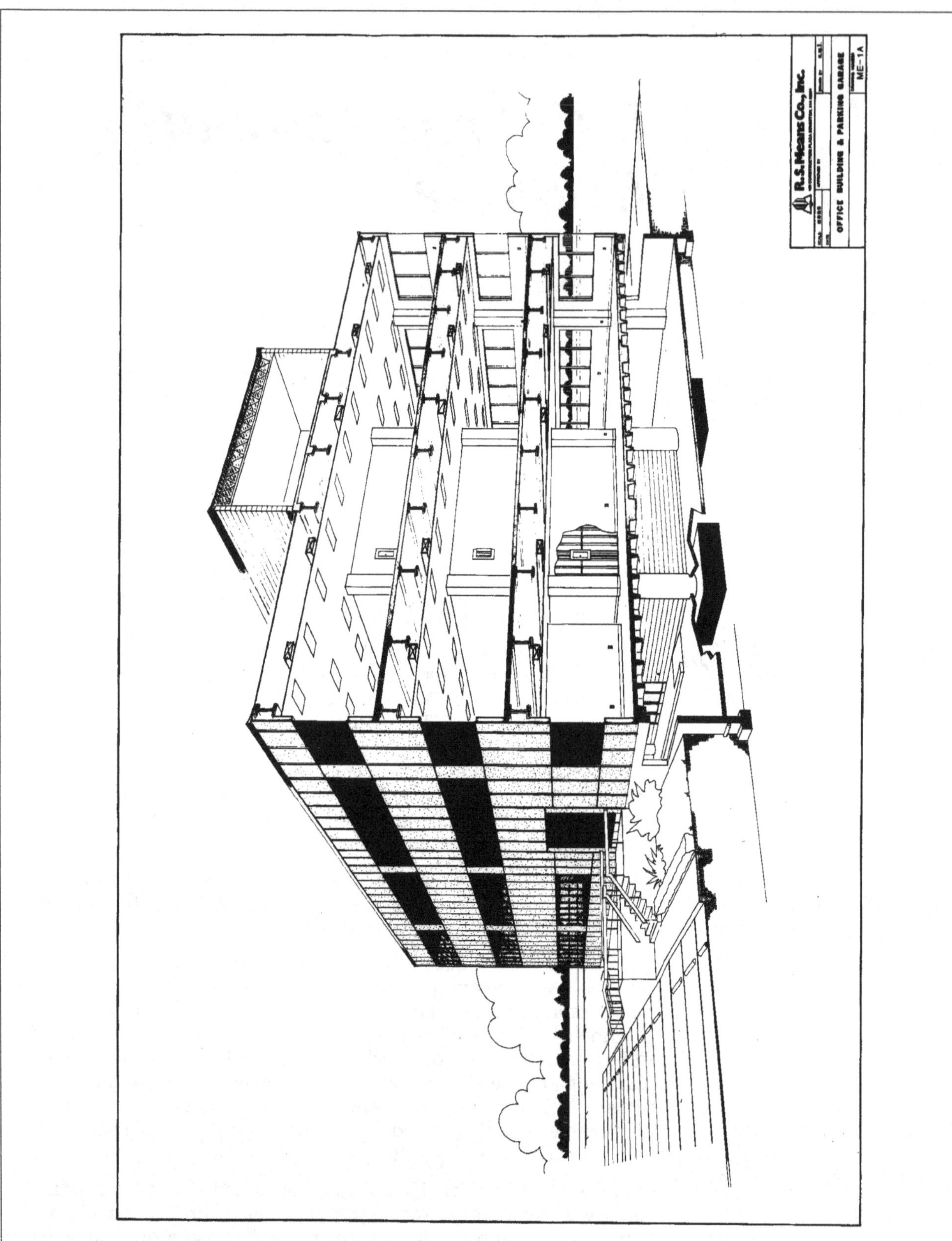

Figure 9.1

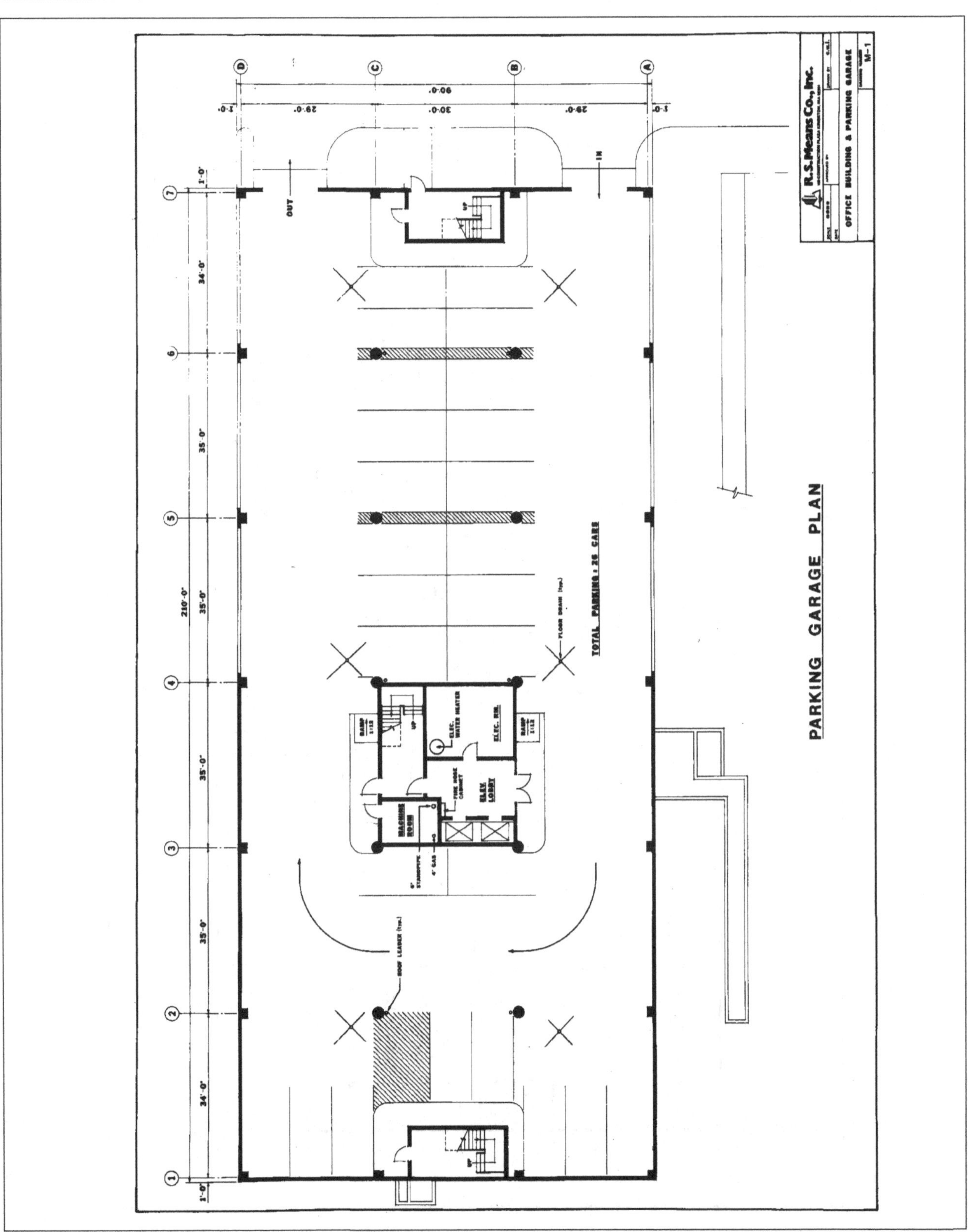
PARKING GARAGE PLAN
TOTAL PARKING : 26 CARS
R.S. Means Co., Inc.
OFFICE BUILDING & PARKING GARAGE
M-1
OUT
IN
ELEC. RM.
ELEC. LOBBY
MACHINE ROOM
ELEC. WATER HEATER
ROOF LEADER (typ.)
FLOOR DRAIN (typ.)
210'-0"
90'-0"

Figure 9.2

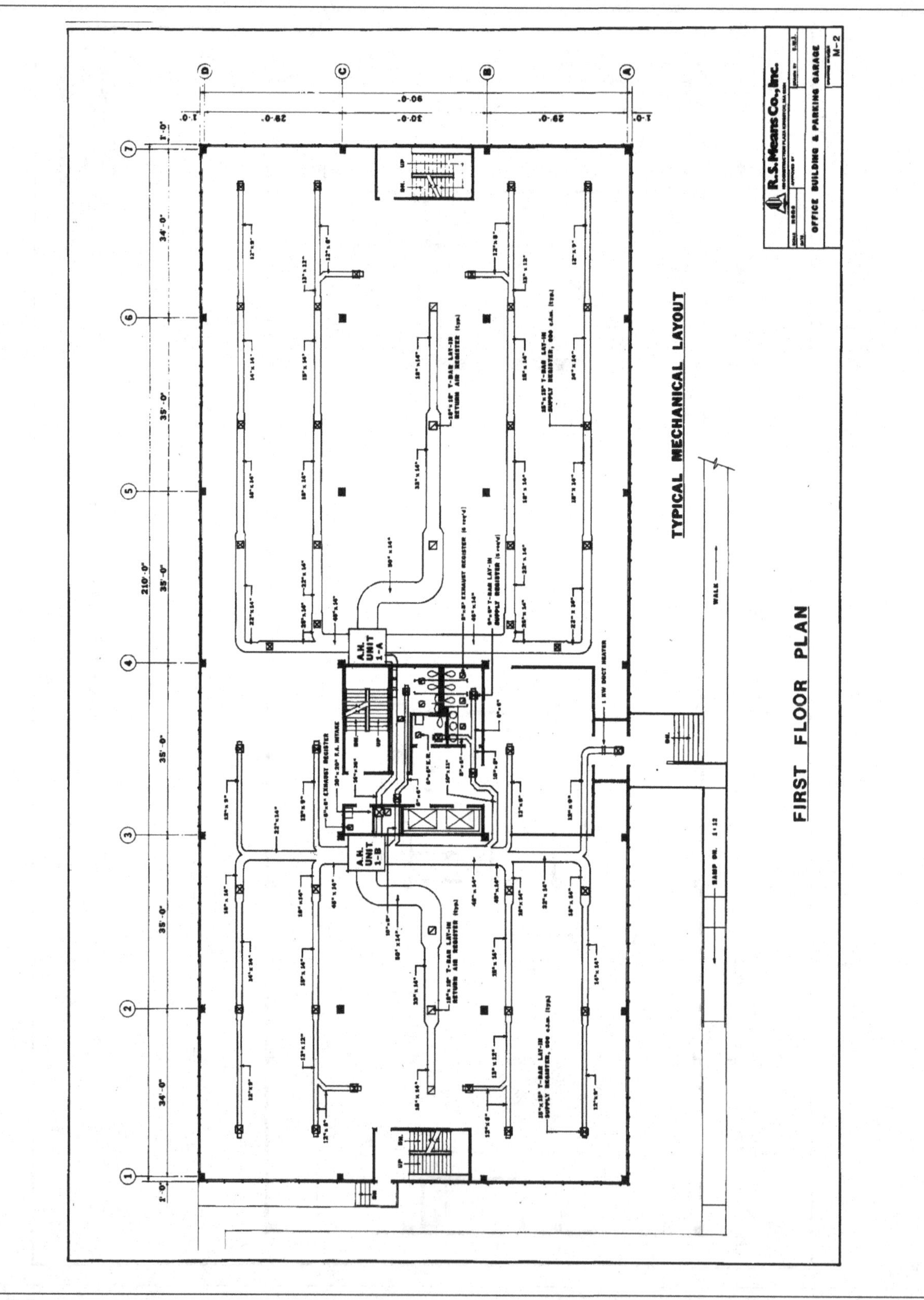

FIRST FLOOR PLAN
TYPICAL MECHANICAL LAYOUT
R.S. Means Co., Inc.
OFFICE BUILDING & PARKING GARAGE
M-2
A.H. UNIT 1-A
A.H. UNIT 1-B
210'-0"
34'-0"
35'-0"
35'-0"
35'-0"
35'-0"
34'-0"
1'-0"
1'-0"
90'-0"
29'-0"
30'-0"
29'-0"
1'-0"
1'-0"
WALK

Figure 9.3

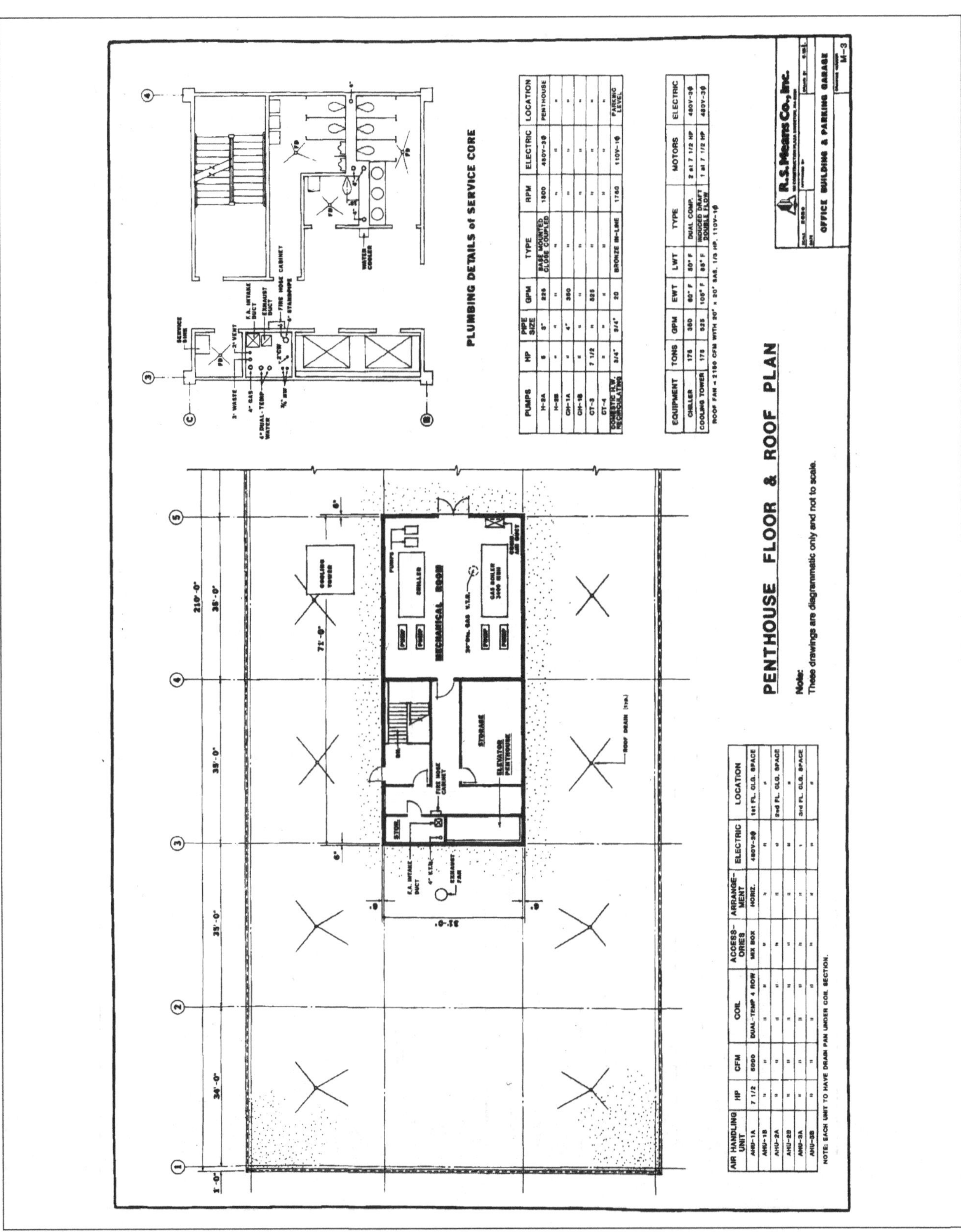
PENTHOUSE FLOOR & ROOF PLAN
Note:
These drawings are diagrammatic only and not to scale.
PLUMBING DETAILS of SERVICE CORE
MECHANICAL ROOM
STORAGE
ELEVATOR PENTHOUSE
COOLING TOWER
CHILLER
GAS BOILER 3400 MBH
PUMPS
FIRE HOSE CABINET
EXHAUST FAN
ROOF DRAIN (typ.)
210'-0"
1'-0" 34'-0" 35'-0" 35'-0" 35'-0"
71'-0"
31'-0"
PUMPS | HP | PIPE SIZE | GPM | TYPE | RPM | ELECTRIC | LOCATION
H-2A | 5 | 3" | 225 | BASE MOUNTED CLOSE COUPLED | 1800 | 480V-3Ø | PENTHOUSE
H-2B
CH-1A | 4" | 350
CH-1B
CT-3 | 7 1/2 | 525
CT-4
DOMESTIC H.W. RECIRCULATING | 3/4" | 3/4" | 20 | BRONZE IN-LINE | 1750 | 110V-1Ø | PARKING LEVEL
EQUIPMENT | TONS | GPM | EWT | LWT | TYPE | MOTORS | ELECTRIC
CHILLER | 175 | 350 | 60° F | 50° F | DUAL COMP. | 2 at 7 1/2 HP | 480V-3Ø
COOLING TOWER | 175 | 525 | 105° F | 85° F | INDUCED DRAFT DOUBLE FLOW | 1 at 7 1/2 HP | 480V-3Ø
ROOF FAN = 2150 CFM WITH 20" x 20" SAS. 1/3 HP. 110V-1Ø
AIR HANDLING UNIT | HP | CFM | COIL | ACCESS-ORIES | ARRANGE-MENT | ELECTRIC | LOCATION
AHU-1A | 7 1/2 | 5000 | DUAL-TEMP 4 ROW | MIX BOX | HORIZ. | 480V-3Ø | 1st FL. CLG. SPACE
AHU-1B
AHU-2A | 2nd FL. CLG. SPACE
AHU-2B
AHU-3A | 3rd FL. CLG. SPACE
AHU-3B
NOTE: EACH UNIT TO HAVE DRAIN PAN UNDER COIL SECTION.
R.S. Means Co., Inc.
OFFICE BUILDING & PARKING GARAGE
M-3

Figure 9.4

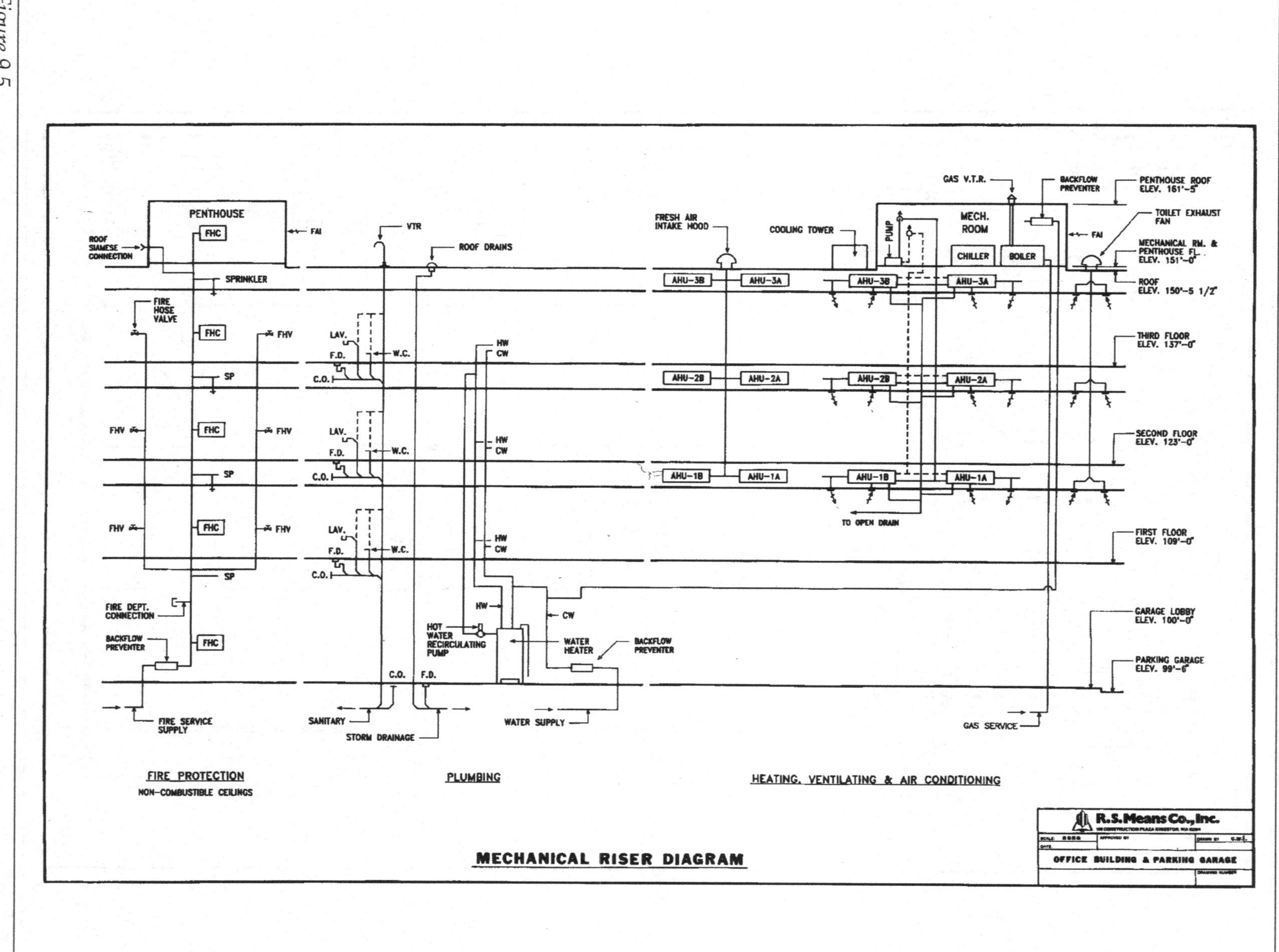
PENTHOUSE
FHC
FAI
ROOF SIAMESE CONNECTION
SPRINKLER
FIRE HOSE VALVE
FHV
SP
FIRE DEPT. CONNECTION
BACKFLOW PREVENTER
FIRE SERVICE SUPPLY
VTR
ROOF DRAINS
LAV.
F.D.
W.C.
C.O.
HW
CW
HOT WATER RECIRCULATING PUMP
WATER HEATER
SANITARY
STORM DRAINAGE
WATER SUPPLY
FRESH AIR INTAKE HOOD
COOLING TOWER
PUMP
MECH. ROOM
CHILLER
BOILER
GAS V.T.R.
AHU-3B
AHU-3A
AHU-2B
AHU-2A
AHU-1B
AHU-1A
TO OPEN DRAIN
GAS SERVICE
PENTHOUSE ROOF ELEV. 161'-5"
TOILET EXHAUST FAN
MECHANICAL RM. & PENTHOUSE FL. ELEV. 151'-0"
ROOF ELEV. 150'-5 1/2"
THIRD FLOOR ELEV. 137'-0"
SECOND FLOOR ELEV. 123'-0"
FIRST FLOOR ELEV. 109'-0"
GARAGE LOBBY ELEV. 100'-0"
PARKING GARAGE ELEV. 99'-6"
FIRE PROTECTION
NON-COMBUSTIBLE CEILINGS
PLUMBING
HEATING, VENTILATING & AIR CONDITIONING
MECHANICAL RISER DIAGRAM
R.S.Means Co., Inc.
OFFICE BUILDING & PARKING GARAGE

Figure 9.5

Getting Started

A good estimator must be able to visualize the proposed plumbing installation from source (i.e., water heaters or meter) to end use (i.e., fixture). This process helps identify each component of the mechanical portion of a building. Before starting the takeoff, the estimator should follow some basic steps:

- Read the plumbing plans and specifications carefully (with special attention given to the General and Special Conditions).
- Examine the drawings for general content: type of structure, number of floors, etc.
- Clarify any unclear areas with the designer, making sure that the scope of work is understood. Addenda to the contract documents may be necessary to clarify certain items of work so that the responsibility for the performance of all work is defined.
- Immediately contact suppliers, manufacturers, and subsystem specialty contractors in order to get their quotations and sub-bids.

Certain information should be taken from the complete architectural plans and specifications if the mechanical work is to be properly estimated. The following items should be given special attention:

- Temporary heat
- Temporary water or sanitary facilities
- Job completion dates (partial occupancy)
- Cleaning

Forms are a necessary tool in the preparation of estimates. This is especially true for the mechanical trades, which include a wide variety of items and materials. Careful measurements and a count of components cannot compensate for an oversight such as forgetting to include pipe insulation. A well designed form acts as a checklist, a guide for standardization, and a permanent record. A typical preprinted form and checklist is shown in Figure 9.6.

Plumbing

As in any estimate, the first step in preparing the plumbing estimate is to visualize the scope of the job. This can be done by scanning the drawings and specifications. The next step is to make up a material "takeoff " sheet. Having read through the plumbing section of the specifications, list each of the item headings on the summary sheet. This will serve as a quantity checklist. Also include labor that will not be specified in the plans, such as cleaning, adjusting, testing, and balancing.

The easiest way to make a material "quantity count" or takeoff is by system, as most of the pipe components of a system will tend to be of the same material, class weight, and grade. A waste system, for example, would consist of pipe varying from 2" to 15" in diameter, but it would all be of service weight cast iron or DWV copper up to a specified size, and extra heavy cast iron for the larger sizes.

Fixture takeoff usually involves nothing more than counting the various types, sizes, and styles, and entering them on a fixture form. It is important, however, that each fixture be fully identified. A common source of error is to overlook what does or does not come with the fixture (i.e., trim carrier, flush valve). Equipment such as pumps, water heaters, water softeners, and all items not previously counted are also

NAME OF JOB: DUE DATE:

LOCATION: BID SENT TO:

ARCHITECT: ENGINEER:

		1	2
	Labor w/Payroll Taxes, Welfare		
	Travel		
	Fixtures		
	Drains, Carriers, etc.		
	Sanitary - Interior		
	Storm - Interior		
	Water - Interior		
	Valves - Interior		
	Water Exterior - Street Connection		
	Water Exterior - Extension to Building		
	Sanitary & Storm Exterior - Street Connection		
	Sanitary & Storm Exterior - Extension to Building		
	Water Meter		
	Gas Service		
	Gas Interior		
	Accessories		
	Flashings for: Vents, Floor & Roof Drains, Pans		
	Hot Water Tanks: Automatic w/trim		
	Hot Water Tanks: Storage, w/stand and trim		
	Hot Water Circ. Pumps, Aquastats, Mag. Starters		
	Sump Pumps and/or Ejectors; Controls, Mag. Starters		
	Flue Piping		
	Rigging, Painting, Excavation and Backfill		
	Therm. Gauges, P.R.V. Controls, Specialties, etc.		
	Record Dwgs., Tags and Charts		
	Fire Extinguishers, Equipment, and/or Standpipes		
	Permits		
	Sleeves, Inserts and Hangers		
	Staging		
	Insulation		
	Acid Waste System		
	Sales Tax		
	Trucking and Cartage		

Figure 9.6

listed at this time. The order of proceeding is not as important as the development of a consistent method that will assist the estimator to improve in speed while minimizing the chances of overlooking any item or class of items. The various counts should then be totaled and transferred to the summary quantity sheet (Figure 9.7). Miscellaneous items that add to the plumbing contract should also be listed. Many of these would have been identified during the preliminary review of the plans and specifications. It is desirable to take off the fixtures and equipment before the piping for several reasons. The fixture and equipment lists can be given to suppliers for pricing while the estimator is performing the more arduous and time-consuming piping takeoff. Taking off the fixtures and equipment first also gives the estimator a good perspective of the building and its systems. Coloring the fixture and equipment as the takeoff is made aids the estimator when he returns for the piping takeoff by acting as targets for piping networks.

Pipe runs for any type of system consist of straight section and fittings of various shapes, styles, and purposes. Depending on job size, separate forms may be used for pipe and fittings, or they may be combined on one sheet. The estimator should start at one end of each system and using a tape, ruler, or wheeled indicator set at the corresponding scale, should measure and record the straight lengths of each size of pipe. When a fitting is encountered, it should be recorded with a check in the appropriate column. The use of colored pencils will aid in marking runs that have been completed, or the termination points on the main line where measurements are stopped, so that a branch may be taken off. Any changes in material should be noted. This would only occur at a joint. Therefore, the estimator should always verify that the piping material going into a joint is the same as that leaving the joint. The plumbing takeoff is shown in Figures 9.8a through 9.8e. Subcontractors (i.e., insulation, controls, etc.) should be notified of the availability of your plans after you have completed your takeoff.

Rounding off piping quantities to the nearest five or ten feet is good practice. In larger projects, rounding might be carried to the nearest hundred or thousand feet. This makes pricing and extending less prone to error. Rounding off pipe quantities is also practical because pipe is purchased in uniform lengths (i.e., rigid copper tubing and steel pipe is rounded off to 20' lengths, cast-iron soil pipe to 5' or 10' lengths, etc.).

The plumbing totals are transferred to cost analysis forms for pricing and extending, as shown in Figures 9.9a through 9.9e.

Units of cost are assigned to each item. These costs may be taken from historical data, or from a reliable reference, such as *Means Plumbing Cost Data*. Be sure to match the description of items to be priced with the historical data as closely as possible when obtaining unit costs. These costs are then extended (multiplied by the total quantity of each item) in order to obtain a total cost per item.

On the final sheet of the plumbing estimate (Figure 9.9e), page subtotals are recorded and totaled to arrive at the "bare" cost for plumbing. These values must then be adjusted for sales tax (if applicable), overhead, and profit on material, labor, and equipment as discussed previously. Finally, the marked up totals are adjusted by the

QUANTITY SHEET

PROJECT: Office Building		SHEET NO.	P 6 of 6
		ESTIMATE NO:	3-33
LOCATION:	ARCHITECT:	DATE:	1 / 4 / XX
TAKE OFF BY: MJM	EXTENSIONS BY: MJM	CHECKED BY:	

DESCRIPTION	NO.	DIMENSIONS				UNIT		UNIT		UNIT		UNIT
Plumbing Fixtures & Appliances												
Water heater, 36 KW, 148 GPH		80 Gal			1	Ea.						
Recirculating pump		3/4"			1	Ea.						
Water closet, white, Wm, w/ flush valve					15	Ea.						
Water closet, white, Fm, (handicap)					3	Ea.						
Urinal, stall type					6	Ea.						
Lavatory, 19" x 16", oval, vanity mount					9	Ea.						
Lavatory, 19" x 17", wall hung					9	Ea.						
Lavatory, 28" x 21", wall hung, (handicap)					3	Ea.						
Service sinks					3	Ea.						
Electric water cooler, wall hung					6	Ea.						

Figure 9.7

PIPING SCHEDULE

JOB: Office Building

SYSTEM: Plumbing - Storm System / Gas Piping

PAGE: P 1 of 6

DATE: 1 / 4 / XX

BY: JJM

	PIPE DIAMETER IN INCHES													
	12	10	8	6	5	4	3 1/2	3	2 1/2	2	1 1/2	1 1/4	1	3/4
Storm system														
Roof drains														
PVC Sch 40								12						
PVC DWV								1100						
DWV 1/4 bends (90)								24						
Floor drains galv.						6								
Cleanout Tee CI						12								
San Y (Tee) CI				1		6								
Pipe CI				220		120								
Gas pipe system														
Pipe BS Sch 40						160							20(vent)	
90 Deg. Ell Blk mall.						7							4	
Tees Blk mall						3								
Caps						3								
Cocks						2								

Figure 9.8a

PIPING SCHEDULE

JOB: Office Building

SYSTEM: Plumbing - Water pipe, Cold water

PAGE: P 2 of 6

DATE: 1 / 4 / XX

BY: JJM

	PIPE DIAMETER IN INCHES												
	12	10	8	6	5	4	3 1/2	3	2 1/2	2	1 1/2	1 1/4	1	3/4
Cold water														
"L" Copper										160			60	460
90 Deg. Ell Copper										6			18	26
Tee Copper										10				
Gate valve Bronze										3				12
Water meter										1				
Hose bibbs														6
Backflow prev.										1(bldg.)			1(boiler rm.)	
Shock absorber														4
Vacuum breaker														4

Figure 9.8b

PIPING SCHEDULE

JOB:	Office Building		DATE:	1 / 4 / XX
SYSTEM:	Plumbing - Water pipe, Hot water		BY:	JJM
			PAGE:	P 3 of 6

			PIPE DIAMETER IN INCHES															
			12	10	8	6	5	4	3 1/2	3	2 1/2	2	1 1/2	1 1/4	1	3/4		
Hot water																		
"L" Copper																220		
90 Deg. Ell Copper																12		
Tee Copper																18		
Check valve Bronze																1		
Relief valve																1		
Unions																2		
Temp. reg/mixing valve																1		
Gate valve Bronze																6		

Figure 9.8c

PIPING SCHEDULE

JOB: Office Building
SYSTEM: Plumbing - Sanitary and Vent.

PAGE: P 4 of 6
DATE: 1 / 4 / XX
BY: JJM

	PIPE DIAMETER IN INCHES													
	12	10	8	6	5	4	3 1/2	3	2 1/2	2	1 1/2	1 1/4	1	3/4
San / Vent														
Hub, Std Wt														
Pipe CI						90		40						
Comb. Y & 1/8 bend						5		2						
Floor Cleanout (FCO)						3		1						
1/4 bend (90)								2						
No Hub, Std Wt														
Pipe CI						170		210						
San Y (Tee)						30		20						
Floor Cleanout (FCO)						3								
Floor drains								12						
Wall cleanout (WCO)						6								
1/4 bend (90)						6								
Vent flashing thru roof						2		2						

Figure 9.8d

PIPING SCHEDULE

JOB: Office Building

SYSTEM: Plumbing - Sanitary and Vent. (cont'd)

PAGE: P 5 of 6

DATE: 1 / 4 / XX

BY: JJM

		PIPE DIAMETER IN INCHES															
		12	10	8	6	5	4	3 1/2	3	2 1/2	2	1 1/2	1 1/4	1	3/4		
San / Vent (cont'd)																	
1/8 bend (45)							11		4								
Coupling St. St.							90		72								
Copper																	
DWV Tube											320						
DWV San Tee											48						
DWV 90 Deg. Ell											24						

Figure 9.8e

COST ANALYSIS

PROJECT: Office Building — CLASSIFICATION: Div 15 Plumbing — SHEET NO. P 1 of 5

LOCATION: — ARCHITECT: — ESTIMATE NO: 3 - 33

DATE: 1/15/XX

TAKE OFF BY: JJM — QUANTITIES BY: JJM — PRICES BY: MJM — EXTENSIONS BY: MJM — CHECKED BY: JJM

DESCRIPTION	SOURCE/DIMENSIONS			QTY	UNIT	MATERIAL		LABOR		EQUIPMENT		SUBCONTRACT		TOTAL	
						COST	TOTAL	COST	TOTAL	COST	TOTAL	COST	TOTAL	COST	TOTAL
Water heater, 36 KW, 80 Gal., 148 GP	15480	200	4200	1	Ea.	3,750.00	3,750	211.00	211						
Circulating pump, 3/4"	15180	200	0640	1	Ea.	120.00	120	35.50	36						
Water closet, white, Wm, flush valve	15418	900	3100	15	Ea.	340.00	5,100	98.50	1,478						
Rough-in, water closet	15418	900	3200	15	Ea.	274.00	4,110	223.00	3,345						
Water closet, white, Fm, (handicap)	15418	900	3380	3	Ea.	355.00	1,065	114.00	342						
Rough-in, water closet	15418	900	3400	3	Ea.	139.00	417	201.00	603						
Urinal, stall type	15411	700	5000	6	Ea.	460.00	2,760	228.00	1,368						
Rough-in, urinal	15411	700	6980	6	Ea.	119.00	714	287.00	1,722						
Lavatory, 20" x 17" oval, vanity mounte	15418	450	0680	9	Ea.	129.00	1,161	89.00	801						
Rough-in, lavatory, vanity mount	15418	450	3580	9	Ea.	63.00	567	248.00	2,232						
Lavatory, 19" x 17", wall mounted	15418	450	4120	9	Ea.	265.00	2,385	71.50	644						
Rough-in, lavatory	15418	450	6960	9	Ea.	220.00	1,980	345.00	3,105						
Lavatory, 28" x 21", Wm, (handicap)	15418	450	6210	3	Ea.	320.00	960	81.50	245						
Rough-in, lavatory, wall mount	15418	450	6960	3	Ea.	220.00	660	345.00	1,035						
Service sink, 24" x 20", wall mounted	15418	600	7100	3	Ea.	465.00	1,395	143.00	429						
Rough-in, service sink, wall mount	15418	600	8980	3	Ea.	282.00	846	440.00	1,320						
Water cooler, wall hung	15413	900	0220	6	Ea.	645.00	3,870	143.00	858						
Rough-in, water cooler, wall mount	15413	900	9800	6	Ea.	42.00	252	143.00	858						
Water cooler, stainless cabinet	15413	900	0640	6	Ea.	99.00	594								
Page P1 subtotal →							32,706		20,630						

Figure 9.9a

COST ANALYSIS

PROJECT: Office Building   CLASSIFICATION: Div 15 Plumbing   SHEET NO. P 2 of 5

LOCATION: ARCHITECT: ESTIMATE NO: 3 - 33

DATE: 1/15/XX

TAKE OFF BY: JJM   UANTITIES BY: JJ   PRICES BY: MJM   EXTENSIONS BY: MJM   CHECKED BY: JJM

DESCRIPTION	SOURCE/DIMENSIONS		QTY	UNIT	MATERIAL		LABOR		EQUIPMENT		SUBCONTRACT		TOTAL	
					COST	TOTAL	COST	TOTAL	COST	TOTAL	COST	TOTAL	COST	TOTAL
PVC														
Roof drain, 3"	15160	4780	12	Ea.	89.00	1,068	40.50	486						
Pipe, 3"	15108	4470	1100	LF	3.19	3,509	10.75	11,825						
90 Deg. ell, 3"	15108	5080	24	Ea.	4.37	105	27.50	660						
Cast Iron, B & S type														
Floor drain, galv., 4"	15150	0440/80	6	Ea.	812.00	4,872	71.50	429						
Tee, clean out, 4"	15150	0240	12	Ea.	140.00	1,680	96.00	1,152						
Sanitary Y (tee), 4"	15107	0620	6	Ea.	25.00	150	71.50	429						
Sanitary Y (tee), 6"	15107	0800	1	Ea.	56.00	56	80.50	81						
Floor clean out, 3"	15150	0120	1	Ea.	112.00	112	39.50	40						
Floor clean outs, 4"	15150	0140	3	Ea.	112.00	336	53.00	159						
Combination Y & 1/8 bend, 2"	15107	1420	2	Ea.	15.75	32	57.00	114						
Combination Y & 1/8 bend, 4"	15107	1520	5	Ea.	33.00	165	71.50	358						
1/4 Bend (90 Deg. ell), 3"	15107	0120	2	Ea.	12.00	24	40.50	81						
Pipe, Cast Iron, B & S, 3"	15107	2140	40	LF	6.50	260	9.50	380						
Pipe, Cast Iron, B & S, 4"	15107	2160	210	LF	8.30	1,743	10.35	2,174						
Pipe, Cast Iron, B & S, 6"	15107	2200	220	LF	13.85	3,047	12.15	2,673						
Cast Iron, No hub														
Pipe, Cast Iron, no hub, 3"	15107	4140	210	LF	7.35	1,544	8.90	1,869						
Pipe, Cast Iron, no hub, 4"	15107	4160	170	LF	9.30	1,581	9.85	1,675						
Sanitary Y (tee), 3"	15107	6470	20	Ea.	9.30	186								
Sanitary Y (tee), 4"	15107	6472	30	Ea.	14.40	432								
Floor clean out, 4"	15150	0140	3	Ea.	112.00	336	53.00	159						
Floor drains, 3"	15150	2040/12	12	Ea.	116.50	1,398	47.50	570						
Wall clean out, 4"	15150	4100	6	Ea.	142.00	852	31.50	189						
Page P2 subtotal →						23,487		25,501						

Figure 9.9b

COST ANALYSIS

PROJECT: Office Building | CLASSIFICATION: Div 15 Plumbing | SHEET NO. P 3 of 5

LOCATION: | ARCHITECT: | ESTIMATE NO: 3 - 33

DATE: 1/15/XX

TAKE OFF BY: JJM | QUANTITIES BY: JJM | PRICES BY: MJM | EXTENSIONS BY: MJM | CHECKED BY: JJM

DESCRIPTION	SOURCE/DIMENSIONS			QTY	UNIT	MATERIAL		LABOR		EQUIPMENT		SUBCONTRACT		TOTAL	
						COST	TOTAL	COST	TOTAL	COST	TOTAL	COST	TOTAL	COST	TOTAL
Cast Iron, no hub (cont'd)															
1/4 Bend (90 Deg. ell), 4"	15107	360	6120	6	Ea.	11.00	66								
1/8 Bend (45 Deg. ell), 3"	15107	360	6212	4	Ea.	6.35	25								
1/8 Bend (45 Deg. ell), 4"	15107	360	6214	11	Ea.	8.05	89								
St. St. Couplings, 3"	15107	360	8650	72	Ea.	8.65	623	15.00	1,080						
St. St. Couplings, 4"	15107	360	8660	90	Ea.	9.80	882	17.30	1,557						
Flashing, for vent thru roof, 3"	15150	900	1450	2	Ea.	17.40	35	18.65	37						
Flashing, for vent thru roof, 4"	15150	900	1460	2	Ea.	19.20	38	19.80	40						
Black Steel															
Pipe, 1"	15107	620	0580	20	LF	2.04	41	6.00	120						
Pipe, 4"	15107	620	0650	160	LF	12.60	2,016	15.85	2,536						
90 Deg. ell Blk. mall., 1"	15107	640	5100	4	Ea.	2.73	11	24.50	98						
90 Deg. ell Blk. mall., 4"	15107	640	5170	7	Ea.	71.50	501	95.00	665						
Tee Blk. mall., 4"	15107	640	5580	3	Ea.	113.00	339	143.00	429						
Caps, 4"	15107	640	5780	3	Ea.	61.50	185	28.50	86		(NOTE: This is 1/2 coupling labor)				
Gas cocks, 4"	15110	600	7030	2	Ea.	335.00	670	190.00	380						
Copper (Sweat joints)															
Type L tube, 3/4"	15107	420	2180	680	LF	1.42	966	4.17	2,836						
Type L tube, 1"	15107	420	2200	60	LF	1.87	112	4.66	280						
Type L tube, 2"	15107	420	2260	160	LF	4.55	728	7.55	1,208						
90 Deg. ell, 3/4"	15107	460	0120	38	Ea.	0.67	25	16.65	633						
90 Deg. ell, 1"	15107	460	0130	18	Ea.	1.65	30	19.80	356						
90 Deg. ell, 2"	15107	460	0160	6	Ea.	7.10	43	29.00	174						
Page P3 subtotal →							7,423		12,514						

Figure 9.9c

COST ANALYSIS

PROJECT: Office Building	CLASSIFICATION: Div 15 Plumbing	SHEET NO. P 4 of 5
		ESTIMATE NO: 3 - 33
LOCATION:	ARCHITECT:	DATE: 1/15/XX

TAKE OFF BY: JJM   QUANTITIES BY: JJM   PRICES BY: MJM   EXTENSIONS BY: MJM   CHECKED BY: JJM

DESCRIPTION	SOURCE/DIMENSIONS			QTY	UNIT	MATERIAL		LABOR		EQUIPMENT		SUBCONTRACT		TOTAL	
						COST	TOTAL	COST	TOTAL	COST	TOTAL	COST	TOTAL	COST	TOTAL
Copper (Sweat joints) (cont'd)															
Tee, 3/4"	15107	460	0500	18	Ea.	1.24	22	26.50	477						
Tee, 2"	15107	460	0540	10	Ea.	13.20	132	45.50	455						
Shock absorber, 3/4"	15140	800	0500	4	Ea.	14.50	58	26.50	106						
Vacuum breaker, 3/4"	15140	600	1080	4	Ea.	28.00	112	15.85	63						
Unions, 3/4"	15107	460	0900	2	Ea.	3.95	8	17.60	35						
Gate valve, 3/4"	15110	160	2940	18	Ea.	18.30	329	15.85	285						
Gate valve, 2"	15110	160	2980	3	Ea.	62.50	188	29.00	87						
Check valve, 3/4"	15110	160	1860	1	Ea.	33.00	33	15.85	16						
Relief valve, 3/4"	15110	160	5640	1	Ea.	74.00	74	11.30	11						
Temp. mixing valve, 3/4"	15110	160	8440	1	Ea.	55.50	56	15.85	16						
DWV Copper (Sweat joints)															
DWV Tube, 2"	15107	420	4140	320	LF	3.29	1,053	7.20	2,304						
Sanitary Tee, 2"	15107	460	2290	48	Ea.	8.75	420	45.50	2,184						
90 Deg. ell, 2"	15107	460	2070	24	Ea.	5.95	143	31.50	756						
Water meter, 2"	15120	940	2360	1	Ea.	415.00	415	53.00	53						
Hose bibbs, 3/4"	15410	300	5000	6	Ea.	4.74	28	13.20	79						
Back flow preventer, 1"	15140	100	1120	1	Ea.	645.00	645	22.50	23						
Back flow preventer, 2"	15140	100	1160	1	Ea.	905.00	905	45.50	46						
Page P4 subtotal →							4,621		6,996						

Figure 9.9d

COST ANALYSIS

PROJECT: Office Building	CLASSIFICATION: Div 15 Plumbing	SHEET NO.	P 5 of 5
LOCATION:	ARCHITECT:	ESTIMATE NO:	3 - 33
TAKE OFF BY: JJM	QUANTITIES BY: JJM	DATE:	1/15/XX
PRICES BY: MJM	EXTENSIONS BY: MJM	CHECKED BY:	MJM

	SOURCE/DIMENSIONS					MATERIAL		LABOR		EQUIPMENT		SUBCONTRACT		TOTAL	
DESCRIPTION				QTY	UNIT	COST	TOTAL	COST	TOTAL	COST	TOTAL	COST	TOTAL	COST	TOTAL
Insulation for Hot/Cold water															
on copper tube															
Fiberglass w / ASJ, 1" wall, 3/4" dia.	15080	600	6840	680	LF		(subcontract)					4.16	2,829		
Fiberglass w / ASJ, 1" wall, 1" dia.	15080	600	6860	60	LF		(subcontract)					4.42	265		
Fiberglass w / ASJ, 1" wall, 2" dia.	15080	600	6890	160	LF		(subcontract)					4.97	795		
Miscellaneous															
Demonstration tests				4	Hrs.			39.60	158						
Sleeving				16	Hrs.			39.60	634						
Warranty				6	Hrs.			39.60	238						
Valves, tags, and charts				4	Hrs.			39.60	158						
Page P5 subtotal									$1,188				$3,889		
Plumbing Subtotals P1							32,706		20,630						
P2							23,487		25,501						
P3							7,423		12,514						
P4							4,621		6,996						
P5									1,188				3,889		
Subtotal Total							$68,237		$66,828				$3,889		
Materials sales tax						0.050	3,412								
Handling, Overhead and Profit markup						0.100	6,824	0.501	33,481			0.100	389		
TOTAL							$78,472		$100,309				$4,278		$183,060
City modifier for Boston						1.002	78,629	1.238	124,183			1.111	4,753		
BOSTON TOTAL							$78,629		$124,183				$4,753		$207,565

Figure 9.9e

appropriate location factor (Boston is used in the example). This completes the plumbing estimate for the office building. The resultant price is what a general contractor would receive as a complete bid for the plumbing work.

Fire Protection

After having scanned the plans and specifications to visualize the scope and type of fire protection proposed for this project, estimate sheets are prepared, as shown in Figure 9.10a through 9.10c. When properly used, such forms will serve as a summary of the specifications.

The specifications should be studied, and all components of fire protection systems should be entered on the takeoff sheets (Figures 9.10a through 9.10c). Material pricing and labor considerations should be listed.

The General Conditions and the Special Conditions should indicate the fire protection contractor's responsibilities regarding excavation and backfill, cutting and patching, masonry and concrete, painting, electrical work, temporary services, and any other items that may affect the estimate. Note on the estimate sheet the services that will be provided "by others." Indicate "labor only" items, such as cleaning, testing, flushing, application of decals, distribution of equipment to be stored in the cabinet, etc.

The material or quantity takeoff should be executed systematically, floor by floor. The fire protection estimator should be familiar with the architectural plans as this may be the only place that equipment such as extinguishers or other specialized equipment will be shown.

Hazardous areas requiring special considerations that may have been overlooked by the designer may be found on the architectural plans. The reflected ceiling plans and ductwork layout should also be studied to identify the degree of coordination required for the sprinkler head layout. The piping takeoff for standpipes should be kept separate from the sprinkler distribution piping. As in other piping takeoffs, pipe should be listed by material and joining method. The sprinkler heads should be taken off first, floor by floor, by type and configuration.

Fire protection materials after the takeoff should be totaled and transferred to a cost analysis form for pricing and extending, as shown in Figures 9.11a through 9.11c. As with the plumbing estimate, bare costs are summarized, totaled, and marked up on the final page (Figure 9.11c) to arrive at the "bid" price for fire protection.

Summarizing the Unit Price Estimate

Overhead and Profit: Since fire protection, and plumbing, contractors do not usually have to purchase large quantities of expensive equipment, their overhead markups are based primarily on labor. In this case, the markup for labor makes up a relatively large percentage of the total cost, and, therefore, the overhead markup percentage will likely be low in relation to total project cost. A contractor whose work involves large equipment purchases and several subcontractors, on the other hand, must include overhead percentages for these items in addition to the labor percentage. Published industry guides such as *Means Plumbing Cost Data* provide useful overhead percentages. However, such books should be used primarily as a reference. Every mechanical task has unique labor and material requirements, and markups for overhead

PIPING SCHEDULE

JOB: Office Building
SYSTEM: Fire Protection - Standpipe / Sprinkler

PAGE: FP 1 of 3
DATE: 1 / 7 / XX
BY: JJM

	PIPE DIAMETER IN INCHES														
	12	10	8	6	5	4	3 1/2	3	2 1/2	2	1 1/2	1 1/4	1	3/4	1/2
Standpipe															
Pipe BS Sch. 40			60	60		300			80						
90 Deg. Ell CI Thrded			3	1		4			16						
Tee CI Thrded			3	5@6x2-1/2 3@6x5		8									
F D Valves									13						
Reducers CI			2												
Roof siamese				1@6x2-1/2x2-1/2											
OS&Y Gate valve			1	3											
Swing check w/ball drip				1											
Wall siamese				1											
Fire hose											500				
Hose adapter									5						
Hose rack									5						
Hose nozzle									5						

Figure 9.10a

PIPING SCHEDULE

JOB: Office Building

SYSTEM: Fire Protection - Standpipe / Sprinkler (cont'd)

PAGE: FP 2 of 3

DATE: 1 / 7 / XX

BY: JJM

	PIPE DIAMETER IN INCHES															
	12	10	8	6	5	4	3 1/2	3	2 1/2	2	1 1/2	1 1/4	1	3/4	1/2	
Standpipe (cont'd)																
Hydrolator									5							
Fire hose cabinet									5							
Sprinkler Distribution																
Alarm valve				1												
Alarm water mtr/gong																Size Std. 1
Check valve				1												
Flow control valve				1												
Pipe BS Grooved Sch 10						50			300	960						
Pipe BS Grooved Sch 40											1150	570	1150			
Tee Grooved						10			30							

Figure 9.10b

PIPING SCHEDULE

JOB: Office Building

SYSTEM: Fire Protection - Sprinkler / Standpipe (cont'd)

PAGE: FP 3 of 3

DATE: 1 / 7 / XX

BY: JJM

	PIPE DIAMETER IN INCHES														
	12	10	8	6	5	4	3 1/2	3	2 1/2	2	1 1/2	1 1/4	1	3/4	1/2
Sprinkler Dist (cont'd)															
Tee Mech joint										80	80	40	@1-1/4x3/4		
										2x1	1-1/2x1				
90 Deg. Ell CI Thrded													50		
Coupling Grooved joint						20			80	20					
90 Deg. Ells Grooved joint										10					
Sprinkler heads															300
Tee CI Thrded													40		

Figure 9.10c

COST ANALYSIS

PROJECT: Office Building | CLASSIFICATION: Div. 15 Fire Protection | SHEET NO. FP 1 of 3

ESTIMATE NO: 3 - 33

LOCATION: | ARCHITECT: | DATE: 1/15/XX

TAKE OFF BY: MJM | QUANTITIES BY: MJM | PRICES BY: MJM | EXTENSIONS BY: MJM | CHECKED BY: JJM

DESCRIPTION	SOURCE/DIMENSIONS			QTY	UNIT	MATERIAL		LABOR		EQUIPMENT		SUBCONTRACT		TOTAL	
						COST	TOTAL	COST	TOTAL	COST	TOTAL	COST	TOTAL	COST	TOTAL
Black Steel (threaded)															
Pipe, S40, 2-1/2"	15107	620	0620	80	LF							24.00	1,920		
Pipe, S40, 4"	15107	620	0650	300	LF							38.00	11,400		
Pipe, S40, 6"	15107	620	0670	60	LF							70.00	4,200		
Pipe, S40, 8"	15107	620	0680	60	LF							89.00	5,340		
90 Deg. ell, CI, 2-1/2"	15107	640	0150	16	Ea.							83.50	1,336		
90 Deg. ell, CI, 4"	15107	640	0180	4	Ea.							211.00	844		
90 Deg. ell, CI, 6"	15107	640	0200	1	Ea.							360.00	360		
90 Deg. ell, CI, 8"	15107	640	0210	3	Ea.							575.00	1,725		
Tee, CI, 4"	15107	640	0620	8	Ea.							315.00	2,520		
Tee, CI, 8"	15107	640	0650	3	Ea.							990.00	2,970		
Tee, CI, 6" x 2-1/2"	15107	640	0672	5	Ea.							795.00	3,975		
Tee, CI, 6" x 5"	15107	640	0672	3	Ea.							795.00	2,385		
Reducer, CI, 8"	15107	640	0699	2	Ea.							1,075.00	2,150		
F. D. Bronze valves, 2-1/2"	13910	800	0090	13	Ea.							124.00	1,612		
Roof siamese, 6" x 2-1/2" x 2-1/2"	13910	800	6060	1	Ea.							315.00	315		
OS&Y gate valve, 6"	15110	200	3700	3	Ea.							1,125.00	3,375		
OS&Y gate valve, 8"	15110	200	3720	1	Ea.							1,725.00	1,725		
Swing check valve w/ball, 6"	13930	400	6540	1	Ea.							780.00	780		
Wall siamese, 6"	13910	400	7370	1	Ea.							870.00	870		
Fire hose, 1-1/2"	13910	400	2260	500	LF							2.08	1,040		
Fire hose adapter, 2-1/2"	13910	400	1600	5	Ea.							8.90	45		
Fire hose rack, 1-1/2"	13910	400	2640	5	Ea.							78.50	393		
Fire hose nozzle, 1-1/2"	13910	400	6700	5	Ea.							28.50	143		
Fire hose coupling, 1-1/2"	13910	400	1410	5	Ea.							31.50	158		
Fire hose rack nipple, 1-1/2"	13910	400	2780	5	Ea.							17.15	86		
Page FP1 subtotal →													51,665		

Figure 9.11a

COST ANALYSIS

PROJECT: Office Building	CLASSIFICATION: Div. 15 Fire Protection	SHEET NO.	FP 2 of 3
LOCATION:	ARCHITECT:	ESTIMATE NO:	3 - 33
		DATE:	1/15/XX
TAKE OFF BY: MJM	QUANTITIES BY: MJM / PRICES BY: MJM / EXTENSIONS BY: MJM	CHECKED BY:	JJM

DESCRIPTION	SOURCE/DIMENSIONS			QTY	UNIT	MATERIAL		LABOR		EQUIPMENT		SUBCONTRACT		TOTAL	
						COST	TOTAL	COST	TOTAL	COST	TOTAL	COST	TOTAL	COST	TOTAL
Hydrolator, 1-1/2"	13910	800	4200	5	Ea.							84.50	423		
Fire hose cabinet	10525	200	4200	5	Ea.							295.00	1,475		
Alarm valve, 6"	13930	400	6300	1	Ea.							1,475.00	1,475		
Water alarm motor/gong	13930	400	1220	1	Ea.							320.00	320		
Check valve, 6"	13930	400	6840	1	Ea.							650.00	650		
Flow control valve, 6"	13930	400	8860	1	Ea.							2,775.00	2,775		
Grooved pipe system, black steel															
Pipe, S10, 2"	15107	690	0550	960	LF							14.00	13,440		
Pipe, S10, 2-1/2"	15107	690	0560	300	LF							17.75	5,325		
Pipe, S10, 4"	15107	690	0590	60	LF							23.00	1,380		
Pipe, S40, 1"	15107	690	1050	1160	LF							9.35	10,846		
Pipe, S40, 1-1/4"	15107	690	1060	580	LF							10.45	6,061		
Pipe, S40, 1-1/2"	15107	690	1070	1160	LF							11.90	13,804		
Tee, grooved, 2-1/2"	15107	690	4780	30	Ea.							47.00	1,410		
Tee, grooved, 4"	15107	690	4800	10	Ea.							83.00	830		
90 deg. ell, grooved, 2"	15107	690	4070	10	Ea.							29.00	290		
Coupling, grooved, 2"	15107	690	4990	20	Ea.							17.55	351		
Coupling, grooved, 2-1/2"	15107	690	5000	80	Ea.							20.50	1,640		
Coupling, grooved, 4"	15107	690	5030	20	Ea.							32.50	650		
Tee, mech joint, 1-1/4" x 1/2"	15107	660	9510	40	Ea.							53.50	2,140		
Tee, mech joint, 1-1/2" x 1"	15107	660	9560	80	Ea.							56.50	4,520		
Tee, mech joint, 2" x 1"	15107	660	9590	80	Ea.							64.50	5,160		
Page FP2 subtotal →													74,965		

Figure 9.11b

COST ANALYSIS

PROJECT: Office Building | CLASSIFICATION: Div. 15 Fire Protection | SHEET NO. FP 3 of 3

LOCATION: | ARCHITECT: | ESTIMATE NO: 3 - 33

DATE: 1/15/XX

TAKE OFF BY: MJM | QUANTITIES BY: MJM | PRICES BY: MJM | EXTENSIONS BY: MJM | CHECKED BY: JJM

DESCRIPTION	SOURCE/DIMENSIONS			QTY	UNIT	MATERIAL		LABOR		EQUIPMENT		SUBCONTRACT		TOTAL	
						COST	TOTAL	COST	TOTAL	COST	TOTAL	COST	TOTAL	COST	TOTAL
(con't)															
90 Deg. ell (threaded), 1"	15107	640	0110	50	Ea.							39.50	1,975		
Tee (threaded), 1"	15107	640	0550	40	Ea.							63.50	2,540		
Sprinkler heads	13930	400	4830	300	Ea.							59.50	17,850		
													22,365		
Fire Protection subtotal, FP 1													51,665		
Fire Protection subtotal, FP 2													74,965		
Fire Protection subtotal, FP 3													22,365		
Subtotal Total →												→	148,994		148,994
Markup, assuming the sprinkler installer is a subcontractor. (Probably to the plumber.)												0.100	14,899		
TOTAL →												→	163,894		
City modifier for Boston												1.111	182,086		
BOSTON TOTAL →												→	182,086		182,086

Figure 9.11c

and profit should ultimately be based on the contractor's individual situation.

The Project Schedule: When the work for all sections is priced, the estimator should complete the project schedule so that time-related costs in the Project Overhead Summary can be determined. When preparing the schedule, the estimator must visualize the entire construction process in order to determine the correct sequence of work. Certain tasks must be completed before others are begun. Different trades will work simultaneously. Material deliveries will also affect scheduling. All such variables must be incorporated into the Project Schedule. An example of a Project Schedule is shown in Figure 9.12. The labor-hour figures, which have been calculated for each section, are used to assist with scheduling. The estimator must be careful not only to use the labor-hours for each section independently, but to coordinate each section with related work.

The schedule shows that the project will last approximately one year. The duration of the plumbing work will be about six months. Time-dependent items, such as equipment rental and superintendent costs, can be analyzed on a Project Overhead Summary form (Figures 9.13a and 9.13b). Some items, such as permits and insurance, are dependent on total job costs. The total direct costs for the project from the Estimate Summary can be used to estimate these costs. These items should be analyzed individually if possible. Alternatively, percentages can be added to material and labor to cover these items. Since the example estimate has been prepared using *Means Plumbing Cost Data,* the percentage method has been used.

The Bottom Line: The estimator is now able to complete the estimate. Appropriate contingency, sales tax, and overhead and profit costs must be added to the direct costs of the project. The overhead and profit percentage for labor has already been added to all labor costs. Wherever possible, contractors should determine appropriate markups for their own companies. Finally, Means prices represent national averages, and should be adjusted with the City Cost Index for your locality.

All totals for our estimate of the office building project have already been marked up to cover these items. Either applying markups individually or at the final summary is acceptable, depending on the estimator's preference.

The prime plumbing contractor may also wish or need to add a final markup for profit if the cost of administering subcontracted items has not been covered previously. Market conditions and project overhead will dictate the amount of final markup, if any, to be applied. The complete plumbing estimate is now ready for submittal to the general contractor.

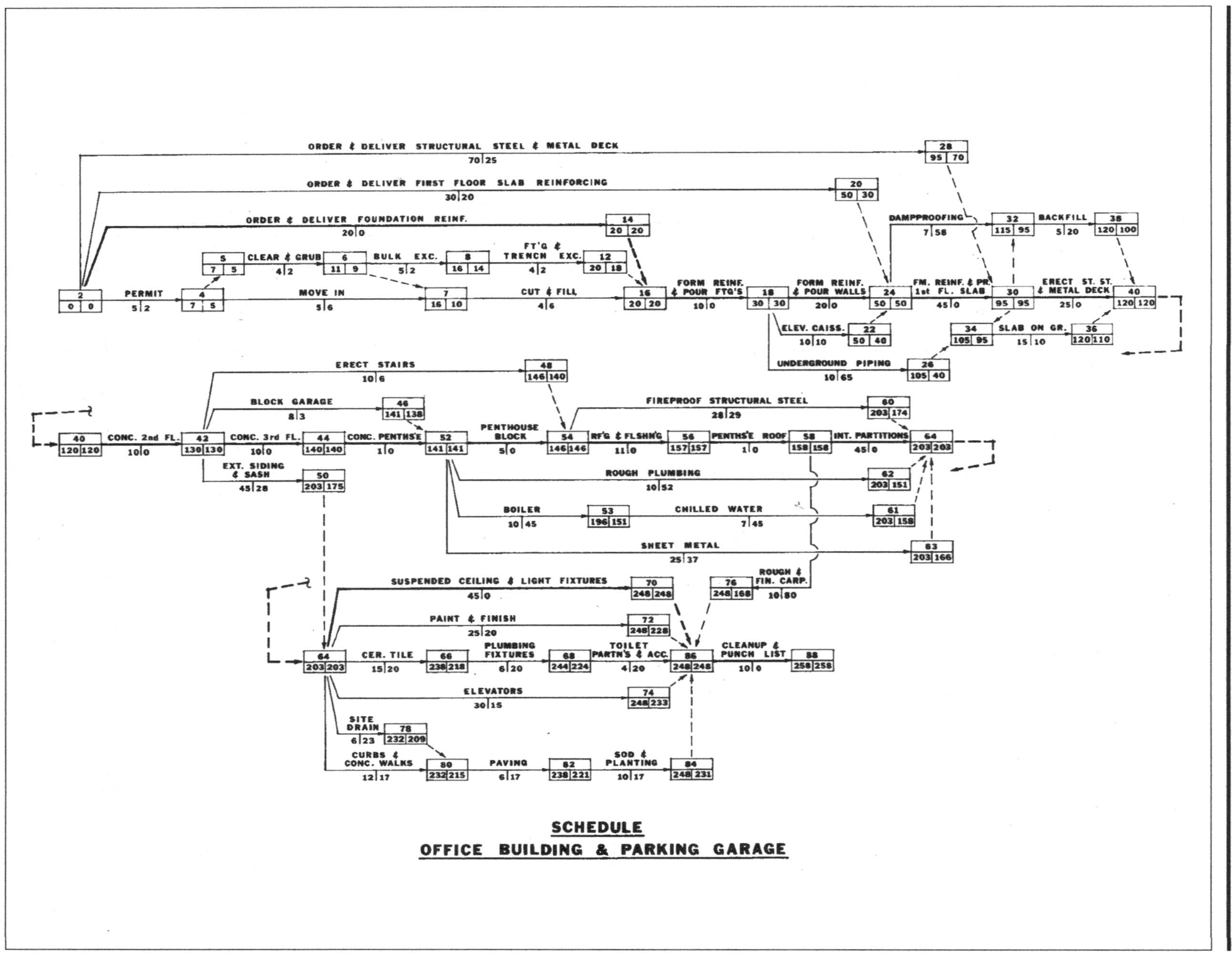
ORDER & DELIVER STRUCTURAL STEEL & METAL DECK
ORDER & DELIVER FIRST FLOOR SLAB REINFORCING
ORDER & DELIVER FOUNDATION REINF.
PERMIT
CLEAR & GRUB
BULK EXC.
FT'G & TRENCH EXC.
MOVE IN
CUT & FILL
FORM REINF. & POUR FTG'S
FORM REINF. & POUR WALLS
FM. REINF. & PR. 1st FL. SLAB
ERECT ST. ST. & METAL DECK
DAMPPROOFING
BACKFILL
ELEV. CAISS.
SLAB ON GR.
UNDERGROUND PIPING
ERECT STAIRS
BLOCK GARAGE
CONC. 2nd FL.
CONC. 3rd FL.
CONC. PENTHS'E
PENTHOUSE BLOCK
RF'G & FLSHN'G
PENTHS'E ROOF
INT. PARTITIONS
FIREPROOF STRUCTURAL STEEL
EXT. SIDING & SASH
ROUGH PLUMBING
BOILER
CHILLED WATER
SHEET METAL
ROUGH & FIN. CARP.
SUSPENDED CEILING & LIGHT FIXTURES
PAINT & FINISH
CER. TILE
PLUMBING FIXTURES
TOILET PARTN'S & ACC.
CLEANUP & PUNCH LIST
ELEVATORS
SITE DRAIN
CURBS & CONC. WALKS
PAVING
SOD & PLANTING
SCHEDULE
OFFICE BUILDING & PARKING GARAGE

Figure 9.12

PROJECT OVERHEAD SUMMARY

SHEET NO.

PROJECT | ESTIMATE NO.

LOCATION | ARCHITECT | DATE

QUANTITIES BY: | PRICES BY: | EXTENSIONS BY: | CHECKED BY:

DESCRIPTION	QUANTITY	UNIT	MATERIAL/EQUIPMENT		LABOR		TOTAL COST	
			UNIT	TOTAL	UNIT	TOTAL	UNIT	TOTAL
Job Organization: Superintendent								
Project Manager								
Timekeeper & Material Clerk								
Clerical								
Safety, Watchman & First Aid								
Travel Expense: Superintendent								
Project Manager								
Engineering: Layout								
Inspection/Quantities								
Drawings								
CPM Schedule								
Testing: Soil								
Materials								
Structural								
Equipment: Cranes								
Concrete Pump, Conveyor, Etc.								
Elevators, Hoists								
Freight & Hauling								
Loading, Unloading, Erecting, Etc.								
Maintenance								
Pumping								
Scaffolding								
Small Power Equipment/Tools								
Field Offices: Job Office								
Architect/Owner's Office								
Temporary Telephones								
Utilities								
Temporary Toilets								
Storage Areas & Sheds								
Temporary Utilities: Heat								
Light & Power								
Water								
PAGE TOTALS								

Page 1 of 2

Figure 9.13a

DESCRIPTION	QUANTITY	UNIT	MATERIAL/EQUIPMENT		LABOR		TOTAL COST	
			UNIT	TOTAL	UNIT	TOTAL	UNIT	TOTAL
Totals Brought Forward								
Winter Protection: Temp. Heat/Protection								
Snow Plowing								
Thawing Materials								
Temporary Roads								
Signs & Barricades: Site Sign								
Temporary Fences								
Temporary Stairs, Ladders & Floors								
Photographs								
Clean Up								
Dumpster								
Final Clean Up								
Punch List								
Permits: Building								
Misc.								
Insurance: Builders Risk								
Owner's Protective Liability								
Umbrella								
Unemployment Ins. & Social Security								
Taxes								
City Sales Tax								
State Sales Tax								
Bonds								
Performance								
Material & Equipment								
Main Office Expense								
Special Items								
TOTALS:								

Page 2 of 2

Figure 9.13b

Appendix

Table of Contents

Plumbing Fixture Symbols

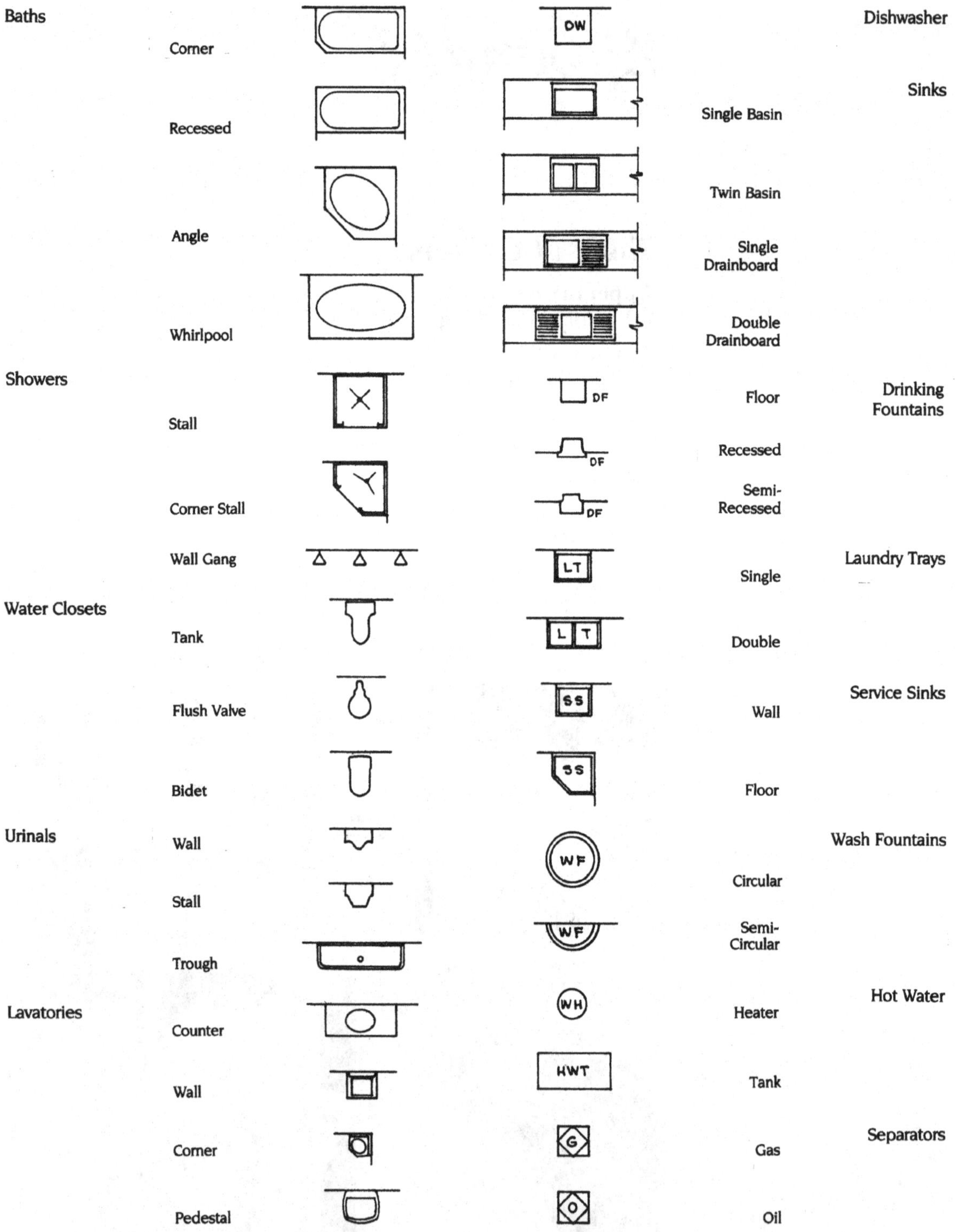

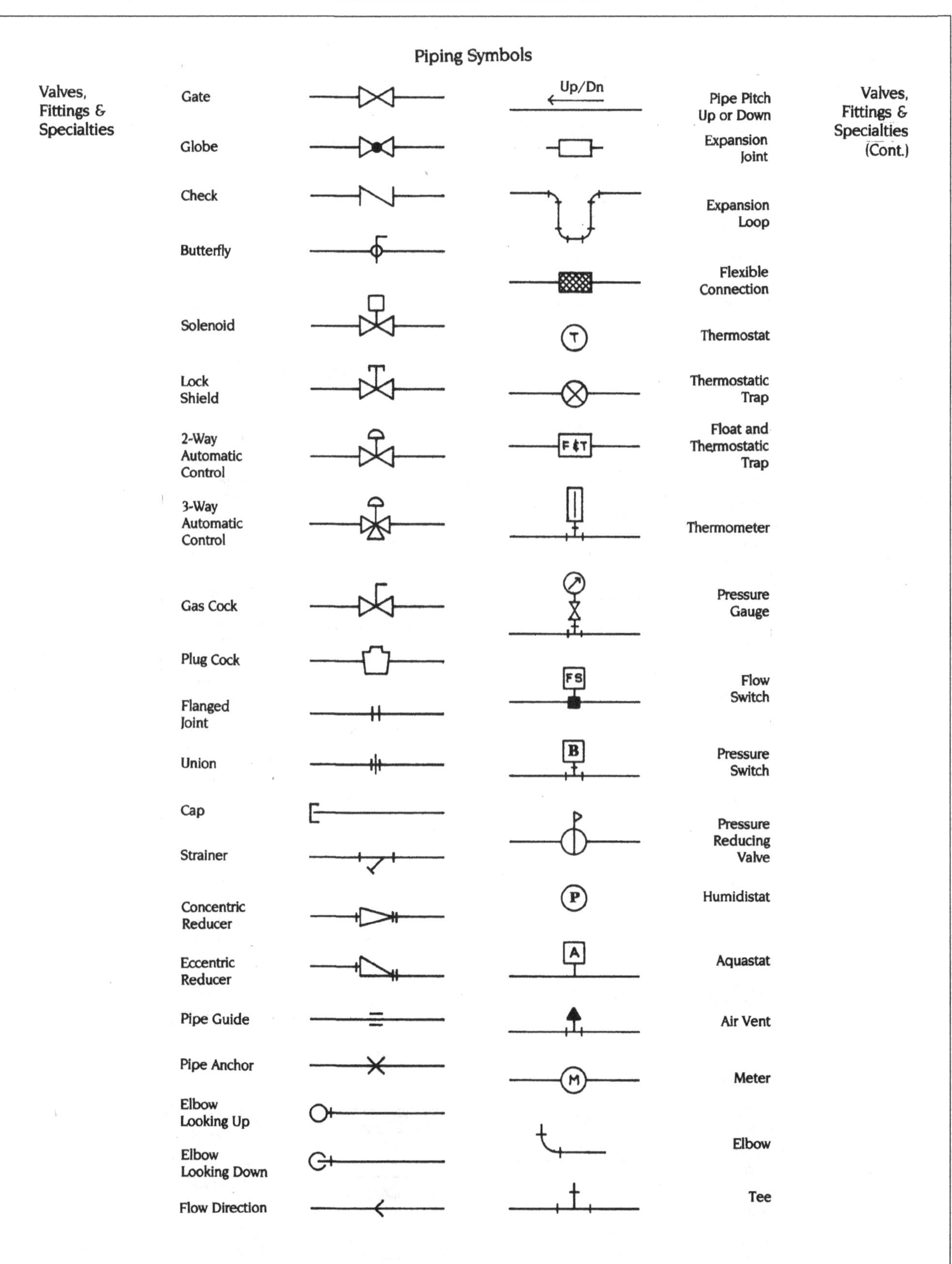
Piping Symbols
Valves, Fittings & Specialties
Gate
Globe
Check
Butterfly
Solenoid
Lock Shield
2-Way Automatic Control
3-Way Automatic Control
Gas Cock
Plug Cock
Flanged Joint
Union
Cap
Strainer
Concentric Reducer
Eccentric Reducer
Pipe Guide
Pipe Anchor
Elbow Looking Up
Elbow Looking Down
Flow Direction
Up/Dn
Pipe Pitch Up or Down
Expansion Joint
Expansion Loop
Flexible Connection
T
Thermostat
Thermostatic Trap
F&T
Float and Thermostatic Trap
Thermometer
Pressure Gauge
FS
Flow Switch
B
Pressure Switch
Pressure Reducing Valve
P
Humidistat
A
Aquastat
Air Vent
M
Meter
Elbow
Tee
Valves, Fittings & Specialties (Cont.)

Piping Symbols (Cont.)

Category	Description	Symbol	Symbol	Description	Category
Plumbing	Floor Drain	▢	——FOG——	Fuel Oil Gauge Line	HVAC (Cont.)
	Indirect Waste	——W——	—o—PD—o—	Pump Discharge	
	Storm Drain	——SD——			
	Combination Waste & Vent	——CWV——	– – – – – –	Low Pressure Condensate Return	
	Acid Waste	——AW——	——LPS——	Low Pressure Steam	
	Acid Vent	– – – AV – – –	——MPS——	Medium Pressure Steam	
	Cold Water	——CW——	——HPS——	High Pressure Steam	
	Hot Water	——HW——			
	Drinking Water Supply	——DWS——	——BD——	Boiler Blow-Down	
	Drinking Water Return	——DWR——	——F——	Fire Protection Water Supply	Fire Protection
	Gas-Low Pressure	——G——			
	Gas-Medium Pressure	——MG——	——WSP——	Wet Standpipe	
	Compressed Air	——A——	——DSP——	Dry Standpipe	
	Vacuum	——V——	——CSP——	Combination Standpipe	
	Vacuum Cleaning	——VC——	——SP——	Automatic Fire Sprinkler	
	Oxygen	——O——	—o——o—	Upright Fire Sprinkler Heads	
	Liquid Oxygen	——LOX——	—●——●—	Pendent Fire Sprinkler Heads	
	Liquid Petroleum Gas	——LPG——			
HVAC	Hot Water Heating Supply	——HWS——		Fire Hydrant	
	Hot Water Heating Return	——HWR——		Wall Fire Dept. Connection	
	Chilled Water Supply	——CHWS——			
	Chilled Water Return	——CHWR——		Sidewalk Fire Dept. Connection	
	Drain Line	——D——			
	City Water	——CW——	FHR	Fire Hose Rack	
	Fuel Oil Supply	——FOS——	FHC	Surface Mounted Fire Hose Cabinet	
	Fuel Oil Return	——FOR——			
	Fuel Oil Vent	——FOV——	FHC	Recessed Fire Hose Cabinet	

HVAC Ductwork Symbols

Supply Duct

Return/ Exhaust Duct

Duct–First Number is Side Shown

10 x 20

Direction of Flow

Lined Ductwork

Inclined Drop in Direction of Air Flow

D

Inclined Rise in Direction of Air Flow

R

Canvas Connection

Access Door

Intake Louver & Screen

Exhaust Louver

20 x 10-L 650 CPM

Splitter Damper

Round Elbow Turning Vanes

M

Automatic Damper

FD

Fire Damper

VD

Volume Damper

SD

Smoke Damper

24" Dia. CD
1250 CFM

Supply Outlet Ceiling Diffuser

20 x 10 CD
500 CFM

Supply Outlet Ceiling Diffuser

84 x 6-LD
375 CFM

Linear Diffuser

Round Elbow

Square Elbow

Square Elbow Turning Vanes

Double Duct Air System

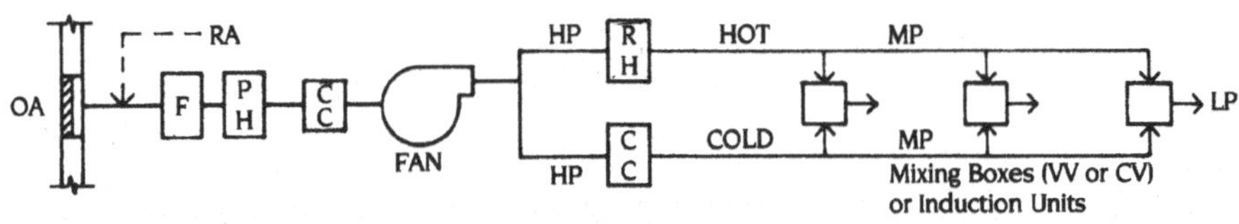

OA = Outside Air
RA = Return Air
F = Filter
PH = Preheat Coil

CC = Cooling Coil
RH = Reheat Coil
HP = High Pressure Duct
MP = Medium Pressure Duct

LP = Low Pressure Duct
VV = Variable Volume
CV = Constant Volume

Labor-Hours to Install Building Piping Systems

This section contains labor-hour units for various system installations, as well as specialties, fixtures, and equipment. The labor unit labor-hours are based on a national average.

The user should evaulate the location and conditions of each individual project and guide their final labor-hour allocations accordingly.

PLUMBING FIXTURES

ITEM DESCRIPTION	LABOR-HOURS EACH
Water Closet (Flush Valve Floor-Mounted)	2.25
(Flush Valve Wall-Hung)	2.00
(Tank Type Floor-Mounted)	2.25
(Tank Type Wall-Hung)	2.25
Urinal (Flush Valve Wall-Hung)	2.00
Lavatory (Wall-Hung)	1.75
Lavatory (Countertop)	2.00
Wash Fountains (54" Diameter)	9.00
Bathtub (C.I. Recessed)	3.50
Bathtub (Steel Recessed)	3.00
Shower (Valve Body and Trim Only)	1.00
Shower (Terrazzo Receptor)	2.50
Shower (Stall w/Fiberglass Walls)	4.00
Sink (Service)	3.00
Sink (Countertop) 1 Compartment	2.25
Sink (Countertop) 2 Compartment	2.75
Floor Sink	2.75
Scrub Sink	3.00
Clinical Sink w/Flushometer and Faucet	4.00
Institutional Bathtub w/Base	8.25
Electric Water Cooler (Recessed)	3.25
(Free Standing)	2.25
(Semi-recessed)	3.00
Eye Wash Fountain	2.50
Floor Drain 2" to 4"	1.00
Area Drain 2" to 4"	1.00
Roof Drain 2" to 4"	1.25
Wall Hydrant	.50
Trench Drain (Light Duty, 10 ft. long)	3.00
Emergency Shower	2.50
Support Carriers	
Water Closet Carrier	2.00
Urinal Carrier	1.50
Lavatory Carrier	2.00
Elec. Water Coolers/Drink. Fount. Carrier	2.00

BUILDING EQUIPMENT

ITEM DESCRIPTION	LABOR-HOURS EACH
Constant Pressure Pumps	
1750 R.P.M. 150 G.P.M.	
5 Horsepower (Duplex)	8.00
7½	8.00
10	10.00
15	12.00
20	14.00
25	16.00
5 Horsepower (Triplex)	10.00
7½	10.00
10	12.00
15	15.00
20	18.00
25	20.00

NOTE:

1) Fixture man-hours include distribution, uncrating, trim and connecting to existing piping and testing.
2) Equipment man-hours include distribution, uncrating, set into place and connecting to existing piping, and testing.

EXCLUSIONS

1) All piping, valves, elec. wiring.

BUILDING EQUIPMENT (continued)	
ITEM DESCRIPTION	LABOR MAN-HOURS EACH
In-line Circulating Pumps	
1/12 Horsepower (Iron or Bronze Body)	1.25
1/6	1.25
¼	1.25
⅓	1.25
½	1.50
¾	2.00
1	2.50
Water Meters	
¾" Disc Type	0.75
1"	1.00
1½"	1.25
2"	1.50
2" Compound Type	1.50
3"	2.50
4"	4.00
6"	6.00
Fire Meters	
3" Detector Type	4.00
4"	6.00
6"	8.00
8"	10.00
Backflow Preventers	
1" Thread 53 G.P.M.	0.50
1½" 100	1.00
2" 160	1.00
2½" Flanged 225	2.50
3" 320	2.75
4" 500	4.00
6" 1000	6.00
8" 1600	8.00
Sewage Ejectors	
6 ft. Shaft Cast-Iron Basin	
2 Horsepower 50 G.P.M. (Duplex)	18.00
2 75	18.00
2 100	18.00
3 125	20.00
3 150	20.00
3 200	22.00
5 250	24.00
7½ 300	28.00
10 350	30.00
15 450	35.00
20 500	40.00
Sump Pumps	
6 ft. Shaft Cast-Iron Basin	
2 Horsepower 30 G.P.M. (Simplex)	16.00
2 40	16.00
2 50	16.00
2 60	16.00
2 75	16.00
2 100	16.00
3 125	18.00
2 30 (Duplex)	18.00
2 40	18.00
2 50	18.00
2 60	18.00
2 75	18.00
2 100	18.00
3 125	20.00
3 150	20.00
3 200	22.00
3 250	24.00
Grease Interceptors	
14 lb. Grease Capacity	3.25
20	3.25
30	3.25
40	4.00
50	4.00
70	5.00
100	5.00
150	6.00
200	6.00
300	7.00
400	7.00
500	8.00
700	8.00
1000	9.00
Oil Interceptors	
10 G.P.M.	3.25
20	3.25
25	3.25
35	4.00
50	4.00
75	5.00
100	5.00
150	6.00
200	6.00
250	7.00
350	7.00
500	8.00

NOTE:

1) Equipment man-hours include distribution, uncrating, set into place, connect to existing piping and testing.

EXCLUSIONS

1) All piping, valves and elec. wiring.

BUILDING EQUIPMENT (continued)	
ITEM DESCRIPTION	LABOR-HOURS EACH
Hot Water Generators (Gas-Fired)	
500 G.P.H. Recovery Rate	14.00
1000	18.00
1500	20.00
2000	24.00
2500	26.00
3000	28.00
Hot Water Generators (Oil-Fired)	
500 G.P.H. Recovery Rate	16.00
1000	20.00
1500	24.00
2000	26.00
2500	28.00
3000	30.00
Steam Hi-Temp Hot Water Generators	
500 G.P.H. Production Rate	20.00
1000	24.00
1500	28.00
2000	30.00
2500	32.00
3000	35.00
Hot Water Generators (240 Volt Electric)	
500 G.P.H. Production Rate 140 Kilowatts	12.00
1000	14.00
1500	16.00
2000	18.00
2500	20.00
3000	24.00
Electric Water Heaters (Residential)	
8 G.P.H. Recovery Rate (Glass-Lined)	1.00
10	1.00
12	1.25
30	3.00
45	3.00
60	3.50
75	4.00
100	5.00
120	6.00
Gas-Fired Water Heaters (Residential)	
8 G.P.H. Recovery Rate (Glass-Lined)	1.50
10	1.50
12	2.00
30	3.50
45	3.50
60	4.00
75	4.50
100	6.00

BUILDING EQUIPMENT (continued)			
ITEM DESCRIPTION			LABOR-HOURS EACH
120			8.00
Acid Neutralizing Tanks			
5 Gal. Capacity			2.50
15			3.50
30			4.00
55			6.00
100			8.00
150			9.00
200			10.00
250			12.00
350			14.00
500			18.00
Air Compressors w/Dryer			
1 Horsepower w/30 Gal. Receiver (Simplex)			18.00
1½			18.00
2	w/60		20.00
3			26.00
1		(Duplex)	20.00
1½			20.00
2			24.00
2	w/80		26.00
3	w/60		28.00
3	w/80		30.00
5	w/100		32.00
10	w/120		38.00
15	w/120		40.00
Vacuum Pumps (Medical Gas System)			
¾ Horsepower w/30 Gal. Receiver (Simplex)			16.00
1			16.00
1½			16.00
2			18.00
3	w/60		24.00
5	w/80		27.00
7½	w/80		32.00
1	w/60	(Duplex)	18.00
1½			18.00
2	w/80		24.00
3			27.00
5	w/100		30.00
10	w/120		36.00
15			38.00

NOTE:

1) Equipment labor-hours include distribution, uncrating, set into place, connect to existing piping and testing.

EXCLUSIONS

1) All piping, valves and elec. wiring.

BUILDING EQUIPMENT (continued)		
ITEM DESCRIPTION		LABOR-HOURS EACH
Fire Pumps (Fire Protection)		
20 Horsepower	500 G.P.M. (Simplex)	16.00
30		18.00
40	750	20.00
50	1000	24.00
Jockey Pump (Fire Protection)		
7½ Horsepower	500 G.P.M.	6.00
10		8.00

MEDICAL GAS SPECIALTIES		
ITEM DESCRIPTION		LABOR-HOURS EACH
5 Cylinder Manifold		7.00
10		9.00
12		10.00
Single Wall Outlet		1.00
Double		1.50
Triple		2.00
Single Ceiling Outlet		3.00
Double		4.00
Triple		5.00
Single Alarm with Pressure Gauge		2.50
Double		3.00
Single Audio Visual Legend		3.00
Double		3.50
Triple		4.50
Single Zone Valve & Box	½″	2.00
	¾″	2.00
	1″	2.25
	1¼″	2.50
	1½″	2.75
Double	½″	3.00
	¾″	3.00
	1″	3.25
	1¼″	3.50
	1½″	3.75

FIRE PROTECTION DEVICES	
ITEM DESCRIPTION	LABOR-HOURS EACH
Sprinkler Heads Pendant Type	0.35
Upright	0.35
Sprinkler Alarm Valves	
4″ Wet Type	8.00
4″ Dry	8.00
6″ Wet	10.00
6″ Dry	10.00
Fire Hose Cabinets (Recessed 125′ Hose)	3.00
(Semi-recessed 125′ Hose)	3.00
Rack (125′ Hose)	1.50
Siamese Connection 2½″ × 2½″ × 4″	3.00
2½″ × 2½″ × 6″	4.00
Roof Manifold 2½″ × 4″	2.50
Fire Dept. Valve 2½″	1.00
Fire Extinguishers	1.00
(in Cabinet)	2.00

MISCELLANEOUS ITEMS	
ITEM DESCRIPTION	LABOR-HOURS EACH
Sheet Lead Flashing 6 S.F. (4 lb.)	1.00
Pressure Gauges	0.24
Thermometers	0.24

NOTE:

1) Labor-hours include distribution, uncrating, set in place, connect to existing piping and test.

EXCLUSIONS

1) All piping, valves and elect. wiring.
2) Setting and connecting gas cylinders

VALVES AND CONTROL DEVICES

ITEM DESCRIPTION: BRONZE GATE, GLOBE, ANGLE, CHECK THREADED JOINT, 125, 150 OR 200 LB. RATING.

SIZE	LH EA.	SIZE	LH EA.	SIZE	LH EA.	SIZE	LH EA.	SIZE	LH EA.	SIZE	LH EA.
¼″	0.35	½″	0.43	1″	0.53	1½″	0.66	2½″	1.53	4″	2.55
⅜″	0.43	¾″	0.47	1¼″	0.63	2″	0.78	3″	1.88		

ITEM DESCRIPTION: BRONZE GATE, GLOBE, ANGLE, CHECK BRAZED OR SOLDER JOINT, 125, 150 OR 200 LB. RATING.

SIZE	LH EA.	SIZE	LH EA.	SIZE	LH EA.	SIZE	LH EA.	SIZE	LH EA.	SIZE	LH EA.
¼″	0.22	½″	0.30	1″	0.41	1½″	0.54	2½″	0.77	4″	1.60
⅜″	0.30	¾″	0.34	1¼″	0.47	2″	0.63	3″	0.98		

ITEM DESCRIPTION: BUTTERFLY VALVES—FLANGED, LEVER HANDLE LUG TYPE, 150 LB. RATING.

SIZE	LH EA.	SIZE	LH EA.	SIZE	LH EA.	SIZE	LH EA.	SIZE	LH EA.	SIZE	LH EA.
2″	0.77	3″	1.02	5″	1.53	8″	2.30	12″	3.44	16″	5.10
2½″	0.89	4″	1.28	6″	1.78	10″	2.80	14″	4.34	18″	5.86

ITEM DESCRIPTION: IRON BODY BRONZE MOUNTED THREADED GATE, GLOBE, CHECK, OS & Y VALVES, 125, 150 OR 250 LB. RATING.

SIZE	LH EA.	SIZE	LH EA.	SIZE	LH EA.	SIZE	LH EA.	SIZE	LH EA.	SIZE	LH EA.
1¼″	0.75	1½″	0.82	2″	0.95	2½″	1.75	3″	2.05	4″	2.80

ITEM DESCRIPTION: IRON BODY BRONZE MOUNTED FLANGED GATE, GLOBE, CHECK, OS & Y VALVES, 150 OR 250 LB. RATING.

SIZE	LH EA.	SIZE	LH EA.	SIZE	LH EA.	SIZE	LH EA.	SIZE	LH EA.	SIZE	LH EA.
1½″	0.9	2½″	1.02	4″	1.79	6″	2.30	10″	3.33	14″	5.14
2″	0.92	3″	1.28	5″	2.09	8″	2.72	12″	4.09	16″	6.05

ITEM DESCRIPTION: IRON BODY OR BRONZE GAS COCKS TEE, LEVER HANDLE OR SQUARE HEAD THREADED, 125 LB. RATING.

SIZE	LH EA.	SIZE	LH EA.	SIZE	LH EA.	SIZE	LH EA.	SIZE	LH EA.	SIZE	LH EA.
¼″	0.40	½″	0.45	1″	0.55	1½″	0.75	2½″	1.49	3″	2.13
⅜″	0.45	¾″	0.51	1¼″	0.68	2″	0.80				

LABOR-HOURS INCLUDE THE FOLLOWING:

DISTRIBUTION, SET IN PLACE, MAKE UP JOINT & TEST.

VALVES AND CONTROL DEVICES

ITEM DESCRIPTION: WRENCH-OPERATED LUBRICATED PLUG VALVES, FLANGED, 150 LB. RATING.

SIZE	LH EA.	SIZE	LH EA.	SIZE	LH EA.	SIZE	LH EA.	SIZE	LH EA.	SIZE	LH EA.
1¼″	0.87	1½″	0.95	2″	1.10	2½″	1.75	3″	2.13	4″	2.76

ITEM DESCRIPTION: BRONZE BALL VALVES, THREADED, 150 LB. RATING.

SIZE	LH EA.	SIZE	LH EA.	SIZE	LH EA.	SIZE	LH EA.	SIZE	LH EA.	SIZE	LH EA.
½″	0.43	¾″	0.47	1″	0.53	1¼″	0.63	1½″	0.66	2″	0.78

ITEM DESCRIPTION: BRONZE BALL VALVES, SOLDERED JOINT, 150 LB. RATING.

SIZE	LH EA.	SIZE	LH EA.	SIZE	LH EA.	SIZE	LH EA.	SIZE	LH EA.	SIZE	LH EA.
½″	0.30	¾″	0.34	1″	0.41	1¼″	0.47	1½″	0.54	2″	0.63

ITEM DESCRIPTION: ASME RATED BRONZE RELIEF VALVES, THREADED.

SIZE	LH EA.	SIZE	LH EA.	SIZE	LH EA.	SIZE	LH EA.	SIZE	LH EA.	SIZE	LH EA.
½″	0.30	¾″	0.30	1″	0.35	1¼″	0.50	1½″	0.55	2″	0.60

ITEM DESCRIPTION: PRESSURE REDUCING VALVES, FLANGED, 125 LB. RATING—200 PSI.

SIZE	LH EA.	SIZE	LH EA.	SIZE	LH EA.	SIZE	LH EA.	SIZE	LH EA.	SIZE	LH EA.
2″	1.70	2½″	1.91	3″	2.10	4″	2.55	6″	3.61	8″	4.50

ITEM DESCRIPTION: BRONZE YELLOW DRAIN VALVES, THREADED.

SIZE	LH EA.	SIZE	LH EA.	SIZE	LH EA.	SIZE	LH EA.	SIZE	LH EA.	SIZE	LH EA.
½″	0.50	¾″	0.50	1″	0.60	1¼″	0.75	1½″	0.82	2″	0.90

LABOR-HOURS INCLUDE THE FOLLOWING:

DISTRIBUTION, SET IN PLACE, MAKE UP JOINT & TEST.

VALVES AND CONTROL DEVICES

ITEM DESCRIPTION: BRONZE BODY "Y" STRAINERS THREADED, 125 LB. RATING.

SIZE	LH EA.	SIZE	LH EA.	SIZE	LH EA.	SIZE	LH EA.	SIZE	LH EA.	SIZE	LH EA.
½"	0.47	1"	0.53	½"	0.66	2½"	1.53	3"	1.88	4"	2.55
¾"	0.47	1¼"	0.63	2"	0.78						

ITEM DESCRIPTION: IRON BODY "Y" STRAINERS FLANGED, 125 LB. RATING.

SIZE	LH EA.	SIZE	LH EA.	SIZE	LH EA.	SIZE	LH EA.	SIZE	LH EA.	SIZE	LH EA.
4"	1.79	5"	2.09	6"	2.30	8"	2.72	10"	3.33	12"	4.09

ITEM DESCRIPTION: VACUUM BREAKERS.

SIZE	LH EA.	SIZE	LH EA.	SIZE	LH EA.	SIZE	LH EA.	SIZE	LH EA.	SIZE	LH EA.
½"	0.47	¾"	0.47	1"	0.53	1¼"	0.63	1½"	0.66	2"	0.78

ITEM DESCRIPTION: EXPANSION JOINTS COPPER BELLOWS TYPE.

SIZE	LH EA.	SIZE	LH EA.	SIZE	LH EA.	SIZE	LH EA.	SIZE	LH EA.	SIZE	LH EA.
½"	0.47	1"	0.53	1½"	0.66	2½"	1.53	4"	2.55		
¾"	0.47	1¼"	0.63	2"	0.78	3"	1.88				

ITEM DESCRIPTION: SHOCK ABSORBERS (PDI)*.*PLUMBING & DRAINAGE INSTITUTE.

SIZE PDI	LH EA.	SIZE PDI	EA.	SIZE PDI	LH EA.	SIZE PDI	LH EA.	SIZE PDI	LH EA.	SIZE PDI	LH EA.
A	0.47	B	0.63	C	0.78	D	0.88	E	1.40	F	1.75

ITEM DESCRIPTION: PLASTIC VALVES, SOCKET AND THREADED, 125 LB. RATED.

SIZE	LH EA.	SIZE	LH EA.	SIZE	LH EA.	SIZE	LH EA.	SIZE	LH EA.	SIZE	LH EA.
½"	0.15	¾"	0.18	1"	0.20	1¼"	0.24	1½"	0.31	2"	0.44

LABOR-HOURS INCLUDE THE FOLLOWING:

DISTRIBUTION, SET IN PLACE, MAKE UP JOINT & TEST.

PIPE INSULATION, SUPPORTS AND SLEEVES

ITEM DESCRIPTION: ½" FIBERGLASS PIPE INSULATION.

SIZE	LH/LF	SIZE	LH/LF	SIZE	LH/LF	SIZE	LH/LF	SIZE	LH/LF	SIZE	LH/LF
½"	.045	1"	.050	1½"	.055	2½"	.067	4"	.073	8"	.080
¾"	.045	1¼"	.050	2"	.065	3"	.070	6"	.075	10"	.094

ITEM DESCRIPTION: HANGER ASSEMBLIES (CLEVIS HANGER, 2'-0" ROD, INSERT W/NUTS & BOLTS).

SIZE	LH EA.	SIZE	LH EA.	SIZE	LH EA.	SIZE	LH EA.	SIZE	LH EA.	SIZE	LH EA.
½"	0.36	1"	0.36	1½"	0.36	2½"	0.36	4"	0.36	8"	0.46
¾"	0.36	1¼"	0.36	2"	.036	3"	0.36	6"	0.41	10"	0.46

ITEM DESCRIPTION: STEEL PIPE SLEEVES.

SIZE	LH EA.	SIZE	LH EA.	SIZE	LH EA.	SIZE	LH EA.	SIZE	LH EA.	SIZE	LH EA.
½"	0.16	1"	0.22	1½"	0.28	2½"	0.40	4"	0.50	8"	0.63
¾"	0.20	1¼"	0.25	2"	0.32	3"	0.45	6"	0.55	10"	0.70

ITEM DESCRIPTION: TOILET AND BATH ACCESSORIES—LABOR-HOURS EACH.

GRAB BAR	TOWEL BAR	TOILET PAPER HOLDER	SOAP DISPENSER SURFACE MTD.	MIRROR AND SHELF	MEDICINE CABINET
2.0	1.0	SURFACE 0.50 RECESSED 0.75	1.0	16" × 20" 1.0 24" × 60" 1.25	SURFACE 1.25 RECESSED 1.50

FACIAL TISSUE HOLDER	SANITARY NAPKIN DISPENSER	TOWEL DISPENSER	PAPER CUP DISPENSER	ELECTRIC HAND DRYER	SOAP DISH
SURFACE .50 RECESSED .75	SURFACE 1.0 RECESSED 1.25	SURFACE .75 RECESSED 1.0	SURFACE 1.0 RECESSED 1.25	1.0	0.50

SHOWER ROD & FLANGES	ASH RECEPTACLE	ASH URN & WASTE RECEPTACLE	TOWEL & ROBE HOOK	JANITORIAL UTILITY SHELF 36" LONG	UTILITY SHELF 5½" × 24"
1.0	0.50	1.0	0.50	1.50	1.0

LABOR-HOURS INCLUDE THE FOLLOWING:

DISTRIBUTION, SET IN PLACE AND INSTALL.

SITE DRAINAGE AND UTILITIES

ITEM DESCRIPTION: EXTRA STRENGTH VITRIFIED CLAY PIPE AND FITTINGS, RING JOINT (ADD FOR EXCAVATION).

SIZE	LH/LF	SIZE	LH/LF	SIZE	LH/LF	SIZE	LH/LF	SIZE	LH/LF	SIZE	LH/LF
4″	0.10	8″	0.12	12″	0.17	18″	0.26	24″	0.43	36″	0.53
6″	0.11	10″	0.14	15″	0.19	21″	0.34	30″	0.62		

ITEM DESCRIPTION: REINFORCED CONCRETE PIPE, CLASS III, RING JOINT (ADD FOR EXCAVATION).

SIZE	LH/LF	SIZE	LH/LF	SIZE	LH/LF	SIZE	LH/LF	SIZE	LH/LF	SIZE	LH/LF
12″	0.10	18″	0.14	24″	0.18	30″	0.32	36″	0.39		
15″	0.12	21″	0.16	27″	0.22	33″	0.35				

ITEM DESCRIPTION: DUCTILE IRON PRESSURE PIPE & FITTINGS, CLASS 150 CEMENT LINED MECHANICAL JOINT.

SIZE	LH/LF	SIZE	LH/LF	SIZE	LH/LF	SIZE	LH/LF	SIZE	LH/LF	SIZE	LH/LF
4″	0.13	6″	0.17	8″	0.20	10″	0.26	12″	0.31	14″	0.37
										16″	0.51

ITEM DESCRIPTION: DUCTILE IRON PRESSURE PIPE & FITTINGS, CLASS 150 CEMENT LINED NEOPRENE SLIP-ON JOINT.

SIZE	LH/LF	SIZE	LH/LF	SIZE	LH/LF	SIZE	LH/LF	SIZE	LH/LF	SIZE	LH/LF
4″	0.11	6″	0.13	8″	0.16	10″	0.21	12″	0.25	14″	0.32
										16″	0.48

ITEM DESCRIPTION: DUCTILE IRON GATE VALVES WITH BOX MECHANICAL & NEOPRENE SLIP-ON JOINTS.

SIZE	LH/LF	SIZE	LH/LF	SIZE	LH/LF	SIZE	LH/LF	SIZE	LH/LF	ITEM	LH EA.
4″	3.0	6″	3.5	8″	4.0	10″	5.0	12″	7.5	FIRE HYD.	5.0
										WET CONN.	3.0
										WATER TAP	2.0

ITEM DESCRIPTION: SCHEDULE 40 STEEL, THREADED MILLWRAP PIPE & FITTINGS.

SIZE	LH/LF	SIZE	LH/LF	SIZE	LH/LF	SIZE	LH/LF	SIZE	LH/LF	SIZE	LH/LF
¾″	0.04	1¼″	0.05	2″	0.08	3″	0.12	5″	0.22	8″	0.38
1″	0.04	1½″	0.06	2½″	0.10	4″	0.18	6″	0.27		

LABOR-HOURS INCLUDE THE FOLLOWING:

DISTRIBUTION, SET IN PLACE AND INSTALL.

1) UP TO 4′ O″ TRENCH DEPTH
2) REQUIRED EQUIPMENT FOR UNLOADING AND PLACING IN TRENCH
   a) VITRIFIED CLAY PIPE 18″ & LARGER
   b) REINFORCED CONCRETE & DUCTILE IRON PIPE—ALL SIZES
3) TESTING
4) CREWS OF THREE WORKERS ARE ASSUMED

EXCLUSIONS

1) DEWATERING
2) EXCAVATION & BACKFILL

LABOR-HOURS TO INSTALL BUILDING PIPING SYSTEMS

ITEM DESCRIPTION:
EXTRA HEAVY CAST IRON PIPE AND FITTINGS
LEAD AND OAKUM JOINT

BELOW GROUND

SIZE	PIPE PER L.F.	ONE JOINT FITTING	TWO JOINT FITTING	THREE JOINT FITTING
2″	.08	.36	.70	1.05
3″	.11	.53	1.04	1.57
4″	.14	.70	1.38	2.07
5″	.18	.87	1.72	2.58
6″	.21	1.06	2.11	3.17
8″	.32	1.57	3.13	4.70
10″	.41	1.95	3.89	5.85
12″	.56	2.74	5.46	8.19
15″	.70	3.42	6.82	10.23

ABOVE GROUND

SIZE	PIPE PER L.F.	ONE JOINT FITTING	TWO JOINT FITTING	THREE JOINT FITTING
2″	.12	.56	1.10	1.67
3″	.17	.83	1.65	2.48
4″	.23	1.10	2.19	3.30
5″	.28	1.38	2.74	4.11
6″	.33	1.65	3.28	4.93
8″	.49	2.44	4.86	7.30
10″	.61	3.04	6.07	9.11
12″	.82	4.11	8.21	12.32
15″	1.03	5.13	10.25	15.38

LABOR-HOURS INCLUDE THE FOLLOWING:

DISTRIBUTION FROM STOCKPILE 100′ DISTANCE
MEASURE AND CUT PIPE
INSTALL PIPE AND FITTING IN PLACE
CAULK JOINT WITH OAKUM AND LEAD
NORMAL LOSS TIME
TESTING
UP TO 3′-0″ TRENCH DEPTH
UP TO 10′-0″ CEILING HEIGHT

EXCLUSIONS

EXCAVATION AND BACKFILL
DEWATERING
HANGERS AND SUPPORTS (BELOW AND ABOVE GROUND)
UNUSUAL JOB CONDITIONS

LABOR-HOURS TO INSTALL BUILDING PIPING SYSTEMS (continued)

ITEM DESCRIPTION:
EXTRA HEAVY CAST IRON PIPE AND FITTINGS
NEOPRENE GASKET JOINT

BELOW GROUND

SIZE	PIPE PER L.F.	ONE JOINT FITTING	TWO JOINT FITTING	THREE JOINT FITTING
2″	.06	.29	.56	.83
3″	.10	.42	.83	1.24
4″	.13	.56	1.10	1.65
5″	.16	.70	1.38	2.06
6″	.19	.83	1.65	2.46
8″	.26	1.21	2.40	3.59
10″	.33	1.53	3.04	4.56
12″	.43	2.07	4.13	6.19
15″	.53	2.99	5.05	7.56

ABOVE GROUND

SIZE	PIPE PER L.F.	ONE JOINT FITTING	TWO JOINT FITTING	THREE JOINT FITTING
2″	.09	.43	.85	1.27
3″	.14	.64	1.26	1.88
4″	.18	.85	1.68	2.52
5″	.22	1.04	2.06	3.08
6″	.26	1.29	2.57	3.84
8″	.36	1.80	3.59	5.37
10″	.46	2.31	4.61	6.90
12″	.65	3.25	6.48	9.71
15″	.81	4.01	8.01	12.00

LABOR-HOURS INCLUDE THE FOLLOWING:

DISTRIBUTION FROM STOCKPILE 100′ DISTANCE
MEASURE AND CUT PIPE
INSTALL PIPE AND FITTING IN PLACE
MAKE UP GASKET JOINT
NORMAL LOSS TIME
TESTING
UP TO 3′-0″ TRENCH DEPTH
UP TO 10′-0″ CEILING HEIGHT

EXCLUSIONS

EXCAVATION AND BACKFILL
DEWATERING
HANGERS AND SUPPORTS (BELOW AND ABOVE GROUND)
UNUSUAL JOB CONDITIONS

LABOR-HOURS TO INSTALL BUILDING PIPING SYSTEMS (continued)

ITEM DESCRIPTION:
SERVICE WEIGHT CAST IRON PIPE AND FITTINGS
LEAD AND OAKUM JOINT

BELOW GROUND					ABOVE GROUND				
SIZE	PIPE PER L.F.	ONE JOINT FITTING	TWO JOINT FITTING	THREE JOINT FITTING	SIZE	PIPE PER L.F.	ONE JOINT FITTING	TWO JOINT FITTING	THREE JOINT FITTING
2″	.06	.32	.66	1.00	2″	.10	.53	1.07	1.61
3″	.10	.49	1.00	1.51	3″	.15	.80	1.61	2.43
4″	.13	.66	1.34	2.02	4″	.21	1.07	2.16	3.25
5″	.16	.83	1.68	2.53	5″	.26	1.34	2.70	4.06
6″	.19	1.03	2.07	3.12	6″	.31	1.62	3.25	4.88
8″	.30	1.54	3.09	4.65	8″	.48	2.41	4.83	7.25
10″	.37	1.92	3.86	5.80	10″	.59	3.01	6.03	9.06
12″	.53	2.70	5.42	8.14	12″	.81	4.08	8.18	12.27
15″	.67	3.38	6.78	10.18	15″	1.01	5.10	10.22	15.33

LABOR-HOURS INCLUDE THE FOLLOWING:

DISTRIBUTION FROM STOCKPILE 100′ DISTANCE
MEASURE AND CUT PIPE
INSTALL PIPE AND FITTING IN PLACE
CAULK JOINT WITH OAKUM AND LEAD
NORMAL LOSS TIME
TESTING
UP TO 3′-0″ TRENCH DEPTH
UP TO 10′-0″ CEILING HEIGHT

EXCLUSIONS

EXCAVATION AND BACKFILL
DEWATERING
HANGERS AND SUPPORTS (BELOW AND ABOVE GROUND)
UNUSUAL JOB CONDITIONS

LABOR-HOURS TO INSTALL BUILDING PIPING SYSTEMS (continued)

ITEM DESCRIPTION:
SERVICE WEIGHT CAST IRON PIPE AND FITTINGS
NEOPRENE GASKET JOINT

BELOW GROUND					ABOVE GROUND				
SIZE	PIPE PER L.F.	ONE JOINT FITTING	TWO JOINT FITTING	THREE JOINT FITTING	SIZE	PIPE PER L.F.	ONE JOINT FITTING	TWO JOINT FITTING	THREE JOINT FITTING
2″	.04	.26	.53	.80	2″	.08	.40	.82	1.23
3″	.09	.39	.80	1.21	3″	.13	.60	1.22	1.84
4″	.12	.53	1.07	1.61	4″	.16	.82	1.65	2.48
5″	.14	.66	1.34	2.02	5″	.20	1.00	2.02	3.04
6″	.17	.80	1.61	2.43	6″	.25	1.26	2.53	3.81
8″	.25	1.17	2.36	3.55	8″	.35	1.77	3.55	5.34
10″	.31	1.50	3.01	4.52	10″	.44	2.28	4.57	6.87
12″	.42	2.04	4.10	6.15	12″	.64	3.21	6.44	9.67
15″	.51	2.50	5.01	7.53	15″	.79	3.98	7.97	11.97

LABOR-HOURS INCLUDE THE FOLLOWING:

DISTRIBUTION FROM STOCKPILE 100′ DISTANCE
MEASURE AND CUT PIPE
INSTALL PIPE AND FITTING IN PLACE
MAKE UP GASKET JOINT
NORMAL LOSS TIME
TESTING
UP TO 3′-0″ TRENCH DEPTH
UP TO 10′-0″ CEILING HEIGHT

EXCLUSIONS

EXCAVATION AND BACKFILL
DEWATERING
HANGERS AND SUPPORTS (BELOW AND ABOVE GROUND)
UNUSUAL JOB CONDITIONS

LABOR-HOURS TO INSTALL BUILDING PIPING SYSTEMS (continued)

ITEM DESCRIPTION:
SERVICE WEIGHT CAST IRON PIPE AND FITTINGS
HUBLESS CLAMP JOINT

BELOW GROUND

SIZE	PIPE PER L.F.	TWO JOINT FITTING	THREE JOINT FITTING	FOUR JOINT FITTING
1½″	.03	.32	.49	.66
2″	.03	.39	.60	.80
3″	.06	.46	.70	.94
4″	.08	.53	.80	1.07
5″	.10	.60	.90	1.21
6″	.13	.66	1.00	1.34
8″	.21	.87	1.31	1.75
10″	.26	1.07	1.61	2.16

ABOVE GROUND

SIZE	PIPE PER L.F.	TWO JOINT FITTING	THREE JOINT FITTING	FOUR JOINT FITTING
1½″	.04	.41	.62	.83
2″	.07	.49	.75	1.00
3″	.10	.58	.88	1.17
4″	.14	.66	1.00	1.34
5″	.17	.75	1.13	1.51
6″	.20	.83	1.26	1.68
8″	.33	1.09	1.64	2.19
10″	.40	1.34	2.02	2.70

LABOR-HOURS INCLUDE THE FOLLOWING:

DISTRIBUTION FROM STOCKPILE 100′ DISTANCE
MEASURE AND CUT PIPE
INSTALL PIPE AND FITTING IN PLACE
MAKE UP CLAMP JOINT
NORMAL LOSS TIME
TESTING
UP TO 3′-0″ TRENCH DEPTH
UP TO 10′-0″ CEILING HEIGHT

EXCLUSIONS

EXCAVATION AND BACKFILL
DEWATERING
HANGERS AND SUPPORTS (BELOW AND ABOVE GROUND)
UNUSUAL JOB CONDITIONS

LABOR-HOURS TO INSTALL BUILDING
PIPING SYSTEMS (continued)

ITEM DESCRIPTION:
IRON ALLOY (SILICON) PIPE AND FITTINGS
LEAD AND OAKUM JOINT

BELOW GROUND						ABOVE GROUND				
SIZE	PIPE PER L.F.	ONE JOINT FITTING	TWO JOINT FITTING	THREE JOINT FITTING		SIZE	PIPE PER L.F.	ONE JOINT FITTING	TWO JOINT FITTING	THREE JOINT FITTING
2″	.09	.41	.80	1.20		2″	.13	.62	1.22	1.84
3″	.13	.60	1.17	1.76		3″	.19	.93	1.84	2.75
4″	.16	.79	1.56	2.35		4″	.25	1.21	2.40	3.60
6″	.24	1.17	2.33	3.49		6″	.37	1.81	3.60	5.41
8″	.37	1.81	3.60	5.41		8″	.55	2.74	5.46	8.19

LABOR-HOURS INCLUDE THE FOLLOWING:

DISTRIBUTION FROM STOCKPILE 100′ DISTANCE
MEASURE AND CUT PIPE
INSTALL PIPE AND FITTING IN PLACE
CAULK JOINT WITH OAKUM AND LEAD
NORMAL LOSS TIME
TESTING
UP TO 3′-0″ TRENCH DEPTH
UP TO 10′-0″ CEILING HEIGHT

EXCLUSIONS

EXCAVATION AND BACKFILL
DEWATERING
HANGERS AND SUPPORTS (BELOW AND ABOVE GROUND)
UNUSUAL JOB CONDITIONS

LABOR-HOURS TO INSTALL BUILDING PIPING SYSTEMS (continued)

ITEM DESCRIPTION: IRON ALLOY (SILICON) PIPE AND FITTINGS LEAD AND OAKUM JOINT

BELOW GROUND						ABOVE GROUND				
SIZE	PIPE PER L.F.	TWO JOINT FITTING	THREE JOINT FITTING	FOUR JOINT FITTING		SIZE	PIPE PER L.F.	TWO JOINT FITTING	THREE JOINT FITTING	FOUR JOINT FITTING
2″	.06	.49	.73	.97		2″	.09	.61	.91	1.21
3″	.09	.56	.83	1.11		3″	.14	.70	1.04	1.38
4″	.12	.65	.96	1.28		4″	.18	.80	1.19	1.58
6″	.17	.80	1.19	1.58		6″	.26	1.00	1.50	2.00
8″	.26	1.04	1.55	2.06		8″	.40	1.29	1.93	2.57

LABOR-HOURS INCLUDE THE FOLLOWING:

DISTRIBUTION FROM STOCKPILE 100′ DISTANCE
MEASURE AND CUT PIPE
INSTALL PIPE AND FITTING IN PLACE
MAKE UP MECHANICAL COUPLING JOINT
NORMAL LOSS TIME
TESTING
UP TO 3′-0″ TRENCH DEPTH
UP TO 10′-0″ CEILING HEIGHT

EXCLUSIONS

EXCAVATION AND BACKFILL
DEWATERING
HANGERS AND SUPPORTS (BELOW AND ABOVE GROUND)
UNUSUAL JOB CONDITIONS

LABOR-HOURS TO INSTALL BUILDING PIPING SYSTEMS (continued)

ITEM DESCRIPTION:
DRAIN-WASTE-VENT COPPER TUBE "DWV"
CAST OR WROUGHT FITTINGS—SOLDER JOINT

SIZE	PIPE PER L.F.	TWO JOINT FITTING	THREE JOINT FITTING	FOUR JOINT FITTING
1¼″	.05	.32	.49	.65
1½″	.06	.36	.54	.71
2″	.06	.41	.62	.82
2½″	.08	.49	.75	1.00
3″	.10	.58	.88	1.16
4″	.12	.83	1.26	1.67
5″	.16	1.26	1.94	2.53
6″	.21	1.68	2.53	3.38

LABOR-HOURS INCLUDE THE FOLLOWING:

DISTRIBUTION FROM STOCKPILE 100′ DISTANCE
MEASURE, CUT AND PREPARE TUBING AND FITTING
INSTALL TUBING AND FITTING IN PLACE
SOLDER JOINT WITH TURBO-TORCH
NORMAL LOSS TIME
TESTING
UP TO 10′-0″ CEILING HEIGHT

EXCLUSIONS

HANGERS AND SUPPORTS
UNUSUAL JOB CONDITIONS

LABOR-HOURS TO INSTALL BUILDING PIPING SYSTEMS (continued)

ITEM DESCRIPTION:
POLYVINYL CHLORIDE (PVC) PLASTIC PIPE—SCHED. 40
DRAIN-WASTE-VENT (DWV) SOLVENT, SOCKET JOINT FITTINGS

BELOW GROUND

SIZE	PIPE PER L.F.	TWO JOINT FITTING	THREE JOINT FITTING	FOUR JOINT FITTING
1¼″	.02	.14	.20	.28
1½″	.03	.18	.26	.36
2″	.04	.26	.38	.52
3″	.05	.38	.58	.77
4″	.06	.46	.70	.93
6″	.11	.58	.88	1.17
8″	.14	.72	1.10	1.47
10″	.19	.91	1.38	1.84
12″	.24	1.14	1.73	2.31

ABOVE GROUND

SIZE	PIPE PER L.F.	TWO JOINT FITTING	THREE JOINT FITTING	FOUR JOINT FITTING
1¼″	.03	.18	.27	.37
1½″	.04	.23	.35	.47
2″	.05	.33	.50	.67
3″	.07	.50	.76	1.01
4″	.08	.60	.91	1.22
6″	.14	.76	1.14	1.52
8″	.19	.95	1.43	1.90
10″	.24	1.19	1.78	2.38
12″	.30	1.49	2.24	2.97

LABOR-HOURS INCLUDE THE FOLLOWING:

DISTRIBUTION FROM STOCKPILE 100′ DISTANCE
MEASURE, CUT AND PREPARE PIPE
INSTALL PIPE AND FITTING IN PLACE
MAKE UP SOLVENT JOINT
NORMAL LOSS TIME
TESTING
UP TO 3′-0″ TRENCH DEPTH
UP TO 10′-0″ CEILING HEIGHT

EXCLUSIONS

EXCAVATION AND BACKFILL
DEWATERING
HANGERS AND SUPPORTS (BELOW AND ABOVE GROUND)
UNUSUAL JOB CONDITIONS

LABOR-HOURS TO INSTALL BUILDING PIPING SYSTEMS (continued)

ITEM DESCRIPTION:
POLYPROPYLENE FUSION JOINT
DRAIN-WASTE-VENT (DWV) SOCKET FUSION TYPE

BELOW GROUND					ABOVE GROUND				
SIZE	PIPE PER L.F.	TWO JOINT FITTING	THREE JOINT FITTING	FOUR JOINT FITTING	SIZE	PIPE PER L.F.	TWO JOINT FITTING	THREE JOINT FITTING	FOUR JOINT FITTING
1½″	.06	.30	.44	.59	1½″	.08	.33	.49	.65
2″	.07	.37	.54	.72	2″	.09	.40	.60	.79
3″	.09	.50	.75	.99	3″	.11	.55	.82	1.10
4″	.10	.65	.98	1.30	4″	.12	.72	1.08	1.44
6″	.11	.84	1.26	1.67	6″	.14	.93	1.39	1.84

LABOR-HOURS INCLUDE THE FOLLOWING:

DISTRIBUTION FROM STOCKPILE 100′ DISTANCE
MEASURE, CUT AND PREPARE
INSTALL PIPE AND FITTING IN PLACE
FUSE JOINT
NORMAL LOSS TIME
TESTING
UP TO 3′-0″ TRENCH DEPTH
UP TO 10′-0″ CEILING HEIGHT

EXCLUSIONS

EXCAVATION AND BACKFILL
DEWATERING
HANGERS AND SUPPORTS (BELOW AND ABOVE GROUND)
UNUSUAL JOB CONDITIONS

LABOR-HOURS TO INSTALL BUILDING PIPING SYSTEMS (continued)

ITEM DESCRIPTION:
STEEL PIPE—THREADED JOINTS
MALLEABLE OR CAST IRON FITTINGS

SCHEDULE 40—STD.

SIZE	PIPE PER L.F.	TWO JOINT FITTING	THREE JOINT FITTING	FOUR JOINT FITTING
½″	.03	.38	.58	.77
¾″	.03	.42	.63	.84
1″	.04	.47	.71	.94
1¼″	.05	.54	.81	1.08
1½″	.06	.57	.86	1.15
2″	.07	.67	1.01	1.35
2½″	.10	1.35	2.03	2.71
3″	.11	1.62	2.44	3.26
4″	.18	2.14	3.20	4.27
5″	.24	2.57	3.84	5.12
6″	.31	3.13	4.68	6.24
8″	.38	3.69	5.52	7.36

SCHEDULE 80—EX. HVY

SIZE	PIPE PER L.F.	TWO JOINT FITTING	THREE JOINT FITTING	FOUR JOINT FITTING
½″	.05	.40	.60	.79
¾″	.05	.43	.65	.86
1″	.07	.50	.75	.99
1¼″	.08	.59	.88	1.16
1½″	.09	.62	.93	1.23
2″	.10	.72	1.08	1.44
2½″	.14	1.50	2.25	3.00
3″	.15	1.81	2.71	3.61
4″	.20	2.36	3.54	4.71
5″	.26	2.82	4.22	5.63
6″	.36	3.50	5.24	6.99
8″	.44	4.13	6.18	8.24

LABOR-HOURS INCLUDE THE FOLLOWING:

DISTRIBUTION FROM STOCKPILE 100′ DISTANCE
MEASURE, CUT AND THREAD PIPE
INSTALL PIPE AND FITTING IN PLACE
MAKE UP THREADED JOINT
NORMAL LOSS TIME
TESTING
UP TO 10′-0″ CEILING HEIGHT

EXCLUSIONS

HANGERS AND SUPPORTS
UNUSUAL JOB CONDITIONS

LABOR-HOURS TO INSTALL BUILDING PIPING SYSTEMS (continued)

ITEM DESCRIPTION:
STEEL PIPE—THREADED
CAST IRON FLANGED FITTINGS

SCHEDULE 40-STD.						SCHEDULE 80-EX. HVY				
SIZE	PIPE PER L.F.	TWO JOINT FITTING	THREE JOINT FITTING	FOUR JOINT FITTING		SIZE	PIPE PER L.F.	TWO JOINT FITTING	THREE JOINT FITTING	FOUR JOINT FITTING
2″	.08	.77	1.16	1.71		2″	.09	.90	1.28	1.92
2½″	.12	.94	1.41	2.08		2½″	.14	1.10	1.56	2.35
3″	.13	1.15	1.72	2.53		3″	.15	1.39	1.90	2.86
4″	.18	1.51	2.26	3.31		4″	.20	1.74	2.49	3.84
5″	.24	1.92	2.87	4.20		5″	.26	2.21	3.15	4.78
6″	.31	2.28	3.41	4.98		6″	.36	2.62	3.74	5.67
8″	.38	3.04	4.56	6.68		8″	.44	3.51	5.01	7.58

LABOR-HOURS INCLUDE THE FOLLOWING:

DISTRIBUTION FROM STOCKPILE 100′ DISTANCE
MEASURE, CUT AND THREAD PIPE
INSTALL PIPE AND FITTING IN PLACE
MAKE UP THREAD FLANGED JOINT AND BOLT-UP
NORMAL LOSS TIME
TESTING
UP TO 10′-0″ CEILING HEIGHT

EXCLUSIONS

HANGERS AND SUPPORTS
UNUSUAL JOB CONDITIONS

LABOR-HOURS TO INSTALL BUILDING PIPING SYSTEMS (continued)

ITEM DESCRIPTION:
COPPER TUBING—95/5 SOLDER JOINT
CAST OR WROUGHT FITTINGS

TYPE "K"						TYPE "L"				
SIZE	PIPE PER L.F.	TWO JOINT FITTING	THREE JOINT FITTING	FOUR JOINT FITTING		SIZE	PIPE PER L.F.	TWO JOINT FITTING	THREE JOINT FITTING	FOUR JOINT FITTING
½″	.03	.29	.43	.57		½″	.03	.29	.43	.57
¾″	.04	.32	.48	.65		¾″	.03	.32	.48	.65
1″	.04	.39	.59	.78		1″	.04	.39	.59	.78
1¼″	.05	.43	.65	.86		1¼″	.05	.43	.65	.86
1½″	.06	.50	.75	1.00		1½″	.06	.50	.75	1.00
2″	.06	.57	.86	1.14		2″	.06	.57	.86	1.14
2½″	.09	.70	1.05	1.39		2½″	.09	.70	1.05	1.39
3″	.12	.89	1.34	1.71		3″	.11	.89	1.34	1.71
4″	.14	1.34	2.01	2.68		4″	.14	1.34	2.01	2.68
5″	.18	1.96	2.95	3.93		5″	.17	1.96	2.95	3.93
6″	.24	2.61	3.91	5.21		6″	.23	2.61	3.91	5.21
8″	.36	3.39	5.09	6.78		8″	.34	3.39	5.09	6.78

LABOR-HOURS INCLUDE THE FOLLOWING:

DISTRIBUTION FROM STOCKPILE 100′ DISTANCE
MEASURE, CUT AND PREPARE TUBING AND FITTING
INSTALL TUBING AND FITTING IN PLACE
SOLDER JOINT WITH TURBO TORCH
NORMAL LOSS TIME
TESTING
UP TO 10′-0″ CEILING HEIGHT
UP TO 2′-0″ TRENCH DEPTH (TYPE "K" ONLY)

EXCLUSIONS

HANGERS AND SUPPORTS
UNUSUAL JOB CONDITIONS

LABOR-HOURS TO INSTALL BUILDING
PIPING SYSTEMS (continued)

ITEM DESCRIPTION:
COPPER TUBING—95/5 SOLDER JOINT
CAST OR WROUGHT FITTINGS
TYPE "M"

SIZE	PIPE PER L.F.	TWO JOINT FITTING	THREE JOINT FITTING	FOUR JOINT FITTING
½"	.03	.29	.43	.57
¾"	.03	.32	.48	.65
1"	.04	.39	.59	.78
1¼"	.05	.43	.65	.86
1½"	.06	.50	.75	1.00
2"	.06	.57	.86	1.14
2½"	.08	.70	1.05	1.39
3"	.10	.89	1.34	1.71
4"	.13	1.34	2.01	2.68
5"	.16	1.96	2.95	3.93
6"	.22	2.61	3.91	5.21
8"	.32	3.39	5.09	6.78

LABOR-HOURS INCLUDE THE FOLLOWING:

DISTRIBUTION FROM STOCKPILE 100′ DISTANCE
MEASURE, CUT AND PREPARE TUBING AND FITTING
INSTALL TUBING AND FITTING IN PLACE
SOLDER JOINT WITH TURBO TORCH
NORMAL LOSS TIME
TESTING
UP TO 10′-0″ CEILING HEIGHT
UP TO 2′-0″ TRENCH DEPTH (TYPE "K" ONLY)

EXCLUSIONS

HANGERS AND SUPPORTS
UNUSUAL JOB CONDITIONS

LABOR-HOURS TO INSTALL BUILDING PIPING SYSTEMS (continued)

ITEM DESCRIPTION:
BRASS PIPE
CAST OR MALLEABLE THREADED FITTINGS

STANDARD—CL. 125 LB.

SIZE	PIPE PER L.F.	TWO JOINT FITTING	THREE JOINT FITTING	FOUR JOINT FITTING
½″	.08	.49	.73	.99
¾″	.08	.53	.78	1.05
1″	.09	.60	.88	1.19
1¼″	.10	.68	1.01	1.36
1½″	.11	.76	1.14	1.53
2″	.12	1.14	1.70	2.28
2½″	.18	1.60	2.39	3.20
3″	.24	2.48	3.71	4.96
4″	.28	2.96	4.43	1.48
5″	.35	3.42	5.12	6.83
6″	.41	3.76	5.63	7.51
8″	.54	4.95	7.41	9.89

EXTRA HEAVY—CL. 250 LB.

SIZE	PIPE PER L.F.	TWO JOINT FITTING	THREE JOINT FITTING	FOUR JOINT FITTING
½″	.08	.49	.73	.99
¾″	.08	.53	.78	1.05
1″	.09	.61	.91	1.22
1¼″	.11	.70	1.04	1.39
1½″	.12	.82	1.22	1.63
2″	.13	1.22	1.83	2.45
2½″	.21	1.75	2.62	3.51
3″	.28	2.74	4.10	5.47
4″	.31	3.30	4.94	6.60
5″	.41	3.83	5.73	7.65
6″	.47	4.32	6.47	8.64
8″	.64	5.70	8.53	11.39

LABOR-HOURS INCLUDE THE FOLLOWING:

DISTRIBUTION FROM STOCKPILE 100′ DISTANCE
MEASURE, CUT AND THREAD PIPE
INSTALL PIPE AND FITTING IN PLACE
MAKE UP THREADED JOINT
NORMAL LOSS TIME
TESTING
UP TO 10′-0″ CEILING HEIGHT

EXCLUSIONS

HANGERS AND SUPPORTS
UNUSUAL JOB CONDITIONS

LABOR-HOURS TO INSTALL BUILDING PIPING SYSTEMS (continued)

ITEM DESCRIPTION:
POLYVINYL CHLORIDE (PVC) PLASTIC PIPE
PRESSURE FITTINGS—SOLVENT SOCKET JOINTS

SCHEDULE 40

SIZE	PIPE PER L.F.	TWO JOINT FITTING	THREE JOINT FITTING	FOUR JOINT FITTING
½″	.02	.13	.20	.26
¾″	.02	.13	.20	.26
1″	.03	.14	.22	.28
1¼″	.03	.18	.27	.36
1½″	.04	.23	.35	.46
2″	.04	.33	.50	.66
2½″	.05	.42	.63	.84
3″	.07	.50	.76	1.00
4″	.08	.60	.91	1.20
5″	.11	.77	1.16	1.54
6″	.15	1.10	1.65	2.20

SCHEDULE 80

SIZE	PIPE PER L.F.	TWO JOINT FITTING	THREE JOINT FITTING	FOUR JOINT FITTING
½″	.02	.14	.21	.28
¾″	.02	.14	.21	.28
1″	.03	.15	.23	.30
1¼″	.04	.19	.28	.38
1½″	.05	.24	.36	.48
2″	.05	.34	.51	.68
2½″	.08	.50	.74	1.00
3″	.09	.60	.88	1.20
4″	.11	.71	1.06	1.42
5″	.14	.91	1.35	1.82
6″	.19	1.28	1.91	2.56

LABOR-HOURS INCLUDE THE FOLLOWING:

DISTRIBUTION FROM STOCKPILE 100′ DISTANCE
MEASURE, CUT AND PREPARE PIPE AND FITTING
INSTALL PIPE AND FITTING IN PLACE
MAKE UP SOLVENT JOINT
NORMAL LOSS TIME
TESTING
UP TO 10′-0″ CEILING HEIGHT

EXCLUSIONS

HANGERS AND SUPPORTS
UNUSUAL JOB CONDITIONS

LABOR-HOURS TO INSTALL BUILDING PIPING SYSTEMS (continued)

ITEM DESCRIPTION: GLASS ACID WASTE AND VENT PIPING CLAMP JOINT FITTINGS

BELOW GROUND

SIZE	PIPE PER L.F.	TWO JOINT FITTING	THREE JOINT FITTING	FOUR JOINT FITTING
1½″	.21	.51	.76	1.02
2″	.28	.61	.92	1.22
3″	.32	.82	1.22	1.63
4″	.43	.90	1.35	1.80
6″	.48	1.31	1.96	2.62

ABOVE GROUND

SIZE	PIPE PER L.F.	TWO JOINT FITTING	THREE JOINT FITTING	FOUR JOINT FITTING
1½″	.24	.54	.82	1.09
2″	.31	.65	.97	1.29
3″	.35	.85	1.27	1.70
4″	.45	.94	1.40	1.87
6″	.51	1.36	2.04	2.72

LABOR-HOURS INCLUDE THE FOLLOWING:

DISTRIBUTION FROM STOCKPILE 100′ DISTANCE
PIPE ENCASED IN STYROFOAM (BELOW GROUND)
UNPACKING PIPE SECTIONS (BELOW GROUND)
MEASURE AND CUT PIPE
INSTALL PIPE AND FITTING IN PLACE
MAKE UP CLAMP JOINTS
COVER FITTING WITH PLASTIC (BELOW GROUND)
NORMAL LOSS TIME
TESTING
UP TO 3′-0″ TRENCH DEPTH
UP TO 10′-0″ CEILING HEIGHT

EXCLUSIONS

EXCAVATION AND BACKFILL
DEWATERING
HANGERS AND SUPPORTS (BELOW AND ABOVE GROUND)
BREAKAGE
UNUSUAL JOB CONDITIONS

Means City Cost Indexes

The following pages contain examples of several different types of Means construction cost indexes for U.S. and Canadian locations. The Historical Cost Index is used to figure construction cost variations by year. The sample City Cost Index page contains the percentages to be used in adjusting installation and labor costs to any of 316 major city locations. The Installing Contractor's Overhead & Profit pages (union and open shop) show the labor rates for different trades according to an average of 30 major city rates. All of these cost pages represent relative costs as of Jan. 1, 2004. Further information on utilizing these indexes may be found in Chapter 8.

The City Cost Indexes serve as factors that can be used to adjust national average costs to a particular location. If, for example, the estimator is pricing a plumbing job in Tampa, Florida, the material costs as shown in Division 15, Mechanical, are the same as the Means national average. The labor costs for plumbing installations in Tampa are 76.5% of the Means national average. The overall percentage factor (the "Total" column) is 89.1%. (The factor in the "Total" column should be used in instances where the costs are not broken down into Labor and Material, such as in square foot estimates.)

Appendix C

Historical Cost Indexes

The table below lists both the Means Historical Cost Index based on Jan. 1, 1993 = 100 as well as the computed value of an index based on Jan. 1, 2004 costs. Since the Jan. 1, 2004 figure is estimated, space is left to write in the actual index figures as they become available through either the quarterly "Means Construction Cost Indexes" or as printed in the "Engineering News-Record." To compute the actual index based on Jan. 1, 2004 = 100, divide the Historical Cost Index for a particular year by the actual Jan. 1, 2004 Construction Cost Index. Space has been left to advance the index figures as the year progresses.

Year	Historical Cost Index Jan. 1, 1993 = 100		Current Index Based on Jan. 1, 2004 = 100		Year	Historical Cost Index Jan. 1, 1993 = 100	Current Index Based on Jan. 1, 2004 = 100		Year	Historical Cost Index Jan. 1, 1993 = 100	Current Index Based on Jan. 1, 2004 = 100	
	Est.	Actual	Est.	Actual		Actual	Est.	Actual		Actual	Est.	Actual
Oct 2004					July 1989	92.1	69.3		July 1971	32.1	24.1	
July 2004					1988	89.9	67.6		1970	28.7	21.6	
April 2004					1987	87.7	65.9		1969	26.9	20.2	
Jan 2004	133.0		100.0	100.0	1986	84.2	63.3		1968	24.9	18.7	
July 2003		132.0	99.2		1985	82.6	62.1		1967	23.5	17.7	
2002		128.7	96.8		1984	82.0	61.6		1966	22.7	17.1	
2001		125.1	94.1		1983	80.2	60.3		1965	21.7	16.3	
2000		120.9	90.9		1982	76.1	57.3		1964	21.2	15.9	
1999		117.6	88.4		1981	70.0	52.6		1963	20.7	15.6	
1998		115.1	86.5		1980	62.9	47.3		1962	20.2	15.2	
1997		112.8	84.8		1979	57.8	43.5		1961	19.8	14.9	
1996		110.2	82.9		1978	53.5	40.2		1960	19.7	14.8	
1995		107.6	80.9		1977	49.5	37.2		1959	19.3	14.5	
1994		104.4	78.5		1976	46.9	35.3		1958	18.8	14.1	
1993		101.7	76.5		1975	44.8	33.7		1957	18.4	13.8	
1992		99.4	74.8		1974	41.4	31.1		1956	17.6	13.2	
1991		96.8	72.8		1973	37.7	28.3		1955	16.6	12.5	
1990		94.3	70.9		1972	34.8	26.2		1954	16.0	12.0	

City Cost Indexes

DIVISION		FLORIDA																	
		MIAMI			ORLANDO			PANAMA CITY			PENSACOLA			ST. PETERSBURG			TALLAHASSEE		
		MAT.	INST.	TOTAL	MAT.	INST.	TOTAL	MAT.	INST.	TOTAL	MAT.	INST.	TOTAL	MAT.	INST.	TOTAL	MAT.	INST.	TOTAL
01590	**EQUIPMENT RENTAL**	.0	89.4	89.4	.0	97.8	97.8	.0	97.8	97.8	.0	97.8	97.8	.0	97.8	97.8	.0	97.8	97.8
02	**SITE CONSTRUCTION**	102.6	73.8	81.1	118.7	86.5	94.6	133.4	84.1	96.5	130.9	86.5	97.7	119.1	86.0	94.3	119.6	86.0	94.4
03100	CONCRETE FORMS & ACCESSORIES	92.5	70.8	73.5	93.9	55.3	60.0	93.0	28.8	36.6	83.9	53.0	56.8	91.6	49.1	54.3	93.8	40.5	47.0
03200	CONCRETE REINFORCEMENT	95.4	74.6	82.8	95.4	80.6	86.5	99.6	51.3	70.4	102.1	51.6	71.6	98.8	59.0	74.8	95.4	51.9	69.1
03300	CAST-IN-PLACE CONCRETE	92.5	71.6	83.8	98.4	70.0	86.5	96.1	35.6	70.8	96.1	56.5	79.5	102.4	57.0	83.4	94.7	49.8	76.0
03	**CONCRETE**	91.0	73.1	81.8	91.8	66.8	79.0	99.2	37.2	67.5	98.1	55.7	76.4	96.4	55.5	75.4	92.2	47.9	69.6
04	**MASONRY**	84.6	70.1	75.5	90.7	64.1	74.1	90.9	28.4	51.8	88.0	51.4	65.1	134.4	50.8	82.1	89.0	40.0	58.3
05	**METALS**	101.4	90.9	97.5	107.3	93.9	102.2	96.4	66.9	85.3	96.4	80.9	90.5	100.2	82.2	93.4	98.1	79.8	91.2
06	**WOOD & PLASTICS**	89.0	68.6	78.0	92.9	53.5	71.6	91.8	28.9	57.9	83.1	53.5	67.2	90.5	49.1	68.2	91.1	38.9	63.0
07	**THERMAL & MOISTURE PROTECTION**	99.4	71.5	85.8	95.7	68.3	82.4	96.0	32.4	65.0	95.7	54.5	75.6	95.3	51.2	73.8	95.7	48.0	72.5
08	**DOORS & WINDOWS**	98.5	66.2	90.3	100.9	58.0	90.0	98.5	27.6	80.5	98.4	51.9	86.6	99.5	48.1	86.4	99.5	42.5	85.0
09200	PLASTER & GYPSUM BOARD	98.2	68.2	77.8	102.1	52.7	68.5	96.9	27.5	49.7	92.8	52.8	65.6	96.4	48.2	63.6	98.7	37.7	57.2
095,098	CEILINGS & ACOUSTICAL TREATMENT	95.0	68.2	76.9	95.0	52.7	66.4	89.8	27.5	47.6	89.8	52.8	64.7	89.8	48.2	61.6	95.0	37.7	56.2
09600	FLOORING	126.0	64.2	110.0	118.6	72.1	106.6	118.0	20.0	92.7	112.2	55.3	97.5	116.9	55.2	101.0	118.6	40.5	98.5
097,099	WALL FINISHES, PAINTS & COATINGS	107.5	51.5	73.9	111.3	57.0	78.7	111.3	25.4	59.8	111.3	57.6	79.1	111.3	48.0	73.3	111.3	40.8	69.1
09	**FINISHES**	107.8	67.1	85.8	108.0	57.7	80.8	107.3	26.3	63.6	104.7	53.8	77.2	105.7	49.9	75.5	107.7	39.7	70.9
10 - 14	**TOTAL DIV. 10000 - 14000**	100.0	89.9	97.8	100.0	79.5	95.6	100.0	48.9	89.0	100.0	54.7	90.3	100.0	57.3	90.8	100.0	64.0	92.2
15	**MECHANICAL**	99.8	76.0	88.9	99.8	56.4	79.8	99.8	26.1	65.9	99.8	51.8	77.7	99.8	51.7	77.6	99.8	41.3	72.9
16	**ELECTRICAL**	98.2	76.2	85.4	98.4	46.3	67.9	96.2	34.7	60.2	100.8	56.0	74.6	98.2	50.0	70.0	98.4	42.9	65.9
01 - 16	**WEIGHTED AVERAGE**	98.6	74.8	87.1	100.6	64.6	83.2	99.8	39.1	70.6	99.5	59.2	80.1	102.2	57.6	80.7	99.1	50.6	75.7

DIVISION		FLORIDA			GEORGIA														
		TAMPA			ALBANY			ATLANTA			AUGUSTA			COLUMBUS			MACON		
		MAT.	INST.	TOTAL	MAT.	INST.	TOTAL	MAT.	INST.	TOTAL	MAT.	INST.	TOTAL	MAT.	INST.	TOTAL	MAT.	INST.	TOTAL
01590	**EQUIPMENT RENTAL**	.0	97.8	97.8	.0	90.2	90.2	.0	92.9	92.9	.0	92.2	92.2	.0	90.2	90.2	.0	102.6	102.6
02	**SITE CONSTRUCTION**	119.3	86.6	94.8	103.6	75.9	82.9	104.6	95.9	98.1	100.9	92.9	94.9	103.5	76.3	83.2	104.3	94.1	96.6
03100	CONCRETE FORMS & ACCESSORIES	95.3	77.7	79.9	93.6	44.4	50.5	91.4	81.4	82.6	88.6	53.9	58.1	94.8	56.9	61.5	92.5	57.5	61.8
03200	CONCRETE REINFORCEMENT	95.4	92.8	93.8	94.8	90.8	92.4	99.1	92.6	95.2	104.8	79.2	89.3	95.4	91.0	92.7	97.8	91.0	93.7
03300	CAST-IN-PLACE CONCRETE	100.1	63.5	84.8	96.7	42.9	74.2	99.9	76.6	90.2	94.4	50.4	76.0	96.3	43.8	74.4	95.0	47.1	75.0
03	**CONCRETE**	95.0	76.7	85.6	93.0	54.3	73.2	98.5	81.7	89.9	94.4	57.9	75.7	93.0	60.2	76.2	92.5	61.7	76.7
04	**MASONRY**	87.6	78.1	81.6	88.7	32.9	53.8	89.8	72.1	78.8	90.0	41.5	59.7	88.8	36.3	55.9	102.6	39.6	63.2
05	**METALS**	101.1	97.4	99.7	95.5	89.4	93.2	91.4	80.7	87.4	89.8	71.0	82.7	95.9	91.1	94.1	90.8	91.5	91.1
06	**WOOD & PLASTICS**	94.2	80.6	86.9	92.4	44.3	66.4	90.5	84.2	87.1	88.1	55.5	70.5	93.8	61.0	76.1	101.1	59.9	78.9
07	**THERMAL & MOISTURE PROTECTION**	95.6	64.3	80.4	95.5	52.2	74.4	91.7	77.5	84.8	91.1	51.4	71.8	95.1	54.8	75.5	93.9	58.6	76.7
08	**DOORS & WINDOWS**	100.9	73.7	94.0	98.5	51.2	86.5	100.0	79.6	94.8	94.1	56.6	84.6	98.5	60.5	88.8	96.9	60.8	87.7
09200	PLASTER & GYPSUM BOARD	98.7	80.6	86.4	98.4	43.3	60.9	117.1	84.2	94.7	116.1	54.7	74.3	98.4	60.5	72.6	107.6	59.3	74.7
095,098	CEILINGS & ACOUSTICAL TREATMENT	95.0	80.6	85.3	93.7	43.3	59.6	96.0	84.2	88.0	97.5	54.7	68.5	93.7	60.5	71.2	87.8	59.3	68.5
09600	FLOORING	118.6	55.2	102.2	118.6	33.8	96.7	84.4	81.1	83.5	83.2	43.1	72.9	118.6	39.4	98.2	92.5	39.7	78.9
097,099	WALL FINISHES, PAINTS & COATINGS	111.3	48.0	73.3	107.5	42.7	68.6	90.4	80.6	84.5	90.4	40.6	60.5	107.5	40.9	67.6	109.2	50.0	73.7
09	**FINISHES**	107.6	71.0	87.8	105.1	41.1	70.5	93.1	81.5	86.8	92.6	50.1	69.6	105.0	51.8	76.2	91.7	53.0	70.8
10 - 14	**TOTAL DIV. 10000 - 14000**	100.0	80.8	95.9	100.0	74.0	94.4	100.0	84.6	96.7	100.0	71.8	93.9	100.0	76.1	94.9	100.0	77.4	95.1
15	**MECHANICAL**	99.8	76.5	89.1	99.8	46.9	75.5	99.9	80.1	90.8	99.9	68.6	85.5	99.8	38.4	71.6	99.8	43.2	73.8
16	**ELECTRICAL**	97.2	50.0	69.6	92.3	62.8	75.1	96.6	85.3	90.0	98.4	51.3	70.8	93.3	40.7	62.6	91.9	62.0	74.4
01 - 16	**WEIGHTED AVERAGE**	99.8	74.8	87.8	97.4	55.6	77.2	96.8	82.0	89.7	95.4	61.1	78.9	97.5	54.2	76.6	95.9	60.7	78.9

DIVISION		GEORGIA						HAWAII			IDAHO								
		SAVANNAH			VALDOSTA			HONOLULU			BOISE			LEWISTON			POCATELLO		
		MAT.	INST.	TOTAL	MAT.	INST.	TOTAL	MAT.	INST.	TOTAL	MAT.	INST.	TOTAL	MAT.	INST.	TOTAL	MAT.	INST.	TOTAL
01590	**EQUIPMENT RENTAL**	.0	91.3	91.3	.0	90.2	90.2	.0	99.3	99.3	.0	101.4	101.4	.0	94.3	94.3	.0	101.4	101.4
02	**SITE CONSTRUCTION**	104.1	77.3	84.0	114.4	76.5	86.1	138.9	106.5	114.7	78.8	103.0	96.9	83.9	96.9	93.7	80.0	103.0	97.2
03100	CONCRETE FORMS & ACCESSORIES	93.3	54.6	59.3	81.0	47.2	51.4	106.3	145.7	140.8	98.4	82.2	84.2	113.7	69.3	74.8	98.3	81.7	83.7
03200	CONCRETE REINFORCEMENT	101.1	79.4	88.0	101.2	45.0	67.3	104.0	120.7	114.1	98.2	77.9	85.9	111.5	96.4	102.4	98.6	77.4	85.8
03300	CAST-IN-PLACE CONCRETE	93.3	52.8	76.4	94.6	53.4	77.4	197.0	130.3	169.1	99.0	93.4	96.6	107.3	88.1	99.3	98.2	93.2	96.1
03	**CONCRETE**	92.0	60.1	75.7	97.8	50.7	73.7	154.7	134.2	144.2	104.7	85.1	94.6	113.1	81.2	96.8	100.9	84.7	92.6
04	**MASONRY**	91.9	54.9	68.7	94.9	49.9	66.7	134.4	131.7	132.7	136.9	76.9	99.3	138.0	87.8	106.5	132.6	68.5	92.5
05	**METALS**	96.2	85.9	92.3	95.7	74.0	87.5	115.7	108.4	112.9	112.8	78.2	99.7	96.5	88.2	93.3	113.1	77.0	99.4
06	**WOOD & PLASTICS**	106.8	52.3	77.4	80.1	43.2	60.2	95.0	149.6	124.4	96.8	80.8	88.2	106.6	62.9	83.0	96.8	80.8	88.2
07	**THERMAL & MOISTURE PROTECTION**	95.5	54.5	75.5	95.3	59.0	77.6	113.2	131.8	122.2	97.1	82.1	89.8	167.6	81.7	125.7	97.3	72.3	85.1
08	**DOORS & WINDOWS**	99.6	52.7	87.7	93.8	40.0	80.1	105.9	137.7	113.9	94.7	77.6	90.4	115.0	68.9	103.3	94.7	70.7	88.6
09200	PLASTER & GYPSUM BOARD	98.7	51.5	66.6	91.3	42.1	57.8	119.3	150.9	140.8	88.2	80.1	82.7	155.2	61.8	91.6	88.2	80.1	82.7
095,098	CEILINGS & ACOUSTICAL TREATMENT	95.0	51.5	65.5	89.8	42.1	57.5	117.3	150.9	140.0	112.4	80.1	90.5	159.6	61.8	93.4	112.4	80.1	90.5
09600	FLOORING	118.6	50.8	101.1	110.4	40.6	92.4	167.8	136.4	159.7	100.3	58.5	89.5	138.4	97.3	127.8	100.6	58.5	89.8
097,099	WALL FINISHES, PAINTS & COATINGS	108.2	50.7	73.7	107.5	37.0	65.2	104.5	149.9	131.8	103.9	53.8	73.9	129.9	76.6	97.9	103.9	56.0	75.2
09	**FINISHES**	105.6	53.2	77.3	101.4	44.2	70.5	135.5	145.9	141.1	99.7	74.8	86.2	168.8	74.0	117.6	99.8	75.0	86.4
10 - 14	**TOTAL DIV. 10000 - 14000**	100.0	74.8	94.6	100.0	73.5	94.3	100.0	122.9	104.9	100.0	91.1	98.1	100.0	93.1	98.5	100.0	91.0	98.1
15	**MECHANICAL**	99.8	49.2	76.5	99.8	43.3	73.8	100.1	120.6	109.5	100.1	79.5	90.6	101.4	86.8	94.7	100.1	79.5	90.6
16	**ELECTRICAL**	94.7	54.6	71.3	90.8	31.4	56.1	113.2	122.9	118.9	84.0	79.7	81.5	85.3	87.7	86.7	86.6	76.8	80.9
01 - 16	**WEIGHTED AVERAGE**	98.1	59.5	79.5	97.5	50.4	74.8	117.6	126.1	121.7	101.6	81.8	92.1	112.2	84.7	99.0	101.3	79.8	90.9

Installing Contractor's Overhead & Profit

Below are the **average** installing contractor's percentage mark-ups applied to base labor rates to arrive at typical billing rates.

Column A: Labor rates are based on union wages averaged for 30 major U.S. cities. Base rates including fringe benefits are listed hourly and daily. These figures are the sum of the wage rate and employer-paid fringe benefits such as vacation pay, employer-paid health and welfare costs, pension costs, plus appropriate training and industry advancement funds costs.

Column B: Workers' Compensation rates are the national average of state rates established for each trade.

Column C: Column C lists average fixed overhead figures for all trades. Included are Federal and State Unemployment costs set at 6.2%; Social Security Taxes (FICA) set at 7.65%; Builder's Risk Insurance costs set at 0.44%; and Public Liability costs set at 2.02%. All the percentages except those for Social Security Taxes vary from state to state as well as from company to company.

Columns D and E: Percentages in Columns D and E are based on the presumption that the installing contractor has annual billing of $4,000,000 and up. Overhead percentages may increase with smaller annual billing. The overhead percentages for any given contractor may vary greatly and depend on a number of factors, such as the contractor's annual volume, engineering and logistical support costs, and staff requirements. The figures for overhead and profit will also vary depending on the type of job, the job location, and the prevailing economic conditions. All factors should be examined very carefully for each job.

Column F: Column F lists the total of Columns B, C, D, and E.

Column G: Column G is Column A (hourly base labor rate) multiplied by the percentage in Column F (O&P percentage).

Column H: Column H is the total of Column A (hourly base labor rate) plus Column G (Total O&P).

Column I: Column I is Column H multiplied by eight hours.

		A		B	C	D	E	F	G	H	I
		Base Rate Incl. Fringes		Work-ers' Comp. Ins.	Average Fixed Over-head	Over-head	Profit	Total Overhead & Profit		Rate with O & P	
Abbr.	Trade	Hourly	Daily					%	Amount	Hourly	Daily
Skwk	Skilled Workers Average (35 trades)	$33.65	$269.20	16.2%	16.3%	13.0%	10%	55.5%	$18.70	$52.35	$418.80
	Helpers Average (5 trades)	24.55	196.40	17.9		11.0		55.2	13.55	38.10	304.80
	Foreman Average, Inside ($.50 over trade)	34.15	273.20	16.2		13.0		55.5	18.95	53.10	424.80
	Foreman Average, Outside ($2.00 over trade)	35.65	285.20	16.2		13.0		55.5	19.80	55.45	443.60
Clab	Common Building Laborers	26.00	208.00	18.5		11.0		55.8	14.50	40.50	324.00
Asbe	Asbestos/Insulation Workers/Pipe Coverers	36.00	288.00	15.2		16.0		57.5	20.70	56.70	453.60
Boil	Boilermakers	41.25	330.00	13.1		16.0		55.4	22.85	64.10	512.80
Bric	Bricklayers	34.25	274.00	15.0		11.0		52.3	17.90	52.15	417.20
Brhe	Bricklayer Helpers	25.60	204.80	15.0		11.0		52.3	13.40	39.00	312.00
Carp	Carpenters	33.00	264.00	18.5		11.0		55.8	18.40	51.40	411.20
Cefi	Cement Finishers	31.55	252.40	9.9		11.0		47.2	14.90	46.45	371.60
Elec	Electricians	39.40	315.20	6.4		16.0		48.7	19.20	58.60	468.80
Elev	Elevator Constructors	41.60	332.80	7.2		16.0		49.5	20.60	62.20	497.60
Eqhv	Equipment Operators, Crane or Shovel	34.80	278.40	10.3		14.0		50.6	17.60	52.40	419.20
Eqmd	Equipment Operators, Medium Equipment	33.65	269.20	10.3		14.0		50.6	17.05	50.70	405.60
Eqlt	Equipment Operators, Light Equipment	32.15	257.20	10.3		14.0		50.6	16.25	48.40	387.20
Eqol	Equipment Operators, Oilers	29.20	233.60	10.3		14.0		50.6	14.80	44.00	352.00
Eqmm	Equipment Operators, Master Mechanics	35.20	281.60	10.3		14.0		50.6	17.80	53.00	424.00
Glaz	Glaziers	32.05	256.40	13.8		11.0		51.1	16.40	48.45	387.60
Lath	Lathers	30.55	244.40	10.7		11.0		48.0	14.65	45.20	361.60
Marb	Marble Setters	31.95	255.60	15.0		11.0		52.3	16.70	48.65	389.20
Mill	Millwrights	34.35	274.80	10.3		11.0		47.6	16.35	50.70	405.60
Mstz	Mosaic & Terrazzo Workers	31.60	252.80	9.6		11.0		46.9	14.80	46.40	371.20
Pord	Painters, Ordinary	29.60	236.80	12.9		11.0		50.2	14.85	44.45	355.60
Psst	Painters, Structural Steel	30.05	240.40	50.0		11.0		87.3	26.25	56.30	450.40
Pape	Paper Hangers	29.35	234.80	12.9		11.0		50.2	14.75	44.10	352.80
Pile	Pile Drivers	32.05	256.40	22.9		16.0		65.2	20.90	52.95	423.60
Plas	Plasterers	29.95	239.60	14.6		11.0		51.9	15.55	45.50	364.00
Plah	Plasterer Helpers	25.80	206.40	14.6		11.0		51.9	13.40	39.20	313.60
Plum	Plumbers	39.60	316.80	7.8		16.0		50.1	19.85	59.45	475.60
Rodm	Rodmen (Reinforcing)	37.10	296.80	24.8		14.0		65.1	24.15	61.25	490.00
Rofc	Roofers, Composition	28.45	227.60	31.8		11.0		69.1	19.65	48.10	384.80
Rots	Roofers, Tile & Slate	28.65	229.20	31.8		11.0		69.1	19.80	48.45	387.60
Rohe	Roofers, Helpers (Composition)	20.95	167.60	31.8		11.0		69.1	14.50	35.45	283.60
Shee	Sheet Metal Workers	38.80	310.40	11.1		16.0		53.4	20.70	59.50	476.00
Spri	Sprinkler Installers	39.25	314.00	9.0		16.0		51.3	20.15	59.40	475.20
Stpi	Steamfitters or Pipefitters	39.75	318.00	7.8		16.0		50.1	19.90	59.65	477.20
Ston	Stone Masons	33.85	270.80	15.0		11.0		52.3	17.70	51.55	412.40
Sswk	Structural Steel Workers	37.15	297.20	38.9		14.0		79.2	29.40	66.55	532.40
Tilf	Tile Layers	31.50	252.00	9.6		11.0		46.9	14.75	46.25	370.00
Tilh	Tile Layers Helpers	24.45	195.60	9.6		11.0		46.9	11.45	35.90	287.20
Trlt	Truck Drivers, Light	25.70	205.60	15.1		11.0		52.4	13.45	39.15	313.20
Trhv	Truck Drivers, Heavy	26.45	211.60	15.1		11.0		52.4	13.85	40.30	322.40
Sswl	Welders, Structural Steel	37.15	297.20	38.9		14.0		79.2	29.40	66.55	532.40
Wrck	*Wrecking	26.00	208.00	40.5		11.0		77.8	20.25	46.25	370.00

*Not included in averages

Installing Contractor's Overhead & Profit

Below are the **average** installing contractor's percentage mark-ups applied to base labor rates to arrive at typical billing rates.

Column A: Labor rates are based on average open shop wages for 7 major U.S. regions. Base rates including fringe benefits are listed hourly and daily. These figures are the sum of the wage rate and employer-paid fringe benefits such as vacation pay, and employer-paid health costs.

Column B: Workers' Compensation rates are the national average of state rates established for each trade.

Column C: Column C lists average fixed overhead figures for all trades. Included are Federal and State Unemployment costs set at 6.2%; Social Security Taxes (FICA) set at 7.65%; Builder's Risk Insurance costs set at 0.44%; and Public Liability costs set at 2.02%. All the percentages except those for Social Security Taxes vary from state to state as well as from company to company.

Columns D and E: Percentages in Columns D and E are based on the presumption that the installing contractor has annual billing of $2,000,000 and up. Overhead percentages may increase with smaller annual billing. The overhead percentages for any given contractor may vary greatly and depend on a number of factors, such as the contractor's annual volume, engineering and logistical support costs, and staff requirements. The figures for overhead and profit will also vary depending on the type of job, the job location, and the prevailing economic conditions. All factors should be examined very carefully for each job.

Column F: Column F lists the total of Columns B, C, D, and E.

Column G: Column G is Column A (hourly base labor rate) multiplied by the percentage in Column F (O&P percentage).

Column H: Column H is the total of Column A (hourly base labor rate) plus Column G (Total O&P).

Column I: Column I is Column H multiplied by eight hours.

		A		B	C	D	E	F	G	H	I
		Base Rate Incl. Fringes		Workers' Comp. Ins.	Average Fixed Overhead	Overhead	Profit	Total Overhead & Profit		Rate with O & P	
Abbr.	Trade	Hourly	Daily					%	Amount	Hourly	Daily
Skwk	Skilled Workers Average (35 trades)	$25.80	$206.40	16.2%	16.3%	27.0%	10%	69.5%	$17.95	$43.75	$350.00
	Helpers Average (5 trades)	19.05	152.40	17.9		25.0		69.2	13.20	32.25	258.00
	Foreman Average, Inside ($.50 over trade)	26.30	210.40	16.2		27.0		69.5	18.30	44.60	356.80
	Foreman Average, Outside ($2.00 over trade)	27.80	222.40	16.2		27.0		69.5	19.30	47.10	376.80
Clab	Common Building Laborers	20.00	160.00	18.5		25.0		69.8	13.95	33.95	271.60
Asbe	Asbestos/Insulation Workers/Pipe Coverers	26.65	213.20	15.2		30.0		71.5	19.05	45.70	365.60
Boil	Boilermakers	30.55	244.40	13.1		30.0		69.4	21.20	51.75	414.00
Bric	Bricklayers	26.70	213.60	15.0		25.0		66.3	17.70	44.40	355.20
Brhe	Bricklayer Helpers	19.95	159.60	15.0		25.0		66.3	13.25	33.20	265.60
Carp	Carpenters	25.75	206.00	18.5		25.0		69.8	17.95	43.70	349.60
Cefi	Cement Finishers	24.60	196.80	9.9		25.0		61.2	15.05	39.65	317.20
Elec	Electricians	31.50	252.00	6.4		30.0		62.7	19.75	51.25	410.00
Elev	Elevator Constructors	31.20	249.60	7.2		30.0		63.5	19.80	51.00	408.00
Eqhv	Equipment Operators, Crane or Shovel	26.80	214.40	10.3		28.0		64.6	17.30	44.10	352.80
Eqmd	Equipment Operators, Medium Equipment	25.90	207.20	10.3		28.0		64.6	16.75	42.65	341.20
Eqlt	Equipment Operators, Light Equipment	24.75	198.00	10.3		28.0		64.6	16.00	40.75	326.00
Eqol	Equipment Operators, Oilers	22.50	180.00	10.3		28.0		64.6	14.55	37.05	296.40
Eqmm	Equipment Operators, Master Mechanics	27.10	216.80	10.3		28.0		64.6	17.50	44.60	356.80
Glaz	Glaziers	25.30	202.40	13.8		25.0		65.1	16.45	41.75	334.00
Lath	Lathers	23.85	190.80	10.7		25.0		62.0	14.80	38.65	309.20
Marb	Marble Setters	24.90	199.20	15.0		25.0		66.3	16.50	41.40	331.20
Mill	Millwrights	26.80	214.40	10.3		25.0		61.6	16.50	43.30	346.40
Mstz	Mosaic & Terrazzo Workers	24.65	197.20	9.6		25.0		60.9	15.00	39.65	317.20
Pord	Painters, Ordinary	23.40	187.20	12.9		25.0		64.2	15.00	38.40	307.20
Psst	Painters, Structural Steel	23.75	190.00	50.0		25.0		101.3	24.05	47.80	382.40
Pape	Paper Hangers	23.20	185.60	12.9		25.0		64.2	14.90	38.10	304.80
Pile	Pile Drivers	25.00	200.00	22.9		30.0		79.2	19.80	44.80	358.40
Plas	Plasterers	23.35	186.80	14.6		25.0		65.9	15.40	38.75	310.00
Plah	Plasterer Helpers	20.10	160.80	14.6		25.0		65.9	13.25	33.35	266.80
Plum	Plumbers	29.30	234.40	7.8		30.0		64.1	18.80	48.10	384.80
Rodm	Rodmen (Reinforcing)	27.85	222.80	24.8		28.0		79.1	22.05	49.90	399.20
Rofc	Roofers, Composition	21.90	175.20	31.8		25.0		83.1	18.20	40.10	320.80
Rots	Roofers, Tile & Slate	22.05	176.40	31.8		25.0		83.1	18.30	40.35	322.80
Rohe	Roofers, Helpers (Composition)	16.15	129.20	31.8		25.0		83.1	13.40	29.55	236.40
Shee	Sheet Metal Workers	28.70	229.60	11.1		30.0		67.4	19.35	48.05	384.40
Spri	Sprinkler Installers	29.05	232.40	9.0		30.0		65.3	18.95	48.00	384.00
Stpi	Steamfitters or Pipefitters	29.40	235.20	7.8		30.0		64.1	18.85	48.25	386.00
Ston	Stone Masons	25.75	206.00	15.0		25.0		66.3	17.05	42.80	342.40
Sswk	Structural Steel Workers	27.85	222.80	38.9		28.0		93.2	25.95	53.80	430.40
Tilf	Tile Layers	24.55	196.40	9.6		25.0		60.9	14.95	39.50	316.00
Tilh	Tile Layers Helpers	19.05	152.40	9.6		25.0		60.9	11.60	30.65	245.20
Trlt	Truck Drivers, Light	20.30	162.40	15.1		25.0		66.4	13.50	33.80	270.40
Trhv	Truck Drivers, Heavy	20.90	167.20	15.1		25.0		66.4	13.90	34.80	278.40
Sswl	Welders, Structural Steel	27.85	222.80	38.9		28.0		93.2	25.95	53.80	430.40
Wrck	*Wrecking	20.00	160.00	40.5		25.0		91.8	18.35	38.35	306.80

*Not included in averages

Codes and Standards

AMERICAN NATIONAL STANDARDS INSTITUTE (ANSI)
1819 L Street, NW
6th floor
Washington, DC 20036
(202) 293-8020
http://www.ansi.org

AMERICAN SOCIETY FOR TESTING AND MATERIALS INTERNATIONAL (ASTM)
100 Barr Harbor Drive
West Conshohocken, PA 19428
(610) 832-9585
http://www.astm.org

AMERICAN SOCIETY OF MECHANICAL ENGINEERS (ASME)
Three Park Avenue
New York, NY 10016
(800) 843-2763
http://www.asme.org

INTERNATIONAL ASSOCIATION OF PLUMBING AND MECHANICAL OFFICIALS (IAPMO)
5001 E. Philadelphia Street
Ontario, CA 91761
(909) 472-4100
http://www.iapmo.org

INTERNATIONAL CODE COUNCIL
5203 Leesburg Pike
Suite 600
Falls Church, VA 22041
(703) 931-4533
http://www.iccsafe.org

NATIONAL FIRE PROTECTION ASSOCIATION (NFPA)
1 Batterymarch Park
Quincy, MA 02169
(617) 770-3000
http://www.nfpa.org

PLUMBING-HEATING-COOLING CONTRACTORS ASSOCIATION (PHCC)
180 S. Washington Street
P.O. Box 6808
Falls Church, VA 22040-6808
(800) 533-7694
http://www.phccweb.org

Abbreviations

A
Area

Ab
Above

Abs
Absolute

ACP
Asbestos cement pipe

AD
Area drain

AFF
Above finished floor

AGA
American Gas Association

AISI
American Iron and Steel Institute

Al
Aluminum

ANSI
American National Standards Institute

API
American Petroleum Institute

ASA
American Standard Association

ASCE
American Society of Civil Engineering

ASHRAE
American Society of Heating, Refrigerating and Air Conditioning Engineers

ASME
American Society of Mechanical Engineers

ASPE
American Society of Plumbing Engineers

ASSE
American Society of Sanitary Engineers

ASTM
American Society for Testing and Materials

AV
Acid vent

Avg
Average

AW
Acid waste

AWWA
American Water Works Association

B&S
Bell and spigot

Bbl
Barrel

BCF
Backfill

BOD
Biochemical oxygen demand

Br
Branch

BT
Bathtub

BTU
British thermal unit

BV
Balancing valve

C
Centigrade

°C
Degrees centigrade

C to C
Center to center

CA
Compressed air

CB
Catch basin

CF
Cubic feet

Cfm
Cubic feet per minute

ChkV
Check valve

CI
Cast iron or cubic inches

Circ.
Circulator/Circulation

CISP
Cast iron soil pipe

CISPI
Cast Iron Soil Pipe Institute

CIWP
Cast iron water pipe

Cl
Chlorine

CLel
Centerline elevation

Clg
Ceiling

CO
Cleanout

CODP
Cleanout deck plate

CS
Cast steel (or commercial standard)

CTE
Connection to existing

Cu
Copper

CW
Cold water

CY
Cubic yard

D
Drain

Deg or °
Degrees

DF
Drinking fountain

DI
Ductile iron/drain inlet

Dn
Down

Dp
Deep

Drg
Drainage

Dwg
Drawing

DWV
Drainage waste and vent

ED
Sewage ejector discharge

Elev.
Elevation

Ell
Elbow

EWC
Electric water cooler

EWF
Eye wash fountain

ES
Emergency shower

Exc
Excavation

F
Fahrenheit

°F
Degrees Fahrenheit

FAI
Fresh air intake

FD
Floor drain

FDV
Fire department valve

Fe
Iron

Fed Spec
Federal specification

FF
Finish floor

FG
Finish grade

FH
Fire hydrant

FHC
Fire hose cabinet

FHR
Fire hose rack

Fig
Figure

Fixt
Fixture

Flr
Floor

FP
Fire plug

FS
Floor sink

FSP
Fire standpipe

FU
Fixture unit

G
Gas

Ga
Gauge

Gal
Gallon (231 CI)

Galv
Galvanized

Gas
Gallons

GC
Gas cock

G1 V
Globe valve

GPID
Gallons per day

GPH
Gallons per hour

GPM
Gallons per minute

GT
Grease trap

GV
Gate valve

H
Hydrogen or handicapped

HB
Hose bibb

HClg
Hung ceiling

Hd
Head

HD
House drain

Hgr
Hanger

HP
Horsepower

Hr
Hour

HT
House trap

Htr
Heater

HW
Hot water

IAPMO
International Association of Plumbing and Mechanical Officials

IB
Iron body

ID
Inside diameter

IE
Invert elevation

In
Inch

IPS
Iron pipe size

Jt
joint

kW
Kilowatt

L
or Ldr Leader

L
or Lth Length

Lav
Lavatory

Lb
Pound

LF
Linear feet

Mal
Malleable

Mat
Material

Max
Maximum

MCAA
Mechanical Contractors Association of America

Mech
Mechanical

MER
Mechanical equipment room

Mfr
Manufacturer

MGD
Million gallons per day

MH
Manhole

MI
Malleable iron

Min
Minimum (or minute)

MS
Milled steel

N_2
Nitrogen

N_20
Nitrous oxide

NAPHCC
National Association of Plumbing, Heating and Cooling Contractors

NBFU
National Board of Fire Underwriters

NBS
National Bureau of Standards

NFPA
National Fire Protection Association

NH
No hub

NPS
Nominal pipe size (Also called IPS)

NTS
Not to scale

O_2
Oxygen

OD
Outside diameter

Oz
Ounce

P
Pump

P&T
Pressure and temperature

Pb
Lead

PD
Pump discharge

PDI
Plumbing and Drainage Institute

PG
Pressure gauge

PH
Hydrogen concentration

PIV
Post indicator valve

PO
Plugged outlet

Ppm
Parts per million

Press
Pressure

PRV
Pressure reducing valve

PSI
Pounds per square inch

PVC
Polyvinylchloride

Qt
Quart

Qty
Quantity

R
Hydraulic radius

Rad
Radius

RCP
Reinforced concrete pipe

RD
Rate of demand (or roof drain)

Red
Reducer

RT
Running trap

RV
Relief valve

S
Soil

S&W
Soil and waste

SA
Shock absorber

San
Sanitary

Sb
Antimony

SC
Sillcock

SE
Sewage ejector

Sec
Second

SF
Square foot

Shwr
Shower

SI
Square inches

Siam.
Conn. Siamese Connection

Sk
Sink

Sn
Tin

Sol
Solder/Solenoid

Sp
Sprinkler

SP
Sump Pump

Spec
Specification

SS
Service sink (slop sink)

St
Storm

Std
Standard

Str
Strainer

Sv
Service

SW
Service weight

T
Temperature (or time)

TD
Trench drain

Therm
Thermometer

Thrd
Threaded

TP
Threadless pipe

UL
Underwriters' Laboratories, Inc.

Ur
Urinal

USASI
USA Standards institute

V
Vent

Vac
Vacuum

Val
Valve

VB
Vacuum breaker

VCP
Vitrified clay pipe

Vel
Velocity

VIV
Valve in vertical

Vol
Volume

VTR
Vent through roof

W
Waste

WC
Water closet

WClr
Water cooler

Wgt
Weight

WH
Wall hydrant

WL
Water level

XH
Extra heavy

XHCI
Extra heavy cast iron

Glossary

Absorption
Immersion in a fluid for a definite period of time. It is usually expressed as a percent of the dry weight.

Addendum
Any change in the drawings or specifications made by the architect or engineer, prior to bid.

Air break
A piping arrangement in which a drain from a fixture appliance or device discharges through an open connection into a receptacle or interceptor at a point above the flood level rim of the receptacle. Also known as an air gap.

Air compressor
The manufactured item of equipment that compresses air so that its expansion may be utilized as a source of power.

Anaerobic
Bacteria living without oxygen.

Anchor
Typically, pieces of metal used to fasten or secure pipes to the building or structure.

Area of circle
To find the area of a circle, multiply the square of the radius by pi. Area = πr^2

Backfill
That portion of a trench excavation that is replaced after the sewer line has been laid. It is the material above the pipe up to the original earth line.

Backflow
The flow of water or other liquids, mixtures, or substances into the distributing pipes of a potable supply of water from any source or sources other than its intended source. Reversal of flow.

Backflow preventer
A device or assembly designed to prevent backflow into the potable water system.

Back-siphonage
The flowing back of used, contaminated, or polluted water from a plumbing fixture or vessel into a water supply system due to a negative pressure in such pipe.

Back vent
An individual vent pipe connected directly into the back of the fixture waste fittings.

Base
The lowest portion or point of a stack of vertical pipe.

Battery of fixtures
Any group of two or more similar adjacent fixtures that discharge into a common horizontal waste or soil branch.

Bell and spigot
A particular type of pipe joint where the spigot or straight end fits into the bell or flared end and is made tight with lead or rubber gaskets.

Below grade
Work below ground level.

Bid
A proposal by a contractor to perform work for a given sum of money.

Branch
The part of a piping system other than the main riser or stack that extends to fixtures on one or two consecutive floors.

Branch vent
A vent connecting one or more individual vents with a vent stack or stack vent.

Building gravity drainage system
A drainage system that drains by gravity into the building house sewer.

Carrier fitting
A manufactured support for a plumbing fixture that is also an integral part of the soil piping system.

Caulking
The operation or method of rendering a joint tight against water or gas by means of plastic substances such as lead and oakum.

Circuit vent
A vent containing a separate vent stack that may or may not extend through the roof independently of the stack vent.

Circumference of a circle
To find the perimeter or circumference of a circle, multiply the diameter of the circle by pi. Circumference $= \pi d$.

Cleanout
A manufactured fitting with a plug end that can be opened to service soil lines that have become clogged.

Color code
The color choices made by an estimator to readily identify piping systems on drawings.

Common vent
A vent connecting at the junction of two fixture drains and serves as a vent for both fixtures.

Compression
Stress that resists the tendency of two forces acting toward each other.

Conductor
A pipe that leads rainwater to a drain.

Continuous vent
A vertical vent that is a continuation of the drain to which it connects.

Cross connection
Any physical connection between a city water supply and any waste pipe, soil pipe, sewer, drain, or private or uncertified water supply. Also any potable water supply outlet that is submerged or can be submerged in waste water and/ or any other source of contamination.

Crude or raw sewage
Untreated sewage.

Dead end
A branch leading from any soil, waste or vent pipe, building drain, or building sewer, which is terminated at a developed distance of two feet or more by means of a cap, plug, or other fitting not used for admitting water or air to the pipe, except branches serving as cleanout extensions.

Developed length
The length along the center line of pipe and fittings, both horizontal and vertical.

Diameter
Unless specifically stated, the term diameter is the nominal diameter as designed commercially.

Direct cost
The cost of a project prior to the addition of any supplementary costs, such as overhead and profit.

Discount
The percentage deduction from a material's list price afforded the contractor.

Drain
Any pipe that carries waste water or water-borne wastes in a building drainage system.

Drains, combined
The portion of the drainage system within a building that carries storm water and sanitary sewage.

Drains, storm
The part of the horizontal piping and its branches that convert subsoil and/or surface drainage from areas, courts, roof, or yards to the building or storm sewer.

Drains, subsoil
The part of the drainage system which conveys the subsoil, ground or seepage water from the footings of walls, or from under buildings, to the building drain, storm water drain, or building sewer.

Drainage system
Includes all the piping within public or private premises that conveys sewage, rainwater, or other liquid wastes to a legal point of disposal.

Dry well
A covered pit constructed to allow the liquid contents to seep into the ground.

Ejector
A mechanical device used to eject or pump sewage.

Erosion
The gradual destruction of metal or other material by the abrasive action of liquids, gases, solids, or mixtures of these materials.

Escutcheon
Chrome-plated ring used to cover a penetration in a wall where pipe is passing through it.

Estimate
An approximate judgment regarding the dollar value of a project.

Estimate form
The form on which all quantities are entered in order that they may be priced.

Excavation
The operation of removing earth for the purpose of installing underground piping.

Fire hose cabinet
The cabinet containing fire hose and a fire department valve, connected to the fire standpipe system.

Fire hydrant
The water supply device on a site used for fire fighting purposes.

Fire line
A system of pipes used exclusively to supply water for extinguishing fires.

Fitting
A manufactured device used to connect pipe.

Fixture branch
A water supply pipe connecting one or more fixtures to the main water supply header or riser.

Fixture unit
A fixture unit is that amount of fixture discharge equivalent to seven and one-half gallons or more, one cubic foot of water per minute.

Fixtures, battery of
An integral unit such as a kitchen sink and a laundry unit.

Fixtures, plumbing
Installed receptacles, devices, or appliances which are supplied with water, or which receive liquids and/or discharge liquids, or liquid-borne wastes, either directly or indirectly into the drainage system.

Flood level rim
The top edge of the receptacle from which water overflows.

Floor drain
A device installed to receive liquid wastes collecting on a floor and to discharge them into the sanitary system.

Fixture supply
A water supply pipe connecting the fixture with the fixture branch.

Flush valve
A device that discharges a predetermined quantity of water to fixtures for flushing purposes and is activated by direct water pressure.

Fresh air inlet pipe
A pipe connected to the building house drain immediately upstream from the house trap to prevent air lock between the fixture trap and the main trap. It also supplies the whole building drainage system with a circulation of fresh air.

Gas meter
The device used to measure gas consumption in a building.

Gas regulator
The device used to regulate gas pressure in a building.

Hubless joint
A type of pipe joint used on no hub cast-iron soil pipe, requiring clamps.

Indirect waste
A drain pipe used to convey liquid wastes that does not connect directly to the drainage system but which discharges into the house drainage system through an air break into a trap, fixture, receptacle, or interceptor.

Interceptor
A device designed to separate and retain harmful, hazardous, or undesirable matter from normal wastes and permit normal sewage or liquid wastes to discharge into the disposal terminal by gravity.

Labor
The physical operation of installing items.

Labor-hour
The unit of time required to install an item.

Labor wage rate
The prevailing hourly pay for a worker.

Leader
A vertical drainage pipe for conveying storm water from roof or gutter drains to the building storm drain.

Low bid
The lowest dollar proposal by a contractor to perform work.

Main
The principal artery or arteries of a piping system to which all branches, risers, and runouts are connected.

Manhole
A device constructed of brick or pre-cast concrete used to service site sewer lines and additionally used as a junction point between two sewer lines.

Master plumber
The master plumber's license grants him the authority to install and to assume responsibility for contractual agreements pertaining to plumbing and to secure any required permits. The journeyman plumber, properly licensed, is allowed to install plumbing only under the supervision of a master plumber.

Material
The item or items on a project requiring installation.

Medical gas outlet
The device used to gain access to medical gases.

Neutralizing basin
The device used to neutralize acid-bearing wastes before their entry into the building drainage system.

Offset
An offset in a line of piping is a combination of pipe, pipes and/or fittings that join two approximately parallel sections of the line of pipe.

Outfall sewers
Sewers receiving sewage from the collection system and carrying it to the point of final discharge or treatment. Usually the largest sewer of the entire system.

Overhead
Business operating expenses of a contractor.

Pipe, horizontal
Any pipe or part thereof which is installed in a horizontal position or which makes an angle of less than 45° with the horizontal.

Pipe supports
The manufactured devices used to support piping from ceilings, walls, floors, or other structural members.

Pipes, water service
The portion of the water piping which supplies one or more structures or premises and which extends from the public or private main in the street, alley, or easement to the meter or, if no meter is to be provided, to the first stop cock or valve inside the premises.

Pitch
The item pitch is used to indicate the amount of slope or grade given to horizontal piping and is expressed in inches of vertically projected drop per foot on a horizontally projected run of pipe.

Plumbing code
The minimum legal standards set forth by municipalities for the installation of plumbing work.

Plumbing inspector
Any person who, under the supervision of the authority having jurisdiction, is authorized to inspect plumbing and drainage as defined in the code for the municipality, and complying with the laws of licensing and/or registration of the state, city, or county.

Potable water
Water free from impurities present in amounts sufficient to cause disease or harmful physiological effects.

Pricing
The physical operation of costing out an estimate.

Profit
Amount of money due the contractor directly for performing work.

Revent
A revent pipe is that part of a vent pipe line which connects directly with any individual waste or group of wastes, underneath or back of the fixture, and extends either to the main or branch vent pipe.

Riser
A pipe extending floor to floor.

Roof drain
A device installed to receive water collecting on the surface of a roof and to discharge it into the leader.

Roughing
The soil, waste, vent, and water piping immediately behind the fixtures and connected to either the waste and vent stacks or branches, or water risers or branches.

Sanitary sewer
A pipe that carries sewage and excludes storm, surface, and ground water.

Scale
A device used to measure scaled drawings. The appropriate reduced dimensions on a drawing.

Septic tank
A receptacle which receives the discharge of a drainage system or part thereof, and is designed and so constructed to separate the solids from the liquid, digest the organic matter through a period of detention, and allow the liquids to discharge into the soil outside of the tank through a system of open-joint or perforated piping, or into a disposal pit.

Sewage
Any liquid waste containing animal or vegetable matter in suspension or solution.

Sewer connection
The physical operation of connecting a sewer on the site.

Sheeting and shoring
The physical operation of bracing earth on trench walls with planks to prevent cave-in.

Site work
That portion of outside work beginning from a point five feet past the building wall.

Sludge
The accumulated suspended solids of sewage deposited in tanks, beds, or basins, mixed with more or less water to form a semi-liquid mass.

Soil pipe
A pipe that conveys sewage containing fecal matter.

Sovent
A special copper D.W.V. fitting that eliminates the need for a vent stack, through a process of aeration and de-aeration.

Specifications
The standards set forth by the design engineer regarding quality of materials and nature of workmanship for a project.

Stack
Any vertical line of soil, waste, vent, or inside leader piping.

Stack venting
A method of venting a fixture or fixtures through the soil or waste stack.

Storm sewer
A sewer used for conveying rainwater, surface water, condensate, cooling water, or similar clear liquid wastes.

Strain
Change of shape or size of a body produced by the action of stress.

Stress
When external forces act on a body they are resisted by reactions within the body, which are called stresses.

Subsoil drain
A drain that receives only subsurface or seepage water and conveys it to a place of disposal.

Summary sheet
The form an estimator uses to summarize a project, indicating all supplementary costs such as overhead and profit.

Sump pump
A mechanical device used to eject liquid waste from a sump pit into the gravity drainage system.

Takeoff
To physically measure and list items shown on drawings.
Takeoff sheets
The sheets used to record quantities while a project is being taken off.

Tension
That stress which resists the tendency of two forces acting away from each other to pull apart two adjoining planes of a body.

Trade price sheet
The published price sheet issued to contractors by manufacturers.

Trap
A waste fitting which provides a liquid seal.

Trap seal
The maximum vertical depth of liquid that a trap will retain, measured between the crown weir and the top of the dip of the trap.

Turbulence
Any deviation from parallel flow in a pipe due to rough inner wall surfaces, obstructions, or directional changes.

Vacuum
Any pressure less than that exerted by the atmosphere and may be termed a negative pressure.

Vacuum pump
The item of equipment in which a partial vacuum can be produced; used on hospital vacuum systems.

Valve
A device used to control the flow of liquids and gases.

Velocity
Time rate of motion in a given direction and sense.

Vent, loop
A vent from a single fixture or battery of fixtures which is connected into the same stack into which the fixtures discharge. If the loop vent serves more than one fixture, it is one type of circuit vent.

Vent, wet
A vent that receives the discharge of wastes other than from water closets.

Vent, yoke
A pipe connecting upward from a soil or waste stack to a vent stack for the purpose of preventing pressure changes in the stacks.

Vent stack
A vertical vent pipe installed primarily for the purpose of providing circulation of air to and from any part of the drainage system. A vent stack or main vent is that part of a venting system to which circuit vents are connected. Branch

vents, revents, or individual vents may connect with a vent stack. The foot of the vent stack may be connected either into a horizontal drainage branch or into a soil or waste stack.

Vent system
Pipe or piping installed to provide a flow of air to or from a drainage system.

Waste pipe
A pipe that conveys only liquid wastes, free of fecal matter.

Water heater
The manufactured item of equipment that generates hot water.

Water meter
The device used to measure water consumption in a building.

Water pipe
Piping that conveys water to the plumbing fixtures and other water outlets.

Index

Printed in the United States
By Bookmasters